CW00705844

Faulting, Fracturing and Igneous Intrusion in the Earth's Crust

Geological Society books refereeing procedures

The Society makes every effort to ensure that the scientific and production quality of its books matches that of its journals. Since 1997, all book proposals have been refereed by specialist reviewers as well as by the Society's Books Editorial Committee. If the referees identify weaknesses in the proposal, these must be addressed before the proposal is accepted.

Once the book is accepted, the Society Book Editors ensure that the volume editors follow strict guidelines on refereeing and quality control. We insist that individual papers can only be accepted after satisfactory review by two independent referees. The questions on the review forms are similar to those for *Journal of the Geological Society*. The referees' forms and comments must be available to the Society's Book Editors on request.

Although many of the books result from meetings, the editors are expected to commission papers that were not presented at the meeting to ensure that the book provides a balanced coverage of the subject. Being accepted for presentation at the meeting does not guarantee inclusion in the book.

More information about submitting a proposal and producing a book for the Society can be found on its web site: www.geolsoc.org.uk.

It is recommended that reference to all or part of this book should be made in one of the following ways:

Healy, D., Butler, R. W. H., Shipton, Z. K. & Sibson, R. H. (eds) 2012. *Faulting, Fracturing and Igneous Intrusion in the Earth's Crust.* Geological Society, London, Special Publications, **367**.

Tavani, S., Fernandez, O. & Muñoz, J. A. 2012. Stress fluctuation during thrust-related folding: Boltaña anticline (Pyrenees, Spain). *In*: Healy, D., Butler, R. W. H., Shipton, Z. K. & Sibson, R. H. (eds) *Faulting, Fracturing and Igneous Intrusion in the Earth's Crust.* Geological Society, London, Special Publications, **367**, 131–140, http://dx.doi.org/10.1144/SP367.9.

GEOLOGICAL SOCIETY SPECIAL PUBLICATION NO. 367

Faulting, Fracturing and Igneous Intrusion in the Earth's Crust

EDITED BY

D. HEALY
University of Aberdeen, Scotland

R. W. H. BUTLER
University of Aberdeen, Scotland

Z. K. SHIPTON
University of Strathclyde, Scotland

and

R. H. SIBSON
University of Otago, New Zealand

2012
Published by
The Geological Society
London

THE GEOLOGICAL SOCIETY

The Geological Society of London (GSL) was founded in 1807. It is the oldest national geological society in the world and the largest in Europe. It was incorporated under Royal Charter in 1825 and is Registered Charity 210161.

The Society is the UK national learned and professional society for geology with a worldwide Fellowship (FGS) of over 10 000. The Society has the power to confer Chartered status on suitably qualified Fellows, and about 2000 of the Fellowship carry the title (CGeol). Chartered Geologists may also obtain the equivalent European title, European Geologist (EurGeol). One fifth of the Society's fellowship resides outside the UK. To find out more about the Society, log on to www.geolsoc.org.uk.

The Geological Society Publishing House (Bath, UK) produces the Society's international journals and books, and acts as European distributor for selected publications of the American Association of Petroleum Geologists (AAPG), the Indonesian Petroleum Association (IPA), the Geological Society of America (GSA), the Society for Sedimentary Geology (SEPM) and the Geologists' Association (GA). Joint marketing agreements ensure that GSL Fellows may purchase these societies' publications at a discount. The Society's online bookshop (accessible from www.geolsoc.org.uk) offers secure book purchasing with your credit or debit card.

To find out about joining the Society and benefiting from substantial discounts on publications of GSL and other societies worldwide, consult www.geolsoc.org.uk, or contact the Fellowship Department at: The Geological Society, Burlington House, Piccadilly, London W1J 0BG: Tel. + 44 (0)20 7434 9944; Fax + 44 (0)20 7439 8975; E-mail: enquiries@geolsoc.org.uk.

For information about the Society's meetings, consult *Events* on www.geolsoc.org.uk. To find out more about the Society's Corporate Affiliates Scheme, write to enquiries@geolsoc.org.uk.

Published by The Geological Society from:
The Geological Society Publishing House, Unit 7, Brassmill Enterprise Centre, Brassmill Lane, Bath BA1 3JN, UK

The Lyell Collection: www.lyellcollection.org
Online bookshop: www.geolsoc.org.uk/bookshop
Orders: Tel. +44 (0)1225 445046, Fax +44 (0)1225 442836

British Library Cataloguing in Publication Data

A catalogue record for this book is available from the British Library.
ISBN 978-1-86239-347-9
ISSN 0305-8719

Distributors
For details of international agents and distributors see:
www.geolsoc.org.uk/agentsdistributors

Typeset by Techset Composition Ltd, Salisbury, UK
Printed by MPG Books Ltd, Bodmin, UK

Contents

Stress, faulting, fracturing and seismicity: the legacy of Ernest Masson Anderson

DAVID HEALY[1]*, RICHARD H. SIBSON[2], ZOE SHIPTON[3] & ROBERT BUTLER[1]

[1]*School of Geosciences, King's College, University of Aberdeen, Aberdeen AB24 3UE UK*

[2]*Department of Geology, University of Otago, P.O. Box 56, Dunedin 9054, New Zealand*

[3]*Department of Civil Engineering, University of Strathclyde, Glasgow, G4 0NG UK*

**Corresponding author (e-mail: d.healy@abdn.ac.uk)*

For as long as geologists have looked at deformed rocks, they have grappled to understand the mechanical origins of deformation. Natural systems are inherently complex so that, for many, purely geometric and kinematic approaches have sufficed. However, we know that stress within the brittle upper crust controls the nucleation, growth and reactivation of faults and fractures, induces seismic activity, affects the transport of magma and modulates structural permeability, thereby influencing the redistribution of hydrothermal and hydrocarbon fluids. An endeavour of structural geology and seismotectonics is therefore to reconstruct states of stress and their evolution over geological time from observations of the final products of rock deformation. Experimentalists endeavour to recreate structures observed in nature under controlled stress conditions. Earth scientists studying earthquakes attempt to monitor or deduce stress changes in the Earth as it actively deforms. All are building upon the pioneering researches and concepts of Ernest Masson Anderson dating back to the start of the 20th century. His insights, encapsulated in a small number of research papers and in the book *The Dynamics of Faulting and Dyke Formation with Applications to Britain*, continue to influence investigations in structural geology, seismology, rock mechanics, processes of hydrothermal mineralization and physical volcanology. This volume celebrates this legacy.

Ernest Masson Anderson, 1877–1960

E. M. Anderson was born in Falkirk, Scotland in 1877. From 1894 to 1898 he attended the University of Edinburgh, gaining 1st Class Honours in Mathematics and Natural Philosophy and a solid grounding in continuum mechanics. Following an early teaching career, he joined the Geological Survey of Great Britain in 1903. Aside from two years of war service during 1916–17 (first as a volunteer in the Highland Light Infantry where he was wounded in action and later with the Royal Engineers), he continued with the Geological Survey until his retirement in 1928 in the role of Senior Geologist (to which we has promoted in 1922). Over the period 1910–30 he contributed to seven Highland, seven Coalfield and four Lowland District Memoirs. He died in Edinburgh in 1960 at the age of 83. He was elected a Fellow of the Geological Society in 1913 and a Fellow of the Royal Society of Edinburgh in 1922. Further recognition came in the form of a DSc from the University of Edinburgh in 1933, the Makdougall Brisbane Prize and Medal for 1934–36 from the Royal Society of Edinburgh, the Clough Medal for 1949–50 from the Edinburgh Geological Society and the award of the Murchison Medal by the Geological Society in 1949.

His first great contribution (Anderson 1905), later elaborated in his book (Anderson 1942/1951), was the recognition of the surface boundary condition of zero shear stress. In a triaxial stress system with principal compressive stresses $\sigma_1 > \sigma_2 > \sigma_3$, just three basic stress regimes should occur in the crust depending whether the vertical stress σ_v is σ_1, σ_2 or σ_3. Combining this boundary condition with what is essentially the Coulomb criterion for brittle shear failure (see Hubbert 1972), whereby shear failure initiates along planes containing the σ_2 axis lying at *c.* 30° to σ_1, he went on to argue that, at first formation in intact crust, there should be three fundamental classes of fault: (1) thrust faults forming with dips of *c.* 30° during crustal compression when $\sigma_v = \sigma_3$; (2) normal faults forming with initial dips of *c.* 60° during crustal extension when $\sigma_v = \sigma_1$; and (3) subvertical wrench (strike–slip) faults developing where $\sigma_v = \sigma_2$. His demonstration in 1905 of the possibility of strike–slip faulting was ahead of its time given the prevailing view that faulting mostly involved vertical displacements. The 1906 San Francisco earthquake in California occurred one year later when a strike–slip rupture with average dextral slip of *c.* 4 m was mapped for 430 km along the San Andreas fault.

No doubt influenced by his extensive field experience in the Inner Hebrides and on the Scottish

From: HEALY, D., BUTLER, R. W. H., SHIPTON, Z. K. & SIBSON, R. H. (eds) 2012. *Faulting, Fracturing and Igneous Intrusion in the Earth's Crust*. Geological Society, London, Special Publications, **367**, 1–6. http://dx.doi.org/10.1144/SP367.1

mainland, he also explored the mechanics of dyke intrusion and demonstrated how they occupy extension fractures controlled by regional and local stress trajectories. Later, more elaborate analyses of curving stress trajectory systems around pressurized igneous intrusions provided explanations for ring-dykes and cone sheets (Anderson 1936). His theoretical analyses were not restricted to the mechanics of brittle structures, however. Along with C. T. Clough he pioneered the systematic recording and interpretation of lineations in metamorphic terrains such as the Dalradian schists of Schiehallion. He published a kinematic analysis which, contrary to conventional wisdom at the time, argued that lineations generally developed parallel to the shearing transport direction (Anderson 1948). In a fitting tribute, Hubbert (1972) wrote: 'At the time when Anderson's book first appeared it had the distinction of being one of the few books, if not the only one, dealing with geologic structures and written by a geologist that was based on valid mechanical premises'.

Anderson's legacy: where has it got us?

Faulting and seismic hazard in regionally homogeneous stress fields

Keeping in mind the important proviso that Anderson's analysis was limited to stress conditions at the time of fault formation (nucleation), modern studies provide ample support for 'Andersonian' structural relationships. In terms of active tectonics, the World Stress Map (a global compilation of *in situ* stress measurements) reveals three basic variants of stress province with associated active faults across the Earth (Heidbach *et al.* 2010). This appears to hold even in complex collision zones (Molnar & Tapponnier 1975). Andersonian-type structures have even been observed extraterrestrially in ice tectonics on the Jovian moon, Europa (Kattenhorn & Marshall 2006). Anderson's three fundamental classes of fault also account for a high proportion of seismologically determined focal mechanisms. Stress inversions from these mechanisms support his inference that one of the principal stresses is generally subvertical (Célérier 2008). In regionally homogeneous stress fields, the Andersonian model provides a good explanation for the nucleation of faults and earthquakes. Sibson *et al.* (2012) show that the M_w7.1 2010 Darfield (Christchurch) earthquake's composite rupture is consistent with the regional stress field in the South Island of New Zealand. The main dextral strike–slip fault responsible for this event is consistent with either a newly formed Andersonian strike–slip fault, or a pre-existing subvertical structure optimally oriented for frictional reactivation. López (2012) uses stress inversion of fault slips and slip tendency analysis of recent deformation in central Costa Rica to show that the seismicity is consistent with Andersonian fault orientations in a convergent stress field.

Structural inheritance

Pre-existing faults and other structural fabrics were not considered by Anderson, yet many subsequent analyses have shown that Andersonian stress fields, with one principal stress vertical, are consistent with reactivation of brittle features or other surfaces formed during earlier events. Faults optimally oriented for frictional reactivation at close to Andersonian orientations for the three stress regimes also appear to govern the strength of much of the upper crust (Townend & Zoback 2000). Sibson (2012) compares optimally (i.e. Andersonian) and non-optimally oriented fault orientations for reverse slip ruptures. The dominant signal on a dip histogram shows a peak in the Andersonian orientation of 30 degree dips, with two subsidiary peaks at higher and lower dips. The reactivation of pre-existing fabrics (faults, bedding planes, etc.) in preference to the nucleation of new Andersonian faults is ascribed to a delicate balance between pore fluid pressure and differential stress levels. Van Noten *et al.* (2012) model the evolution of successive generations of quartz veins invoking a transitional strike–slip stress regime between extensional and contractional deformation episodes. An analysis of the 3D brittle failure mode is used to show that this transitional strike–slip stress state should always occur during basin inversion, that is, in the change-over from extension to shortening. MacDonald *et al.* (2012) show how pre-existing normal faults striking parallel to the continental margin in the Bight Basin (offshore Australia) are optimally oriented for reactivation by ridge-push stresses. Inversion anticlines associated with this reactivation may yet be prospective for hydrocarbon exploration, although some potential traps are shown to have been breached. Tassone *et al.* (2012) show how similar plate boundary stresses have affected the continental interior of Australia. Integrating data from seismicity records, exhumation measurements and structural restorations, these authors describe Miocene–Pliocene inversion of the southern Australian margin with exhumation in the region of 1 km, in an Andersonian framework of reactivated pre-existing faults.

Faulting and fracturing in locally perturbed stress fields

Stress fields in the crust are not always homogeneous, however, and several authors address perturbations in stress fields. Three papers in this volume deal with stress perturbations around magma bodies, folds or salt diapirs. Gerbault (2012) models the wall rocks of a magma chamber with elastoplasticity theory and, by incorporating terms for the gravitational load, shows that wall rocks will fail in shear and not tension. Complex arrays of eccentric shear bands (i.e. 'faults') match predictions from engineering plasticity and are consistent with patterns of fractures mapped around exhumed intrusions. Anderson also recognized the significance of local perturbations to the stress field in his work on cone sheets and ring dykes (Anderson 1936). Stresses also vary with position around folds and this has an effect on fold-related fractures (Tavani *et al.* 2012). In detail, during fold growth stresses vary with spatial position and time; Tavani *et al.* (2012) assign observed sets of extensional and contractional fractures to distinct phases of these 'stress fluctuations'. There is an implicit issue of scale in this analysis, and the field relationships of the secondary fractures with respect to the primary fold architecture are critical. King *et al.* (2012) describe evidence from petroleum wells for stress deflections (perturbations) around salt diapirs in the Gulf of Mexico. Maximum horizontal stress orientation is shown to deviate from a regional margin-parallel trend towards parallelism with the local salt–sediment interface. These rotations are attributed to contrasts in geomechanical properties between the salt and the hosting sediments and gravitational collapse of deltaic sediments down the flanks of the diapirs.

Low-angle structures and the nature of stress variations with depth

Stress perturbations away from homogeneity are not restricted to intrusions of salt or magma or folding. Several authors consider the nature of stress in the vicinity of low-angle normal faults or detachments. These structures have often been presented as a paradox, with slip (and nucleation) considered to be difficult or impossible in these non-optimal Andersonian orientations due to the high value of normal stress acting across the fault plane. A thorough recent review of the issues surrounding low-angle normal faulting is presented by Collettini (2011).

Tingay *et al.* (2012) use data from petroleum wells in the Nile Delta to describe non-Andersonian fault orientations in rocks lying above an evaporitic detachment. Present-day stress orientations from geomechanical analysis are consistent with Andersonian fault orientations in rocks below the detachment, or in areas where the salt layer is absent. This stress field in the rocks above the detachment is distinct from that below. The concept of significant stress rotations across layers or zones of mechanically distinct (weak?) rocks is not new (e.g. Casey 1980; Mount & Suppe 1987), but the observation of this behaviour across a subhorizontal detachment is entirely novel. The observation of vertical partitioning in the crustal stress field orientation prompts questions about the nature of the stress-depth function, often shown simply as the *magnitudes* of Sv, SH or Sh v. depth, and highlights a significant contrast to standard Andersonian stress states. Tuitt *et al.* (2012) use finite element models of a sediment wedge above a detachment to investigate the relative importance of pore fluid pressure, basal friction, wedge rigidity and wedge angle on sliding along the detachment. For hydrostatic pore fluid pressures, the rigidity of the wedge sediments and the wedge angle have little effect on sliding. These models could be extended to analyse in more detail the situation reported in Tingay *et al.* (2012), exploring the parameter space for controlling stress rotations above and below a detachment within the sediment package. Bistacchi *et al.* (2012) develop a novel model of low-angle detachment instability due to anisotropic coefficients of sliding friction. Many low-angle detachments contain zones of strongly aligned fabrics defining an intense structural anisotropy, with these local fabrics oriented parallel to the detachment. Such assemblages are likely to display anisotropy in their frictional behaviour. Bistacchi *et al.* (2012) extend the slip tendency analysis of Morris *et al.* (1996) to rocks with directional variations in frictional resistance to sliding. Improved laboratory characterization of rock mechanical properties is the basis for several recent approaches to brittle fracture in the crust, representing another departure from Andersonian orthodoxy.

The role of pore fluids in fault stability and seismicity

The role of anisotropy in rock properties is further developed by Healy (2012). Patterns of fluid-saturated microcracks in the damage zones of faults are modelled using equations of anisotropic poroelasticity. Changes in pore fluid pressure in the wall rocks of faults, either natural or induced by engineering activity such as hydrocarbon extraction or water injection, can alter the stresses acting on the fault zone. Predicted changes in effective stress are significantly different for anisotropic pores (i.e. cracks) in comparison to the common assumption of spherical pores. The dilatancy-diffusion hypothesis in seismic precursors is assessed by Main *et al.* (2012). Changes in pore volume and pore pressure have been observed in association with seismic events in laboratory experiments, but the field evidence from natural earthquakes is less compelling. Simple scaling between laboratory and field behaviours is unlikely for several reasons, even though the fracture systems that can dominate the fluid transport properties of the subsurface may themselves be scale invariant. The concept and consequences of effective stress are not explicit in the orthodox Andersonian analysis, and the role of subsurface pore pressure heterogeneity in space and time provides a rich field for further work into fault stability and seismic hazard.

Outstanding issues

Field geologists commonly find that low-displacement faults fit into one or the other of Anderson's three classes, and that conjugate brittle faults typically intersect each other at 50–60° as he suggested (e.g. Kelly *et al.* 1998). However, there are notable exceptions from both field datasets and laboratory experiments (Aydin & Reches 1982; Reches & Dieterich 1983). Polymodal fractures, with quadrimodal fractures forming a subset, are distinctly non-Andersonian. Recent work based on field data and numerical models has sought to explain these shear fracture patterns in terms of interacting and coalescing tensile cracks (Healy *et al.* 2006*a*, *b*). A review of field data suggests that tensile fractures associated with shear fractures are often oblique, as predicted by the new non-Andersonian model (Blenkinsop 2008). The utility of the Andersonian approach in predicting fracture orientations and seismic hazard for certain stress states is amply demonstrated by many of the papers in this volume. However, fracture prediction generally requires more than just the orientations of fractures; fracture size (length or area) and spacing (intensity or density) are also critical, especially for industrial applications. Fluid transport properties in fractured rock are functions of all of these fracture and fracture pattern attributes. Curved fractures also fall outside the scope of the orthodox Andersonian model, and yet these can be found in such diverse settings as wall rock fractures around magma chambers (Gerbault 2012), near-surface fractures in exhumed terrain (Martel 2011) and vein systems around faults. A key area of outstanding research lies in deciphering the role of pre-existing structural anisotropy on fracture nucleation; in essence, the upscaling rules of Griffith's (1924) notions of microstructural fracture processes have to be established. Field studies show that structural anisotropy can condition complex fault patterns (e.g. Butler *et al.* 2008). The detailed relationships between these phenomena, the origin of frictional or elastic mechanical anisotropy, their relationship to failure criteria and the imposed stress field remain fertile areas for further investigation, however. The papers contained in this volume are a testament to the enduring power of E. M. Anderson's ideas which continue to yield profound insights and further stimuli for research into the mechanical behaviour of the Earth's crust.

This volume originated from a conference called to commemorate the contributions of E. M. Anderson on the 50th anniversary of his death. 'Stress controls on faulting, fracturing and igneous intrusion in the Earth's crust' was held in Glasgow in September 2010. Over 120 delegates from academic institutes and industry attended 3 days of presentations. The volume editors would like to thank all authors and reviewers for their time, effort and patience in producing this collection. Special thanks go to: T. Anderson (no relation to Ernest Masson so far as we are aware) at the Geological Society of London for her tireless efforts in nursing manuscripts through to production; to A. Macpherson at the Edinburgh Geological Society for the decision to allow a facsimile of Anderson's original paper from 1905 to be included in this volume; and to H. Moir for helping to organize the conference.

References

ANDERSON, E. M. 1905. The dynamics of faulting. *Transactions of the Edinburgh Geological Society*, **8**, 387–402.

ANDERSON, E. M. 1936. The dynamics of formation of cone sheets, ring-dykes, and cauldron subsidence. *Proceedings of the Royal Society of Edinburgh*, **56**, 128–163.

ANDERSON, E. M. 1942/1951. *The Dynamics of Faulting and Dyke Formation with Application to Britain.* Oliver & Boyd, Edinburgh.

ANDERSON, E. M. 1948. On lineation and petrofabric structure and the shearing movement by which they have been produced. *Quarterly Journal of Geological Society*, **104**, 99–126.

AYDIN, A. & RECHES, Z. 1982. Number and orientation of fault sets in the field and in experiments. *Geology*, **10**, 107.

BISTACCHI, A., MASSIRONI, M., MENEGON, L., BOLOGNESI, F. & DONGHI, V. 2012. On the nucleation of non-Andersonian faults along phyllosilicate-rich mylonite belts. *In*: HEALY, D., BUTLER, R. W. H., SHIPTON, Z. K. & SIBSON, R. H. (eds) *Faulting, Fracturing, Igneous Intrusion in the Earth's Crust.* Geological Society, London, Special Publications, **367**, 185–199.

BLENKINSOP, T. G. 2008. Relationships between faults, extension fractures and veins, and stress. *Journal of Structural Geology*, **30**, 622–632.

BUTLER, R. W. H., BOND, C. E., SHIPTON, Z. K., JONES, R. R. & CASEY, M. 2008. Fabric anisotropy controls faulting in the continental crust. *Journal of Geological Society, London*, **165**, 449–452.

CASEY, M. 1980. Mechanics of shear zones in isotropic dilatant materials. *Journal of Structural Geology*, **2**, 143–147.

CÉLÉRIER, B. 2008. Seeking Anderson's faulting in seismicity: a centennial celebration. *Reviews of Geophysics*, **46**, RG4001, doi: 10.1029/2007RG000240.

COLLETTINI, C. 2011. The mechanical paradox of low-angle normal faults: Current understanding and open questions. *Tectonophysics*, **510**, 253–268.

GERBAULT, M. 2012. Pressure conditions for shear and tensile failure around a circular magma chamber; insight from elasto-plastic modelling. *In*: HEALY, D., BUTLER, R. W. H., SHIPTON, Z. K. & SIBSON, R. H. (eds) *Faulting, Fracturing and Igneous Intrusion in the Earth's Crust.* Geological Society, London, Special Publications, **367**, 111–130.

GRIFFITH, A. A. 1921. The phenomena of rupture and flow in solids. *Philosophical Transactions of the Royal Society of London*, **A221**, 163–198.

HEALY, D. 2012. Anisotropic poroelasticity and the response of faulted rock to changes in pore-fluid pressure. *In*: HEALY, D., BUTLER, R. W. H., SHIPTON, Z. K. & SIBSON, R. H. (eds) *Faulting, Fracturing and Igneous Intrusion in the Earth's Crust.* Geological Society, London, Special Publications, **367**, 201–214.

HEALY, D., JONES, R. R. & HOLDSWORTH, R. E. 2006*a*. Three-dimensional brittle shear fracturing by tensile crack interaction. *Nature*, **439**, 64–67.

HEALY, D., JONES, R. R. & HOLDSWORTH, R. E. 2006*b*. New insights into the development of brittle shear fractures from a 3-D numerical model of microcrack interaction. *Earth and Planetary Science Letters*, **249**, 14–28.

HEIDBACH, O., TINGAY, M., BARTH, A., REINECKER, J., KURFEβ, D. & MÜLLER, B. 2010. Global crustal stress pattern based on the World Stress Map database release 2008. *Tectonophysics*, **482**, 3–15.

HUBBERT, M. K. 1972. *Foreword to the 1972 Facsimile Reprint of the 1951 Edition of The Dynamics of Faulting and Dyke Formation with Applications to Britain by E. M. Anderson.* Hafner Publishing Company, New York, v–xii.

KATTENHORN, S. A. & MARSHALL, S. T. 2006. Fault-induced perturbed stress fields and associated tensile and compressive deformation at fault tips in the ice shell of Europa: implications for fault mechanics. *Journal of Structural Geology*, **28**, 2204–2221.

KELLY, P. G., SANDERSON, D. J. & PEACOCK, D. C. 1998. Linkage and evolution of conjugate strike–slip fault zones in limestones of Somerset and Northumbria. *Journal of Structural Geology*, **20**, 1477–1493.

KING, R., BACKÉ, G., TINGAY, M., HILLIS, R. & MILDREN, S. 2012. Stress deflections around salt diapirs in the Gulf of Mexico. *In*: HEALY, D., BUTLER, R. W. H., SHIPTON, Z. K. & SIBSON, R. H. (eds) *Faulting, Fracturing and Igneous Intrusion in the Earth's Crust.* Geological Society, London, Special Publications, **367**, 141–153.

LÓPEZ, A. 2012. Andersonian and Coulomb stresses in Central Costa Rica and its fault slip tendency potential: new insights into their associated seismic hazard. *In*: HEALY, D., BUTLER, R. W. H., SHIPTON, Z. K. & SIBSON, R. H. (eds) *Faulting, Fracturing and Igneous Intrusion in the Earth's Crust.* Geological Society, London, Special Publications, **367**, 19–37.

MACDONALD, J., BACKÉ, G., KING, R., HOLFORD, S. & HILLIS, R. 2012. Geomechanical modelling of fault reactivation in the Ceduna Sub-basin, Bight Basin, Australia. *In*: HEALY, D., BUTLER, R. W. H., SHIPTON, Z. K. & SIBSON, R. H. (eds) *Faulting, Fracturing and Igneous Intrusion in the Earth's Crust.* Geological Society, London, Special Publications, **367**, 71–89.

MAIN, I. G., BELL, A. F., MEREDITH, P. G., GEIGER, S. & TOUATI, S. 2012. The dilatancy–diffusion hypothesis and earthquake predictability. *In*: HEALY, D., BUTLER, R. W. H., SHIPTON, Z. K. & SIBSON, R. H. (eds) *Faulting, Fracturing and Igneous Intrusion in the Earth's Crust.* Geological Society, London, Special Publications, **367**, 215–230.

MARTEL, S. J. 2011. Mechanics of curved surfaces, with application to surface-parallel cracks. *Geophysical Research Letters*, **30**, L20303, doi: 10.1029/2011GL049354.

MOLNAR, P. & TAPPONNIER, P. 1975. Cenozoic tectonics of Asia: effect of a continental collision. *Science*, **189**, 419–425.

MORRIS, A., FERRILL, D. A. & HENDERSON, D. B. 1996. Slip-tendency analysis and fault reactivation. *Geology*, **24**, 275.

MOUNT, V. S. & SUPPE, J. 1987. State of stress near the San Andreas fault. *Geology*, **15**, 1143–1146.

RECHES, Z. & DIETERICH, J. H. 1983. Faulting of rocks in three-dimensional strain fields I. Failure of rocks in polyaxial, servo-control experiments. *Tectonophysics*, **95**, 111–132.

SIBSON, R. H. 2012. Reverse fault rupturing: competition between non-optimal and optimal fault orientations. *In*: HEALY, D., BUTLER, R. W. H., SHIPTON, Z. K. & SIBSON, R. H. (eds) *Faulting, Fracturing and Igneous Intrusion in the Earth's Crust.* Geological Society, London, Special Publications, **367**, 39–50.

SIBSON, R. H., GHISETTI, F. C. & CROOKBAIN, R. A. 2012. Andersonian wrench faulting in a regional stress field during the 2010–2011 Canterbury, New Zealand, earthquake sequence. *In*: HEALY, D., BUTLER, R. W. H., SHIPTON, Z. K. & SIBSON, R. H. (eds) *Faulting, Fracturing and Igneous Intrusion in the Earth's Crust.* Geological Society, London, Special Publications, **367**, 7–17.

TASSONE, D. R., HOLFORD, S. P., HILLIS, R. R. & TUITT, A. K. 2012. Quantifying Neogene plate-boundary controlled uplift and deformation of the southern Australian margin. *In*: HEALY, D., BUTLER, R. W. H., SHIPTON, Z. K. & SIBSON, R. H. (eds) *Faulting, Fracturing and Igneous Intrusion in the Earth's Crust.* Geological Society, London, Special Publications, **367**, 91–110.

TAVANI, S., FERNANDEZ, O. & MUÑOZ, J. A. 2012. Stress fluctuation during thrust-related folding: Boltaña anticline (Pyrenees, Spain). *In*: HEALY, D., BUTLER, R. W. H., SHIPTON, Z. K. & SIBSON, R. H. (eds) *Faulting, Fracturing and Igneous Intrusion in the Earth's Crust.* Geological Society, London, Special Publications, **367**, 131–140.

TINGAY, M., BENTHAM, P., DE FEYTER, A. & KELLNER, A. 2012. Evidence for non-Andersonian faulting above evaporites in the Nile Delta. *In*: HEALY, D., BUTLER, R. W. H., SHIPTON, Z. K. & SIBSON, R. H. (eds) *Faulting, Fracturing and Igneous Intrusion in the Earth's Crust.* Geological Society, London, Special Publications, **367**, 155–170.

TOWNEND, J. & ZOBACK, M. D. 2000. How faulting keeps the crust strong. *Geology*, **28**, 399–402.

Tuitt, A., King, R., Hergert, T., Tingay, M. & Hillis, R. 2012. Modelling of sediment wedge movement along low-angle detachments using ABAQUS™. *In*: Healy, D., Butler, R. W. H., Shipton, Z. K. & Sibson, R. H. (eds) *Faulting, Fracturing and Igneous Intrusion in the Earth's Crust.* Geological Society, London, Special Publications, **367**, 7–17.

Van Noten, K., Van Baelen, H. & Sintubin, M. 2012. The complexity of 3D stress-state changes during compressional tectonic inversion at the onset of orogeny. *In*: Healy, D., Butler, R. W. H., Shipton, Z. K. & Sibson, R. H. (eds) *Faulting, Fracturing and Igneous Intrusion in the Earth's Crust.* Geological Society, London, Special Publications, **367**, 51–69.

Andersonian wrench faulting in a regional stress field during the 2010–2011 Canterbury, New Zealand, earthquake sequence

R. H. SIBSON[1]*, F. C. GHISETTI[2] & R. A. CROOKBAIN[3]

[1]*Department of Geology, University of Otago, P.O. Box 56, Dunedin 9054, New Zealand*

[2]*Terrageologica, 129 Takamatua Bay Rd., RD1, Akaroa 7581, New Zealand*

[3]*Shell Exploration NZ Ltd, 167 Devon Street West, New Plymouth 4310, New Zealand*

**Corresponding author (e-mail: rick.sibson@otago.ac.nz)*

Abstract: The initial M_w7.1 Darfield earthquake sequence was centred west of Christchurch City in the South Island of New Zealand but aftershocks, including a highly destructive M_w6.3 event, eventually extended eastwards across the city to the coast. The mainshock gave rise to right-lateral strike-slip of up to 5 m along the segmented rupture trace of a subvertical fault trending 085 ± 5° across the Canterbury Plains for *c*. 30 km, in agreement with teleseismic focal mechanisms. Near-field data however suggest that the mainshock was composite, initiating with reverse-slip north of the surface rupture. Stress determinations for the central South Island show maximum compressive stress σ_1 to be horizontal and oriented 115 ± 5°. The principal dextral rupture therefore lies at *c*. 30° to regional σ_1, the classic 'Andersonian' orientation for a low-displacement wrench fault. An aftershock lineament trending *c*. 145° possibly represents a conjugate left-lateral strike-slip structure. This stress field is also consistent with predominantly reverse-slip reactivation of NNE–NE faults along the Southern Alps range front. The main strike-slip fault appears to have a low cumulative displacement and may represent either a fairly newly formed fault in the regional stress field, or an existing subvertical fault that happens to be optimally oriented for frictional reactivation.

The M_w7.1 Darfield earthquake and ensuing aftershocks were initially largely restricted to the upper crust below the Canterbury Plains lying between Christchurch and the foothills of the Southern Alps in the South Island of New Zealand. The epicentral region lies about 100 km SE of the dextral-reverse Alpine Fault trending 050° and nearly the same distance SSE from the *c*. 070° trending dextral Hope Fault, which are principal fast-moving (20–30 mm a^{-1}) components of the Pacific–Australia plate boundary fault system (Fig. 1; note that all listed horizontal trends are given as 000–360° azimuthal bearings). The tectonic setting is one of continental convergence with local geology comprising a basement of Mesozoic greywackes (Rakaia Terrane) overlain unconformably by a Late Cretaceous–Tertiary cover sequence about 1 km in thickness, capped in turn by an upper Quaternary sequence of post-glacial alluvial gravels up to a few hundred metres thick (Forsyth *et al.* 2008). Following the earthquake, a subvertical strike-slip rupture with dextral displacements of up to 5 m was mapped east–west across the Canterbury Plains for *c*. 30 km (Quigley *et al.* 2010, 2012) (Fig. 2). The causative fault structures for the Darfield sequence had no prior topographic expression and were unrecognized before the event although parallel east–west trending faults, some clearly Holocene-active, had been recognized to the north, south and offshore to the east (e.g. Wood & Herzer 1993; Forsyth *et al.* 2008). In this note we explore the relationships between the principal strike-slip rupture and the tectonic stress field in the northern South Island.

The 2010 Darfield earthquake sequence

The M_w7.1 mainshock of the 2010 Darfield earthquake sequence occurred at 04:35 local time on 4 September 2010 (16:35 UTC on September 3), initiating at a depth of 11 km below an extremely low-relief portion of the Canterbury Plains (Fig. 2). The epicentre was *c*. 6 km ESE from the town of Darfield some 38 km west of Christchurch city, and about 18 km SE of the foothills of the Southern Alps (Gledhill *et al.* 2011).

Surface rupture

A segmented strike-slip rupture trace with right-lateral displacement averaging 2.5 m but ranging up to 5 m (now termed the Greendale Fault) was mapped east–west for *c*. 30 km across the Canterbury Plains (Quigley *et al.* 2010, 2012). The surface expression of the rupture in the alluvial gravels was that of a dextral Riedel shear array, typical of

From: Healy, D., Butler, R. W. H., Shipton, Z. K. & Sibson, R. H. (eds) 2012. *Faulting, Fracturing and Igneous Intrusion in the Earth's Crust*. Geological Society, London, Special Publications, **367**, 7–18. http://dx.doi.org/10.1144/SP367.2

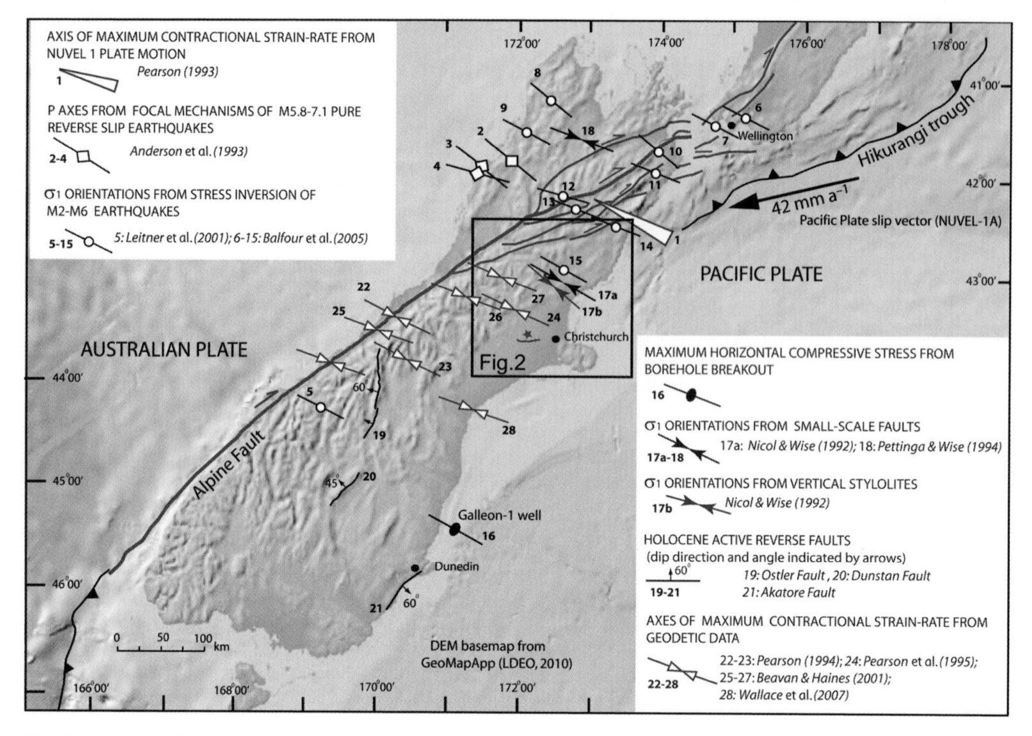

Fig. 1. Setting of the 2010–2011 Canterbury earthquake sequence west of Christchurch (red star denotes epicentre of the 4 September M_w7.1 Darfield mainshock; irregular bold red line is the surface rupture along the Greendale Fault) in relation to the main plate boundary fault system in the South Island of New Zealand. Insets list data sources for different stress indicators.

a buried right-lateral strike-slip fault propagating up through alluvial cover (cf. Tchalenko 1970). Measured displacements were consistent with pure dextral strike-slip on a subvertical fault with an enveloping surface for the left-stepping rupture segments striking 085°, on average. Notably, however, the surface rupture trace lies *c.* 8 km south of the well-constrained seismological epicentre (Gledhill *et al.* 2011).

Seismological characteristics

Teleseismic moment tensor analyses from long-period waves (US Geological Survey; Harvard Global CMT Catalogs; http://earthquake.usgs.gov) yield focal mechanisms consistent with near-vertical dextral strike-slip on a rupture paralleling the mapped surface trace of the Greendale Fault. Near-field seismological and geodetic data however suggest that the mainshock was composite. Both first motion and regional moment tensor analyses yielded initial mechanisms involving reverse-slip on NE–SW striking planes lying north of the strike-slip surface trace (Gledhill *et al.* 2011). This initial reverse-slip is believed to have contributed to the high vertical accelerations (0.5–1.0 g) measured in the epicentral area.

Aftershocks

Seven months after the M_w7.1 earthquake on 4 September 2010, nearly 5000 aftershocks had been recorded. These include one M_w6.3 event, 16 shocks in the range $5.9 > M_w > 5.0$ and 202 in the range $4.9 > M_w > 4.0$ (Fig. 2). A more complete structural analysis of the still ongoing sequence is given by Sibson *et al.* (2011). The great majority of the aftershocks were located at depths less than 15 km. Note that a strike-slip environment has the singular advantage that epicentre alignments can be used to map out subsidiary vertical faults if such occur (e.g. Fukuyama *et al.* 2003). Most of the aftershocks occur in an east–west swathe around the surface rupture trace of the Greendale Fault (Fig. 2), but a number of subsidiary clusters and lineaments are also evident (Gledhill *et al.* 2011). A cluster west of the western curving termination of the surface rupture abuts the foothills of the Southern Alps and is dominated by reverse-slip events. More diffuse activity occurs along the foothills to the NE.

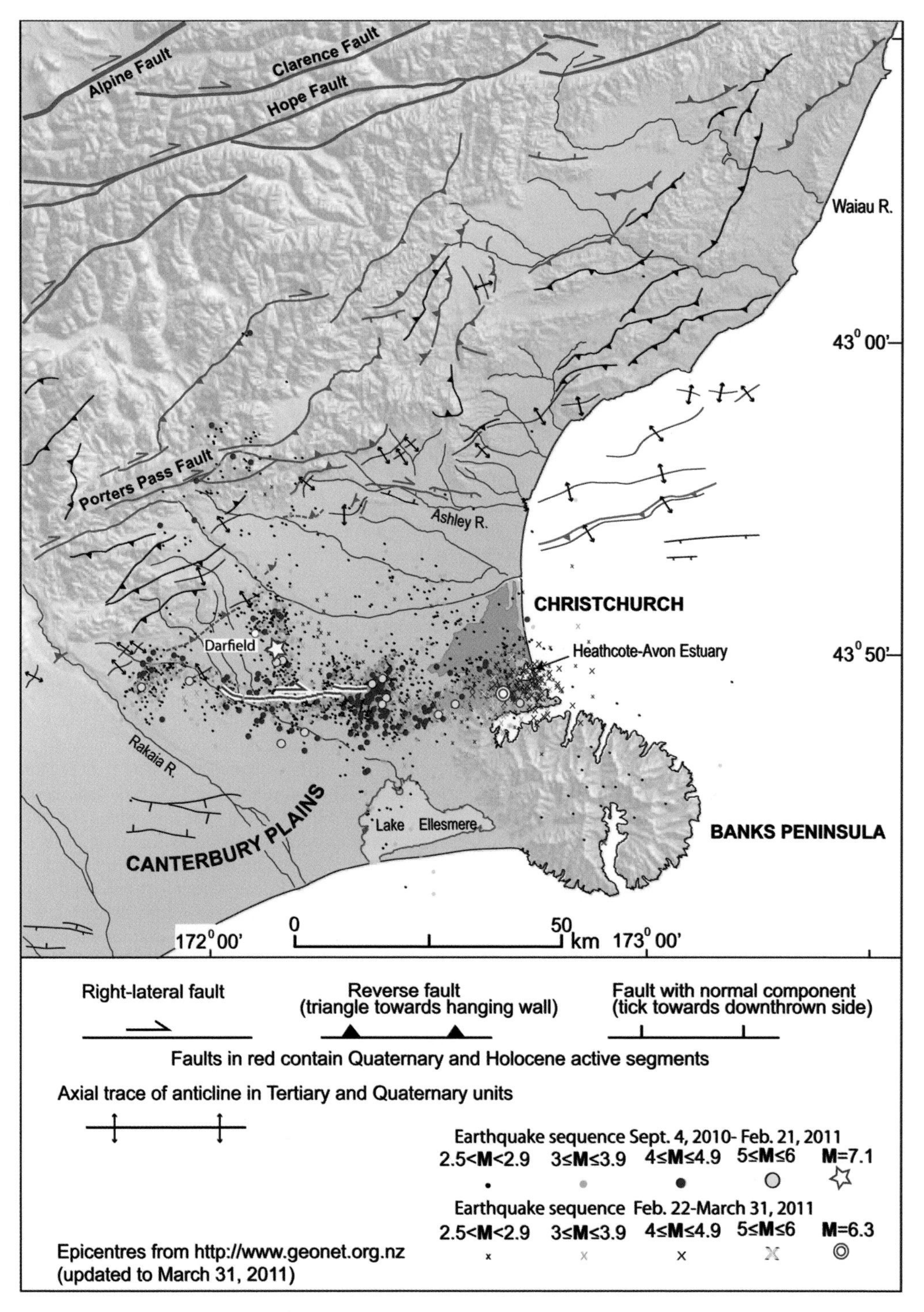

Fig. 2. Local setting of the 2010–2011 Darfield–Christchurch earthquake sequence in relation to mapped structures (after Rattenbury *et al.* 2006; Forsyth *et al.* 2008). Surface rupture accompanying the Darfield earthquake from Quigley *et al.* (2010) with distribution of mainshock and aftershock epicentres updated to 31 March 2011 from GEONET (http://www.geonet.org.nz). DEM from USGS/NASA Shuttle Radar Topography Mission v. 4 (http://srtm.csi.cgiar.org).

A subsidiary belt of activity trends NNW from the mainshock epicentre towards the foothills and the Porters Pass system of strike-slip faults. Of particular interest is a diffuse aftershock lineament developed south of the surface rupture in the first two weeks of the sequence, trending *c.* 145° towards the mouth of Lake Ellesmere (Fig. 2). A strong concentration of activity including five of the $M_w > 5.0$ events occurs at the eastern end of the surface rupture and just south of it, associated with a stepover (dilational?) to an ENE-trending en echelon aftershock lineament that continues eastwards (Fig. 2).

On 22 February (23:52 UTC on 21 February), central and eastern Christchurch were devastated by an M_w6.3 aftershock associated with the eastwards extension of this lineament. From focal mechanism and geodetic analyses, dextral-reverse oblique slip occurred over a rupture with dimensions 15 × 7 km oriented *c.* 060°/60°SE beneath the Heathcote–Avon estuary at depths from 1–7 km (Ristau *et al.* 2011). No surface fault break was observed but intensified aftershock activity continues, extending out to sea north of Banks Peninsula (Fig. 2).

Regional stress in the South Island

A variety of seismological, geodetic and geological indicators suggest that regional maximum compressive stress σ_1 is horizontal and oriented WNE–ESE through much of the South Island (Fig. 1). The different indicators are listed on an azimuthal plot (Fig. 3) for comparison, with line solidity giving a qualitative indication of their reliability. Data are grouped for three domains: a Northern Domain covering approximately the top quarter of the South Island including the Marlborough faults; a Central Domain including north and south Canterbury (including Christchurch and the Darfield epicentral area) together with the central Southern Alps extending across the Alpine Fault into Westland; and a Southern Domain embracing Otago and Southland.

Stress inversions and focal mechanisms

Stress inversions from crustal focal mechanisms led McGinty *et al.* (2000) to infer subhorizontal σ_1 stress trajectories oriented 120 and 118° in northern and southern Marlborough, respectively. Likewise, Balfour *et al.* (2005) carried out a comprehensive analysis of stress in the northern South Island and around the Marlborough fault system based on stress inversions from focal mechanisms, yielding a 'best-fit' horizontal σ_1 stress trajectory trending 115 ± 16° which is remarkably uniform across the northern South Island. These trajectories are consistent with *P*-axes determined for the 1968 M_w7.1 Inangahua earthquake and the 1991 M_w5.8 Hawks Crag I and M_w6.0 Hawks Crag II earthquakes in the northern South Island, all of which involved near-pure reverse-slip mechanisms (Anderson *et al.* 1993). In a study of seismicity around the Alpine Fault (mostly concentrated in its hanging wall below the Southern Alps) Leitner *et al.* (2001) carried out a series of stress inversions for various domains, finding a fairly constant subhorizontal σ_1 stress trajectory oriented 110–120°.

Galleon-1 borehole breakout

Borehole breakouts, commonly assessed from dipmeter logs, are widely employed to determine the contemporary horizontal stress orientations (Zoback 2007). Generally, a vertical borehole coincides with one of the three principal compressive stresses. Initial elastic distortion of the approximately circular cross-section by the stress differential between the two horizontal principal compressive stresses ($\sigma_{H1} > \sigma_{H2}$) is amplified by induced shear fractures and spalling of the borehole walls because of stress concentrations at the areas of greatest curvature. This creates an oval cross-section to the borehole with the long axis at right angles to the maximum horizontal compressive stress, σ_{H1}. For an Andersonian strike-slip regime, the orientations of σ_{H1} and σ_{H2} correspond to σ_1 and σ_3, respectively, while for a thrust regime they correspond to σ_1 and σ_2, respectively.

The Galleon-1 well, drilled 25 km offshore from the south Canterbury coast in 1985 to a total depth of 3086 m (Wilson *et al.* 1985), provides the only breakout stress determination available to us. Systematically aligned breakouts were recorded in Late Cretaceous–Eocene strata from depths of 3018 to 2149 m. A high-resolution dipmeter tool (HDT) log was run in the 12.25″ diameter hole and a stratigraphic HDT log (SHDT) in the 8.5″ hole. Caliper and orientation data were available digitally for the HDT, while the pertinent SHDT data were digitized from a scan of the field print. For the purpose of this analysis the well is essentially vertical, deviating a maximum of 4° from the vertical along an azimuth of 246° at 2918 m. Azimuths derived from both logging tools were corrected for a magnetic declination of 23°E. Breakouts were analysed using the criteria proposed by Reinecker *et al.* (2003); the data are summarized in Table 1.

Note that the breakout azimuths vary slightly between the 12.25″ diameter and 8.5″ diameter sections of the borehole. Table 1 lists the mean breakout azimuth and standard deviation for the two hole diameters which are then aggregated over the entire logged interval. Cumulative breakout length totalled 268 m over six separate zones. The average breakout azimuth was 024 ± 9°, implying

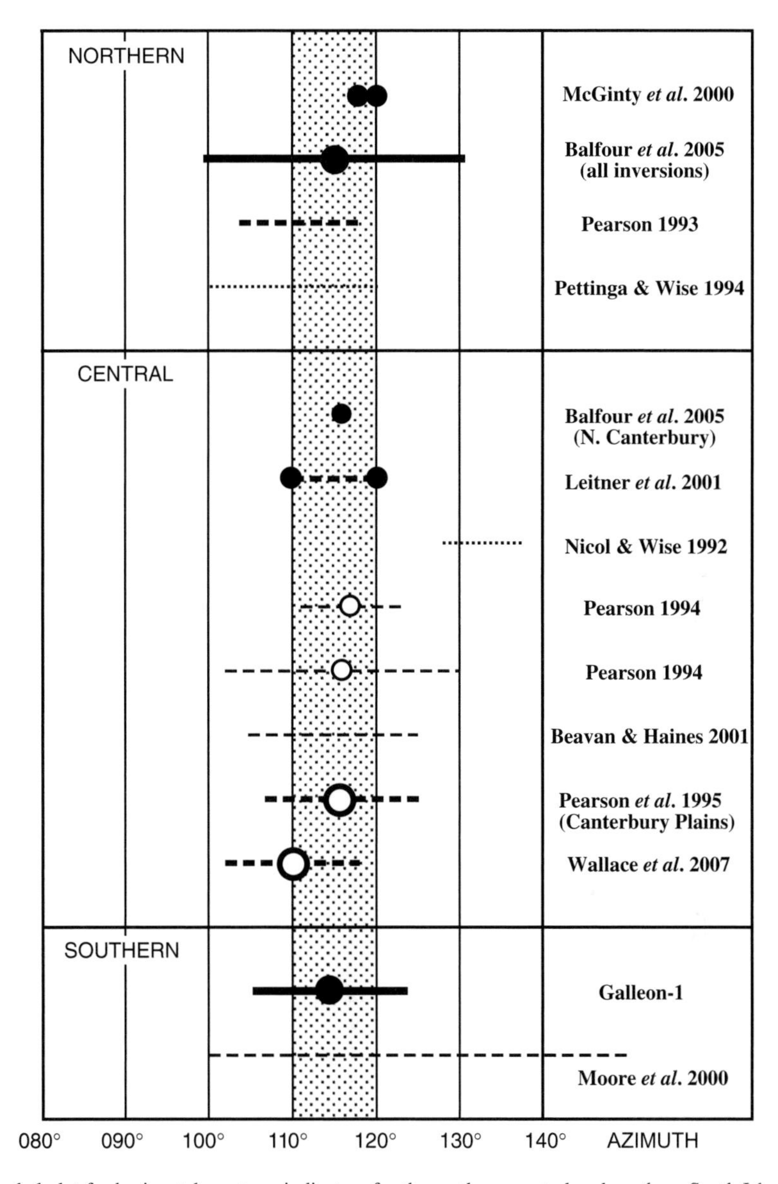

Fig. 3. Azimuthal plot for horizontal σ_1 stress indicators for the northern, central and southern South Island. Line thickness is a qualitative assessment of reliability. Solid circles, stress inversions from focal mechanisms or borehole breakouts (solid bars represent uncertainties where given); open circles, trends for maximum contractional strain rate (dashed bars represent uncertainties where given; dashed lines terminated by circles are range estimates); dotted lines, geological estimates of palaeostress orientation.

σ_{H1} oriented $114 \pm 9°$. Using criteria from the World Stress Map (Heidbach *et al.* 2010) this ranks as a B-quality determination on a 5-point scale.

Geodetic strain rate determinations

Axes of maximum contractional strain rate have comparable trends to those inferred for σ_1 trajectories throughout much of the South Island (Figs 1 & 3). This parallelism is explicable if, over short time intervals without significant strain release from large earthquakes, maximum contractional strain rate is regarded as a measure of maximum incremental shortening subparallel to σ_1. Comparable relationships have been noted elsewhere (e.g. Keiding *et al.* 2009), but are not invariably the case

Table 1. *Frequency weighted breakout azimuths in the Galleon-1 well. Depths are recorded in metres measured along the hole below the rotary table (MAHBRT). Cumulative BO refers to the cumulative breakout length (m); R refers to the mean resultant length of the breakout sample unit vectors and is an indicator of angular dispersion (Fisher 1996)*

Logged interval	Depth to top (m)	Depth to base (m)	Cumulative BO (m)	*R*	Mean BO azimuth	Standard deviation of BO azimuth
12.25″ hole	1930.1	2721.1	71	0.98	031.7°	5.7°
8.5″ hole	2721.1	3078.0	197	0.97	019.1°	7.1°
Total	1930.1	3078.0	268	0.95	023.9°	9.1°

(Townend & Zoback 2006). Thus, Reilly (1990) inferred a regional compressive stress oriented at 110° from consideration of shear strain rates along the Hikurangi Margin. Similarly, Pearson (1993) extrapolated from the NUVEL-1 plate model (De Mets *et al.* 1990) to find a maximum contractional strain rate averaging 114–118° through the northern South Island. Wallace *et al.* (2007) employed a rotating elastic block model to describe active deformation in the South Island, determining a maximum contractional strain rate oriented 110 ± 8° within the Canterbury/Otago block which includes the area of the Darfield earthquake sequence.

In more local studies, Pearson (1994) found comparably oriented axes of maximum contractional strain rate from areas west and east of the Alpine Fault in the central South Island: 117 ± 6° for the Okarito network and 116 ± 14° for the Godley Valley region. In the area of the Canterbury Plains NW of Christchurch, Pearson *et al.* (1995) found the maximum contractional strain rate to be oriented 116 ± 9° just north of the Darfield epicentral area. From inversion of modern GPS networks in the central Southern Alps further west, Beavan & Haines (2001) determined maximum contractional strain rates trending from 105 to 115°. Strain rates are significantly lower in the southern South Island (Otago and Southland) and are less well constrained. From repeated triangulation surveys, Walcott (1984) found maximum contractional strain rates in Southland oriented 103 ± 18° while Moore *et al.* (2000) found an azimuthal range of 100–150°. Given these uncertainties, the Galleon-1 borehole determination seems likely to be the most reliable indicator of regional stress in the southern South Island.

Geological indicators of regional stress

Comparatively few palaeostress determinations have been carried out from neotectonic studies in the South Island and the age of the deformation is often not well constrained. Nicol & Wise (1992) infer a subhorizontal σ_1 palaeostress trending 133 ± 5° from a systematic set of subvertical stylolitic solution surfaces in Oligocene limestone plus related minor structures in the Doctor's Dome area of north Canterbury, which they infer to be probably of Late Pliocene–Early Pleistocene age. From slickenline analysis of minor faults of probable Miocene–Pliocene age around the Waimea Fault near Nelson, Pettinga & Wise (1994) deduced an overall subhorizontal σ_1 palaeostress trending 110–120°. Because of the age uncertainty, these palaeostress indicators are ranked low in terms of reliability.

Crude constraints on present stress-field orientation come from the lack of obvious strike-slip on several major reverse faults that are Holocene-active, restricting the σ_1 stress trajectory to lie nearly orthogonal to strike (Fig. 1). The Ostler Fault strikes 010–030° through the McKenzie Basin south of Mount Cook (Ghisetti *et al.* 2007), implying a σ_1 trajectory oriented 100–120°. Likewise, Quaternary active traces of the Dunstan Fault in Central Otago, which Beanland *et al.* (1986) regard as a pure reverse-slip structure, strike 033° on average, implying σ_1 trending 123°. Further south, the Akatore Fault strikes 035–040° (Litchfield & Norris 2000) implying σ_1 oriented 125–130°. The uncertainties, however, are considerable. For a fault with dip δ and a horizontal slip vector oblique to the fault dip direction by an angle ϕ, the ratio of horizontal to vertical displacement is $H/V = \tan\delta \tan\phi$. On the grounds that a strike-slip component will only become geomorphically obvious when the horizontal displacement exceeds the vertical displacement (i.e. $H/V > 1$), this allows the obliquity of the slip vector to range ±45° from the strike-normal direction for $\delta = 45°$. Because of the large uncertainties, these estimates of possible σ_1 trends are therefore omitted from Figures 1 and 3. Principal Horizontal Shortening (PHS) directions obtained by Berryman (1979) from H/V analysis of active faults throughout the South Island are not greatly dissimilar from the σ_1 trajectories illustrated in Figures 1 and 3, however.

Estimated stress field for the Darfield–Christchurch earthquake sequence

While recognizing that local stress heterogeneity may occur, a remarkably consistent picture of the regional stress field emerges from Figures 1 and 3 for the northern and central South Island especially if one accepts axes of maximum contractional strain rate as proxies for σ_1 trajectories. We therefore adopt a σ_1 trajectory trending $115 \pm 5^\circ$ as the best estimate for the regional stress field in the area of the Darfield earthquake sequence. This is in fact nearly identical to the σ_1 trajectory derived for north Canterbury by Balfour *et al.* (2005) in the southernmost of their stress inversions from focal mechanisms (Fig. 3). It is also comparable with the stress orientations for the northern South Island listed in the 2008 World Stress Map that are derived entirely from focal mechanism inversions (Heidbach *et al.* 2010). Note that this σ_1 orientation is also consistent with predominantly reverse-slip reactivation of structures trending NNE–NE along the Southern Alps range front. Given the mixture of strike-slip and reverse faulting in the Darfield sequence, it seems likely that the stress field is of the form $\sigma_1 > \sigma_v = \sigma_2 \sim \sigma_3$ with local variance between $\sigma_v = \sigma_2$ and $\sigma_v = \sigma_3$.

Stress controls on the initiation and reactivation of strike-slip faults

Initiation of Andersonian wrench (strike-slip) faults

Anderson (1905, 1951) recognized three fundamental stress regimes within the crust, depending which of the principal compressive stresses ($\sigma_1 > \sigma_2 > \sigma_3$) lies vertical in accordance with the boundary condition imposed by the Earth's free surface. He then argued that this should give rise to three basic classes of fault (normal faults when $\sigma_v = \sigma_1$; wrench faults when $\sigma_v = \sigma_2$; thrust faults when $\sigma_v = \sigma_3$) forming in accordance with the Coulomb criterion for brittle shear failure which, for a fluid-saturated rock mass with pore-fluid pressure P_f may be written:

$$\tau = C + \mu_i\sigma_n' = C + \mu_i(\sigma_n - P_f) \qquad (1)$$

where τ and σ_n are the shear and normal stress components on the failure plane, respectively, and the intact rock is characterized by a cohesive strength C and a coefficient of internal friction $\mu_i = \tan\phi_i$ (the friction angle ϕ_i being the slope of the failure envelope on a Mohr diagram; Jaeger & Cook 1979). Shear failure then occurs on planes containing the σ_2 direction oriented at an angle to the maximum compression σ_1 given by:

$$\theta_i = 45^\circ - \frac{\phi_i}{2} = 0.5\tan^{-1}\left(\frac{1}{\mu_i}\right). \qquad (2)$$

From experimental rock mechanics, internal friction generally lies in the range $0.5 < \mu_i < 1.0$ (Jaeger & Cook 1979) so that new faults should lie at $32^\circ > \theta_i > 22^\circ$ to σ_1 and contain the σ_2 axis. For the particular case where $\sigma_v = \sigma_2$, wrench (strike-slip) faults should develop along vertical planes lying at 22–32° to horizontal σ_1, with the possibility of conjugate wrench faults forming at a dihedral angle of 44–64° (Fig. 4). While rock anisotropy may cause significant deviations from these idealized relationships, they often appear to hold for brittle strike-slip faults with low cumulative displacement (e.g. Anderson 1951; Kelly *et al.* 1998).

Frictional reactivation of strike-slip faults

Reshear of an existing fault retaining some cohesive strength c (generally $c < C$, but may be vanishingly small at low normal stresses; Byerlee 1978) is of similar form:

$$\tau = c + \mu_s\sigma_n' = \mu_s(\sigma_n - P_f) \qquad (3)$$

where τ and σ_n are again the shear and normal stress components on the fault plane, P_f is the pore-fluid pressure and μ_s is the coefficient of frictional sliding on the existing plane. Consider the case of an existing fault containing the σ_2 axis oriented at a reactivation angle θ_r to σ_1 (Fig. 4). The optimal orientation of the fault with respect to σ_1 for frictional reactivation (when the differential stress required for reshear is a minimum) is given by:

$$\theta_r^* = 0.5\tan^{-1}\left(\frac{1}{\mu_s}\right). \qquad (4)$$

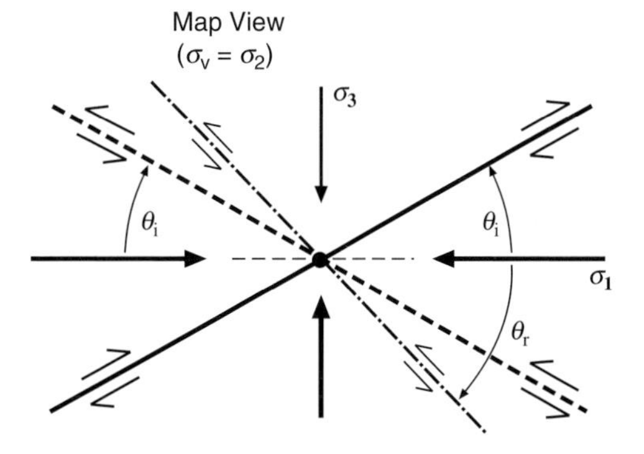

Fig. 4. Cartoon map illustrating a regional 'wrench fault' stress regime with $\sigma_v = \sigma_2$ and the expected orientations of a new-forming Andersonian dextral strike-slip fault and conjugate sinistral fault at $\pm\theta_i$ to σ_1. Also shown is the reactivation angle θ_r defined with respect to σ_1 for an existing vertical fault containing the σ_2 axis (dash-dot line).

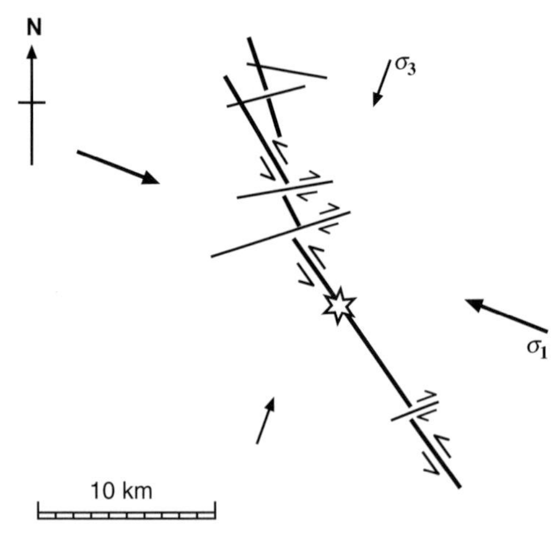

Fig. 5. 2000 M_w6.6 Western Tottori earthquake and inferred stress field (after Ohmi *et al.* 2002; Fukuyama *et al.* 2003). Star, instrumental epicentre; bold lines, main left-lateral strike-slip rupture defined by aftershocks; thinner lines, conjugate right-lateral faults defined by aftershock lineaments.

Note that θ_r^* is always $\leq 45°$. In an extensive set of experiments, Byerlee (1978) demonstrated that for most rocks $0.6 < \mu_s < 0.85$ so that (generally) $30° > \theta_r^* > 25°$. In fact, θ_r^* remains within $30 \pm 5°$ over the broad range of friction coefficients of $0.36 < \mu_s < 0.84$. Faults therefore reactivate most easily when oriented close to their original Andersonian attitudes. While reactivation may occur at $\theta_r < \theta_r^*$ and $\theta_r > \theta_r^*$, frictional lock-up occurs at $\theta_r = 2\theta_r^*$, prohibiting reshear at higher reactivation angles unless the tensile overpressure condition $P_f > \sigma_3$ is met (Sibson 1985).

2000 M_w6.6 Western Tottori earthquake: Andersonian wrench faulting in action

The previously unrecognized fault that ruptured in the 2000 M_w6.6 Western Tottori earthquake in SW Honshu, Japan, is an example of an active strike-slip fault (probably of low cumulative displacement) that approximates to Andersonian character (Fig. 5). No definite surface rupturing was recognized, but focal mechanism analysis and a well-defined planar distribution of aftershocks showed that left-lateral strike-slip rupturing with maximum slip ranging up to *c.* 5 m occurred for over 30 km along strike and to a depth of about 12 km, on a subvertical fault striking 145–150° on average (Ohmi *et al.* 2002; Fukuyama *et al.* 2003; Yukutake *et al.* 2007). The high-resolution aftershock study also shows that the principal NNW–SSE fault is cross-cut and dextrally offset along a set of subsidiary aftershock lineaments defining conjugate right-lateral strike-slip faults at dihedral angles ranging between 45 and 75° (Fig. 6). Although some of these dihedral angles are somewhat in excess of those anticipated from experimental rock

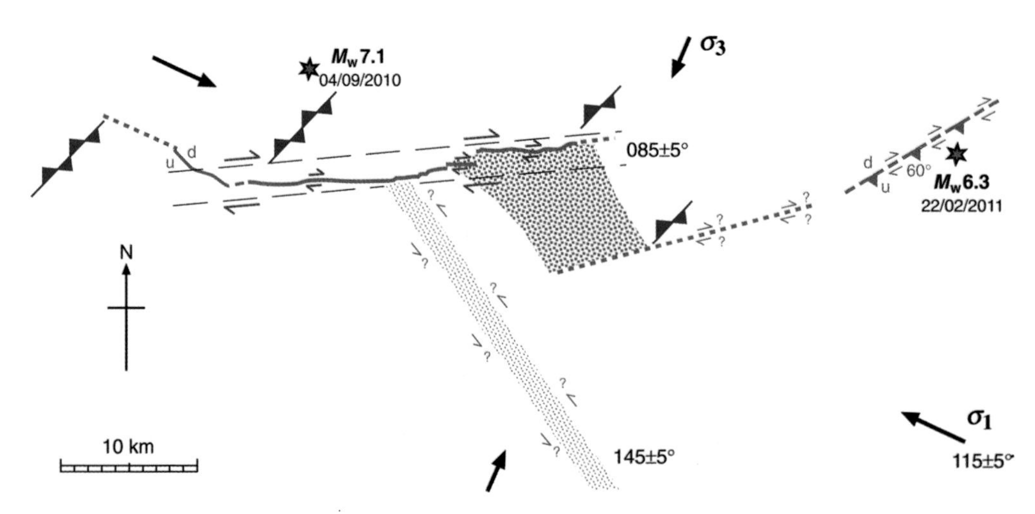

Fig. 6. Seismotectonic cartoon for the 2010–2011 Darfield–Christchurch earthquake sequence. Stars, instrumental epicentres for major shocks; bold red lines, dextral strike-slip surface rupture along Greendale Fault accompanying the 4 September M_w7.1 earthquake (after Quigley *et al.* 2010); dotted red lines, inferred subsurface extensions of surface rupture and aftershock lineaments; dashed toothed red line, inferred dextral-reverse rupture for 22 February M_w6.3 earthquake (Ristau *et al.* 2011); double-toothed blue lines, inferred strike traces of subsurface reverse-slip ruptures (dips uncertain); light stippled band, Norwood–Doyleston–Ellesmere aftershock lineament with inferred left-lateral strike-slip; heavy stippled area, concentrated aftershock activity within possible dilational stepover.

deformation (44–64°), this could result from finite deformation and/or the presence of inherited structures favourably oriented for reactivation. Some stress field heterogeneity is evident locally, but stress inversions from aftershock focal mechanisms define an Andersonian wrench regime with σ_2 subvertical and subhorizontal σ_1 trending 115 ± 10° (Fukuyama *et al.* 2003; Yukutake *et al.* 2007). The vertical left-lateral strike-slip fault therefore lies at 30 ± 10° to regional σ_1 trajectories. Together, the vertical fault plane containing σ_2, the reactivation angle of *c.* 30° and the set of conjugate dextral faults are all consistent with wrench fault development as envisaged by Anderson (1905).

Stress and large-displacement strike-slip faults

Simple Andersonian relationships between regional σ_1 trajectories and fault orientation do not seem to hold for large-displacement continental strike-slip faults. Although some stress heterogeneity is apparent, horizontal σ_1 trajectories lie at up to 85° to the trace of the San Andreas fault in central California and at 60–70° to the fault in southern California (Mount & Suppe 1987; Zoback *et al.* 1987; Townend & Zoback 2004). Likewise, stress inversions in the northern South Island suggest horizontal σ_1 trajectories lying at *c.* 60° to the main strands of the Marlborough fault system (Balfour *et al.* 2005). These relationships may be explained, at least in part, by the significant block rotations expected to occur about vertical axes in areas of distributed continental deformation (Nur *et al.* 1986; Lamb 1988). Nonetheless, continued slip activity on these structures, oriented at beyond frictional lock-up for Byerlee's (1978) friction values, implies extreme weakness of the large-displacement faults that is not easily accounted for either by the presence of low-friction fault material or by near-lithostatic fluid overpressures.

Regional stress and the 2010 Darfield earthquake sequence

A seismotectonic cartoon (Fig. 6) illustrates the geometrical relationships between the regional stress field (horizontal σ_1 trending 115 ± 5°) and the principal subvertical strike-slip rupture striking 085 ± 5° in the Darfield sequence (Greendale Fault), along with other subsidiary features defined by the aftershock distribution and focal mechanisms provided by Gledhill *et al.* (2011). The subvertical dextral strike-slip rupture lies at 30 ± 10° to inferred σ_1, the expected relationship for a new-formed strike-slip fault (Anderson 1905) or for an existing fault at the optimal orientation for frictional reactivation in a stress field with $\sigma_v = \sigma_2$.

Note that the western termination of the Greendale Fault surface rupture, which exhibits a component of normal slip (Quigley *et al.* 2010), curves NW to lie subperpendicular to the inferred σ_3 direction. At the eastern termination of the Greendale surface rupture, a rhomboidal area of concentrated aftershock activity with mixed strike-slip and extensional normal faulting (Gledhill *et al.* 2011) defines a stepover structure to an en echelon microearthquake lineament trending ENE towards the Heathcote–Avon estuary. This is the typical expression of a dilational stepover in a dextral strike-slip fault system (Sibson 1986); such features are commonly associated with time-dependent slip transfer to the en echelon fault segment. During the 22 February M_w6.3 aftershock, dextral-reverse oblique slip occurred on a more northerly striking (*c.* 060°) fault dipping *c.* 60°SE from below the Heathcote–Avon estuary (Ristau *et al.* 2011). While this major aftershock likely involves reactivation of an inherited fault structure, the kinematics remain compatible with the inferred stress field (Fig. 6).

Within such a regional stress regime, a conjugate left-lateral strike-slip fault would be expected to be subvertical and strike *c.* 145 ± 5°, which is the approximate trend of an aftershock lineament (Norwood–Doyleston–Ellesmere lineament) discernible south of the Greendale rupture running SE towards the mouth of Lake Ellesmere (Figs 2 & 6). Left-lateral strike-slip along a lineament of this orientation is compatible with a number of the focal mechanisms provided for this area by Gledhill *et al.* (2011). If this 145° trending lineament is indeed a left-lateral strike-slip fault, it confirms the inferred 115° trend for the regional σ_1 trajectory in the area of the Darfield earthquake sequence. A degree of aftershock clustering occurs around the intersection of the two lineaments. It seems likely that sinistral strike-slip along this subsidiary lineament may be interfering with the main dextral fault structure, offsetting it left-laterally to create a local contractional jog as occurs further east along the segmented rupture trace. A mesh structure of conjugate right- and left-lateral faults may contribute to areas of distributed aftershock activity.

Predominantly reverse-slip rupturing on NNE–NE striking planes in the area of the mainshock epicentre south of Darfield and immediately SW of the western termination of the Greendale rupture is also broadly compatible with the inferred σ_1 trajectory, but with $\sigma_v = \sigma_3$. Given the preponderant mixture of strike-slip and reverse fault aftershocks in the Darfield sequence, it seems likely that the stress field is transitional between a wrench and a thrust regime with $\sigma_1 > \sigma_v = \sigma_2 \sim \sigma_3$.

Conclusions

Not withstanding the evidence of reverse-slip associated with the nucleation of the mainshock rupture (Gledhill *et al.* 2011), right-lateral strike-slip rupturing along a subvertical plane oriented at 30 ± 10° to regional σ_1 trajectories trending 115 ± 5° together with possible conjugate left-lateral faults at *c.* 60° to the principal rupture (Fig. 6) all indicate a classic Andersonian wrench regime with $\sigma_v = \sigma_2$ (Anderson 1905, 1951). The structural geometry is comparable to the evidence for conjugate Andersonian wrench faulting visible in the aftershock distribution of the 2000 M_w6.6 Western Tottori earthquake in central Honshu, Japan, where the main left-lateral strike-slip fault structure is offset by a set of conjugate right-lateral strike-slip faults oriented at 50–70° to the main structure (Ohmi *et al.* 2002; Fukuyama *et al.* 2003).

Orientation of the main dextral structure (the Greendale Fault) at the optimal angle for reactivation coupled with the evidence of persistent activity on conjugate structures makes it likely that the cumulative strike-slip on the Greendale Fault is comparatively low (<1 km?). It could represent either a fault that is fairly new-formed in the contemporary stress field or an existing subvertical fault that is optimally oriented for reshear in the stress field. From the complexity of the subsidiary aftershock trends and their disparate mechanisms it appears probable that the structure is growing by amalgamating inherited basement structures, perhaps largely derived from Late Cretaceous–Paleogene extensional tectonics (Field *et al.* 1989), with optimally oriented strike-slip segments formed in the contemporary stress field. With time and increasing displacement, the east–west Greendale Fault will likely evolve into a 'smoother' less complex strike-slip fault (Wesnousky 1988) and will perhaps rotate progressively towards parallelism with the ENE-trending faults to the north and become integrated into the Marlborough fault system.

We express our thanks to OMV New Zealand LTD and their joint venture partners Mitsui E&P Australia Pty Ltd and PTTEP New Zealand LTD for permission to publish the results of the Galleon-1 breakout analysis and to Occam Technology PTY for digitization of HDT and SHDT field data. Thanks also to M. Reyners, J. Beavan and J. Townend for helpful discussions on the aftershock distribution and focal mechanisms and on stress/strain rate relationships. Helpful comments from M. Tingay and an anonymous reviewer significantly improved the manuscript.

References

ANDERSON, E. M. 1905. The dynamics of faulting. *Transactions of the Edinburgh Geological Society*, **8**, 387–402.

ANDERSON, E. M. 1951. *The Dynamics of Faulting and Dyke Formation with Application to Britain*. 2nd edn. Oliver & Boyd, Edinburgh.

ANDERSON, H., WEBB, T. & JACKSON, J. 1993. Focal mechanisms of large earthquakes in the South Island of New Zealand: implications for the accommodation of Pacific-Australia plate motion. *Geophysical Journal International*, **115**, 1032–1054.

BALFOUR, N. J., SAVAGE, M. K. & TOWNEND, J. 2005. Stress and crustal anisotropy in Marlborough, New Zealand: evidence for low fault strength and structure-controlled anisotropy. *Geophysical Journal International*, **163**, 1073–1086.

BEANLAND, S., BERRYMAN, K. R., HULL, A. G. & WOOD, P. R. 1986. Late Quaternary deformation at the Dunstan Fault, Central Otago, New Zealand. *Royal Society of New Zealand Bulletin*, **24**, 293–306.

BEAVAN, J. & HAINES, J. 2001. Contemporary horizontal velocity and strain-rate fields of the Pacific-Australia plate boundary zone through New Zealand. *Journal of Geophysical Research*, **106**, 741–770.

BERRYMAN, K. 1979. Active faulting and derived PHS directions in the South island, New Zealand. *Royal Society of New Zealand Bulletin*, **18**, 29–34.

BYERLEE, J. D. 1978. Friction of rocks. *Pure & Applied Geophysics*, **116**, 615–626.

DE METS, C., GORDON, R. G., ARGUS, D. F. & STEIN, S. 1990. Current plate motions. *Geophysical Journal International*, **101**, 425–478.

FIELD, B. D., BROWNE, G. H. ET AL. 1989. *Cretaceous and Cenozoic Sedimentary Basins and Geological Evolution of the Canterbury Region, South Island, New Zealand. New Zealand Geological Survey Basin Studies 2*. Department of Scientific and Industrial Research, Wellington.

FISHER, N. I. 1996. *Statistical Analysis of Circular Data*. Cambridge University Press, UK.

FORSYTH, P. J., BARRELL, D. J. A. & JONGENS, R. 2008. *Geology of the Christchurch area. Geological Map 16, 1:250,000*. GNS Science, Lower Hutt, New Zealand.

FUKUYAMA, E., ELLSWORTH, W. L., WALDHAUSER, F. & KUBO, A. 2003. Detailed fault structure of the 2000 Western Tottori, Japan, earthquake sequence. *Bulletin of the Seismological Society of America*, **93**, 1468–1478.

GHISETTI, F. C., GORMAN, A. R. & SIBSON, R. H. 2007. Surface breakthrough of a basement fault by repeated seismic slip episodes: the Ostler Fault, South Island, New Zealand. *Tectonics*, **26**, TC6004, doi: 10.1029/2007TC002146.

GLEDHILL, K., RISTAU, J., REYNERS, M., FRY, B. & HOLDEN, C. 2011. The Darfield (Canterbury, New Zealand) M_w 7.1 earthquake of September 4 2010: a preliminary seismological report. *Seismological Research Letters*, **82**, 378–386.

HEIDBACH, O., TINGAY, M., BARTH, A., REINECKER, J., KURFEß, D. & MÜLLER, B. 2010. Global crustal stress pattern based on the World Stress Map database release 2008. *Tectonophysics*, **482**, 3–15.

JAEGER, J. C. & COOK, N. G. W. 1979. *Fundamentals of Rock Mechanics*. 3rd edn. Chapman & Hall, London.

KEIDING, M., LUND, B. & ÁRNADÓTTIR, T. 2009. Earthquakes, stress, and strain along an obliquely divergent plate boundary: Reykjanes Peninsula, southwest Iceland. *Journal of Geophysical Research*, **114**, B09306, doi: 10.1029/2008JB006253.

KELLY, P. G., SANDERSON, D. J. & PEACOCK, D. C. 1998. Linkage and evolution of conjugate strike-slip fault zones in limestones of Somerset and Northumbria. *Journal of Structural Geology*, **20**, 1477–1493.

LAMB, S. H. 1988. Tectonic rotations about vertical axes during the last 4 Ma in part of the New Zealand plate-boundary zone. *Journal of Structural Geology*, **10**, 875–893.

LEITNER, B., EBERHART-PHILLIPS, D., ANDERSON, H. & NABELEK, J. 2001. A focused look at the Alpine Fault, New Zealand: seismicity, focal mechanisms and stress observations. *Journal of Geophysical Research*, **106**, 2193–2220.

LITCHFIELD, N. J. & NORRIS, R. J. 2000. Holocene motion on the Akatore fault, south Otago coast, New Zealand. *New Zealand Journal of Geology and Geophysics*, **43**, 405–418.

MCGINTY, P., REYNERS, M. & ROBINSON, R. 2000. Stress directions in the shallow part of the Hikurangi subduction zone, New Zealand, from the inversion of earthquake first motions. *Geophysical Journal International*, **142**, 339–350.

MOORE, M. A., ANDERSON, H. J. & PEARSON, C. 2000. Seismic and geodetic constraints on plate boundary deformation across the northern Macquarie Ridge and southern South Island of New Zealand. *Geophysical Journal International*, **143**, 847–880.

MOUNT, V. S. & SUPPE, J. 1987. State of stress near the San Andreas fault. *Geology*, **15**, 1143–1146.

NICOL, A. & WISE, D. U. 1992. Paleostress adjacent to the Alpine Fault of New Zealand: fault, vein, and stylolite data from the Doctors Dome area. *Journal of Geophysical Research*, **97**, 17 685–17 692.

NUR, A., RON, H. & SCOTTI, O. 1986. Fault mechanics and the kinematics of block rotations. *Geology*, **14**, 746–749.

OHMI, S., WATANABE, K., SHIBUTANI, T., HIRANO, N. & NAKAO, S. 2002. The 2000 Western Tottori earthquake – seismic activity revealed by the regional seismic networks. *Earth Planets Space*, **54**, 819–830.

PEARSON, C. 1993. Rate of coseismic strain release in the northern South Island, New Zealand. *New Zealand Journal of Geology and Geophysics*, **36**, 161–166.

PEARSON, C. 1994. Geodetic strain determinations from the Okarito and Godley-Tekapo regions, central South Island, New Zealand. *New Zealand Journal of Geology and Geophysics*, **37**, 309–318.

PEARSON, C., BEAVAN, J., DARBY, D., BLICK, G. H. & WALCOTT, R. I. 1995. Strain distribution across the Australian-Pacific plate boundary in the central South Island, New Zealand, from 1992 GPS and earlier terrestrial observations. *Journal of Geophysical Research*, **100**, 22 071–22 081.

PETTINGA, J. R. & WISE, D. U. 1994. Paleostress adjacent to the Alpine fault: broader implications from fault analysis near Nelson, South Island, New Zealand. *Journal of Geophysical Research*, **99**, 2727–2736.

QUIGLEY, M., VILLAMOR, P. *ET AL.* 2010. Previously unknown fault shakes New Zealand's South Island. *EOS, Transactions American Geophysical Union*, **91**, 469–472.

QUIGLEY, M., VAN DISSEN, R. *ET AL.* 2012. Surface rupture during the 2010 M_w 7.1 Darfield (Canterbury) earthquake: implications for fault rupture dynamics and seismic-hazard analysis. *Geology*, **40**, 55–58.

RATTENBURY, M. S., TOWNSEND, D. B. & JOHNSTON, M. R. 2006. *Geology of the Kaikoura Area. Geological Map 13, 1:250,000*. GNS Science, Lower Hutt, New Zealand.

REILLY, W. I. 1990. Horizontal crustal deformation on the Hikurangi Margin. *New Zealand Journal of Geology and Geophysics*, **33**, 393–400.

REINECKER, J., TINGAY, M. & MÜLLER, B. 2003. *Borehole breakout analysis from four-arm caliper logs*. World Stress Map Project Stress Analysis Guidelines (available online at www.world-stress-map.org).

RISTAU, J., BANNISTER, S. *ET AL.* 2011. The M_w6.3 Christchurch, New Zealand, earthquake of 22 February 2011: preliminary seismic and geodetic results (Abstract). *Seismological Society of America 2011 Annual Meeting*, April 13–15, Memphis Tennessee.

SIBSON, R. H. 1985. A note on fault reactivation. *Journal of Structural Geology*, **7**, 751–754.

SIBSON, R. H. 1986. Rupture interaction with fault jogs. *In:* DAS, S., BOATWRIGHT, J. & SCHOLZ, C. H. (eds) *Earthquake Source Mechanics*. American Geophysical Union, Washington, Monograph 37 (Maurice Ewing Series 6), 157–167.

SIBSON, R. H., GHISETTI, F. C. & RISTAU, J. 2011. Stress control of an evolving strike-slip fault system during the 2010-2011 Canterbury, New Zealand, earthquake sequence. *Seismological Research Letters*, **82**, 824–832.

TCHALENKO, J. S. 1970. Similarities between shear zones of different magnitudes. *Geological Society of America Bulletin*, **81**, 1625–1640.

TOWNEND, J. & ZOBACK, M. D. 2004. Regional tectonic stress near the San Andreas fault in central and southern California. *Geophysical Research Letters*, **31**, L15S11, doi: 10.1029/2003GL018918.

TOWNEND, J. & ZOBACK, M. D. 2006. Stress, strain, and mountain building in central Japan. *Journal of Geophysical Research*, **111**, B03411, doi: 10.1029/2005JB003759.

WALCOTT, R. I. 1984. The kinematics of the plate boundary zone through New Zealand: a comparison of short-and long-term deformations. *Geophysical Journal of the Royal Astronomical Society*, **79**, 613–633.

WALLACE, L. M., BEAVAN, J., MCCAFFREY, R., BERRYMAN, K. & DENYS, P. 2007. Balancing the plate motion budget in the South Island, New Zealand, using GPS, geological and seismological data. *Geophysical Journal International*, **168**, 332–352.

WESNOUSKY, S. G. 1988. Seismological and structural evolution of strike-slip faults. *Nature*, **335**, 340–343.

WILSON, I. R., RENTON, P. H., MOUND, D. G. & GRANT, J. 1985. *Galleon-1 Geological Completion Report PPL 38203*, BP Shell Todd (Canterbury) Services. Unpublished Petroleum Report PR 1146 (http://www.crownminerals.govt.nz).

WOOD, R. A. & HERZER, R. H. 1993. The Chatham Rise, New Zealand. *In:* BALLANCE, P. F. (ed.) *South Pacific Sedimentary Basins, Sedimentary Basins of the World 2*. Elsevier, Amsterdam, 329–349.

YUKUTAKE, Y., IIO, Y., KATAO, H. & SHIBUTANI, T. 2007. Estimate of the stress field in the region of the 2000 Western Tottori earthquake: using numerous aftershock focal mechanisms. *Journal of Geophysical Research*, **112**, B09306, doi: 10.1029/2005JB004250.

ZOBACK, M. D. 2007. *Reservoir Geomechanics*. Cambridge University Press, Cambridge.

ZOBACK, M. D., ZOBACK, M. L. *ET AL.* 1987. New evidence on the state of stress on the San Andreas fault system. *Science*, **238**, 1105–1111.

Andersonian and Coulomb stresses in Central Costa Rica and its fault slip tendency potential: new insights into their associated seismic hazard

ALLAN LÓPEZ

Ingeniería Geológica, ICE, Centro de Investigaciones en Ciencias Geológicas-UCR, Universidad Latina de Costa Rica (e-mail: allan.lopez@geologos.or.cr)

To the memory of Jacques Angelier, 1947–2010

Abstract: Plate boundary forces acting within the Cocos Plate that is being subducted at a rate of 8.5–9.0 cm a^{-1} towards N32°E below the Caribbean Plate and the Panama microplate are found responsible for contemporaneous superimposed compressive, wrench and extensive fault patterns in Central Costa Rica. The stress inversion of fault-slip planes and focal mechanisms reveals a prevailing convergence-imposed N20°–45°E almost horizontal compression. Ellipsoid R values $[R = (\sigma_1 - \sigma_2)/(\sigma_2 - \sigma_3)]$ in the range of 0.3–0.05 and 0.8–0.93 are responsible for the permutation of σ_2 to σ_3 and σ_2 to σ_1, respectively, and show typical Andersonian configurations with one stress axis vertical or close to it. Coulomb failure stress (CFS) analysis reveals that up to 5 bars (0.5 MPa) of tectonic loading are being imposed on east–west thrusts and on critically oriented conjugate NW- and NE-trending strike-slip faults. Non-optimally oriented structures are potential targets for reactivation even with 2 bars (0.2 MPa) of load. Triggering and interaction with volcanic activity is highly suspected in one documented recent case. When the regional fault population was tested for its slip tendency (τ/σ_n), a good correlation with CFS results was found.

The central region of Costa Rica, where the majority of the population and the social and economic infrastructure are concentrated, is affected by continuous seismic activity and has been the scene of several destructive earthquakes in the last centuries and years. For this reason, it is vital to know the geometry and mechanical properties of active and potentially active faults acting as seismic sources to adequately assess the seismic hazard and its impact on national development.

Montero & Morales (1990), López (1999), Montero (2001) and Lewis *et al.* (2008) have modelled the regional stress and strain fields with different databases and identified the main regional faults with evidence of recent activity. These studies have revealed a number of important structures, for example the Aguacaliente fault which has hosted earthquakes that destroyed the city of Cartago on several occasions (the last in 1910). However, a geometric and numerical analysis on the reactivation potential of such faults has been lacking. The purpose of this publication is to help overcome this objective constraint through the joint integrated implementation, for the first time, of several analytical methods that have proven valuable analytical tools for the appropriate assessment of the seismic hazard and risk. The implemented analysis is described below.

Tectonic setting

Costa Rica is located at the SW corner of the Caribbean Plate and to the NE of the Middle America trench. There the rough and smooth sea bottom of the Cocos Plate subducts beneath the Caribbean Plate along the 5 Ma old Middle America trench at a velocity of 8.3 and 9.0 cm a^{-1} in the NW and SE, respectively, towards N32°E (DeMets *et al.* 1994; Protti *et al.* 1996). The bathymetry of the northern part is smooth, created at the East Pacific Rise, while the rough southern sector comes from the Cocos–Nazca spreading centre and is thermally and chemically affected at it passes over the Galapagos hotspot. This interaction has created several inactive volcanic ridges, almost perpendicular to the trench and as high as 2000 m above sea bottom (Fig. 1). This morphology and the relative upper plate buoyancy has a variable influence on the current deformation response and seismicity rates along the erosive convergent margin, trench and Benioff zone (Von Huene *et al.* 2000; Ranero & Von Huene 2000; Arroyo *et al.* 2009; Linkimer *et al.* 2010) and induces indenter tectonics in the central and south sectors of the Caribbean Plate in Costa Rica (Gardner *et al.* 1992; López 1999). In the northern part, the seismicity is not locked due to its steeper subduction angle and smooth seafloor

From: Healy, D., Butler, R. W. H., Shipton, Z. K. & Sibson, R. H. (eds) 2012. *Faulting, Fracturing and Igneous Intrusion in the Earth's Crust.* Geological Society, London, Special Publications, **367**, 19–38. http://dx.doi.org/10.1144/SP367.3

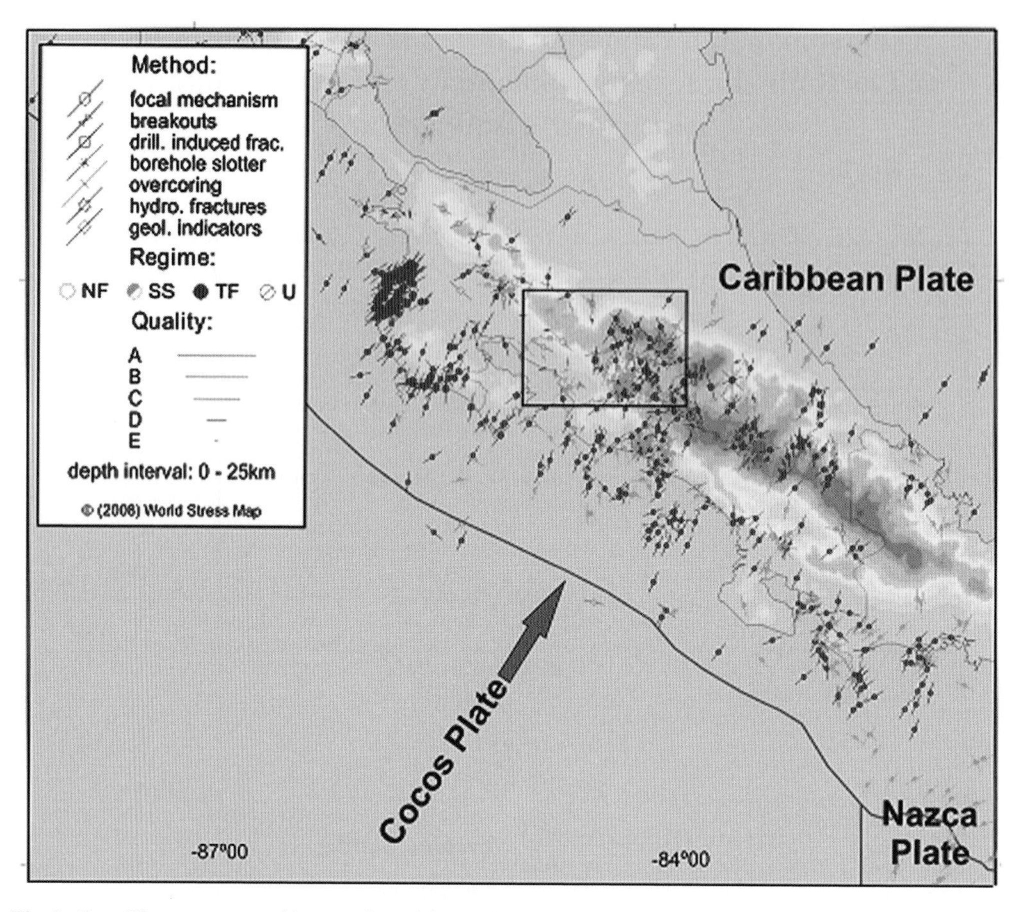

Fig. 1. Costa Rica stress map and its tectonic setting showing the location of the Cocos Plate, which is being subducted at a rate of 8.5–9.0 cm a^{-1} towards N32°E below the Caribbean Plate and the Nazca Plate. The plotted data is larger than the WSM available sets and is currently being enlarged by the author. The box symbols and their shading refer to the WSM applied methods.

with background activity and undetermined recurrence intervals for highly destructive earthquakes. Recent studies (Climent *et al.* 2008) also suggest high seismic hazard to Central Costa Rica and in southern Costa Rica–northern Panamá where the triple junction of the Nazca–Cocos–Caribbean Plates creates a complex and active seismotectonic setting (DeMets 2001).

Central Costa Rica lies halfway along the igneous arc backbone of this segment of the Caribbean Plate, located some 150 km from the subduction zone, and is characterized by a range of volcanic styles (Fernãndez *et al.* 1998). Abundant acidic batholiths and extrusive deposits occur to the SE in the extinct Cordillera de Talamanca arc (Denyer & Arias 1991). Volcanism is active to the NE with numerous Neogene to modern volcaniclastic and sedimentary basins that record repeated phases of intrusion, uplift, erosion, shearing and thrusting (Alvarado *et al.* 1988; Arias & Denyer 1991*a*, *b*, 1994; Barquero *et al.* 1995; Colombo 1997). This region, referred to by some workers as the Panamá block or the Panamá microplate, constitutes the boundary between blocks along the western margin of the Caribbean Plate (Montero & Alvarado 1995; Fernández & Pacheco 1996, 1998; Fernández & Montero 2002; Montero 1999, 2001). An ongoing debate regarding its existence and the evolution of the concept has been summarized by Fernández (2011). There is agreement on the existence of the Central Costa Rica Deformed Belt (CCRDB) (Marshall 2000; Montero 2001) as a system of NW dextral and NE sinistral shears, and this model can be traced back to Alvarado Villalón (1984) who was the first to decipher this framework in the southern part of the central valley. The CCRDB is an arcuate zone of diffuse strain at the map scale that crosses the Isthmus from west to east along

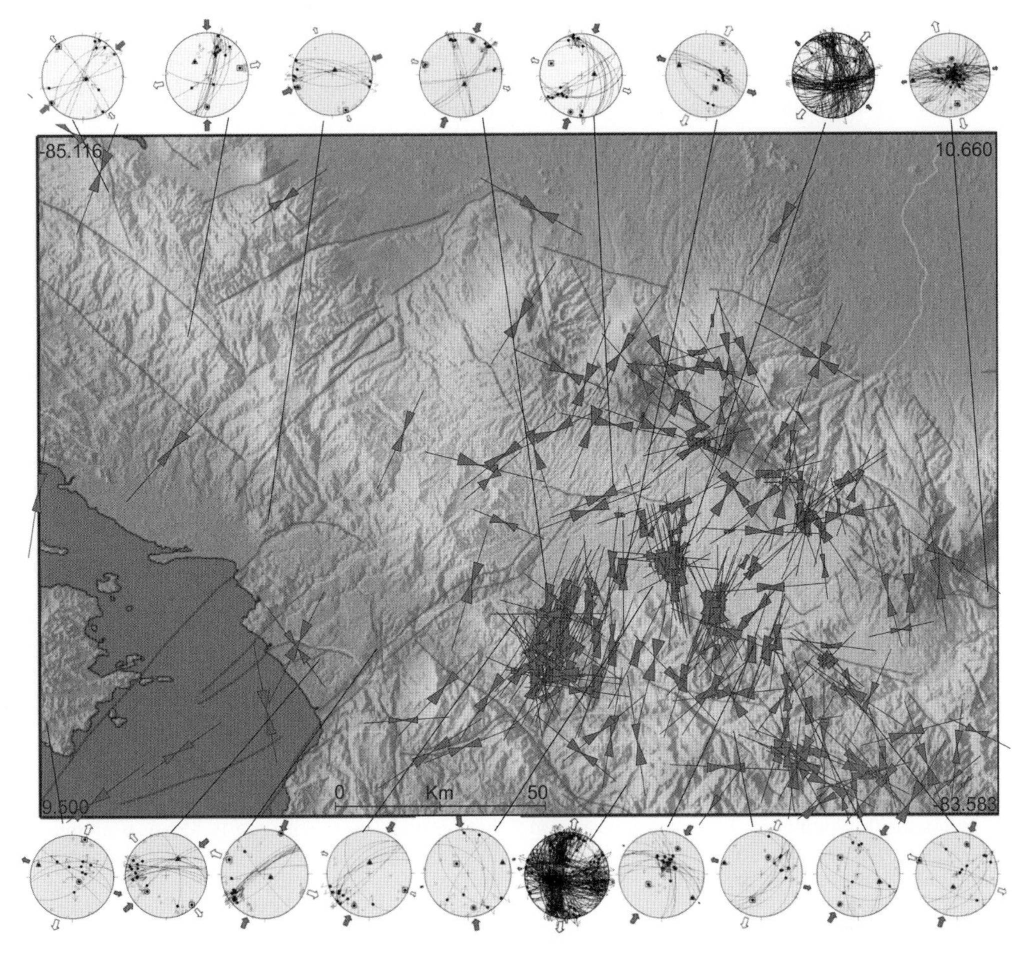

Fig. 2. Main tectonic elements of Central Costa Rica illustrating the changes in direction of the SH (arrows) gathered from the different stress methods and revealing the existence of three different contemporaneous and partially superimposed stress fields. The first order is controlled by the plate-scale boundary force of the Cocos Plate convergence and the second and third by neotectonic regional faulting and density contrast between the volcanic backbone and the intermontane basins and subordinated local inclusions and detachment horizons.

middle Costa Rica (Fig. 1). López (1999) explains the broad development of this zone at different scales and locations in response to stresses driven by the Cocos ridge collision–indentation relations, referring to it as an X fault pattern due to its conjugate style. Fisher *et al.* (1994) and Gardner *et al.* (2000) presented similar views for the Central Pacific as did Denyer & Arias (1994) who showed several examples with similar kinematics and trends.

This study covers the area between latitudes 9.400 and 10.500 and longitudes −85.000 and −83.450, an area of 12 500 km^2 (Fig. 2). All the structures shown are modified from the Atlas Tectónico de Costa Rica (Denyer *et al.* 2003) with superposed indicators of maximum horizontal compression orientations (SH) and a few representative stress tensor fault-slip stations (Fig. 2). The analytical tools are described in the following section.

Neotectonic and present-day stress field

A database of 483 focal mechanism solutions available from the author and 57 outcrop-scale fault kinematic observations were compiled and used to decipher the contemporaneous and recent stress fields, respectively. The focal mechanisms were carefully selected and only reliable data with good seismograph coverage, relevant magnitudes and a good distribution of polarities were used. Datasets were provided by the Red Sismológica Nacional (RSN) de Costa Rica, Harvard Centroid Moment Tensor (CMT) and US Geological Survey (USGS) solutions and the data of Quintero & Güendel (2000) were also employed. The mesoscopic fault data were collected in Upper Tertiary and Quaternary outcrops where the kinematics are unambiguously preserved. The tensor results were subject to the regulations of the World Stress Map (http://dc-app3-14.gfz-potsdam.de/). The fault data were provided by López (1999), Marshall (2000) and López *et al.* (2008*a*, *b*). Special attention was paid to discarding artificial tensors during the inversions and to use only results from the recent tectonic interval. In general, the inversions produced reliable tensors regardless of the size of the input dataset as long as the quality was good. Poor-quality input data were avoided as they tended to produce unreliable tensors.

This combination of results is relevant since they are the only tectonic stress sources and the output of the inversion of focal mechanisms or of their selected nodal planes does not need to be rotated. This is not the case for mesoscopic fault-slip data, which have been retilted in several cases to avoid the misleading effect of the active tectonism characteristics of the Caribbean Plate interaction on the overall rock mass. Celerier (2008) investigated the genetic relationships between Andersonian stress, reactivation, friction laws and the use of geometrical plotting of focal mechanisms to isolate the different regimes as a means of detecting anomalous behaviours. A preliminary scenario of the SH is obtained from the orientation of the P axes of every thrust and strike-slip focal mechanisms solutions and B axes of normal faults. Within the region covered by this study, they show a general NE trend with local fluctuations trending, in some cases, NW and NWW as depicted in Figure 2.They also follow the regional trend observed in Figure 1 (López *et al.* 2008*a*, *b*).

The dip-pitch graph of De Vicente (1988) was applied to identify the active nodal plane geometry for each focal solution. This method has proven to be very reliable and its strength lies in its ability to select the true moving plane from a given population event by event, independent of the whole set as for other methods whose results may change if the original dataset is increased or reduced. Aftershock lineaments also confirm its robustness and tectonic logic. Those planes belonging to well-constrained crustal subvolumes, ranging from 2 to 27 km depth, were inverted in a similar fashion as the fault data to obtain representative stress tensors.

These analyses are only spatial since time data are relatively few for formal inversions and their output confirmed the SH trends. The importance of such differentiation needed to understand stress changes after large earthquakes is however not underestimated, although its application at this moment is limited by the fact that the database of instrumental focal mechanisms only covers the last 30 years.

The geometrical analysis was performed plotting all the focal mechanisms in triangular plots and separating them according to the World Stress Map (WSM) guidelines. Well-constrained sets were obtained for three tectonic regimes and an unknown set for those focal mechanism solutions with inclined axes not belonging to any of the other groups, although they plot according to less restrictive criteria (Fig. 3). The identified datasets consist of 83 pure normal seismic faults with one group indicating NW–SE to NNE–SSW striking extension and another NS–NE oriented for NWW–SEE extension. Two groups of 104 strike-slip fault planes are present; one corresponds to a symmetrical conjugate set with NE–SW oriented SH and a second to a much less populated set with SH towards NW. Two contrasting groups are also depicted by 102 reverse nodal planes, sharing the above-mentioned contrasting SH. Truly transtensive and transpressive populations are rather small and vanished during the inversions; perhaps they are not well constrained for central Costa Rica in the same manner as Celerier found for another context. Those mechanisms with all three axes having moderate plunges between 25° and 45° or both *P* and *T* axes with nearly identical plunges in the range of 40–50° (Zoback 1992; Zoback & Burke 1993; Zobach & Mooney 2003; Zoback & Zoback 1989, 1991, 2002) belong to the unknown regime. This group consists of 104 planes which are useless for stress regimes and tensor determinations, yet they have a WNW SH trend consistent with the third rank. In general, every tectonic stress regime shows Andersonian configurations; one of the stress axes is vertical or close to it within a small cluster and the related tensors are even more consistent despite the depth range (Fig. 3).

The results (Table 1) are consistent with the Cocos Plate convergence N32°E direction, but the local deflections in the SH trajectories at intersections with major tectonovolcanic edifices and regional faults is also evident (Fig. 2). The standardized quality ranking scheme of the WSM was applied providing mainly A, B and C indicators (Sperner *et al.* 2003; Heidbach *et al.* 2007, 2010), and was used during the inversions utilizing Win-Tensor software of D. Delvaux (2011). The *R* value ($R = \sigma_1 - \sigma_2/\sigma_2 - \sigma_3$) is the stress ellipsoid shape value, N_{original} is the original number of data for each population, N_{final} is the remaining data after the inversion and S_{Hmax} is the horizontal projected trend of the maximum compression. It is

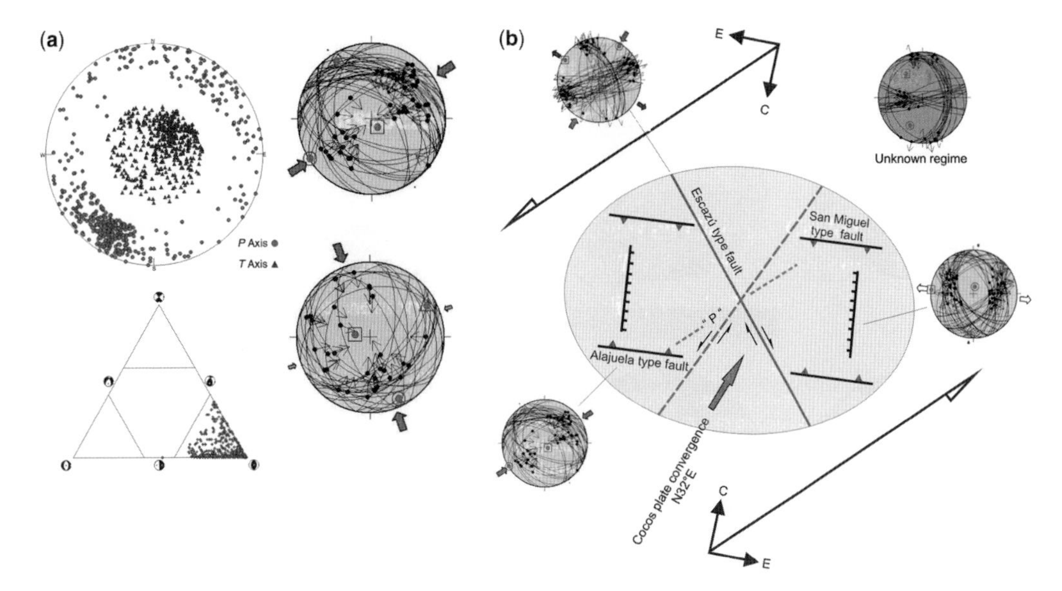

Fig. 3. (**a**) P–T axes distribution and triangular plotting for the 102 reverse selected nodal planes. They have the highest M_w magnitudes because of its genetic link with the Cocos Plate convergence. The stress tensor stereograms clearly depict the first NE–SW SH rank and the NW-trending SH. In most of the populations of the other stress regimes, the Andersonian configurations prevail with one of the *P*, *B* or *T* axes within a small vertical cluster; in the inversions, this mechanical conduct is very well constrained for all depth ranges. (**b**) The facts confirm the theory in this synthetic strain couple with nodal strike-slips parallel to the Escazú type faults as well as its conjugate set with Alajuela and San Miguel-type reverse structures perpendicular to the compressive vector and this one at 11° off the Cocos convergence direction plus orthogonal normal faulting and 'P' fractures.

found that Anderson's theory satisfactorily explains the results of the nodal planes inversion with one of the stress axes nearly vertical, even at hypocentre depths up to 25 km.

Slip tendency

This analytical tool is based on the Amonton law which relates fault stability to the ratio of shear

Table 1. *Tensor parameters for the most representative stress configurations. The first two letters denote the stress regime (normal, Strike-slip, thrust faulting, U is unknown) and the third, fourth and fifth denote their orientations. The* R *value* ($R = \sigma_1 - \sigma_2/\sigma_2 - \sigma_3$) *is the stress ellipsoid shape value, QWSM is the quality ranking system of the WSM (A being the highest quality and E the lowest). A quality means that the orientation of the maximum horizontal compressional stress SH is accurate to within* ±15°*, B quality to within* ±20°*, C quality to within* ±25°*, and D quality to within* ±40°*.* $N_{original}$ *is the original number of data for each population,* N_{final} *is the remaining data after the inversion and* S_{Hmax} *is the horizontal projected trend of the maximum compression*

Tensor	σ_1	σ_2	σ_3	R	QWSM	N_{original}	N_{final}	S_{Hmax}
1NS	201/05	97/68	293/21	0.40	A	58	55	022
NF NNE	343/81	107/04	197/07	0.31	D	83	17	110
NF NNW	022/89	197/00	287/00	0.26	C	83	46	018
SS NE	031/07	165/79	300/07	0.38	C	104	60	027
SS NW	334/10	122/77	243/06	0.06	D	104	27	149
TF NE	237/01	328/11	143/78	0.52	B	102	48	056
TF NNW	154/10	061/13	278/73	0.70	C	102	32	159
U						104	104	078

The R value ($R = \sigma_1 - \sigma_2/\sigma_2 - \sigma_3$) is the stress ellipsoid shape value, N_{original} is the original number of data for each population, N_{final} is the remaining data after the inversion and S_{Hmax} is the horizontal projected trend of the maximum compression.

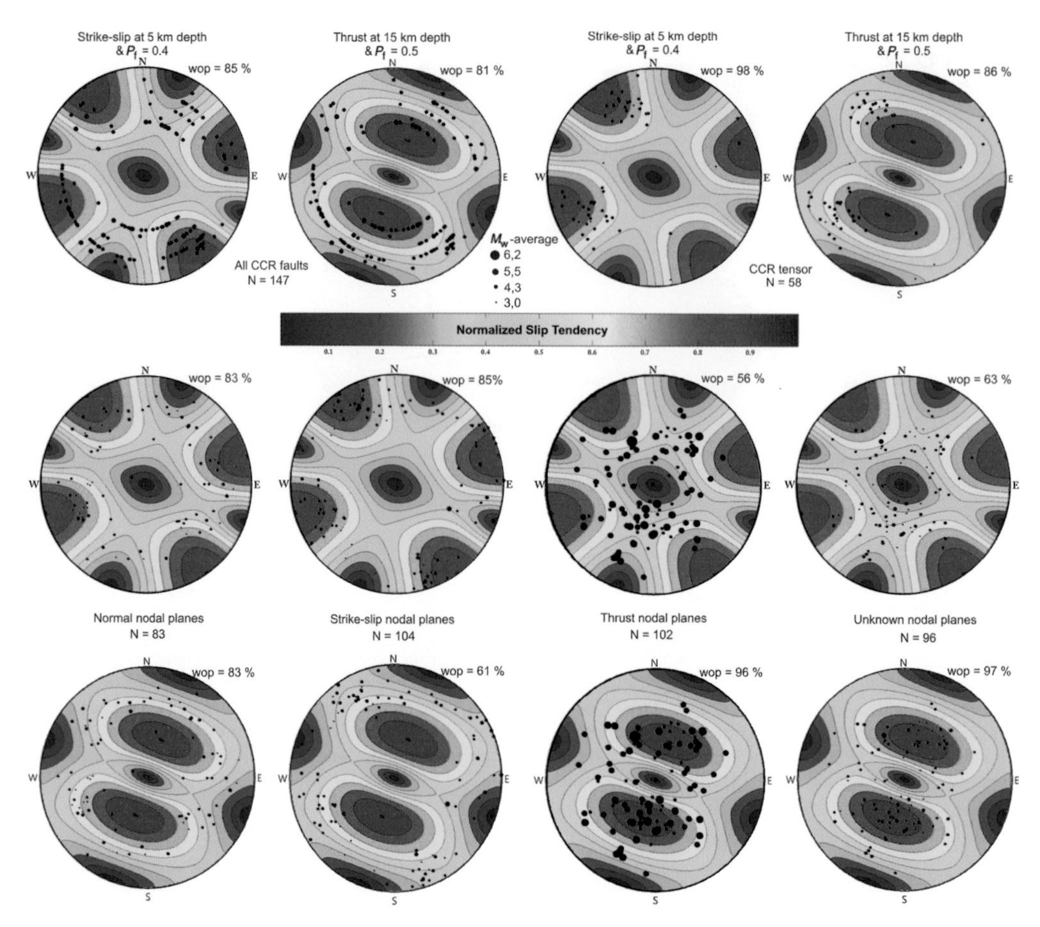

Fig. 4. Normalized T_s stereograms for different stress regimes and depths illustrating the fault plane pole distribution for the total 147 CCR structures with 85% of well-oriented planes for the strike-slip scenario and 81% for the compressive scenario at 5 and 15 km and 0.4 and 0.5 effective pore pressure, respectively. The same parameters are applied to the main sets of focal mechanisms rendering medium to very high T_s values in general with wrench and reverse faults tending to reactivate again with the same motion as noted when contrasting their respective T_s. The unknown stress regime set also depicts very high T_s under the thrust regime. Note the high wop (well-oriented planes) percentage values for the majority of cases, probing its robust mechanical concept.

stress to effective normal stress resolved on a plane of weakness-fault plane. Slip tendency T_s is defined as τ/σ_n, where τ is shear stress and σ_n is normal stress, and was fully described by Jaeger (1979) and Morris *et al.* (1996). When the sliding friction exceeds that ratio, the structure is said to be in the stable field; the opposite applies for unstable cases. Colletini & Trippeta (2007) discussed in detail its geomechanical assumptions, restrictions and especially the application to aftershock-triggered seismicity and associated focal mechanism solutions. They examined the reliability of the method and the seismotectonic logic of the results for contrasting crust types and stress fields. They also highlighted the correlations with poro-elastic effects, pore-fluid diffusion, viscoelastic stress transfer and also with *in situ* stress measurements, shock rupture planes, fault-slip analysis and static stress transfer (the latter two items are applied in this research at several scales of stress order). Here the method is also applied to regional faults and selected focal mechanism nodal planes to test their frictional fault reactivation potential under known and/or assumed values for the input parameters.

The T_s stereoplots (Fig. 4) correspond to the contouring of the normalized slip tendency with the pole plotting of the selected nodal planes and regional faults within the studied region using the stress shape ratio, pore pressure, depth and stress tensor axes orientation determined for the central Costa Rica-1NS tensor. The analysis follows the method proposed by Colletini & Trippeta (2007) which overcomes the poorly constrained fluid pressure P_f and differential stress $\sigma_1 - \sigma_3$ within seismogenic volumes. In the same fashion, the normalized slip tendency is defined as NTs $\leq T_s/T_{smax}$ with well-oriented fault planes in the range: $0.5 <$ NTs ≤ 1.0 while misoriented structures are fall within the range $0 \leq$ NTs ≤ 0.5.

Coulomb static stress transfer

There is growing evidence from the analysis of seismic sequences in California, Anatolia, Kobe, Chile (Lin & Stein 2004), and quite recently in Japan (Toda *et al.* 2011*a–c*), that the alteration of shear and normal stresses conditions in the vicinity of active faults during earthquakes induces increments in seismicity rate. There is also evidence that stress transfer promotes or delays new activity in the surrounding structures and that their mutual interaction is favoured by such changes, even having a role in magmatic systems and eruptions. Although these changes are small, according to Stein (1999) they are not sufficient to cause earthquakes but only trigger them. He found remarkably good correlation between the model calculations and observations of seismicity. It is therefore considered that the small stress changes calculations are sufficient from which to learn something important because they are sudden, cause large changes in the seismicity rate and thus likelihood of earthquakes. So even though the perturbation is small, its effect is large. He also defines the Coulomb failure stress (CFS) change (Okada 1992) in its simplest form as $\Delta\sigma_f$ (also written ΔCFS or ΔCFF) as:

$$\Delta\sigma_f = \Delta\tau + \mu(\Delta\sigma_n + \Delta P)$$

where $\Delta\tau$ is the shear stress change on a fault (reckoned positive in the direction of fault slip), $\Delta\sigma_n$ is the normal stress change (positive if the fault is unclamped), ΔP is the pore pressure change in the fault zone (positive in compression) and μ is the friction coefficient (with range 0–1). Failure is encouraged if $\Delta\sigma_f$ is positive and discouraged if negative; both increased shear and unclamping of faults promote failure. The tendency of ΔP to counteract $\Delta\sigma_n$ is often incorporated into Equation (1) by a reduced 'effective' friction coefficient, μ'.

Being an intraplate setting with strong subduction influence and with several active volcanoes, Central Costa Rica constitutes an interesting scenario in which to explore the behaviour of Coulomb static stress transfer. From this perspective three key regional faults, one of them believed to be responsible for historic catastrophic earthquakes, are tested for their Coulomb stresses. Only a few decades of seismic records are available and little can be inferred from the past seismic history except the supposed magnitudes. These representative structures are the Aguacaliente, Escazú and Alajuela faults of known geometry and enough trace length to be considered dangerous with high seismic hazard potential, in particular because of their close location to civilian infrastructure. These three structures are therefore tested for their CFS. A shear modulus $G = 3.2 \times 10^5$ bars, Young's Modulus $E = 8 \times 10^5$ bars, Poisson's ratio PR $= 0.25$ and a friction coefficient $\mu = 0.4$ were used for the CFS modelling. Since no directly related focal mechanism solutions to the selected analysed regional structures exist, the geometry and other seismotectonic parameters are derived from the empirical relationships of Wells & Coppersmith (1994). The input stress tensor for the optimally oriented planes, those with the most positive Coulomb stress change resolved (Toda *et al.* 2005) during the CFS calculations, is the central Costa Rica-1NS which is the most stable. It has properties $R = 0.40$ and QWSM = A and less loss of original nodal plane data (55 out of 58), also with a logic SH $= 22°$ subparallel to the N32°E convergent direction of the Cocos Plate. The description and setting of the analysed faults are discussed in the following sections.

Aguacaliente fault

Cartago City in SW Central Costa Rica was destroyed in 1834, 1841 and most recently in 1910 (Ms 6.1), an event accompanied by a local seismicity which lasted until 1912. The Aguacaliente fault (hot water) has been identified (Montero *et al.* 2005) as the source fault (Figs 2 & 5). This structure displays a surface length trace of more than 25 km with a width of 10–100 m and unknown core zone. Its kinematics has been inferred from morphotectonic indicators and no conclusive evidence of the real motion has yet been found in old and recently excavated trenches. The strong effects of tropical weathering and tectonic shearing combine to produce complex near-surface exposures with medium-to-poor interpretation possibilities. Some of these excavations exposed parts of the main core and deformed zone of 40 m but the sense of motion and orientations are based on the presence of: small faceted and elongated pressure ridges; grouped hot springs aligned with tensional fractures in a E–W strain couple fashion at both sides of the main trace; tilted and faulted terraces; sinistral modelled displacements of rivers and creeks; concordance of groundwater flow directions with an E–W trace; and detailed magnetic and geoelectric trends with good correlation with the proposed location of the main core (Montero & Kruse 2006). All these features suggest an active fault with additional constraints from a

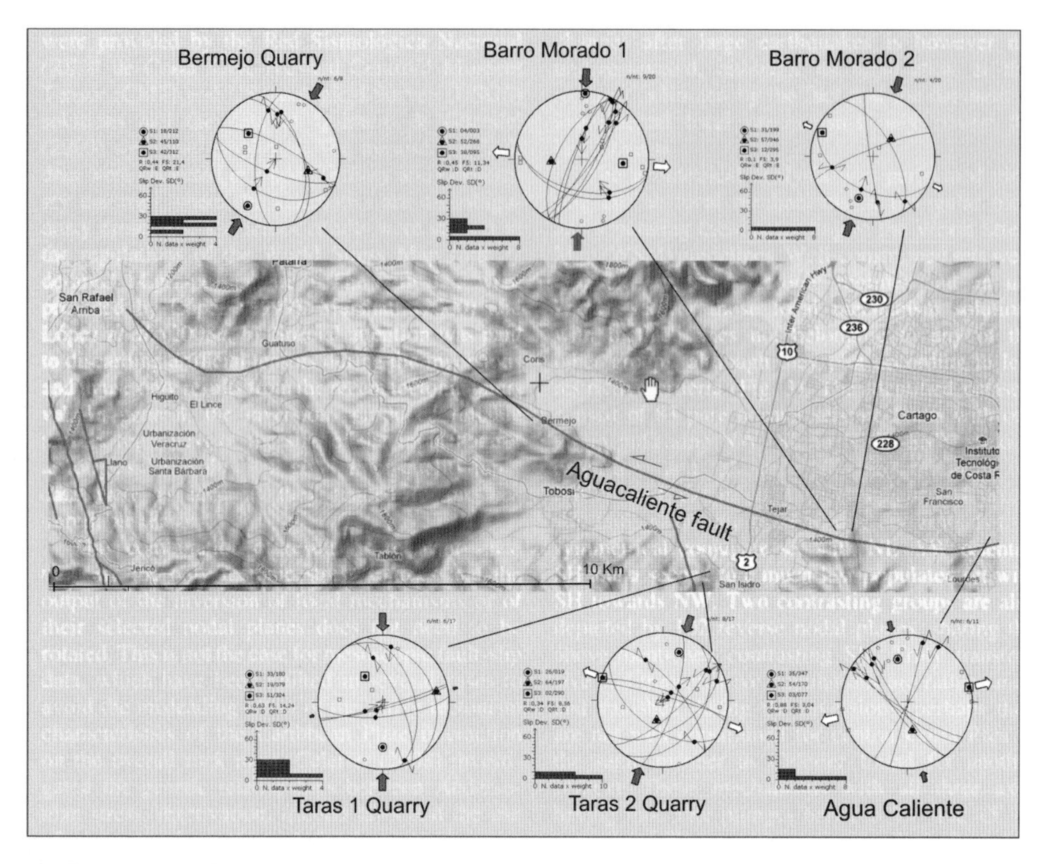

Fig. 5. Mesoscale fault sets support the sinistral motion proposed for the Aguacaliente fault although the whole scenario is more complicated; not morphologically represented NE–SW striking dextral and sinistral tears affect the deformation zone. The fault should have a reverse component attending the interpreted first rank SH; interestingly, this structure does not fit with the strain ellipse shown in Figure 3 unless it is a reverse fault.

carbon sample found in a displaced horizon dated 3665 ± 145 years BC (Woodward-Clyde 1993).

Escazú fault

This structure is the NNW continuation of the Aguacaliente system in the western middle part of the central Costa Rica. It bounds the eastern edge of San José City where the majority of the socio-economic infrastructure of Costa Rica is located. Its stress transfer potential is vital because it is also a link with the Alajuela fault and the active Poás volcano edifice (Fig. 2). Its existence was suggested by Woodward-Clyde (1993) and Montero *et al.* (2005) further studied its interaction with 300 Ka old Ignimbrites which generate some complex landforms despite being mostly covered by other Quaternary volcanics and recent alluvial deposits with a non-continuous morphotectonic signature. Two en echelon NNW segments are identified, but for stress transfer modelling it is represented as a single plane. Its length is 15 km with a width of 3 km and capable of M_w 6.4 earthquakes.

Alajuela fault

A regional prominent east–west scarp 150–200 m high with a small half-trace WNW bending, mainly due to outcrop pattern, is the morphotectonic signature of the Alajuela fault. It is a faceted reverse structure running 25 km between Alajuela and Grecia Cities (Fig. 2) (Montero *et al.* 2011). The involved lithologies comprise debris flows, recent ignimbrites and lavas from the nearby Barva volcano and alluvial and lacustrine deposits, all of them affected by the fault-bend deformation with varying dips at both sides of the frontal step and facing towards SW, E and NE. Soto (2005) studied an outcrop that displays vertical and slightly tilted tuffs and ignimbrites in contrast with the almost flat horizons of the same sequence, locally broken by subvertical cataclastic filled metre-scale-width fracture bands interpreted as the core of the partially blind faulted antiform axis. Neotectonic and modern displacement is apparent from faulted recent alluvial units and tectonic interaction with 322 Ka dated ignimbrites and 40 Ka young lavas.

Discussion

The modern stress field

A joint correlation and interpretation of the neotectonic and present-day stress field along with the slip tendency and stress transfer of regional structures is an innovative and original approach with direct implications for the seismic hazard. Central Costa Rica is a good source of data and provides a challenging seismotectonic setting to apply such new tools to distinguish between dormant and active or potentially active scenarios under known or assumed geomechanical parameters and nucleation of displacements on both old and/or new faults.

The modern stress field as modelled from focal mechanisms clearly indicates the presence of three main trends for the SH. The main trend is of first order trending NNE parallel to and governed by the Cocos Plate convergence direction. This SH covers the entire Central Costa Rica with local fluctuations less than 15° deviating at its intersection with large faults and volcanic edifices (Fig. 1). The second-order SH clearly trends NW and is by far much less developed than the former; it is concentrated in the central part of the study area, sometimes overlapping with the first. The third SH is poorly developed, sparse and isolated across Central Costa Rica but still a recognizable NWW trend. The whole attitude can hardly be attributed to the presence of different tectonic phases during a period of less than 40 years and, without enough large earthquakes which could have significantly altered those three ranks of SH, as can be accepted in principle when fault-slip data are collected at outcrops representing different levels of the regional stratigraphic column and contrasting palaeostress orientations are found.

These are the cases worked out by Villegas (1997) who presented an integrated model for the most westerly part of Central Costa Rica in volcanics with Upper Miocene–Lower Pliocene NW compression. During the Upper Pliocene–Lower Pleistocene it rotates to an almost EW compression, finally assuming a NNE trend for the Upper Pleistocene to recent interval, coinciding in space and time with the modern first-order SH as calculated from the focal mechanism database. Arias & Denyer (1994) suggested the existence of a regional north–south-directed compression and associated extension from fractures and kinematic indicators at a single station in volcanics. Locally this is true, but cannot be accepted for a regional context and does not have a tectonic explanation from known interactions of the Cocos and Caribbean plates or local blocks.

The results of Marshall & Fisher (2000) and their reformatted and reinterpreted data in this study clearly indicate the presence of variable SH orientations at 37 representative sampling stations throughout the volcano–sedimentary sequence. The same applies to the results obtained by L. Campos (pers. comm., 2003) and by graduate students at some other isolated outputs. López (1999) sampled hundreds of fault-slips at the Central Costa Rica oceanic basalts basement and its surrounding sedimentary cover in the southern edge of studied area. He found that the most recent conjugate strike-slip faults correlate very well with the ongoing seismotectonic first-order NE SH trend, and also with the NW dextral and NE sinistral conjugate set so well bisected by the main SH and parallel to the convergence direction of the Cocos Plate.

Stress patterns

The tectonic logic and consistence of first-order SH is fully justified from the kinematic and stress approaches when the NW SH trends obtained by the inversions are used to design a typical strain couple. There the strike-slip faults parallel the abundant NW dextral Escazú type faults and its conjugate counterpart and less-developed NE sinistral shears are very well represented; the Alajuela and San-Miguel-like reverse structures are orthogonal to that SH and the normal faulting perpendicular to this compression is fairly developed. The difference between the Cocos Plate N32°E convergence direction and the couple compressive vector is 11°, a value which is subparallel to that obtained from the inversions as observed when this vector is compared with the individual compressive trends of each regime tensor (Fig. 3). Even the 'P' shear is present, and its assigned nodal planes fit very well with the first-order NE SH.

This view is consistent with and better explains the presence of second- and third-order stress patterns (Heidbach *et al.* 2007; Tingay *et al.* 2006) which do not fit the discussed strain ellipse but are sustained by five tensors. Indeed, this kind of regional and local stress trajectory-field fluctuations and deviations as envisaged in this study for Central Costa Rica have been detected at varying scales in different plate tectonic settings. They have been explained as induced and controlled by active faults with stress changes genetically linked to seismic–volcanic activity and local density contrasts. These recognized factors precisely characterize the recent and current upper crust properties within and along the North Panamá Deformed Belt (NPDB) and the Caribbean Plate in Central Costa Rica, where the present activity of the Irazú, Turrialba and Poás volcanoes compete with the neo and present-day deformation imposed by displacements on quite active seismic sources. The density contrasts are also present at deep-rooted contacts between those igneous massifs and the surrounding Upper Tertiary volcaniclastic–sedimentary shelves. In this scenario, the NE-trending SH trajectories represent the first-order stress field imposed directly by the Cocos–Caribbean Plate boundary while the NW-oriented trajectories are the second-order stress field. Third-order elements are present in the form of NWW-trending trajectories.

The permutation of principal stress axes is usual inside Central Costa Rica; although a rather common phenomenon it has not yet been given enough attention. It is controlled according to the work of Hu & Angelier (2004) and Angelier (1984, 1989, 1990, 1994) by the relative heterogeneity and contrast of the brittle deformation in compartmented settings such as the partitioned blocks of Central Costa Rica. They are accompanied by changes in the anisotropy of the mechanical properties, resulting from high rates of seismotectonic activity inducing geomechanical imbalances and thus variations in the stress ellipsoid stability and exchange in the orientations and magnitudes of the stress axes and the related tectonic regimes. In Central Costa Rica, the interpreted ellipsoid R values ($R = \sigma_1 - \sigma_2 / \sigma_2 - \sigma_3$) are in the range 0.3–0.05 and 0.8–0.93 which are responsible for the permutation of σ_2 to σ_3 or σ_2 to σ_1, respectively. The result is a change from transpression to axial compression or from transtension to almost axial extension within the neovolcanic edifices with stable pure normal, strike-slip and reverse tensors everywhere. These findings are important for a realistic interpretation of the successive deformation stages in this SW corner of the Caribbean Plate.

All this evidence supports three distinct SH configurations that related to surface structures on three different scales. The surface traces of known structures in central Costa Rica correlate with the first-order pattern of crustal stress at scales of 300 km, the second-order pattern at scales of 50–75 km and the third order pattern at scales of 20–30 km. This scale-dependent property appears to reflect the regional and local complexity of the Panamá microplate–Caribbean Plate boundaries.

Coulomb stress and kinematics

The correlation of the discussed stress outputs with mesoscopic findings at outcrops of the three studied regional faults yields interesting interpretations. Along the Aguacaliente fault, within and outside the deformation band and separated by a few to hundreds of kilometres, the fault slips resemble to some extent the prevailing fault kinematic model (Fig. 5). Local slip indicators from representative outcrops are extrapolated and sustained for the whole structure attending that no segments are identified. Its expected displacement should be reverse with sinistral component attending the properties of the stress field and the regional trend of the first-order SH already discussed. The stress tensors at these sites also confirm the NE, N and NNE σ_1 orientation for populations with WSM qualities ranging from D to E (i.e. from poor to unaccepted ellipsoids). The former are presented only to illustrate the type of movements. The WNW–ESE trace of the Aguacaliente fault is apparent in mesoscale fault exposures dipping SSW. This trend is systematically cross-cut by NW-striking dextral shears and NE-striking sinistral shears, as in other many localities in Central Costa Rica. Nevertheless, such orientations are not locally evident and geomorphic elements cannot be associated. The almost-pure strike-slip movement seems difficult to support due to the apparent lack of a regional WNW SH trend of first or second order in the vicinity.

Figure 6 displays the positive and negative stress shadows for a Aguacaliente fault length of 24.51 km with a width of 10.09 km at a depth of 6 km, the classical interpretation based on morphometry and scarce data from a few trenches for a typical strike-slip motion calculated as a specified fault oriented 95°/75°/10°. The trace is taken as a straight line (although in reality some bends are recognized), but the results for a more complicated outcrop pattern are similar. This is a conceptual average fault trace representative of the Aguacaliente fault system which comprises at least two main EW to WNW–NNW sinuosing branches. Increased seismic hazard depicted as red lobes is depicted towards NWW and S at the western end of the trace; at the eastern end the stress change petals are oppositely oriented. This configuration can explain seismic swarms active around Puriscal village during early 1990 which culminated with the 1990 Piedras Negras earthquake M_s 5.7. The northwards-directed lobe with 2 bars points directly to the active Irazú volcano magmatic chamber, whose depth is estimated at an upper limit of 5 km according to detected seismic activity. The eastern tip generates a positive 2–3 bars stress shadow into the Tucurrique region where seismic activity is recurrent (Fernández 1995). The scenario is interesting since a water reservoir is well covered by that shadow under the present assumptions. It is outwith the scope of this study to analyse the potential interactions of its hydrostatic column and related pore pressures and frictions with that stress transfer in a future earthquake in the same source

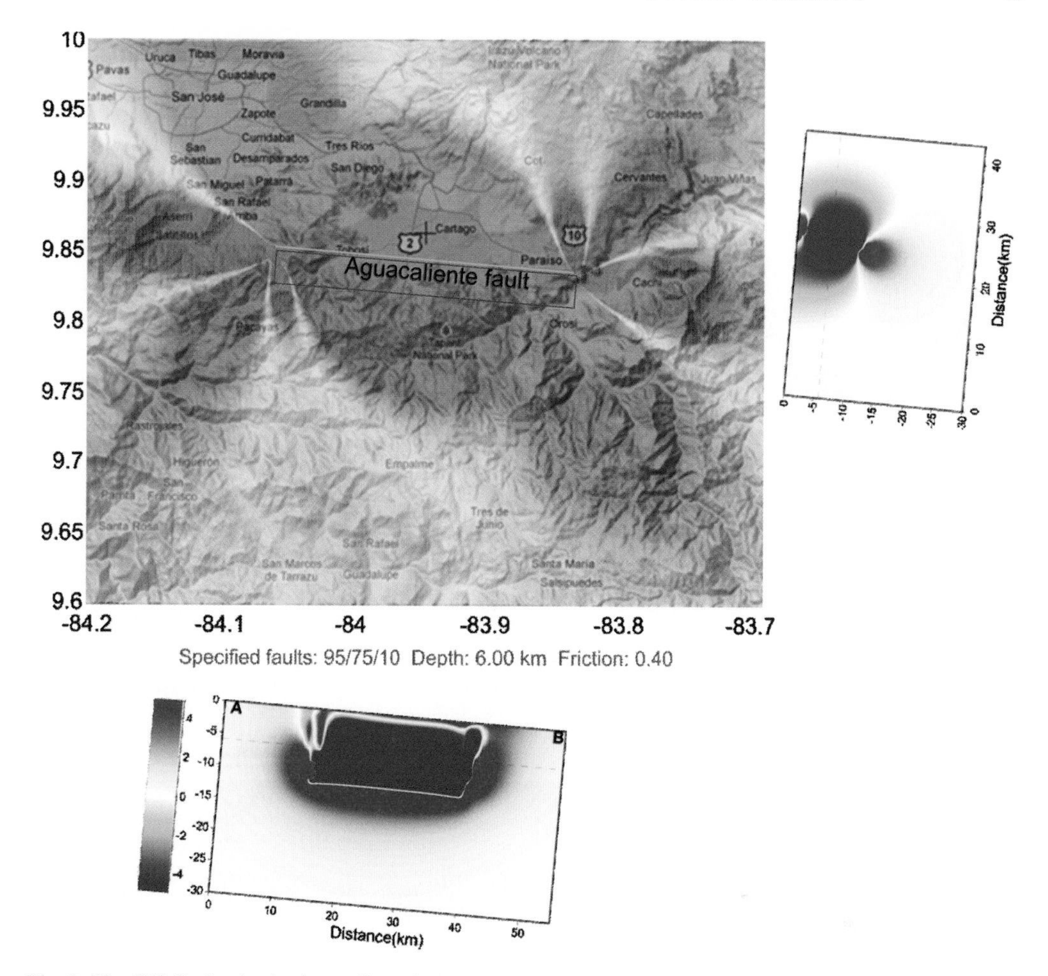

Fig. 6. The CFS display for the Aguacaliente fault at 6 km depth explains recent seismic swarms at both tip ends. One swarm is located to the west around Puriscal village during 1990 which culminated with the Piedras Negras earthquake M_s 5.7 while continuous tremors affect the eastern part. One positive 2 bars (0.2 MPa) lobe increasing the seismic hazard points towards the Irazú volcano without cause–effect reported links so far.

or other ones in the vicinity and the possibility to trigger seismicity. This must be undertaken during the next stage of research because this artificial lake did not exist at the time of the 1910 earthquake. T_s values of 0.60 for strike-slip and 0.80 for thrusting suggest a medium–high reactivation potential. No time–space relations with the Irazú volcano were reported after the mentioned historical events.

The Escazú fault displays an exceptional outcrop (Fig. 7) with more than 200 m of exposures located within the western damaged zone. These offer hundreds of valuable kinematic indicators whose stress inversion indicates that the NNE-striking mesoscale faults are of an almost-pure sinistral nature with an EW conjugate set similar to that found along the Alajuela fault. Only a few structures resemble the reverse character assigned by Montero *et al.* (2005) to this fault, probably based on a single small-magnitude focal mechanism with a nodal plane coincident with its trace. The model SH trends correspond to the second- and first-order ranks. These facts can be jointly interpreted as the representative modern stress regime for the structure after having previously acted as a tear. This view is supported mechanically by the low value of the ellipsoid shape factor ($R = 0.02$), indicating a shift from transpression to axial compression and reverse faulting (i.e. the focal solution interpretation). The other separated phase with an $R = 0.92$ points to transtension, a logical temporal relaxation behaviour between the change of stress regimes. When the total fault dataset is mixed with suspected shear and compressive joints, the overall result is quite similar. These

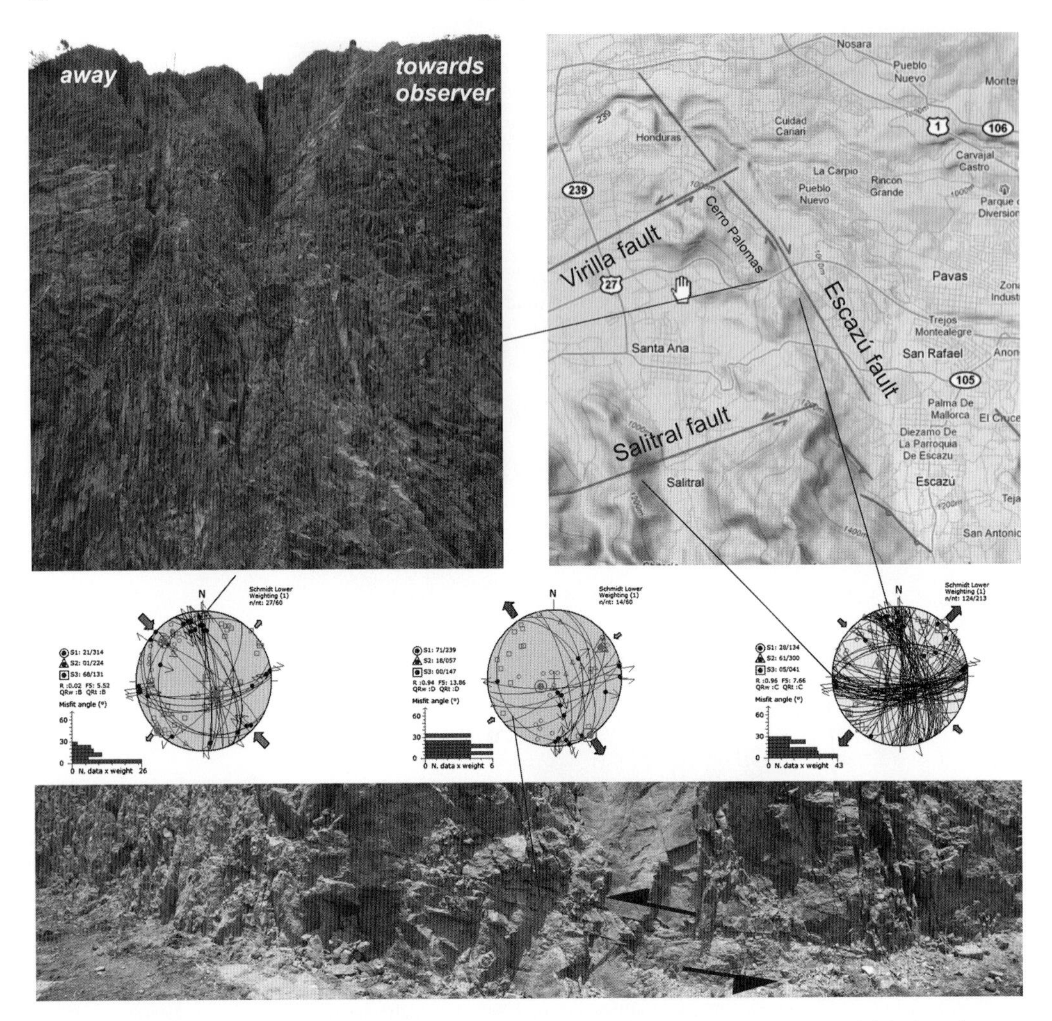

Fig. 7. Top-quality outcrop at Cerro Palomas within the Escazú fault deformation zone. Dozens of right-lateral kinematic indicators and a few reverse indicators leaves no doubt about its overall motion, which should also apply to the nearby Virilla and Salitral faults. The former was the source of the Piedras Negras earthquake, modelled with a right-lateral focal mechanism. As mentioned in Figure 6, the Aguacaliente CFS probably overloaded its hypocentre volume.

repeated time–space instabilities represented by the rather common permutations is herewith valuated as an important palaeostress property to be taken into consideration on any tectonic research on the evolution of Central Costa Rica.

The model stress shadows generated by this fault during its undated last displacement are similar to those produced by the Aguacaliente sector and are based on assigning a length and width of 15 and 3 km respectively and testing the Coulomb stress at 6 km depth, a critical depth for the seismic hazard (Fig. 8). Here the northern tip loads the active Alajuela fault with up to 3 bars in its middle sector, and even 1–2 bars to the southern flank of the Poás volcano and its magma chamber. This load may overlap with that imposed by the Cinchona earthquake (M_w 6.2) in the eastern part of the volcano in 2009. One petal pointing SW interferes directly with the Virilla fault, responsible for the 1990 M_S 5.7 Piedras Negras earthquake. In the southern extreme another stress lobe imposes 2 bars to the Jaris fault, the source of the 1990 Puriscal swarms. From the same tip, another 3-bar shadow projects into San José City. A cross-section along the fault plane imposes a potential slip at 12 km depth in its middle section and at shallower depths close to the tips as described above. The blue low-seismic-hazard shadows could in fact be

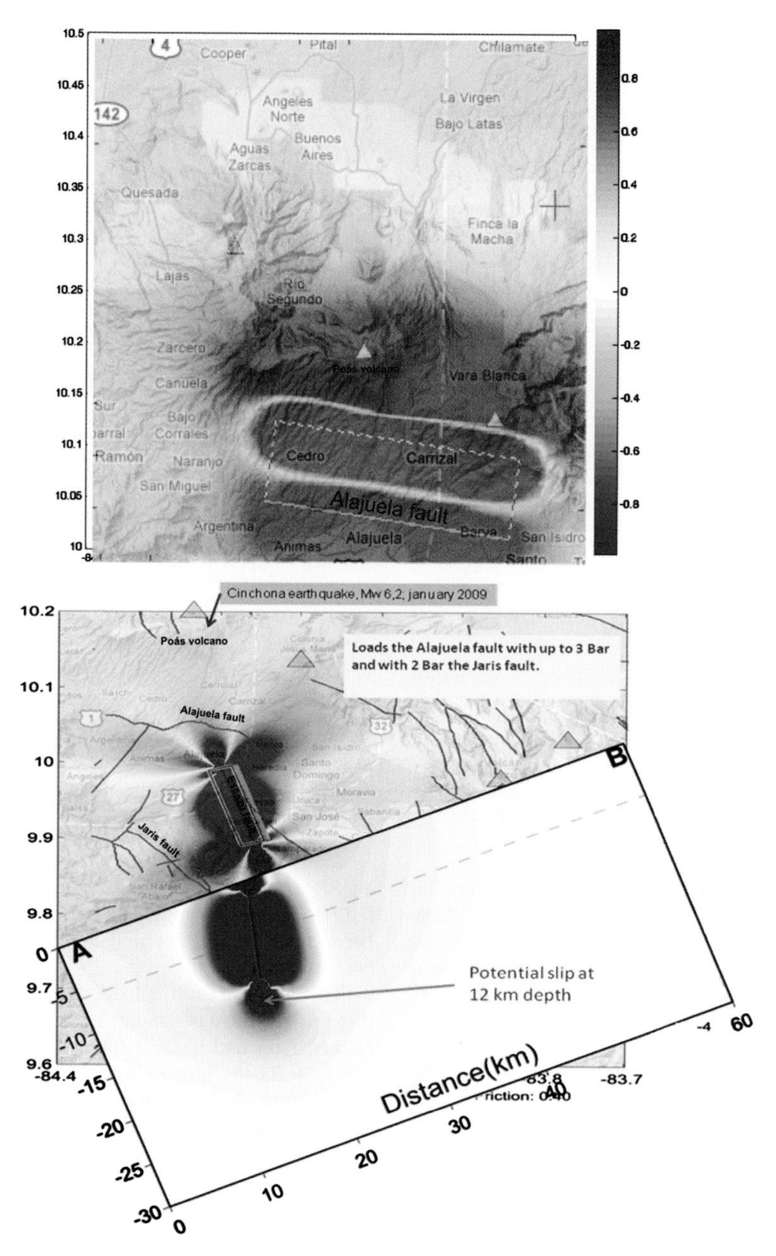

Fig. 8. The Alajuela reverse fault with sinistral component lies at an angle of 75° with respect to the first-order SH, being an attractive target for fault reactivation. It was also affected by the recent Cinchona earthquake M_w 6.2; 8 January 2009. The northern tip of the Escazú CFS induces a shadow on the Alajuela fault of up to 3 bars in its middle sector; 1–2 bars are affecting the southern flank of the Poás volcano and its magma chamber. The Virilla fault, responsible for the 1990 Piedras Negras earthquake, the Jaris fault and the source of the Puriscal swarms the same year, can be explained by the imposed 2 bars by the SW shadow. The T_s value for this structure is 0.83 and 0.45 for the wrench and thrust scenarios, respectively.

positive if the Aguacaliente western tip remains active. The T_s value for this structure is 0.83 and 0.45 for the wrench and thrust scenarios, respectively. These values must be taken into account for more detailed reactivation studies along with its well-oriented plane with respect to the first rank SH, making it a good candidate for an unknown future displacement.

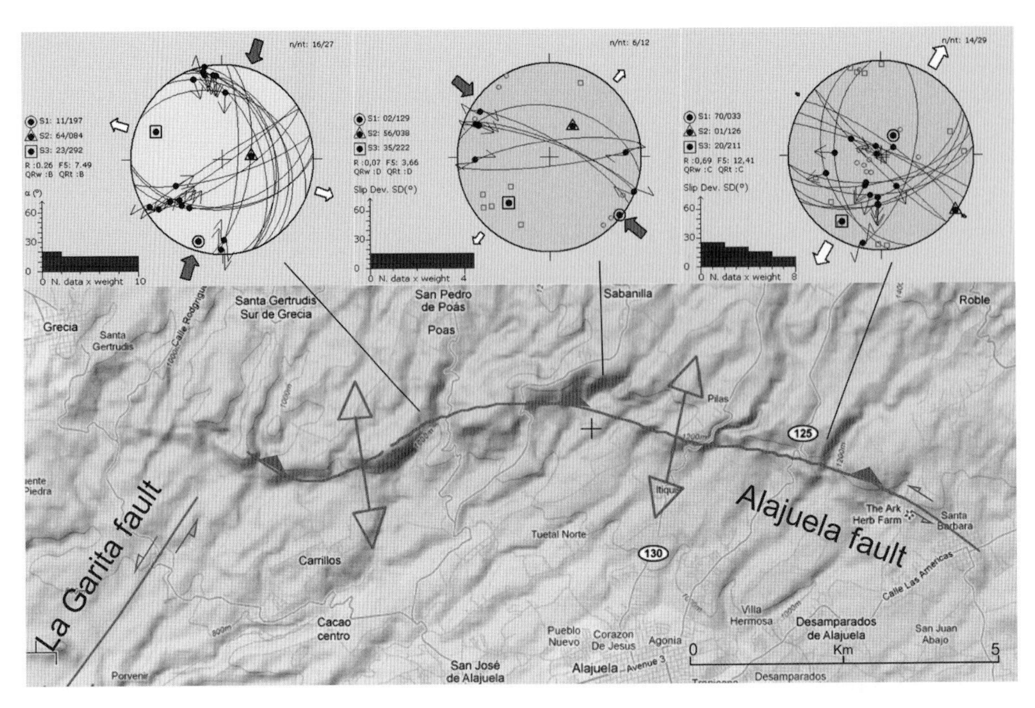

Fig. 9. Along strike (west to east) neostress stations on the Alajuela fault at Prendas, Chilamate and Las Americas quarries shows typical first- (NE) and second-order (NW) SH trends with important sinistral component on EW and NW tears affected by strong tendency for the s_2/s_3 permutation (i.e. transforming the strike-slip regime into axial compression). This structure is herewith interpreted as having an important sinistral component for its confirmed main reverse displacement. See discussion in the text.

At the Prendas quarry in the Alajuela fault deformation zone, M. Meschede (pers. comm. 2008) measured high-quality (B quality rank of the WSM) fault-slip populations (Fig. 9) interpreting reverse kinematics with important sinistral component on EW and NW tears; a few reverse data were rejected during the inversion, which produced a typical NNE trend for the local first-order SH. Since this outcrop is highly representative of and close to the Alajuela fault scarp, and without evidence of segmentation, this result is considered consistent and in principle representative of the entire structure. Marshall & Fisher (2000) data were reprocessed for stress determination, indicating the presence of the second-order NW compression at the nearby Chilamate and Las Americas quarries with lesser qualities (D and C) and a unstable ellipsoid ($R = 0.07$) and hence a strong tendency for the σ_2/σ_3 permutation (i.e. transforming the strike-slip regime into axial compression). A less-developed similar stress solution is detected at Las Animas site and Las Americas is slightly transtensive with value $R = 0.69$. By accepting these results, the east–west to NE–SW strike of the Alajuela fault is confirmed but dipping southwards opposite to the current model; the NE-dipping faults are dextral slips. A solution is to interpret this fault as part of a conjugate system, but more population analysis and detailed mapping is needed to confirm this.

The Alajuela fault is loaded with at least 3 bars (Fig. 8) by the last undated displacement of the Escazú fault and a T_s of 0.65 (strike-slip outline) and 0.36 for the thrust situation. Its EW strike lies at an angle of 75° with respect to the first-order SH, being an attractive target for fault reactivation. These results are offset by the doubt that this structure was the source of the I. Boschini and W. Rojas (pers. comm. 2011) M_w 5.2 and 6.2 earthquakes, both being epicentres inside the southern positive lobe. Furthermore, the recent Cinchona earthquake M_w 6.2 which occurred on 8 January 2009 loaded this region with 1 bar, so it is reasonable to accept that the entire volume is being increasingly overloaded (Barquero 2009).

A CFS simulation assigning a length trace of 24 km to its average plane and hence a theoretical M_w 6.1 indicates increased seismic hazard toward the active Poás volcano massif to the north, a combination which should be investigated since the positive lobe could interact with the magma chamber as proposed for the Vesuvius volcano

Table 2.

Fault	Strike	Dip	Rake	Synthetic FM from T_s value	Fault name
1	165	75	−172		Chiripa
2	133	45	142		Cote-Arenal
3	3	70	−148		Danta
4	61	70	−11		Peñas Blancas
5	72	70	−2		Javillos
12	126	45	132		San Miguel 4
15	175	70	−161		Zarcero 1
16	148	70	169		Zarcero 2
17	163	70	−175		Zarcero 3
18	9	70	−133		Congo
19	175	70	−161		Porvenir
20	164	70	−174		Carbonera
22	157	70	179		Alto Grande
23	163	70	−176		Blanquito
24	40	80	−32		Barranca 1
25	51	80	−21		Barranca 2
26	52	80	−20		Barranca 3
27	77	80	−8		Barranca 4
29	50	80	17		Jesús María
32	77	45	25		Tivives 1
33	42	75	15		Tivives 2
34	52	75	16		Bajamar
36	68	75	20		Tárcoles 1
37	177	70	175		Tárcoles 2
39	179	70	−154		Tárcoles 3
40	168	70	−170		Tárcoles 4
41	76	70	1		Tronco Negro
42	56	70	−15		Camaronal 1
43	146	77	173		Delicias
47	147	70	168		Tronco Negrito
49	41	78	18		Tulin
50	74	78	21		Tulin
51	53	78	17		Candelaria 1
52	48	78	17		Candelaria 2
53	33	78	25		Candelaria 3
55	76	45	23		Candelaria 4
56	104	45	60		Picagres 1
58	159	70	−180		Picagres 2
60	96	70	44		La Garita 1
62	166	70	171		La Garita 2
63	163	70	−175		La Garita 3
67	79	70	26		La Garita 4

(*Continued*)

Table 2. *Continued*

Fault	Strike	Dip	Rake	Synthetic FM from T_s value	Fault name
69	50	70	13		La Garita 5
70	143	70	162		Alajuela 1
71	158	70	169		Alajuela 2
72	170	70	−168		Alajuela 3
73	2	70	−149		Jaris 1
76	105	45	62		Frailes
77	69	45	16		Frailes Madre-1
78	95	45	47		Río Azul
79	83	45	32		Cipreses
80	105	45	61		Lara
81	68	70	20		Cangreja
83	45	70	−27		Agua Caliente 1
84	47	70	−24		Agua Caliente 2
93	75	45	22		Agua Caliente 3
94	66	45	12		Navarro-3
100	52	45	−6		Navarro-2
102	55	60	10		Navarro-1
108	64	45	10		Orosí
110	97	70	45		Duan
111	42	60	−1		Viejo-A Zarcas 1
112	51	70	13		Viejo-A Zarcas 2
114	56	70	−16		Flor-Pal-Jam 1
115	55	70	−16		Flor-Pal-Jam 2
116	77	45	25		Pta Gigante
118	131	70	153		Cedro 1
121	79	60	28		Cedro 2
122	73	60	23		Jeringa-1
123	50	60	6		Jeringa-2
124	69	60	20		Curú
125	132	60	146		Artieda
126	129	70	151		Pógeres
127	65	70	18		Chiquero
129	163	70	−176		La Mesa 1
130	56	70	15		La Mesa 2
131	5	80	−149		Resbalón
132	46	70	−26		Tablazo
133	152	60	165		Alumbre 1
134	43	70	−29		Patio de Agua
136	146	70	166		Morote
138	98	45	51		Tempisque
141	21	79	101		Manzanillo-1
142	154	79	170		Manzanillo-2
143	161	70	170		Coyolito

(Nostro *et al.* 1998). The fault network around the main crater can also be reactivated under appropriate hydrothermal conditions (Figs 2 & 8).

Slip tendency behaviour

The applied T_s method imputing the Central Costa Rica-1NS stress tensor is robust since only 15% of the active mapped fault poles for the total 147 structures in Central Costa Rica and their segments are misoriented with individual values lower than 0.5 (Fig. 4). This is also true for both nodal planes of the population used to calculate that tensor, with only 2% for the shallow strike-slip determination and 14% for the compressive scenario at 15 km depth and an estimated fluid pressure of 0.5. The other regimes show 17, 15, 44 and 37% for the shallow normal, strike-slip, thrust and unknown regimes, respectively. The subduction influence begins weakly below 15 km depth and hence a compressive scenario for the respective T_s was implemented, adopting 17, 39, 4 and 3% for the above-mentioned regimes and assuming that the regional structures could be reactivated at those seismogenic depths. Regarding the M_w magnitudes, it is noted that the thrust nodal planes have higher magnitudes in the range 3.0–6.2 while the other sets range 3.0–4.5; this highlights the prevalence of compressive stresses commanded by the SH controlled by the Cocos Plate convergence.

The T_s stereogram for the 147 faults covering the entire region have a multidirectional trending with dips clustering at 45° and 80° because of the assigned inclinations (Fig. 4). Structures dipping between 25° and 80° and striking NW–SE and NE–SW are part of the 85% well-oriented planes, have a high slip potential and are prone to reactivate within the strike-slip regime at 5 km depth. In the compressive scenario however, below 15 km, at increasing subduction levels and assuming 81% propagate at medium-to-high T_s, almost every plane orientation is likely to be reactivated if it dips between 20° and 75° across NW–SE strikes. The Aguacaliente, Escazú and Alajuela faults report T_s values of 0.6, 0.83 and 0.65 in the strike-slip regime and 0.81, 0.47 and 0.36 in the thrust regime. Table 2 presents the geometry and related synthetic focal mechanism in the case they reactivate (Neves *et al.* 2009) for 86 of those regional faults with $T_s > 0.5$.

Conclusions

The analysed focal mechanism solutions and fault-slip populations for the Central Costa Rica region indicate that the recent and present stress states in Central Costa Rica depict an Andersonian configuration with one principal stress axis vertical or close to it. Important permutations are detected, shifting the stable thrust, normal and strike-slip regimes into transpresion, axial compression, transtension and axial extension depending on the extreme values of the stress ellipsoid R ($R = \sigma_1 - \sigma_2 / \sigma_2 - \sigma_3$) which fluctuate between 0.3–0.05 and 0.8–0.93, explaining the permutation of σ_2 to σ_3 and σ_2 to σ_1. Unknown stress regimes with a particular range of inclined axes or identical plunges are frequent; they may be the result of inversions made with unreliable focal mechanism solution nodal planes, despite the care of the selection procedures, and represent 29% of the database. For mesoscale fault populations, the explanation could be an inaccurate retilting before the inversions due to the lack of an adequate reference horizon. Three orders of contemporaneous superimposed SH trajectories are clearly detected, and its spatial attitude is used to justify known or assumed displacements on regional faults and their local segments. The first-order SH trend is rather similar to and imposed by the N22°E convergence direction of the Cocos Plate on the Caribbean Plate. Refraction of the SH does occur across major volcanotectonic features, where its trajectories locally deviate and return to the original trend after crossing them. These changes do not confirm the existence of the proposed contorted and diffuse Panamá microplate boundary with the Caribbean Plate within Central Costa Rica, since a larger contrast was expected for such a major tectonic frontier; it does not constitute definite evidence against its existence, however. The inversion of the focal mechanism and fault kinematic data yield results that are broadly consistent with regional patterns compiled on the World Stress Map.

The static stress transfer changes modelled for the Aguacaliente, Escazú and Alajuela faults are congruent with recent seismic swarms and justify them as sources of historic destructive earthquakes. This in turn fits with the modelled slip tendency determinations, yielding values from 0.36 to 0.83 depending on the adopted stress regime. These stress states control the deformation which is being accommodated in a complex seismotectonic array synthesized in a regional left-lateral strain couple, which demonstrates a good fit with both known regional faults and nodal planes. The Aguacaliente fault plane cannot be explained by this ellipse unless it has an important reverse component, a proposal supported by the first- and third-order SH trajectories.

This integrated seismotectonic analysis methodology not only demonstrates the validity of the Andersonian postulates within a rather complex and unstable inter- and intra-plate setting in the vicinity of a triple junction, but also its continuous applicability to new analytical tools such as the slip tendency and the Coulomb stress transfer. At the same time, this method provides greater insight into the seismic hazard studies as well as new perspectives into the sequence of deformation necessary to objectively reconstruct the tectonic history of the SW corner of the Caribbean Plate.

This research was supported by the Costa Rica Ministry of Science and Technology (MICIT) and the National Council for Science and Technology (CONICIT). The Instituto Costarricense de Electricidad (ICE) and DAAD are acknowledged and thanked for their support during previous stages. D. Delvaux, F. Trippeta, V. Sevilgen, R. Stein, S. Toda, J. Lin, B. Müller, O. Heidbach, J. Angelier, W. Montero, M. L. Zoback, M.C. Neves, R. Barquero, A. Olaiz, G. de Vicente, A. Cerdas, D. Jiménez, R. Quintero, M. Fernández, W. Rojas, J. Marshall, M. Meschede, W. Frisch, S. Chiesa, C. Redondo, L. Campos, A. Climent, J. Martínez, J. Alvarez, L.G. Obando, S. Carboni, M. Sancho, A. Moya, G. Soto, E. Louvari, G. E. Alvarado, Z. Reches, and R. E. Holdsworth provided software, additional data, advice and field assistance. I am grateful to D. Healy for his patience. The revisions by M. Tingay, an anonymous reviewer and J. Lewis greatly improved the text and content.

References

ALVARADO, G., MORALES, L. D. ET AL. 1988. Aspectos sismológicos y morfotectónicos en el extremo occidental de la Cordillera Volcánica Central de Costa Rica. *Revista Geológica de América Central*, **9**, 75–98.

ALVARADO VILLALÓN, F. 1984. *Geología estructural y tectónica al sur del Valle Central, Costa Rica (Tarbaca)*. Tésis Lic, Escuela Centroamericana de Geología, UCR.

ANGELIER, J. 1984. Tectonic analysis of fault slip data sets. *Journal of Geophysical Research*, **89**, 5835–5848.

ANGELIER, J. 1989. From orientation to magnitudes in paleostress determination using fault slip data. *Journal of Structural Geology*, **11**, 37–50.

ANGELIER, J. 1990. Inversion of field data in fault tectonics to obtain the regional stress tensor–III. A new rapid direct inversion method by analytical means. *Geophysical Journal International*, **103**, 363–376.

ANGELIER, J. 1994. Palaeostress analysis of small-scale brittle structures. Chapter 4. *In*: HANCOCK, P. L. (ed.) *Continental Deformation*. Pergamon Press, 53–120.

ARIAS, O. & DENYER, P. 1991*a*. Aspectos neotectónicos y geológicos de Puriscal y alrededores, Costa Rica. *Revista Geológica de América*, **12**, 83–95.

ARIAS, O. & DENYER, P. 1991*b*. Estructura geológica de la región comprendida en las hojas topográficas Abra, Caraigres, Candelaria y Río Grande, Costa Rica. *Revista Geológica de América Central*, **12**, 61–74.

ARIAS, A. & DENYER, P. 1994. Compresión y dilatación norte-sur en el suroeste del valle central, Costa Rica. *Revista Geológica de América Central*, **17**, 85–94.

ARROYO, I. G., HUSEN, S. ET AL. 2009. Three dimensional P-wave velocity structure on the shallow part of the Central Costa Rican Pacific margin from local earthquake tomography using off-and onshore networks. *Geophysical Journal International*, **175**, 827–849, doi: 10.111/246x.2009. 04342.x.

BARQUERO, R. (ed.) 2009. *El terremoto de Cinchona del 8 de enero de 2009*. RSN, Costa Rica.

BARQUERO, R., LESAGE, P., ET AL. 1995. La crisis sísmica en el volcán Irazú en 1991 (Costa Rica). *Revista Geológica de América Central*, **18**, 5–18.

CELERIER, B. 2008. Seeking Anderson's faulting in seismicity: a centennial celebration. *American Geophysical Union; Review of Geophysics*, **46**, RG4001.

CLIMENT, L. 2008. *In*: CLIMENT, A., ROJAS, W., ALVARADO, G. E. & BENITO, B. (eds) *PROYECTO RESIS II. Evaluación de la Amenaza Sísmica en Costa Rica.* RSN, UPM, I.C.E., NORSAR.

COLLETINI, C. & TRIPPETA, F. 2007. A slip tendency analysis to test mechanical and structural control on after-shock rupture planes. *Earth and Planetary Science Letters*, **255**, 402–413.

COLOMBO, D. 1997. Application of pattern recognition techniques to long-term earthquake prediction in Central Costa Rica. *Engineering Geology*, **48**, 7–18.

DE VICENTE, G. 1988. *Análisis poblacional de fallas. El Sector de enlace del Sistema Central-Cordillera Ibérica.* PhD thesis, Universidad Complutense, Madrid.

DELVAUX, D. 2011. Win-Tensor software (http://users.skynet.be/damien.delvaux/Tensor/WinTensor/win-tensor_download.html).

DEMETS, C. 2001. A new estimate for present-day Cocos-Caribbean Plate motion: implications for slip along the Central American volcanic arc. *Geophysical Research Letters*, **28**, 4043–4046.

DEMETS, C., GORDON, R. G., ARGUS, D. F. & STEIN, S. 1994. Effect of recent revisions to the geomagnetic reversal timescale on estimates of current plate motions. *Geophysical Research Letters*, **21**, 2191–2194.

DENYER, P. & ARIAS, O. 1991. Estratigrafía de la región central de Costa Rica. *Revista Geológica de América Central* **12**, 1–59.

DENYER, P., MONTERO, W. & ALVARADO, G. 2003. *Atlas Tectónico de Costa Rica.* Ed Universidad de Costa Rica.

FERNÁNDEZ, M. 1995. Evaluación del hipotético sistema de falla transcurrente Este-Oeste de Costa Rica. *Revista Geológica de América Central*, **19/20**, 57–54.

FERNÁNDEZ, M. 2011. *Límite tectónico hipotético, deformación y sismotectónica del sector central de Costa Rica entre el Golfo de Nicoya y Limón.* PhD thesis, University of Costa Rica.

FERNÁNDEZ, M. & PACHECO, J. 1996. Complejidad de la estructura sísmica de la región central de Costa Rica según un análisis multifractal. *Revista Geológica de América Central*, **19/20**, 29–36.

FERNÁNDEZ, M. & PACHECO, J. 1998. Sismotectónica de la región central de Costa Rica. *Revista Geológica de América Central*, **21**, 5–23.

FERNÁNDEZ, M. & MONTERO, W. 2002. Fallamiento y sismicidad del área entre Cartago y San José, Valle Central de Costa Rica. *Revista Geológica de América Central*, **26**, 25–37.

FERNÁNDEZ, M., MORA, M. & BARQUERO, R. 1998. Los procesos sísmicos en el volcán Irazú (Costa Rica). *Revista Geológica de América Central*, **21**, 47–59.

FISHER, D., GARDNER, T. W., MARSHALL, J. S. & MONTERO, W. 1994. Kinematics associate with late Cenozoic deformation in central Costa Rica: Western boundary of the Panamá microplate. *Geology*, **22**, 263–266.

GARDNER, T. W., VERDONCK, D., PINTER, N. M., SLINGERLAND, R., FURLONG, K. P., BULLARD, T. F. & WELLS, S. G. 1992. Quaternary uplift astride the aseismic Cocos Ridge, Pacific coast, Costa Rica. *Geological Society of America Bulletin*, **104**, 219–232.

GARDNER, T. W., FISHER, D. M., MARSHALL, J. S. & SAK, P. 2000. Fore arc deformation, segmentation, and drainage evolution in response to seamount subduction along an eroding margin, Pacific coast, Costa Rica. *Proceedings of the American Geophysical Union*, **81**, F1206.

HEIDBACH, O., REINECKER, J., TINGAY, M., MÜLLER, B., SPERNER, B., FUCHS, K. & WENZEL, F. 2007. Plate boundary forces are not enough: second-and third-order stress patterns highlighted in the World Stress Map database. *Tectonics*, **26**, TC6014, doi: 10.1029/2007 TC2133.

HEIDBACH, O., TINGAY, M., BARTH, A., REINECKER, J., KURFEβ, D. & MÜLLER, B. 2010. Global crustal stress patterns based on the World Stress Map database release 2008. *Tectonophysics*, **482**, 3–15.

HU, J. & ANGELIER, J. 2004. Stress permutations: three-dimensional distinct element analysis accounts for a common phenomenon in brittle tectonics. *Journal of Geophysical Research*, **109**, b09403, doi: 10.1029/2003JB002616.

JAEGER, J. C. & COOK, N. G. W. 1979. *Fundamentals of Rock Mechanics*. 3rd edn. Chapman and Hall, London.

LEWIS, J., BOOZER, A. C., LOPEZ, A. & MONTERO, W. 2008. Collision *v.* sliver transport in the hanging wall at the Middle America subduction zone: constraints from background seismicity in Central Costa Rica. *Geochemistry, Geophysics, Geosystems*, **9**, 1–12.

LIN, J. & STEIN, R. S. 2004. Stress triggering in thrust and subduction earthquakes, and stress interaction between the southern San Andreas and nearby thrust and strike-slip faults. *Journal of Geophysical Research*, **109**, B02303, doi: 10.1029/2003JB002607.

LINKIMER, L., BECK, S., SCHWARTZ, S., ZANDT, G. & VADIM, L. 2010. Nature of crustal terranes and the Moho in Northern Costa Rica from receiver function analysis. *Geochemistry, Geophysics, Geosystems G3 AGU*, **11**, 1–24.

LÓPEZ, A. 1999. *Neo-and paleostress partitioning in the SW corner of the Caribbean Plate and its fault reactivation potential*. PhD thesis, Tübingen, Germany.

LÓPEZ, ET AL. 2008*a*. *The Costa Rica Stress Map*. IX Congreso Geológico de América Central, 2008.

LÓPEZ, ET AL. 2008*b*. *Costa Rican Stresses: the Sigma Project*. 3rd World Stress Map conference, GFZ, Potsdam.

MARSHALL, J. S. 2000. *Active tectonics and Quaternary landscape evolution across the western Panama block, Costa Rica, Central America*. Unpublished PhD thesis, University Park, Pennsylvania State University.

MARSHALL, J. & FISHER, D. 2000. Central Costa Rica deformed belt: kinematics of diffuse faulting across the western Panamá block. *Tectonics*, **19**, 468–492.

MESCHEDE, M., BARCKHAUSEN, U. & WORM, H. 2000. Desarrollo del centro de dispersión entre las placas de Cocos y Nazca y los trazos de los puntos calientes. *Revista Geológica de América Central*, **23**, 5–16.

MONTERO, W. 1999. El terremoto del 4 de marzo de 1924 (Ms 7,0): un gran temblor interplaca asociado al límite incipiente entre la Placa Caribe y la microplaca de Panamá? *Revista Geológica de América Central*, **22**, 25–62.

MONTERO, W. 2001. Neotectónica de la región central de Costa Rica: frontera Oeste de la Microplaca de Panamá. *Revista Geológica de América Central*, **24**, 29–56.

MONTERO, W. & MORALES, L. D. 1990. Deformación y esfuerzos neotectónicos en Costa Rica. *Revista Geológica de América Central*, **11**, 69–87.

MONTERO, W. & ALVARADO, G. 1995. El terremoto de Patillos del 30 de diciembre 1952 (Ms 5,9) y el contexto neotectónico de la región del volcán Irazú. *Revista Geológica de América Central*, **18**, 25–42.

MONTERO, W. & KRUSE, S. 2006. Neotectónica y Geoísica de la falla Aguacaliente en los valles Coris y El Guarco; Costa Rica. *Revista Geológica de América Central*, **34–35**, 43–58.

MONTERO, W., BARAHONA, M., ROJAS, W. & TAYLOR, M. 2005. Los sistemas de falla Aguacaliente y Río Azul y relevos compresivos asociados, Valle central de Costa Rica. *Revista Geológica de América Central*, **33**, 7–27.

MONTERO, W., SOTO, G., ALVARADO, G. E. & ROJAS, W. 2010. División del deslizamiento tectónico y transtensión en el macizo del volcán Poás (Costa Rica), basado en estudios neotectónicos y de sismicidad histórica. *Revista Geológica de América Central*, **43**, 13–36.

MORRIS, A., FERRILL, D. & HENDERSON, D. 1996. Slip-tendency analysis and fault reactivation. *Geology*, **24**, 275–278, doi: 10.1130/0091-7613.

NEVES, M. C., PAIVA, L. T. & LUIS, J. 2009. Software for slip-tendency analysis in 3D: a plug-in for Coulomb. *Computers & Geosciences*, **35**, 2345–2352, doi: 10.1016/j.cageo.2009.03.008.

NOSTRO, C., STEIN, R. S., COCCO, M., BELARDINELLI, M. E. & MARZOCCHI, W. 1998. Two-way coupling between Vesuvius eruptions and southern Apennine earthquakes (Italy) by elastic stress transfer. *Journal of Geophysical Research*, **103**, 24 487–24 504.

OKADA, Y. 1992 Internal deformation due to shear and tensile faults in half space. *Bulletin of the Seismological Society of America*, **82**, 1018–1040.

PROTTI, M., SCHWARTZ, S. Y. & ZANDT, G. 1996. Simultaneous inversion for earthquake location and velocity structure beneath central Costa Rica. *Bulletin of the Seismological Society of America*, **86**, 19–31.

QUINTERO, R. & GÜENDEL, F. 2000. Stress field in Costa Rica, Central America. *Journal of Seismology*, **4**, 297–319.

RANERO, C. R. & VON HUENE, R. 2000. Subduction erosion along the Middle America convergent margin. *Nature*, **404**, 748–752.

SOTO, G. J. (ed.) 2005. Geología del cantón de Poás y estudios adicionales. FUNDEVI, University of Costa Rica.

SPERNER, B., MÜLLER, B., HEIDBACH, O., DELVAUX, D., REINECKER, J. & FUCHS, K. 2003. Tectonic stress in the Earth's crust: advances in the World Stress Map project. *In*: Nieuwland, D. A. (ed.) *New Insights into Structural Interpretation and Modelling*. Geological Society, London, Special Publications, **212**, 101–116.

Stein, R. S. 1999. The role of stress transfer in earthquake occurrence. *Nature*, **402**, 605–609, doi:pi10.1038/45144.

Tingay, M. R. P., Muller, B., Reinecker, J. & Heidbach, O. 2006. *State and origin of the present-day stress field in sedimentary basins: New results from the World Stress Map Project, paper presented at Golden Rocks 2006. The 41st US Symposium on Rock Mechanics (USRMS): 50 Years of Rock Mechanics Landmarks and Future Challenges*. American Rock Mechanics Association, Virginia.

Toda, S., Stein, R. S., Richards-Dinger, K. & Bozkurt, S. 2005. Forecasting the evolution of seismicity in southern California: animations built on earthquake stress transfer. *Journal of Geophysical Research*, **B05s16**, 314–334, doi: 10.1029/2004JB 003415.

Toda, S., Lin, J. & Stein, R. 2011*a*. Using the 2011 M = 9.0 Tohoku earthquake to test the Coulomb stress triggering hypothesis and to calculate faults brought closer to failure. *Earth Planets and Space*, **63**, 725–730. (Revised 10 May 2011 for the Tohoku Earthquake Special Issue of Earth Planets Space.)

Toda, S., Stein, R. & Lin, J. 2011*b*. Widespread seismicity excitation throughout central Japan following the 2011 M = 9.0 Tohoku earthquake, and its interpretation by Coulomb stress transfer. *Geophysical Research Letters*, **38**, L00G03, doi: 10.1029/2011GL047834.

Toda, S., Stein, R. S., Sevilgen, V. & Lin, J. 2011*c*. *Coulomb 3.3 graphic-rich deformation and stress-change software: user guide*. US Geological Survey, Open-File Report 2011–1060.

Villegas, A. 1997. *Relación entre el vulcanismo y la evolución geodinámica de la Cordillera de Guanacaste*. MSc thesis, Escuela Centromericana de Geología, Universidad de Costa Rica.

Von Huene, R., Ranero, C., Weinrebe, W. & Hinz, K. 2000. Quaternary convergent margin tectonics of Costa Rica, segmentation of the Cocos Plate, and Central American volcanism. *Tectonics*, **19**, 314–334.

Wells, D. & Coppersmith, K. 1994. New empirical relationships among magnitude, rupture length, Rupture width, Rupture Area, and Surface Displacement. *Bulletin of the Seismological Society of America*, **84**, 974–1002.

WOODWARD-CLYDE CONSULTANTS 1993. *A preliminary evaluation of earthquake and volcanic hazards significant to the major population centers of the Valle Central, Costa Rica*. Ret Corporation.

Zoback, M. D. & Zoback, M. L. 1991. Tectonic stress field of North America and relative Plate motions, *In*: Slemmons, D. B. et al. (eds) *Neotectonics of North America, Decade Map vol. I*. Geological Society of America, Boulder, CO, 339–366.

Zoback, M. D. & Zoback, M. L. 2002. State of stress in the Earth's lithosphere, *In*: Lee, W. H. K., Jennings, P. C. & Kanamori, H. (eds) *International Handbook of Earthquake and Engineering Seismology, International Geophysics Series*. Academic Press, Amsterdam, 559–568.

Zoback, M. L. & Zoback, M. D. 1989. Tectonic stress field of the conterminous United States. *In*: Pakiser, L. C. & Mooney, W. D. (eds) *Geophysical Framework of the Continental United States*. Geological Society of America, Boulder, Memoirs, **172**, 523–539.

Zoback, M. L. 1992. First-and second-order patterns of stress in the lithosphere: the World Stress Map project. *Journal of Geophysical Research*, **97**, 11 703–11 728.

Zoback, M. L. & Burke, K. 1993. Lithospheric stress patterns: a global view. *Eos Transactions*, **74**, 609–618.

Zoback, M. L. & Mooney, W. D. 2003. Lithospheric buoyancy and continental intraplate stresses. *International Geological Review*, **45**, 95–118.

Reverse fault rupturing: competition between non-optimal and optimal fault orientations

R. H. SIBSON

[1]*Department of Geology, University of Otago, P.O. Box 56, Dunedin 9054, New Zealand*

(e-mail: rick.sibson@otago.ac.nz)

Abstract: A dip histogram for intracontinental $M > 5.5$ reverse-slip ruptures reveals a trimodal distribution with a dominant Andersonian peak (fault dip, $\delta = 30 \pm 5^\circ$) flanked by subsidiary clusters at $\delta = 10 \pm 5^\circ$ and $50 \pm 5^\circ$, and no dips greater than 60°. For a simple compressional regime ($\sigma_v = \sigma_3$), the dominant peak is in accord with the reshear of optimally oriented faults with a friction coefficient of $\mu_s = 0.6 \pm 0.2$, implying frictional lock-up at $\delta = 60 \pm 10^\circ$ consistent with the observed upper dip bound. The low-dip cluster ($\delta = 10 \pm 5^\circ$) is dominated by thrusting in the frontal Himalaya and may incorporate staircase thrust systems in cover sequences with deflections along bedding anisotropy. The cluster of moderate-to-steep reverse fault ruptures ($\delta = 50 \pm 5^\circ$) is likely dominated by compressional inversion of inherited normal faults. In both circumstances, however, there appears to be competition between Andersonian thrusts in various stages of development and non-optimal failure planes dipping at either high or low angles. A delicate balance between levels of differential stress and fluid-pressure determines whether or not a poorly oriented thrust or reverse fault reactivates in preference to the development of new, favourably oriented Andersonian thrusts.

Tectonic style varies considerably in areas of continental contraction from thin-skinned fold-thrust belts, largely restricted to the sedimentary cover sequence, to thick-skinned assemblages where generally steeper thrust and reverse fault systems involve crystalline basement and cut across at least the entire upper half of the crust before flattening into décollement shear zones at the base of the brittle crust (Pfiffner 2006). To some extent, the structural literature is biased by the detailed attention paid to thin-skinned foreland fold-thrust belts because of their economic significance as hosts for oil–gas fields (e.g. McClay 2004; Cooper 2007). In addition, the development of Coulomb wedge models for accretionary prisms in subduction settings and the extension of such analyses to other settings have emphasized the role of décollements associated with low-angle thrusting in developing critical wedge tapers (Davis *et al.* 1983; Dahlen 1990). This bias has been redressed, at least in part, over the past 20 years or so by the increasing recognition of inversion assemblages involving the compressional reactivation of moderate-to steep faults within the basement (sometimes extending into the cover sequence) inherited from earlier phases of extensional tectonics (Williams *et al.* 1989; Buchanan & Buchanan 1995).

Here frictional fault mechanics is used to analyse the dip distribution of close-to-pure reverse-slip ruptures on active thrust-reverse faults within the seismogenic zone that commonly occupies the top 10–20 km of deforming continental crust (e.g. Nazareth & Hauksson 2004). The analysis follows on from E. M. Anderson's (1905, 1951) seminal work on the initial formation of faults in different tectonic settings.

Andersonian thrust faulting in a compressional regime

Anderson (1905, 1951) discussed the initial formation of faults in homogeneous isotropic intact rock. He argued first that one of the principal compressive stresses ($\sigma_1 > \sigma_2 > \sigma_3$) should generally be subvertical because of the boundary condition imposed by the Earth's free surface, giving rise to three fundamental stress regimes depending on whether $\sigma_v = \sigma_1$, σ_2 or σ_3. By adopting the equivalent of the Coulomb criterion for brittle shear failure, he argued that new faults should typically form along planes containing the σ_2 axis lying at about 30° to the maximum compression, σ_1 (Fig. 1a). Three fundamental classes of faults should therefore be associated with the different stress regimes: normal faults dipping at *c.* 60° where $\sigma_v = \sigma_1$; vertical wrench (strike-slip) faults where $\sigma_v = \sigma_2$; and thrust faults dipping at *c.* 30° where $\sigma_v = \sigma_3$.

Thrust fault initiation

Taking account of effective stress, the Coulomb criterion for brittle shear failure in a saturated rock

From: Healy, D., Butler, R. W. H., Shipton, Z. K. & Sibson, R. H. (eds) 2012. *Faulting, Fracturing and Igneous Intrusion in the Earth's Crust*. Geological Society, London, Special Publications, **367**, 39–50. http://dx.doi.org/10.1144/SP367.4

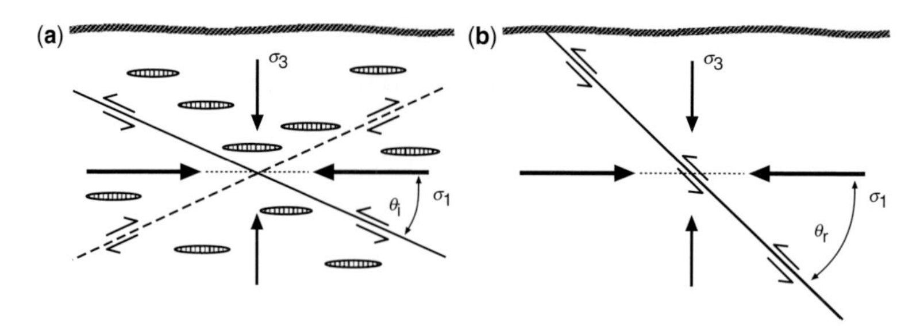

Fig. 1. (**a**) Formation of a new Andersonian thrust fault in a standard compressional regime with $\sigma_v = \sigma_3$ (potential orientation of extension fractures and conjugate thrust fault denoted by crosshatched ellipses and dashed line, respectively); and (**b**) reactivation of existing fault containing σ_2 axis and oriented at θ_r to horizontal σ_1.

mass with pore-fluid pressure P_f may be written:

$$\tau = C + \mu_i \sigma_n' = C + \mu_i(\sigma_n - P_f) \qquad (1)$$

where τ and σ_n are the shear and normal stress components on the failure plane, respectively, and the intact rock is characterized by a cohesive strength C and a coefficient of internal friction $\mu_i = \tan \phi_i$ where the friction angle ϕ_i is the slope of the failure envelope on a Mohr diagram. Shear failure then occurs on planes containing σ_2 oriented at an angle $\theta_i = 45 - \phi_i/2 = 0.5 \tan^{-1}(1/\mu_i)$ to σ_1. From experimental rock mechanics, internal friction generally lies in the range $0.5 < \mu_i < 1.0$ (Jaeger & Cook 1979) so the corresponding initial dips of reverse faults in an Andersonian compressional regime with $\sigma_v = \sigma_3$ should always be less than 45° and occupy the range $32° > \delta = \theta_i > 22°$ (Fig. 1a). Conjugate thrust faults would then be expected to intersect at 44–64°. Various forms of anisotropy in the rock mass (bedding, foliation, etc.) may cause significant departures from these idealized relationships, but they are often observed to hold for at least low-displacement brittle faults. Some major structures such as the Wind River Thrust in Wyoming (Brewer *et al.* 1980; Lynn *et al.* 1983) and the Outer Hebrides Thrust zone together with parallel structures in NW Scotland (Jehu & Craig 1925; Brewer & Smythe 1984) also appear to resemble transcrustal Andersonian thrusts dipping at 25–35° through crystalline basement assemblages.

Reactivation of existing faults in a compressional regime

The condition for reshear of an existing fault retaining some cohesive strength, $c < C$, is simply:

$$\tau = c + \mu_s \sigma_n' = c + \mu_s(\sigma_n - P_f) \qquad (2)$$

where τ and σ_n are again the shear and normal stress components acting on the fault plane and, for most rocks, the coefficient of sliding friction lies in the range $0.6 < \mu_s < 0.85$ (Byerlee 1978). When the fault lacks cohesive strength, this reduces to:

$$\tau = \mu_s \sigma_n' = \mu_s(\sigma_n - P_f). \qquad (3)$$

In a compressional regime, the effective vertical stress is:

$$\sigma_v' = \sigma_v - P_f = \rho g z(1 - \lambda_v) = \sigma_3' \qquad (4)$$

where ρ is average rock density, g is gravitational acceleration and λ_v is the pore-fluid factor, defined $\lambda_v = (P_f/\sigma_v)$. For the particular case of a cohesionless fault containing σ_2 and oriented at a reactivation angle θ_r to σ_1 (Fig. 1b), Equation (3) may be rewritten in terms of the principal stresses as:

$$\sigma_1 - \sigma_3 = \mu_s \left(\frac{\tan \theta_r + \cot \theta_r}{1 - \mu_s \tan \theta_r} \right) \sigma_3' \qquad (5)$$

or

$$\sigma_1 - \sigma_3 = \mu_s \left(\frac{\tan \theta_r + \cot \theta_r}{1 - \mu_s \tan \theta_r} \right) \rho g z(1 - \lambda_v) \qquad (6)$$

allowing the differential stress for reactivation of reverse-slip faults to be plotted as a function of the reactivation angle θ_r and the fluid-pressure state λ_v (Sibson 1985, 1990). Note that for an ideal Andersonian compressional regime with horizontal σ_1, the dip of the reverse fault δ is equal to the reactivation angle θ_r (Fig. 1).

The reactivation diagram in Figure 2 is plotted from Equation (6) for a friction coefficient $\mu_s = 0.6$ at the lower end of Byerlee's (1978) range and at a representative focal depth of 10 km. This selected value for μ_s is in accord with recent findings from experimental studies on gouge friction which demonstrate that only high-strength gouges with $\mu_s \sim 0.6$ possess the velocity-weakening

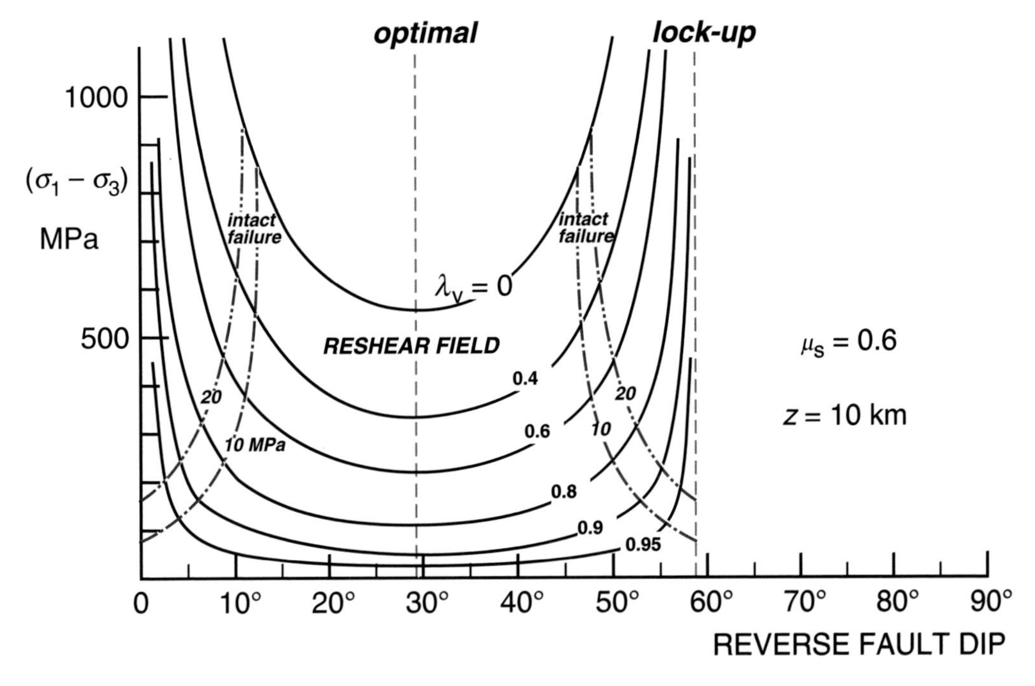

Fig. 2. Two-dimension reactivation plot of differential stress plotted against dip for a cohesionless pure reverse fault with $\mu_s = 0.6$ at 10 km depth, for different values of the pore-fluid factor, $\lambda_v = P_f/\sigma_v$. The $\lambda_v = 0.4$ line equates approximately with hydrostatic fluid pressure. Red dashed lines define dip values for optimal reactivation and frictional lock-up. Blue dash-dot lines approximate shear failure conditions for intact crust with tensile strengths, $T = 10$ and 20 MPa (after Sibson 2009).

characteristics needed for seismic slip instability (Ikari *et al.* 2011). Optimal orientation for reactivation, when the differential stress required for reshear is a minimum, is given by $\theta_r^* = 0.5 \tan^{-1} (1/\mu_s)$. Values of θ_r^* are listed in Table 1 for different friction coefficients. When $\mu_s = 0.6$, $\theta_r^* = 29.5°$ but note that θ_r^* remains at $30 \pm 5°$ over the broad range of friction values of $0.4 < \mu_s < 0.8$. Faults are therefore easiest to reactivate when oriented at close to their original Andersonian attitudes.

Table 1. *Optimal reactivation angles (θ_r^*) and angles of frictional lock-up (θ_r^*) corresponding to different friction coefficients μ*

μ	θ_r^* (°)	$2\theta_r^*$ (°)
0.0	45	90.0
0.1	42.1	84.3
0.2	39.4	78.7
0.3	36.7	73.3
0.4	34.1	68.2
0.5	31.7	63.4
0.6	29.5	59.0
0.7	27.5	55.0
0.8	25.7	51.3
0.9	24.0	48.0
1.0	22.5	45.0

The differential stress needed for reshear is at a minimum when the fault is optimally oriented but rises towards infinity as the reactivation angle either decreases towards zero or increases towards frictional lock-up which, when $\mu_s = 0.6$, gives $\delta = 2\theta_r^* \sim 59°$ (Fig. 2). At higher pore-fluid pressures with $\lambda_v \rightarrow 1$, the reactivation curves flatten to cover more of the allowable dip range ($0 < \delta < 59°$) at low differential stress. Reactivation beyond lock-up ($\delta > \theta_r = 2\theta_r^*$) is possible but only under special circumstances when the *tensile overpressure condition* $P_f > \sigma_3$ is met, as is the case in some mineralized reverse fault systems (Sibson *et al.* 1988). Note that if the existing fault retains some cohesive strength, $c < C$, the reshear curves remain symmetric about $\theta_r^* = 29.5°$ but are displaced upwards and tighten (Sibson 2009). As fault orientation departs from optimal, the situation may also arise where a new, optimally oriented thrust fault may form in intact crust in preference to the reactivation of the existing, badly oriented fault. The two sets of lines labelled with tensile strength values $T = 10$ and 20 MPa approximate the shear failure conditions for intact rock of moderate and high competence (Lockner 1995), respectively, at 10 km depth and different fluid-pressure values assuming a composite Griffith–Coulomb failure envelope (internal friction $\mu_i = 0.75$; Sibson 2000). These curves for intact rock failure restrict the reshear field to the centre of the reactivation plot, except at high fluid overpressures with $\lambda_v \rightarrow 1$.

Stress trajectory issues

Critical to the analysis that follows is the assumption of subhorizontal σ_1 trajectories (to within, say, $\pm 5°$) in areas under compression. Irregular topography will deviate stress trajectories in the near-surface but these deviations tend to die out at depths greater than the wavelength of the irregularities (Zang & Stephansson 2010). The uniformity of stress trajectory determinations over vast regions of continental interiors (Zoback 1992) points to the prevalence of predominantly Andersonian stress states and trajectories. Palaeostress determinations from fault slip data yield a preponderance of near-Andersonian stress states with one principal stress subvertical (Lisle *et al.* 2006), as does an analysis of the Global Centroid Moment Tensor catalogue for crustal earthquakes (Célérier 2008). Note, however, that potentially significant deviations of σ_1 trajectories from the horizontal have been postulated for accretionary prisms and fold-thrust belts where non-cohesive Coulomb wedge theory is applicable (Dahlen 1990). Development of a ductile décollement at the base of the seismogenic upper crust in areas of active 'flake tectonics' (Oxburgh 1972; Armstrong & Dick 1974) also requires σ_1 and σ_3 stress trajectories to deviate at depth from the horizontal and vertical.

Dip compilation of reverse fault ruptures

Figure 3a is a global compilation of 65 dip values (δ) for $M > 5.5$ intracontinental earthquakes involving reverse-slip rupturing where the slip vector lies within $\pm 30°$ of the dip direction. Such earthquakes likely have rupture dimensions greater than *c.* 5 km. Subduction thrust ruptures are excluded although it can be argued that the main Himalayan thrust interface, dipping at a very low angle, resembles many subduction interfaces (Seeber & Armbruster 1981; Ni & Barazangi 1984). Data sources are listed in Sibson & Xie (1998), Collettini & Sibson (2001) and Sibson (2004, 2009), with one additional dip estimate of 11° from Cotton *et al.* (1996) for the 1991 *M* 6.8 Uttarkashi earthquake. Dip values are estimated from first motion arrivals, body-wave modelling or centroid moment tensor (CMT) focal mechanisms where the rupture plane can be positively distinguished by either: (1) the presence of a surface rupture; (2) the aftershock distribution; (3) analysis of surface deformation; or (4) topography (mountains assumed to occupy the hanging wall) in the case of earthquakes in the Himalayan foothills. Note that while the first arrival mechanisms yield estimates of rupture orientation during nucleation at the hypocentre, the CMT mechanisms provide an estimate for the average dip of the rupture plane. Uncertainties in such dip estimates are likely to be at least $\pm 5°$. It also has to be borne in mind that rupture dips from focal mechanisms may give a misleadingly simple picture. For example, the 1999 *M* 7.6 Chi-Chi thrust rupture in Taiwan appears from its focal mechanism as an out-of-sequence Andersonian thrust with a dip of *c.* 25° that projects to the surface rupture along the Chelungpu Fault (Kao & Chen 2000). Structural analyses however point to the possibility of a much more complicated thrust system incorporating subhorizontal detachments (Yue *et al.* 2005; Lee & Chan 2007). Focal depth determinations for the different ruptures are far from uniformly reliable, but the majority probably nucleated at depths greater than 5 km within crystalline basement.

The distribution in Figure 3a is trimodal with a dominant peak at $\delta = 30 \pm 5°$ (*c.* 40%) flanked by two subsidiary clusters at $\delta = 10 \pm 5°$ (*c.* 20%) and $50 \pm 5°$ (*c.* 29%), and with no dips greater than 60°. On the assumption of horizontal σ_1, the dominant peak at $\delta = 30 \pm 5°$ can be interpreted as reflecting optimal orientation of existing reverse faults with a friction coefficient $\mu_s = 0.6$ at the lower end of Byerlee's (1978) range for rock friction (Fig. 4). The lack of reverse-slip ruptures at $\delta > 60°$ is then in agreement with the expectation of frictional lock-up at twice this optimal orientation for reactivation. Note also that there is no obvious correlation of a dip cluster with the plane of maximum shear stress (τ_{max}), as might be expected if plastic slip-planes were controlling fault orientation.

The lower dip cluster ($\delta = 10 \pm 5°$) is dominated by thrust ruptures in the fold-thrust belt of the frontal Himalaya (Fig. 3b). Data are mostly from Molnar & Lyon-Caen (1989) with the likely rupture planes discriminated on topographic grounds, but also include dip estimates for the large 1991 *M* 6.8 Uttarkashi earthquake (Cotton *et al.* 1996), the 1999 *M* 6.8 Chamoli earthquake (Kayal *et al.* 2003) and the 2005 *M* 7.6 Kashmir earthquake (Avouac *et al.* 2006). While low-dipping thrusts predominate, the dip distribution is distinctly bimodal with occasional ruptures also occurring on out-of-sequence Andersonian ($\delta \sim 30°$) thrusts, as in the case of the 2005 *M* 7.6 Kashmir earthquake (Avouac *et al.* 2006). Most of the low-dipping ruptures seem likely to have occurred on, or near-parallel to, the low-dipping Main Himalayan Thrust (MHT) interface (Avouac 2003), perhaps combined

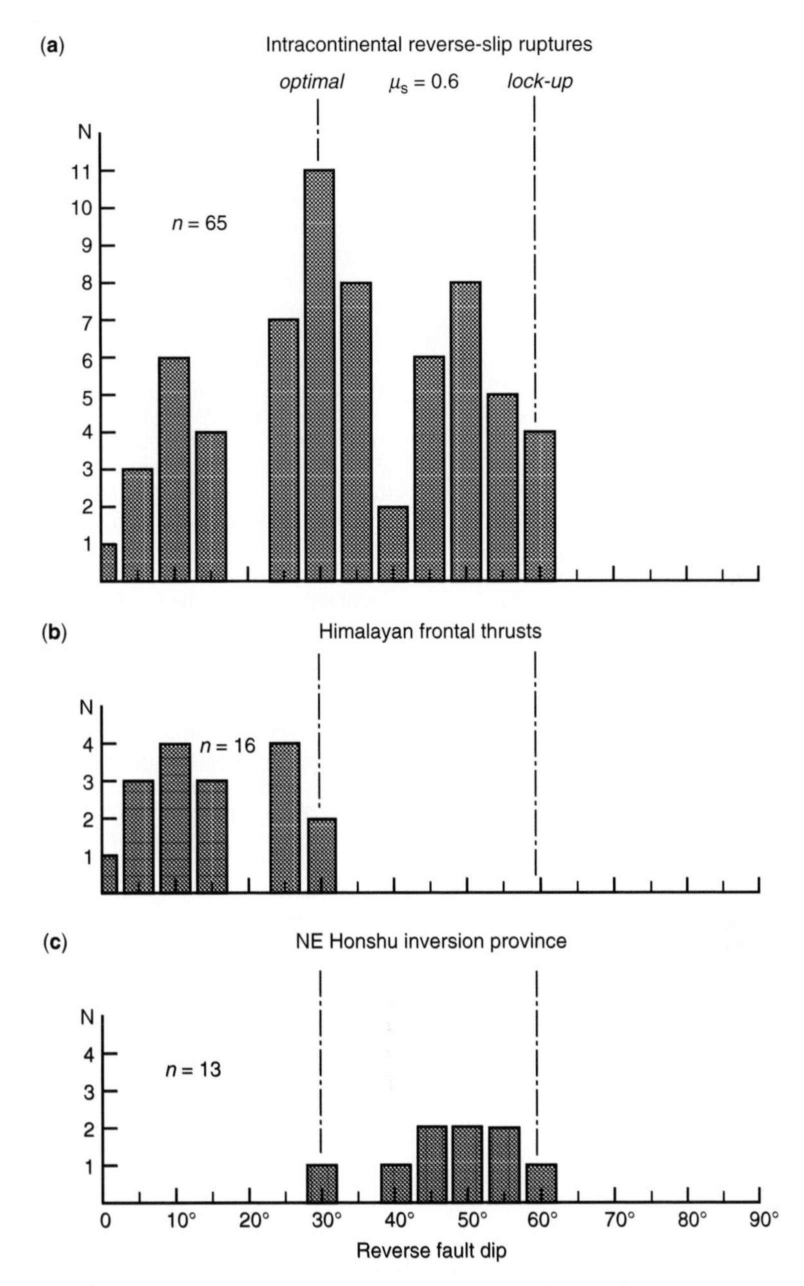

Fig. 3. (**a**) Global dip distribution for intracontinental $M > 5.5$ mainshocks involving reverse-slip rupturing where the slip-vector rakes within $\pm 30^\circ$ of the dip direction ($n = 65$); (**b**) dip distribution for thrusts in the frontal Himalaya; and (**c**) dip distribution for reverse faults in NE Honshu (all updated from Sibson 2009).

with the development of 'staircase' systems within the cover sequence involving preferential deflection of thrusts along bedding (Fig. 5a). Given the regional topographic gradient and the results of Coulomb wedge analysis (Dahlen 1990), it could be argued that here the σ_1 trajectories are likely to be plunging at shallow angles towards the range front; this is belied, at least locally, by the occurrence of typical Andersonian thrusts dipping at *c.* 30°.

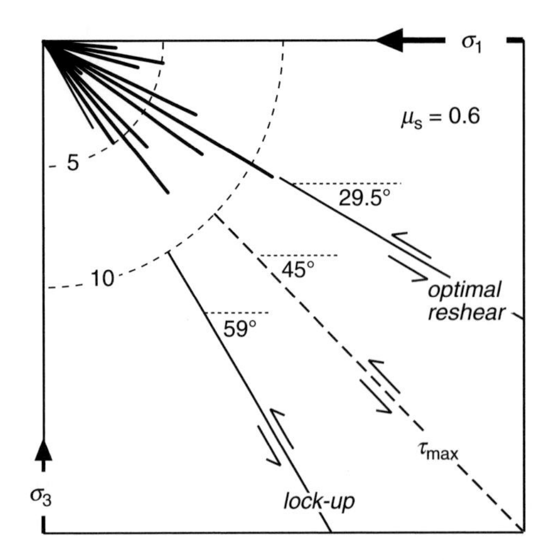

Fig. 4. Rose histogram of reverse-slip rupture dips in relation to a standard Andersonian compressional regime showing orientations of optimal reshear and lock-frictional lock-up for $\mu_s = 0.6$, plus maximum shear stress orientation.

The cluster of moderate-to-steeply dipping reverse-slip ruptures with $\delta > 45°$ in Figure 3a likely comprises a mixture of active reverse faults arising from compressional reactivation of inherited normal faults (compressional inversion; Fig. 5b), plus reverse faults that have undergone 'domino-steepening' from initially lower dips through bulk shortening in areas of continental collision (Sibson & Xie 1998). Recent seismic activity onshore in the magmatic arc of NE Honshu has been dominated by $6 < M < 7.5$ rupturing on moderate-to-steep reverse faults and provides an example of ongoing compressional inversion (Fig. 3c). Stress inversions from focal mechanisms are generally consistent with an Andersonian compressional regime roughly orthogonal to the magmatic arc (Townend & Zoback 2006). On structural and stratigraphic grounds, these structures became active as reverse faults during Pliocene–Quaternary compressional inversion of steep normal faults inherited from Miocene rifting (Sato & Amano 1991; Kato *et al.* 2006). However, there are also hints of a bimodal dip distribution with occasional Andersonian thrusts such as that responsible for the 1896 *M* 7.2 Riku-u earthquake, along with evidence of subsidiary Andersonian thrust ruptures associated with some of the compressional inversion sequences (Sibson 2007, 2009). The NW South Island of New Zealand provides a comparable area of active compressional inversion with competition between reverse-slip reactivation of steep inherited normal faults and development of new Andersonian thrusts (Ghisetti & Sibson 2006). Notably, both these active inversion provinces occupy crust overlying subducting and dehydrating

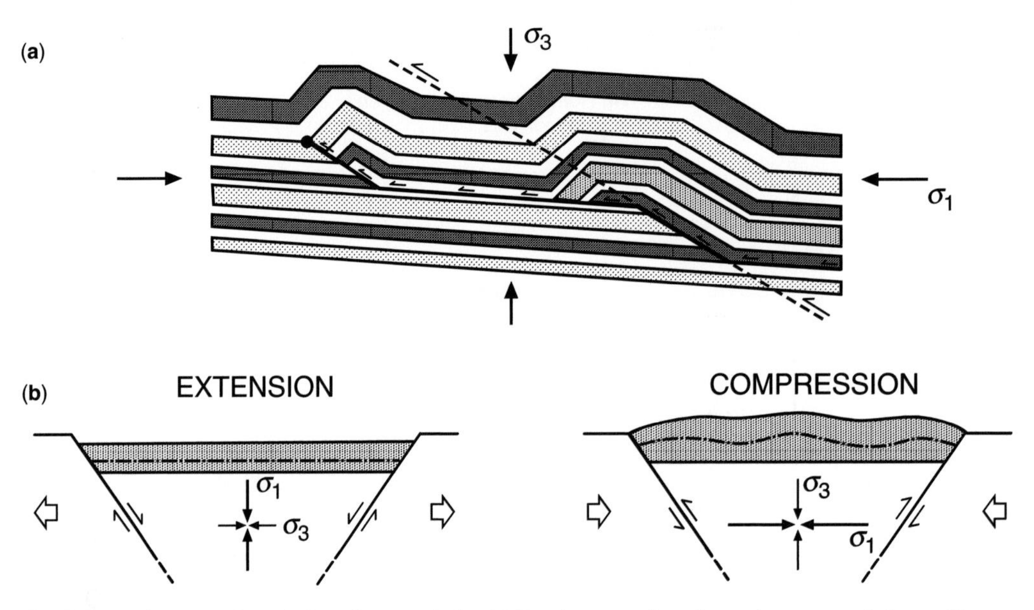

Fig. 5. Tectonic settings for misoriented reverse faults. (**a**) Development of a staircase thrust system influenced by bedding anisotropy comprising flats along bedding linked by cross-cutting ramps, with the development of associated fault-bend and fault-propagation folds (dashed line represents a possible propagation path for an out-of-sequence Andersonian thrust). (**b**) Compressional inversion of a former rift basin leading to the development of steep reverse faults.

lithosphere, with geophysical evidence for the transfer into the crust of slab-derived hydrothermal fluids (Hasegawa *et al.* 2005; Wannamaker *et al.* 2009).

Competition between optimal and non-optimal reverse faults

On the assumption of horizontal σ_1 trajectories, competition between optimal ($\delta \sim 30°$) and non-optimal existing reverse faults can be assessed in terms of frictional mechanics. For $\mu_s = 0.6$, the differential stress needed for reactivation as a function of the reactivation angle θ_r (Equation 5) is normalized against that required for reactivation at optimal orientation to produce the depth-independent plot in Figure 6 (the shaded columns represent the three dominant dip clusters). The differential stress required for reactivation of non-optimal reverse faults at $\delta = 10 \pm 5°$ and $50 \pm 5°$ approaches twice that required for optimal reactivation, other factors (depth, fluid pressure, etc.) being equal. Also plotted is the pore-fluid factor ($\lambda_v = P_f/\sigma_v$) needed for the non-optimally oriented reverse faults to be reactivated at the same levels of differential stress as optimally oriented faults under hydrostatic fluid pressure ($\lambda_v \sim 0.38$). Significant overpressuring ($\lambda_v > 0.6$) is required for this to occur. Note the critical point that unfavourably oriented reverse faults are only likely to reshear in the absence of through-going favourably oriented (i.e. Andersonian) thrusts which, if present, would always be preferentially reactivated.

The balance between the reactivation of a non-optimally oriented reverse fault and the formation within intact crust of a new favourably oriented Andersonian thrust (cf. Fig. 2) is illustrated on a brittle failure mode plot of differential stress ($\sigma_1 - \sigma_3$) versus effective vertical stress $\sigma_v{}'$ (Sibson 2000). For illustrative purposes, the shear failure condition for 'generic' intact crust of moderate competence is represented by a composite Griffith–Coulomb envelope with internal friction $\mu_i = 0.75$ normalized to a tensile strength, $T = 10$ MPa (Fig. 7). The reshear condition for an existing cohesionless fault with $\mu_s = 0.6$ oriented at $\delta = \theta_r = 50°$ to horizontal σ_1 is plotted from Equation 4 (and is close to that for $\delta = \theta_r = 10°$). Shear failure along anisotropy in the cover sequence can also be considered by treating bedding as a potential failure plane with a cohesive strength c significantly lower than the surrounding rock. Reshear lines are then displaced upwards to an intercept $-c/\mu_s$ on the σ_v' axis (Sibson 2009). The example in Figure 7 is for $\delta = 10°$ with $\mu_s = 0.6$ and $c = 3$ MPa, giving an intercept at $\sigma_v' = -5$ MPa. At shallow depth bedding planes may also have lowered friction because of high clay mineral content, but this is not considered further in the analysis.

The intercept between the reshear condition for $\delta = 50°$ and the failure curve for intact rock at $\sigma_v' = 62$ MPa (which, coincidentally, approximates the intercept with the reshear line for $\delta = 10°$ and $c = 3$ MPa) defines two different failure regimes. At $\sigma_v' < 62$ MPa, less differential stress is required to reactivate the steep 50°-dipping reverse fault

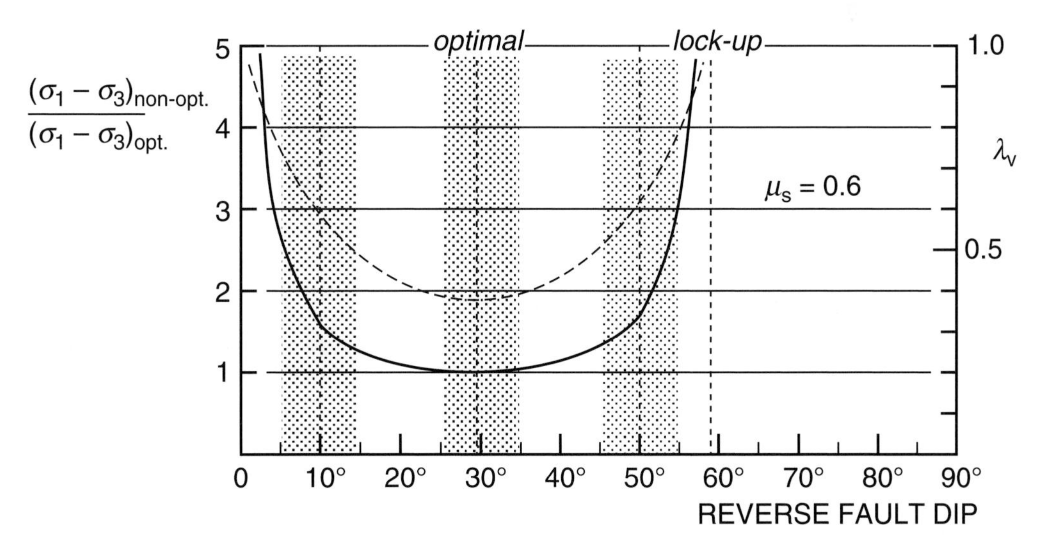

Fig. 6. Differential stress required for frictional reactivation of a non-optimally oriented reverse fault with $\mu_s = 0.6$ normalized against that required for an optimally oriented fault, plotted as a function of fault dip. Dashed curve illustrates the value of λ_v required for the non-optimal reverse fault to fail at the same differential stress as an optimally oriented fault under hydrostatic fluid pressure (i.e. $\lambda_v = 0.38$).

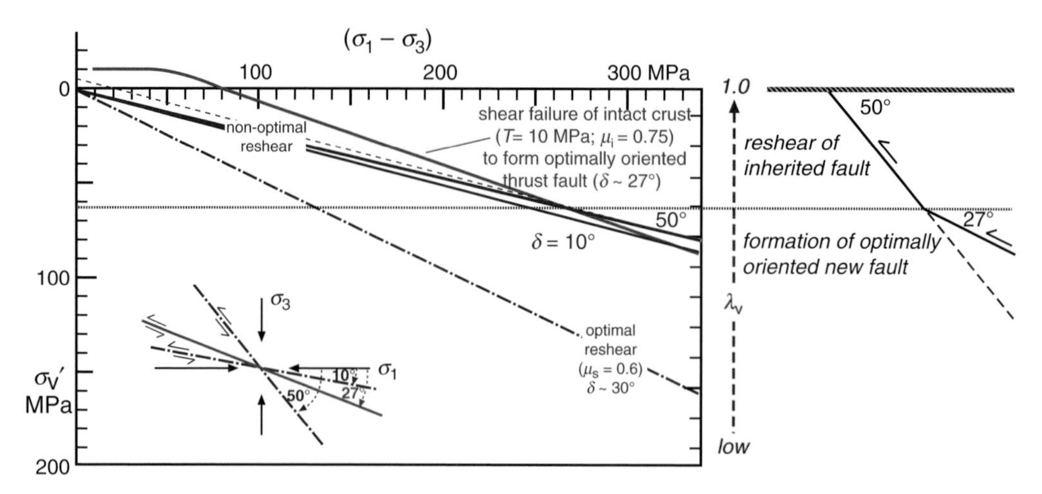

Fig. 7. Brittle failure mode plot comparing the stress conditions for reshear of existing cohesionless reverse faults with dips of 10° and 50° in a compressional regime, with the failure condition of intact rock with tensile strength $T = 10$ MPa. The plot shows the transition from preferential reshear of the steep reverse fault at shallow depth (or high λ_v) to the formation of a new Andersonian thrust at greater depth (or low λ_v).

than to form a new optimally oriented thrust fault; at $\sigma_v' > 62$ MPa however, it becomes easier to develop a new optimal thrust fault than to reactivate the steep existing structure. For this situation, the changeover value $\sigma_v' = 62$ MPa is equivalent to a depth of *c*. 3.85 km under hydrostatic fluid-pressure conditions ($\lambda_v = 0.38$) but equates to a depth of *c*. 11.92 km at a fluid overpressure with $\lambda_v = 0.8$. Provided stress trajectories remain Andersonian, the plot demonstrates: (1) a general tendency for Andersonian thrusts to become dominant over misoriented faults with increasing overburden pressure at depth unless fluid overpressures are localized around the unfavourably oriented structures; (2) non-optimally oriented low-angle thrusts ($\delta = 10 \pm 5°$) and steep reverse faults ($\delta = 50 \pm 5°$) are only likely to remain active in areas of crust that are fluid-overpressured and lack favourably oriented Andersonian thrusts; and (3) downdip segmentation of steep reverse faults by lower-dipping Andersonian thrusts (as observed in some mineralized fault systems; Fig. 8) becomes likely when fluid overpressuring can no longer be maintained.

Evidence for fluid overpressuring around non-optimal reverse faults

It is not uncommon for pore-fluid pressure to rise above hydrostatic values (i.e. $\lambda_v = P_f/\sigma_v > 0.4$) at depths of more than a few kilometres in sedimentary basins infilled with low-permeability sequences, and rise towards lithostatic values ($\lambda_v \rightarrow 1.0$) with increasing depth. Hubbert & Rubey (1959) first associated low-angle overthrusting with fluid overpressuring in sedimentary basins in areas of compressional tectonics, an association since substantiated by direct borehole measurements during oil–gas exploration around the world. Cello & Nur (1988) showed that the loading imposed by an advancing thrust sheet itself contributes to the generation of overpressure. The compressional stress field also plays an important role helping to contain the overpressure, especially when the faults are poorly oriented for reactivation (Sibson 2003). Coulomb wedge analysis of accretionary prisms and foreland fold-thrust belts suggests extensive but variable overpressuring (Davis *et al.* 1983) which is borne out by borehole investigations, geological and geophysical characterization and modelling (Moore & Vrojlik 1992; Tobin & Saffer 2009).

Overpressuring is not necessarily restricted to the sedimentary cover but may also affect fault structures in crystalline basement. On the basis of associated extension vein arrays requiring the hydraulic fracture criterion $P_f > \sigma_3$ to be met, near-lithostatic overpressures are inferred to be associated with the formation of some thrust faults (Nguyen *et al.* 1998) and the reactivation of severely misoriented steep reverse faults in crystalline basement (Sibson *et al.* 1988). Moreover, in the active compressional inversion province of NE Honshu, a wealth of geophysical evidence (local occurrence of bright-spot S-wave reflectors, seismic low-velocity zones, anomalous Vp/Vs ratios and zones of high electrical conductivity) supports the existence of a fluid-rich, variably overpressured mid-crust which extends well into the lower seismogenic

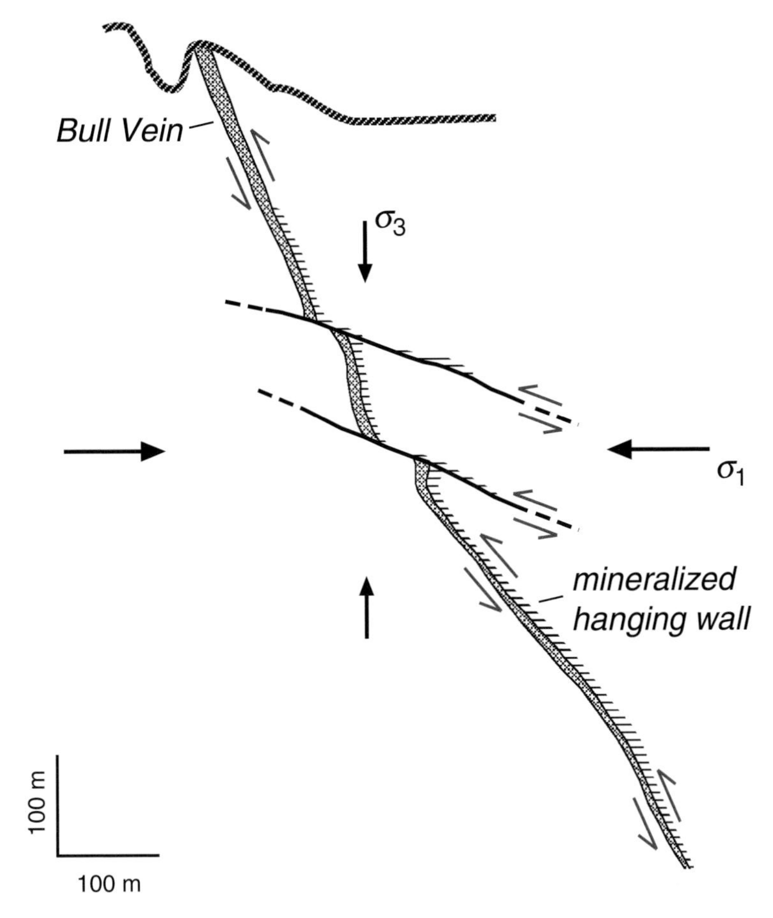

Fig. 8. Sketch cross-section of fault-hosted mesozonal Au-quartz mineralization in the Carson Hill Mine, Calaveras County, in the Mother Lode belt of the Sierra Nevada foothills, California (after Burgess 1948) shown in relation to the inferred stress system at the time of mineralization. A mesozonal Au-quartz lode (the Bull Vein) hosted on a steep reverse fault was mined to over 900 m depth, though disrupted by late Andersonian thrusts. Disseminated mineralization also occurs in the immediate hanging wall of the Bull Vein.

zone, especially in the vicinity of the active fault systems (Hasegawa *et al.* 2005; Okada *et al.* 2006; Sibson 2009).

The inference from frictional mechanics that reactivation of low-angle thrusts and high-angle reverse faults that depart significantly from optimal orientation involves fluid overpressuring towards lithostatic values is thus supported by circumstantial evidence. Outstanding questions include the extent to which overpressures are localized in the vicinity of the active fault structures, and whether the faults themselves serve as the dominant pathways for migrating overpressured fluids at seismogenic depths.

Supplanting misoriented reverse faults

Strongly misoriented faults are only likely to remain active as long as high fluid overpressures are maintained. If overpressuring is not maintained, Andersonian thrusts will develop through the rock mass and supplant the badly oriented structures which then become inactive. In the case of foreland fold-thrust belts, 'staircase' thrust systems evolve initially as bedding-parallel 'flats' linked by Andersonian thrust ramps across thicker competent units (Fig. 5a). If the levels of fluid overpressure needed for continued reactivation of the flats cannot be maintained, an Andersonian thrust is likely to propagate from one of the ramps to form an out-of-sequence thrust that may ultimately transect the entire upper crustal seismogenic zone. The *c.* 30°-dipping thrust that broke to the surface from *c.* 11–15 km depth in the 2005 *M* 7.6 Kashmir earthquake (Avouac *et al.* 2006) could represent rupturing along one such structure.

In areas of compressional inversion, new-forming Andersonian thrusts may likewise cut across and supplant steep inversion structures as overpressures wane. Evidence for this comes from steep reverse faults hosting mesozonal mineralization derived from episodic discharge of near-lithostatically overpressured hydrothermal fluids through fault–valve action (Sibson *et al.* 1988). Not infrequently, these mineralized steep reverse faults are disrupted by sets of late-stage Andersonian thrusts that mark the cessation of the overpressure conditions, allowing reactivation of the steep reverse faults (Fig. 8). The same may occur on a regional scale. One possible example is the Carmel Head Thrust of Caledonian age on the Isle of Anglesey, north Wales. With a present average dip of *c.* 25°N along its better-exposed western trace (Greenly 1919), this structure emplaces rocks of the Mona Complex rocks over younger Lower Paleozoic sequences. The Carmel Head Thrust slices through a series of downfaulted Ordovician outliers, partly flanked by steep reverse faults (some hydrothermally mineralized) active during compressional inversion that predated the Carmel Head Thrust (Bates 1974).

A general expectation is that, because of their lesser ability to contain fluid overpressures, areas with active Andersonian thrust faulting likely have lower prevailing fluid pressures than areas where misoriented reverse faults (either low- or high-angle) remain active.

Conclusions

The frequent occurrence of very low-angle thrusts subparallel to bedding recognized by field mapping in foreland fold-thrust belts is not matched by the dip distribution of reverse-slip earthquake ruptures. A dip histogram for reverse-slip ruptures shows a predominance of Andersonian thrusts dipping at $30 \pm 5^\circ$ in accord with the presumption of fault formation or reactivation at optimal orientations with respect to horizontal σ_1 in a simple compressional regime. For horizontal σ_1, the spread in dip values ($0 < \delta < 60^\circ$) and the peak at $30 \pm 5^\circ$ are both consistent with a fault sliding friction $\mu_s = 0.6 \pm 0.2$, towards the lower end of Byerlee's (1978) experimentally determined range for sliding friction in rocks. Two subordinate dip clusters occur at $\delta = 10 \pm 5^\circ$ and $50 \pm 5^\circ$, symmetrically disposed at $\pm 20^\circ$ from the dominant peak. With the recognition that misoriented faults in compressional regimes provide optimum conditions for 'holding-in' fluid overpressure (Sibson 2003), the preferred interpretation for these subordinate dip clusters is for fluid overpressure preferentially activating shear failure along bedding anisotropy in foreland fold-thrust belts or reactivating an existing fabric of steep faults. As in NE Honshu, the latter may be inherited from an earlier phase of extensional tectonics or may arise from domino-steepening of originally low-angle thrusts in collision zones (Sibson & Xie 1998). In all such settings, there is evidence of competition between the poorly oriented faults and Andersonian thrusts, consistent with mechanical analyses. This suggests that the relative levels of differential stress and fluid overpressure determine whether the reactivation of strongly misoriented reverse faults (either low- or high-angle) occurs in preference to the formation of new, optimally oriented thrust faults.

Alternative explanations could include varying fault friction or plunging stress trajectories. For the subordinate dip cluster at $\delta = 10 \pm 5^\circ$ to represent optimal thrust orientation under horizontal σ_1 requires $\mu_s = 2.75$, which is well outside Byerlee's (1978) range and unrealistically large. The subordinate cluster at $\delta = 50 \pm 5^\circ$ is beyond the possible range of optimal orientation ($\theta_r^* = 45^\circ - \phi_s/2$) for horizontal σ_1. From Coulomb wedge analysis of foreland fold-thrust belts, there is certainly the possibility of σ_1 plunging at shallow angles towards the range front (Dahlen 1990); optimal orientation of thrusts with $\delta = 10 \pm 5^\circ$ would however require σ_1 to plunge at a rather substantial angle of *c.* 20°. In addition, the local occurrence of *c.* 30° dipping thrust ruptures along parts of the frontal Himalaya suggest subhorizontal σ_1, at least locally. Reverse-slip rupturing with $\delta = 50 \pm 5^\circ$ could likewise be accounted for by near-optimal fault reactivation in the presence of σ_1 stress trajectories plunging at *c.* 20° in the dip directions of the faults. However, extreme stress heterogeneity would then be required in areas of active compressional inversion such as NE Honshu) where reverse-slip rupturing occurs on opposite-facing steep reverse faults (Sibson 2009). Moreover, stress inversions from focal mechanisms throughout the region tend to yield standard compressional stress fields with σ_3 subvertical and σ_1 subhorizontal (Townend & Zoback 2006).

Special thanks to F. Ghisetti and J. Townend for much helpful advice and to J. Streit and an anonymous reviewer for constructive criticism and suggestions.

References

ANDERSON, E. M. 1905. The dynamics of faulting. *Transactions of the Edinburgh Geological Society*, **8**, 387–402.

ANDERSON, E. M. 1951. *The Dynamics of Faulting and Dyke Formation with Application to Britain*. 2nd edn. Oliver & Boyd, Edinburgh.

ARMSTRONG, R. L. & DICK, H. J. B. 1974. A model for the development of thin overthrust sheets of crystalline rock. *Geology*, **2**, 35–40.

AVOUAC, J.-P. 2003. Mountain building, erosion, and the seismic cycle in the Nepal Himalaya. *In*: DMOWSKA, R. (ed.) *Advances in Geophysics*, **46**, Elsevier, Amsterdam.

AVOUAC, J.-P., AYOUB, F., LEPRINCE, S., KONCA, O. & HELMBERGER, D. V. 2006. The 2005, Mw 7.6 Kashmir earthquake: sub-pixel correlation of ASTER images and seismic waveforms analysis. *Earth and Planetary Science Letters*, **249**, 514–528.

BATES, D. E. B. 1974. The structure of the Lower Paleozoic rocks of Anglesey, with special reference to faulting. *Geological Journal*, **9**, 39–66.

BREWER, J. A. & SMYTHE, D. K. 1984. MOIST and the continuity of crustal reflector geometry along the Caledonian–Appalachian orogen. *Journal of the Geological Society, London*, **141**, 105–120.

BREWER, J. A., SMITHSON, S. B., OLIVER, J. E., KAUFMAN, S. & BROWN, L. D. 1980. The Laramide Orogeny; evidence from COCORP deep crustal seismic profiles in the Wind River Mountains, Wyoming. *Tectonophysics*, **62**, 165–189.

BUCHANAN, J. G. & BUCHANAN, P. G. (eds) 1995. *Basin Inversion*. Geological Society, London, Special Publications, **88**.

BURGESS, J. A. 1948. Mining on Carson Hill, California. *California Division of Mines Bulletin*, **141**, 88–90.

BYERLEE, J. D. 1978. Friction of rocks. *Pure & Applied Geophysics*, **116**, 615–626.

CÉLÉRIER, B. 2008. Seeking Anderson's faulting in seismicity: a centennial celebration. *Reviews of Geophysics*, **46**, RG4001, doi: 10.1029/2007RG000240.

CELLO, G. & NUR, A. 1988. Emplacement of foreland thrust systems. *Tectonics*, **7**, 261–271.

COLLETTINI, C. & SIBSON, R. H. 2001. Normal faults, normal friction? *Geology*, **29**, 927–930.

COOPER, M. 2007. Structural style and hydrocarbon prospectivity in fold and thrust belts: a global review. *In*: RIES, A. C., BUTLER, R. W. H. & GRAHAM, R. H. (eds) *Deformation of the Continental Crust: The Legacy of Mike Coward*. Geological Society, London, Special Publications, **272**, 447–472.

COTTON, F., CAMPILLO, M., DESCHAMPS, A. & RASTOGI, B. K. 1996. Rupture history of the 1991 Uttarkashi, Himalaya earthquake. *Tectonophysics*, **258**, 35–51.

DAHLEN, F. A. 1990. Critical taper model of fold-and-thrust belts and accretionary wedges. *Annual Reviews of Earth & Planetary Sciences*, **18**, 55–99.

DAVIS, D., SUPPE, J. & DAHLEN, F. A. 1983. Mechanics of fold and thrust wedges and accretionary wedges. *Journal Geophysical Research*, **88**, 1153–1172.

GHISETTI, F. & SIBSON, R. H. 2006. Accommodation of compressional inversion in northwestern South Island (New Zealand): old faults v. new? *Journal of Structural Geology*, **28**, 1994–2010.

GREENLY, E. 1919. The Geology of Anglesey. Memoir of the Geological Survey of Great Britain, 2.

HASEGAWA, A., NAKAJIMA, J., UMINO, N. & MIURA, S. 2005. Deep structure of the northeastern Japan arc and its implications for crustal deformation and shallow seismic activity. *Tectonophysics*, **403**, 59–75.

HUBBERT, M. K. & RUBEY, W. W. 1959. Role of fluid pressure in the mechanics of overthrust faulting. *Geological Society of America Bulletin*, **70**, 115–166.

IKARI, M. J., MARONE, C. & SAFFER, D. M. 2011. On the relation between fault strength and frictional stability. *Geology*, **39**, 83–86.

JAEGER, J. C. & COOK, N. G. W. 1979. *Fundamentals of Rock Mechanics*. 3rd edn. Chapman & Hall, London.

JEHU, T. J. & CRAIG, R. M. 1925. Geology of the Outer Hebrides: Part II – South Uist and Eriskay. *Transactions of the Royal Society of Edinburgh*, **53**, 615–641.

KAO, H. & CHEN, W.-P. 2000. The Chi-Chi earthquake sequence: active, out-of-sequence thrust faulting in Taiwan. *Science*, **288**, 2346–2349.

KATO, N., SATO, H. & UMINO, N. 2006. Fault reactivation and active tectonics on the fore-arc side of the back-arc rift system, NE Japan. *Journal of Structural Geology*, **26**, 1127–1136.

KAYAL, J. R., RAM, S., SINGH, O. P., CHAKRABORTY, P. K. & KARUNAKAR, G. 2003. Aftershocks of the March 2009 Chamoli earthquake and seismotectonic structure of the Himalaya. *Bulletin of the Seismological Society of America*, **93**, 109–117.

LEE, J.-C. & CHAN, Y.-C. 2007. Structure of the 1999 Chi-Chi earthquake rupture and interaction of thrust faults in the active fold belt of western Taiwan. *Journal of Asian Earth Sciences*, **31**, 226–239.

LISLE, R. J., ORIFE, T. O., ARLEGUI, L., LIESA, C. & SRIVASTAVA, D. C. 2006. Favoured states of paleostress in the Earth's crust: evidence from fault-slip data. *Journal of Structural Geology*, **28**, 1051–1066.

LOCKNER, D. A. 1995. Rock failure. *In*: AHRENS, T. J. (ed.) *Rock Physics and Phase Relations: A Handbook of Physical Constants*. American Geophysical Union, Washington, Reference Shelf, **3**, 127–147.

LYNN, H. B., QUAM, S. & THOMPSON, G. A. 1983. Depth migration and interpretation of the COCORP Wind River, Wyoming, seismic reflection data. *Geology*, **11**, 462–469.

MCCLAY, K. R. 2004. Thrust tectonics and hydrocarbon systems; Introduction. *In*: MCCLAY, K. R. (ed.) *Thrust Tectonics and Hydrocarbon Systems*. American Association of Petroleum Geologists, Boulder, Memoir, **82**, ix–xx.

MOLNAR, P. & LYON-CAEN, H. 1989. Fault plane solutions of earthquakes and active tectonics of the Tibetan Plateau and its margins. *Geophysical Journal International*, **99**, 123–153.

MOORE, J. C. & VROJLIK, P. 1992. Fluids in accretionary prisms. *Reviews in Geophysics*, **30**, 113–135.

NAZARETH, J. & HAUKSSON, E. 2004. The seismogenic thickness of the southern California crust. *Bulletin of the Seismological Society of America*, **94**, 940–960.

NGUYEN, P. T., COX, S. F., HARRIS, L. B. & POWELL, C. MCA. 1998. Fault–valve behaviour in optimally oriented shear zones: an example at the Revenge gold mine, Kambalda, Western Australia. *Journal of Structural Geology*, **20**, 1625–1640.

NI, J. & BARAZANGI, M. 1984. Seismotectonics of the Himalayan collision zone: geometry of the underthrusting Indian plate beneath the Himalaya. *Journal of Geophysical Research*, **89**, 1147–1163.

OKADA, T., YAGINUMA, T., UMINO, N., MATSIZAWA, T., HASEGAWA, A., ZHANG, H. & THURBER, C. H. 2006.

Detailed imaging of the fault planes of the 2004 Niigata-Chuetsu, central Japan, earthquake sequence by double-difference tomography. *Earth & Planetary Science Letters*, **244**, 32–43.

Oxburgh, E. R. 1972. Flake tectonics and continental collision. *Nature*, **239**, 202–204.

Pfiffner, O. A. 2006. Thick-skinned and thin-skinned styles of continental contraction. *In*: Mazzoli, S. & Butler, R. W. H. (eds) *Styles of Continental Contraction*. Geological Society of America, Boulder, Special Paper, **414**, 153–177.

Sato, H. & Amano, K. 1991. Relationship between tectonics, volcanism, sedimentation and basin development, Late Cenozoic, central part of northern Honshu, Japan. *Sedimentary Geology*, **74**, 323–343.

Seeber, L. & Armbruster, J. G. 1981. Great detachment earthquakes along the Himalayan arc and long-term forecasting. *In*: Simpson, D. W. & Richards, P. G. (eds) *Earthquake Prediction: An International Review*. Maurice Ewing Series, American Geophysical Union, Washington D.C., Vol. 4, 259–279.

Sibson, R. H. 1985. A note on fault reactivation. *Journal of Structural Geology*, **7**, 751–754.

Sibson, R. H. 1990. Rupture nucleation on unfavourably oriented faults. *Bulletin of the Seismological Society of America*, **80**, 1580–1604.

Sibson, R. H. 2000. A brittle failure mode plot defining conditions for high-flux flow. *Economic Geology*, **95**, 41–48.

Sibson, R. H. 2003. Brittle failure controls on maximum sustainable overpressure in different tectonic regimes. *AAPG Bulletin*, **87**, 901–908.

Sibson, R. H. 2004. Frictional mechanics of seismogenic thrust systems in the upper continental crust: implications for fluid overpressures and redistribution. *AAPG Memoir*, **82**, 1–17.

Sibson, R. H. 2007. An episode of fault–valve behaviour during compressional inversion? The 2004 MJ6.9 Mid-Niigata Prefecture, Japan, earthquake. *Earth & Planetary Science Letters*, **257**, 404–416.

Sibson, R. H. 2009. Rupturing in overpressured crust during compressional inversion: the case from NE Honshu, Japan. *Tectonophysics*, **473**, 404–416.

Sibson, R. H. & Xie, G. 1998. Dip range for intracontinental reverse fault ruptures: truth not stranger than friction? *Bulletin of the Seismological Society of America*, **88**, 1014–1022.

Sibson, R. H., Robert, F. & Poulsen, K. H. 1988. High-angle reverse faults, fluid pressure cycling, and mesothermal gold-quartz deposits. *Geology*, **16**, 551–555.

Tobin, H. J. & Saffer, D. M. 2009. Elevated fluid pressure and extreme mechanical weakness of a plate boundary thrust, Nankai Trough subduction zone. *Geology*, **37**, 679–682.

Townend, J. & Zoback, M. D. 2006. Stress, strain, and mountain building in central Japan. *Journal of Geophysical Research*, **111**, B03411, doi: 10.1029/2005JB003759.

Wannamaker, P. E., Caldwell, T. G. *et al.* 2009. Fluid and deformation regimes of an advancing subduction system at Marlborough, New Zealand. *Nature*, **460**, 733–737.

Williams, G. D., Powell, C. M. & Cooper, M. A. 1989. Geometry and kinematics of inversion tectonics. *In*: Cooper, M. A. & Williams, G. D. (eds) *Inversion Tectonics*. Geological Society, London, Special Publications, **44**, 3–15.

Yue, L.-F., Suppe, J. & Hung, J.-H. 2005. Structural geology of a classic thrust belt earthquake: the 1999 Chi-Chi earthquake, Taiwan ($M_w = 7.6$). *Journal of Structural Geology*, **27**, 2058–2083.

Zang, A. & Stephansson, O. 2010. *Stress Field of the Earth's Crust*. Springer, Dordrecht.

Zoback, M. L. 1992. Stress field constraints on intraplate seismicity in eastern North America. *Journal of Geophysical Research*, **97**, 11 761–11 782.

The complexity of 3D stress-state changes during compressional tectonic inversion at the onset of orogeny

KOEN VAN NOTEN[1,2], HERVÉ VAN BAELEN[1] & MANUEL SINTUBIN[1]*

[1]*Geodynamics and Geofluids Research Group, Department of Earth and Environmental Sciences, Katholieke Universiteit Leuven, Celestijnenlaan 200E, B-3001 Leuven, Belgium*

[2]*Present address: Seismology Section, Royal Observatory of Belgium, Ringlaan 3, B-1180 Brussels, Belgium*

**Corresponding author (e-mail: manuel.sintubin@ees.kuleuven.be)*

Abstract: Compressional tectonic inversions are classically represented in 2D brittle failure mode (BFM) plots that illustrate the change in differential stress ($\sigma_1 - \sigma_3$) versus the pore-fluid pressure during orogenic shortening. In these BFM plots, the tectonic switch between extension and compression occurs at a differential stress state of zero. However, mostly anisotropic conditions are present in the Earth's crust, making isotropic stress conditions highly questionable. In this study, theoretical 3D stress-state reconstructions are proposed to illustrate the complexity of triaxial stress transitions during compressional inversion of Andersonian stress regimes. These reconstructions are based on successive late burial and early tectonic quartz veins which reflect early Variscan tectonic inversion in the Rhenohercynian foreland fold-and-thrust belt (High-Ardenne Slate Belt, Belgium, Germany). This theoretical exercise predicts that, no matter the geometry of the basin or the orientation of shortening, a transitional 'wrench' tectonic regime should always occur between extension and compression. To date, this intermediate regime has never been observed in structures in a shortened basin affected by tectonic inversion. Our study implies that stress transitions are therefore more complex than classically represented in 2D. Ideally, a transitional 'wrench' regime should be implemented in BFM plots at the switch between the extensional and compressional regimes.

Extension (Mode I) fractures are classically interpreted to form at low differential stresses less than four times the tensile strength (T) of rock (e.g. Secor 1965). They open perpendicular to the minimum principal stress (σ_3) and propagate in the plane of the maximum (σ_1) and intermediate (σ_2) principal stresses. The alignment of fractures has therefore often been used in orogenic belts to determine the orientation of the palaeostress field at the time of fracturing (Cox *et al.* 2001; Gillespie *et al.* 2001; Laubach *et al.* 2004). If veins which initiated as extension fractures are uniform in trend over a large area, then they can also be used as such a palaeostress indicator (e.g. Boullier & Robert 1992; Cosgrove 2001; Gillespie *et al.* 2001; Mazzarini *et al.* 2010; Van Noten & Sintubin 2010). Different cross-cutting or successive vein generations are even more interesting because they have the potential (1) to serve as a tool to decipher changes in the overall stress regime during multiple phases of fracturing and (2) to define the P–T conditions at which these stress changes occurred. On the one hand, different cross-cutting fracture and/or vein generations have been used in many studies to deduce local stress field rotation in a consistent remote stress regime (Jackson 1991; Crespi & Chan 1996; Stowell *et al.* 1999; Laubach & Diaz-Tushman 2009; Wiltschko *et al.* 2009). On the other hand, successive cross-cutting veins (i.e. vein sets that are oriented differently with respect to bedding) have often been used to determine a transition in regional stress regime from extension to compression or reversely (Manning & Bird 1991; Boullier & Robert 1992; Teixell *et al.* 2000; Hilgers *et al.* 2006*b*; Van Noten *et al.* 2008, 2011; Laubach & Diaz-Tushman 2009). Alternatively, cross-cutting vein sets are observed in close relationship to faults in which they represent fluctuating fluid pressures during fault–valve activity and are related to short-lived local stress transitions in a consistent remote stress regime (Boullier & Robert 1992; Sibson 1995; Nielsen *et al.* 1998; Muchez *et al.* 2000; Collettini *et al.* 2006). Local mutually cross-cutting vein sets may also develop in the presence of mechanical anisotropy around fault planes in a consistent remote stress field (Healy 2009; Fagereng *et al.* 2010). In all these studies, transitions between regional stress regimes are defined as tectonic inversions; a positive (compressional) tectonic inversion corresponds to a transition from an extensional to a compressional stress regime, while a negative (extensional) tectonic inversion reflects the transition from compression to extension.

From: HEALY, D., BUTLER, R. W. H., SHIPTON, Z. K. & SIBSON, R. H. (eds) 2012. *Faulting, Fracturing and Igneous Intrusion in the Earth's Crust*. Geological Society, London, Special Publications, **367**, 51–69. http://dx.doi.org/10.1144/SP367.5

The conditions of brittle failure of an isotropic rock have classically been visualized in a Mohr–Coulomb diagram, relating the shear (τ) and the normal stress (σ_n) on a particular plane of failure with a particular orientation to the stress state (represented by principal stresses). Although Mohr–Coulomb diagrams provide an excellent way to illustrate both the relationship between stress magnitudes and differential stress, and which type of fracture will develop (i.e. extension, extensional-shear or shear fractures) (Cosgrove 1995), they are limited to illustrate a specific stress state at a specific moment. Unfortunately, the evolution of the fluid pressure (P_f) during orogeny can only be illustrated by means of successive Mohr circle reconstructions. Sibson (1998, 2000) introduced the concept of 2D brittle failure mode plots (BFM plots; Fig. 1) which allow a more dynamic and evolutionary analysis of brittle structures than possible in a classical Mohr–Coulomb diagram. In these BFM plots the transition of Andersonian stress regimes at a certain depth is visualized by means of the change of differential stress versus the evolution of the pore-fluid factor (λ_v) and the effective vertical stress ($\sigma'_V = \sigma_V - P_f$). Moreover, they illustrate a change in Andersonian stress orientations during

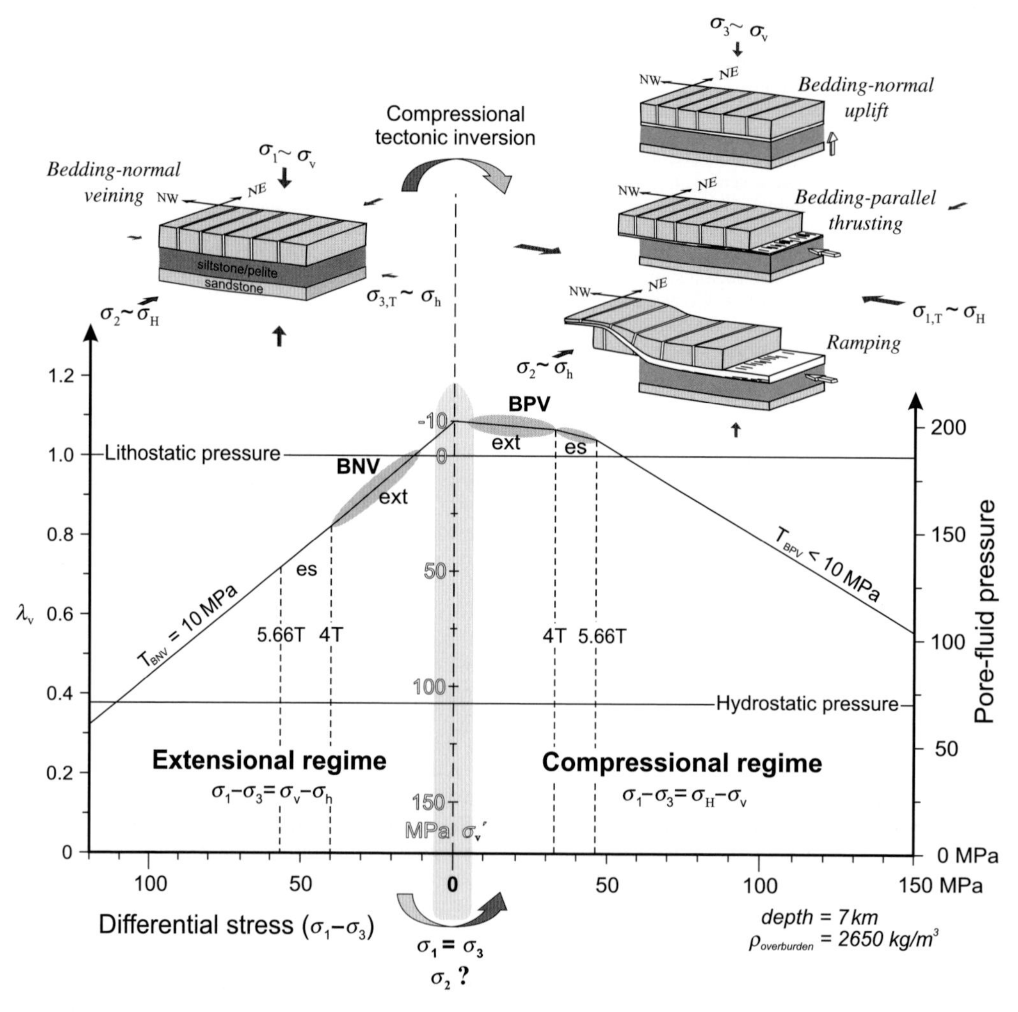

Fig. 1. Example of a 2D brittle failure mode plot (after Sibson 2004) which illustrates the evolution of fluid (over)pressure during compressional tectonic inversion, applied on the emplacement of two successive vein sets oriented normal and parallel to bedding in the High-Ardenne case study (after Van Noten *et al.* 2011). The effect of σ_2 at the time of the tectonic inversion ($\sigma_1 - \sigma_3 = 0$; grey area) is not illustrated in these mode plots and forms the research question of this study.

tectonic inversion in which a vertical σ_1 switches to horizontal during compressional inversion, while σ_3 has an opposite reorientation from horizontal to vertical. BFM plots are useful to visualize parameters such as the tensile strength (T) of rock or the friction along faults (μ) which strongly influence the maximum overpressure that can be built up during tectonic inversion (Cox 2010). The layout of BFM plots is designed to illustrate contrasting failure conditions which may be expected from the transitions between two stress regimes (Sibson 2004). Although these plots are originally constructed to understand the development and evolution of fault systems with their associated vein sets and are used in a wide range of tectonic settings, they are also very useful to define the relative timing of vein formation in terms of changing differential stress during tectonic inversion (Van Noten *et al.* 2011). Sibson (2004) and Cox (2010) moreover illustrate that the interaction between changes in the tectonic stress regime, fluid-pressure changes and fluid migration during tectonic inversion are important for the genesis of ore deposits (e.g. Tunks *et al.* 2004). A better understanding of the link between stress-state transitions during mountain-building processes and ore genesis could therefore have major economic implications (cf. Blundell 2002).

As illustrated on the BFM plot and shown by several studies (e.g. Sibson 1995, 2004), fluid overpressures are likely to develop during the transition from extension to compression and are easier to maintain at low differential stress in the compressional regime than in the extensional regime. Due to low differential stress during tectonic inversion, fluid overpressures are mostly captured in extensional or in extensional-shear veins that accompany the tectonic inversion. These extension veins have a vertical orientation in the extensional regime and a horizontal orientation in the compressional regime at the time of fracturing. The tectonic switch, which is illustrated on the BFM plot mode by the line at zero differential stress (see Fig. 1), reflects the specific stress state at which σ_1 equals σ_3 (so that $\sigma_1 - \sigma_3 = 0$) without considering the role of σ_2. A zero differential stress-state can therefore only be reached in a state of isotropic pressure ($\sigma_1 = \sigma_2 = \sigma_3$) during the tectonic switch. The chances that a stress state equals a pure pressure state in the brittle upper crust are, however, questionable (e.g. Healy 2009). Because of the assumption of an isotropic pressure, the stress-state evolution illustrated in BFM plots is actually an oversimplification of the true 3D stress-state evolution in the Earth's crust. In more realistic anisotropic conditions, the transition from extensional to compressional might therefore be more complex due to the fact that a triaxial stress state might remain present during inversion. Sibson (1998) mentioned that the brittle failure curves for an intermediate wrench regime ($\sigma_2 \sim \sigma_V$) lie anywhere at the zero differential stress line between the extensional and compressional plots, depending on the value of σ_2 with respect to σ_1 and σ_3. To date, however, this intermediate regime has been ignored in any tectonic inversion model.

In this study, we analyse possible stress-state evolutions to illustrate the complexity of 3D stress transitions during compressive tectonic inversion, based on the geometric and kinematic analysis of late-burial bedding-normal and successive early tectonic bedding-parallel quartz veins which occur in the frontal part of the Rhenohercynian foreland fold-and-thrust belt (High-Ardenne Slate Belt, Belgium, Germany). These successive vein sets indicate an important fluid-assisted deformation during the Early Variscan (Early Carboniferous) tectonic inversion affecting the Ardenne-Eifel basin at the onset of Variscan Orogeny. By means of different scenarios, it is illustrated that both the intermediate principal stress σ_2 and the increasing tectonic stress σ_T plays an important role during tectonic inversion in anisotropic stress conditions, and that a zero differential stress state is not necessarily reached during inversion. It is shown that stress transitions are commonly more complex than classically interpreted. Although the starting point of this theoretical exercise is based on a case study, the application of 3D stress transitions under triaxial stress conditions may have a wide range of implications with respect to the evolution of dynamic permeability during orogenic shortening.

Geological framework: the High-Ardenne case study

Geological setting

The dataset on which the present exercise on the complexity of 3D stress transitions during tectonic inversion is based, comprises two successive sets of extensional quartz veins that occur in Lower Devonian siliciclastic multilayers. The study area is situated in the frontal part of the Rhenohercynian foreland fold-and-thrust belt, more specifically in the High-Ardenne slate belt (Belgium, Germany), which forms together with the Dinant fold-and-thrust belt and several Lower Palaeozoic (Cambro-Ordovician) inliers, part of the Ardenne allochton (Fig. 2). This allochton has been thrust over the Brabant parautochton during the latest Asturian stage of the Variscan Orogeny (Late Carboniferous, Meilliez & Mansy 1990). The host-rock lithologies, competent sandstones, psammites and quartzites, alternating with incompetent siltstone and pelites,

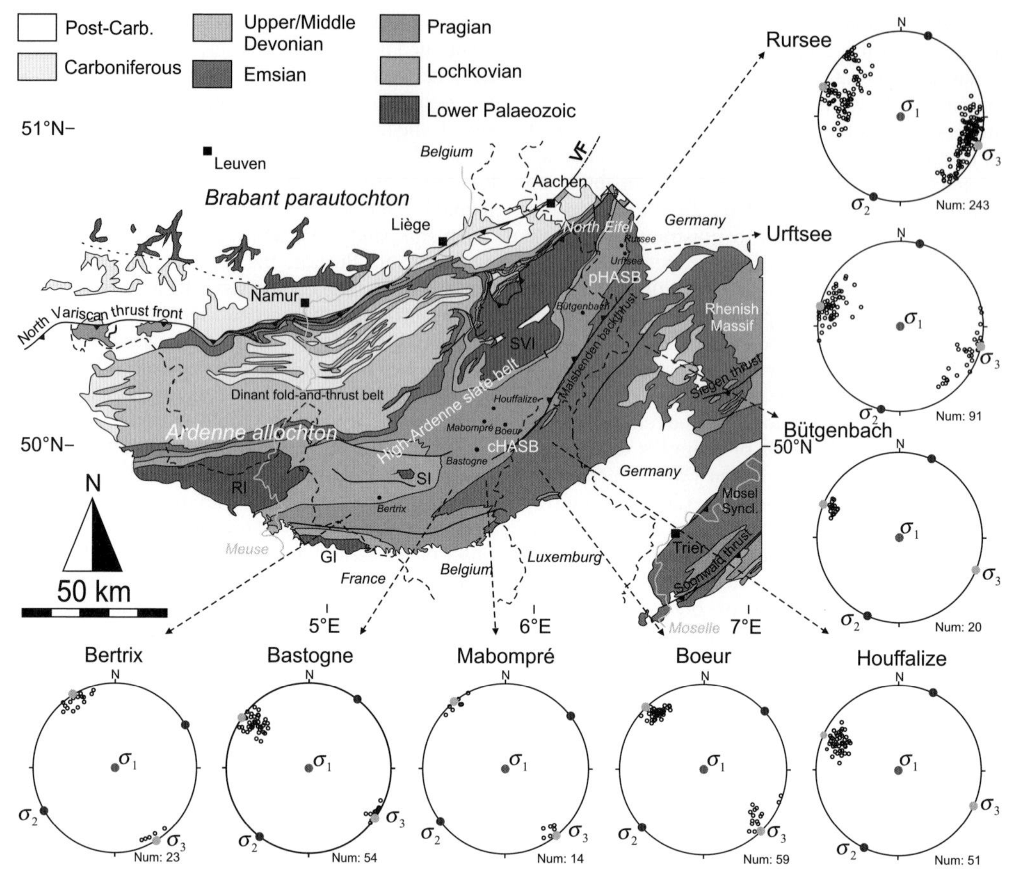

Fig. 2. Structural map and palaeostress analysis of bedding-normal quartz veins in the High-Ardenne slate belt (Belgium, Germany). The lower-hemisphere, equal-area stereographic projections show the original orientation of the veins prior to folding and reflect a consistent extensional stress field at the time of veining. The slight rotation of the $\sigma_1 - \sigma_2$ plane from NE–SW in the SW to NNE–SSW in the NE of the slate belt can be attributed to postveining oroclinal bending of the slate belt related to the NW–SE-directed Variscan compression.

reflect a Lower Devonian shallow marine deltaic deposition (Stets & Schafer 2009) in the Ardenne-Eifel basin that developed on the passive margin of the Rhenohercynian Ocean. The rocks are affected by very-low-grade to low-grade metamorphism that was pre- to synkinematic to the prograding Variscan deformation but is considered to be primarily of burial origin (Fielitz & Mansy 1999). Metamorphic conditions vary from epizonal in the deepest part (10 km) of the shortened sedimentary Ardenne-Eifel basin (Kenis *et al.* 2002, 2005) to anchizonal in the shallower parts that reflect burial conditions of 7 km (von Winterfeld 1994; Van Noten *et al.* 2011). The deepest parts are currently exposed in the central part of the High-Ardenne Slate Belt, while the higher structural levels are present in the North Eifel in the north-eastern periphery of the slate belt (Fig. 2).

Successive vein sets

Two successive quartz vein sets are considered to reflect the Early Variscan (Early Carboniferous) compressional tectonic inversion of the Ardenne-Eifel sedimentary basin. A first prefolding and precleavage set of veins is oriented perpendicular to bedding and is regionally distributed in the Lower Devonian sequences of the High-Ardenne slate belt. These prefolding bedding-normal veins are mostly confined to competent layers (Fig. 3a) and are regionally consistent in orientation, after restoring the beds to their original orientation prior to the formation of folds (Fig. 2). They are determined to be extension veins that formed in the extensional regime during the latest part of the burial. A limited compressibility of the competent host-rock at the time of veining is exemplified by (1)

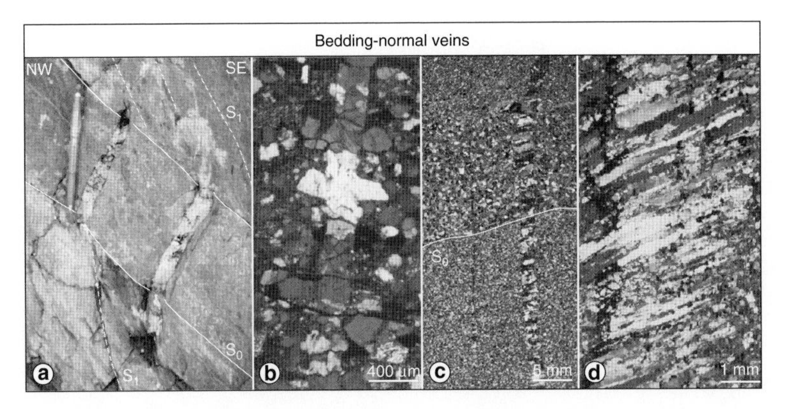

Fig. 3. Characteristics of bedding-normal quartz veins. (**a**) Veins oriented perpendicular to bedding and refracted at the sandstone–siltstone interface, similar to cleavage. (**b**) Transgranular, millimetre-thin vein fracturing through the host-rock grains. (**c**) Millimetre-thin vein with fibrous quartz infill. (**d**) Fibrous quartz crystals spanning the vein walls. Ataxial vein growth.

transgranular microveins with a low tortuosity (Fig. 3b), demonstrating that fracturing is not affected by the grain-scale heterogeneity of the competent host-rock and (2) the absence of buckled veins which suggests that subsequently, no postveining compaction occurred (Fig. 3c). The vein infill mostly contains a fibrous fabric (Fig. 3d) in which crack-seal microstructures are often present. Furthermore, it has been demonstrated by means of a combined (micro)structural and microthermometric study of fluid inclusions in vein quartz that these extension veins initiated at elevated, up to lithostatic, fluid pressures under low differential stress and that they result from several phases of fracturing and sealing (Van Noten *et al.* 2009, 2011). Maximum pore-fluid overpressures of *c.* 190 MPa are measured for the bedding-normal veins in the North Eifel, exceeding a lithostatic pressure of *c.* 185 MPa and reflecting the pressure at a burial depth of 7 km (Van Noten *et al.* 2011). This regional occurrence of overpressures at the time of vein formation suggests that the Ardenne-Eifel basin can be considered as an exposed example of a 350 Ma fractured reservoir, composed of sequences of interbedded high-pressure compartments (Hilgers *et al.* 2006*a*; Kenis & Sintubin 2007). An initial increasing tectonic stress σ_T at the onset of Variscan Orogeny has been invoked as the driving mechanism (1) to decrease the differential stress at the latest stages of the burial and (2) to generate lithostatic overpressures, both of which are necessary to form vertical extension veins in the extensional regime. The initial tectonic stress may be considered as the driving force for the stagnating extension at the end of the extensional regime.

In the higher structural levels of the High-Ardenne slate belt, currently exposed in the North Eifel (Germany; Fig. 2), bedding-normal veins are succeeded and cross-cut by bedding-parallel quartz veins which reflect bedding-normal uplift (Fig. 4a) and bedding-parallel thrusting (Fig. 4b) at the onset of folding during the initial stages of the compressional tectonic regime. Microstructural analysis of their typical laminated fabric (i.e. blocky and crack-seal quartz laminae intercalated by host-rock inclusion lines; Fig. 4c) reveals that they have been formed during successive cycles of uplift and collapse. Repetitive host-rock inclusion bands parallel to bedding (in the crack-seal laminae; Fig. 4c) result from repetitive fracturing and sealing by the crack-seal mechanism (Ramsay 1980) and are supportive for defining these bedding-parallel veins as extension veins. While the orientation of repetitive quartz laminae parallel to bedding (Fig. 4c) clearly reflect bedding-normal uplift, bedding-parallel stylolites (Fig. 4d) reflect bedding-normal collapse between the multiple phases of fracturing and sealing that can possibly be linked to the earthquake cycle (cf. Cox 1987) and to fault–valving during vein development (Sibson 1990; Cox 1995). Additionally, the pronounced bedding-parallel fabric of the bedding-parallel veins shows robust evidence for bedding-parallel shearing prior to folding. This is indicated by bedding-parallel veins cross-cutting layers in small ramps (Fig. 4e). These veins which bear evidence of thrusting are therefore classified as extensional-shear veins which reflect bedding-parallel movements during progressive orogenic shortening. Microthermometric analysis of fluid inclusions shows that these bedding-parallel veins are induced at a supralithostatic fluid pressure (*c.* 205 MPa) (Van Noten 2011; Van Noten *et al.* 2011).

The presence of these vein sets illustrates an important fluid-assisted deformation in the brittle

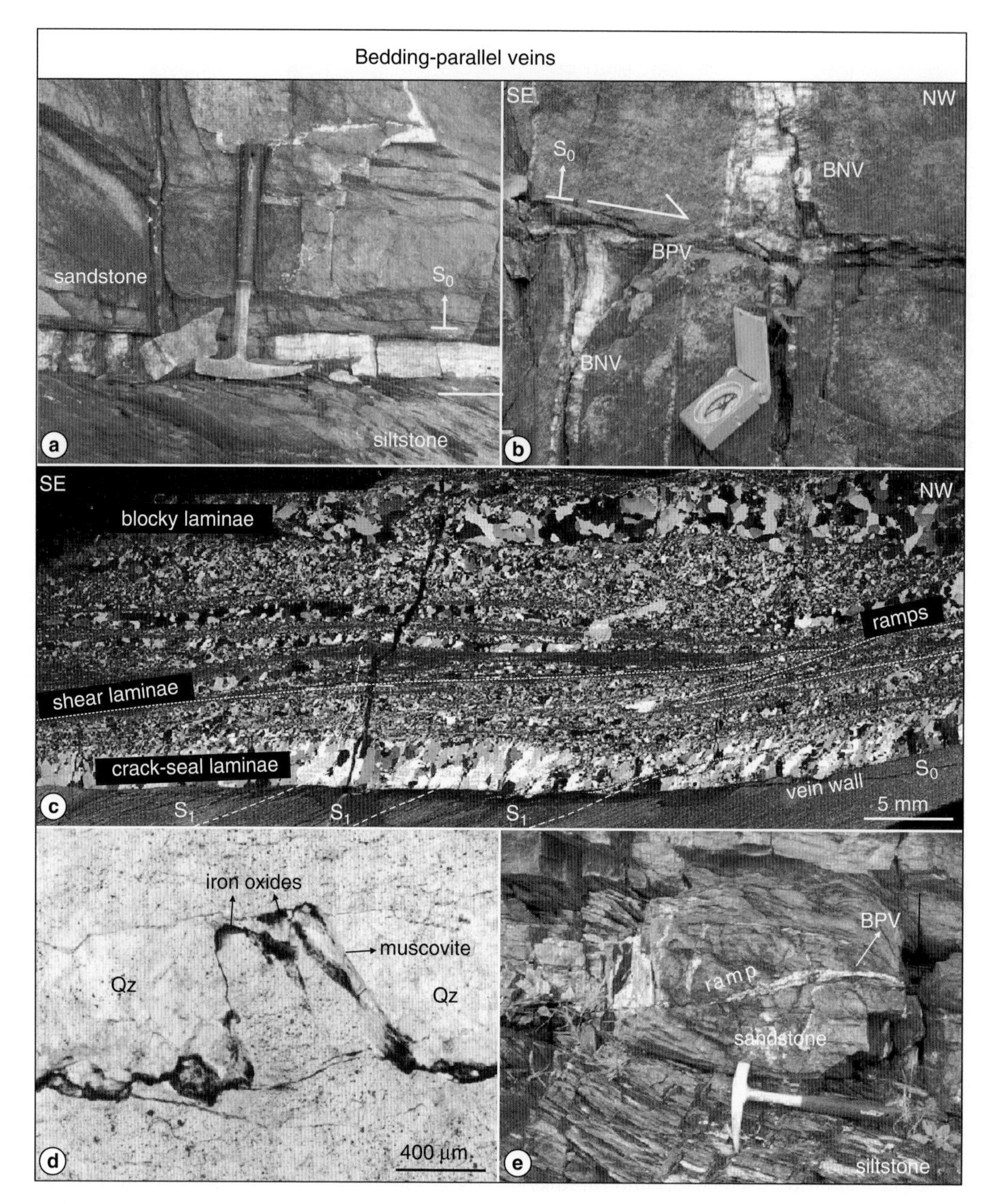

Fig. 4. Characteristics of bedding-parallel quartz veins (BPV). (**a**) Vein situated between sandstone and siltstone. (**b**) Vein cross-cutting and offsetting a bedding-normal vein. (**c**) Blocky and crack-seal quartz laminae intercalated with host-rock inclusions lines and shear laminae in a centimetre-thin laminated vein. Laminae show a pronounced top-to-the-NW bedding-parallel shear along small-scale ramps. (**d**) Anastomosing stylolite observed in the vein containing muscovite and iron oxides. (**e**) Small-scale thrust accompanied by a quartz vein cross-cutting a coarse-grained sandstone.

upper crust during the Early Carboniferous tectonic inversion affecting the Ardenne-Eifel basin at the onset of Variscan Orogeny. In contrast to these two important vein-forming events at the onset of the orogeny, quartz veining seems to have occurred rather occasionally during the main compressional stage of the Variscan Orogeny and is restricted to faulting and local boudinage of competent layers (e.g. Van Baelen & Sintubin 2008). The Late Variscan negative (extensional) tectonic inversion is locally exemplified in the High-Ardenne slate belt (Bertrix, Herbeumont, Belgium) by subvertical discordant extension quartz veins that formed closely after the tectonic inversion and underwent a shape modification during progressive shearing, reflecting the destabilization of the Variscan Orogen (Van Baelen 2010). It has been proven by several microthermometric analyses of fluid inclusions that the geothermal gradient – and thus the temperature–depth conditions during the positive tectonic inversion (Kenis *et al.* 2005; Van Noten 2011), the main Variscan compression (Fielitz & Mansy 1999) and during the Variscan destabilization after negative inversion (Van Baelen 2010) – remained constant. This implies that (1) no significant subsidence or uplift of the shortened Ardenne-Eifel basin occurred during the whole Variscan Orogeny and (2) the magnitude of σ_V remained constant during both positive and negative Variscan tectonic inversion.

Palaeostress analysis

The regional occurring bedding-normal veins are determined to be extension veins that formed in the extensional regime at the latest stages of the burial. If rotated to their original prefolding orientation (cf. method of Debacker *et al.* 2009) they are more or less consistent in orientation across the whole High-Ardenne slate belt (Fig. 2). The veins, which were nearly vertical at the time of vein formation, originally range from NE–SW in the SW and central part of the High-Ardenne slate belt (Bertrix, Bastogne, Mabompré, Boeur; Fig. 2) to NNE–SSW towards the NE-part (Houffalize, Bütgenbach, Urftsee, Rursee; Fig. 2). This structural change is probably related to a post-veining oroclinal bending of the slate belt during the main Variscan contraction because of the presence of a rigid basement in the north (cf. Brabant parautochton in Fig. 2; Fielitz 1992; Mansy *et al.* 1999). Notwithstanding this oroclinal bending, the extension veins can be used as a regional palaeostress indicator for the stress state during the extensional regime. The orientation of the principal stress axes is derived from the distribution of extension veins that are assumed to form perpendicular to σ_3 and align in the $\sigma_1 - \sigma_2$ plane. Veins are therefore formed in a consistent anisotropic Andersonian stress field that is characterized by a vertical σ_1, corresponding to the overburden stress σ_V, and two well-defined horizontal principal stresses (σ_H and σ_h), reflecting a triaxial stress state during bedding-normal veining.

After tectonic inversion, bedding-normal vein formation is followed by bedding-parallel veining in the upper structural levels of the High-Ardenne slate belt. In the tectonic switch σ_3 changes to vertical and corresponds to the overburden stress σ_V, while σ_1 now corresponds to σ_H and reflects regional shortening in the compressional regime (Fig. 1). Similar to the bedding-normal veins, the extensional bedding-parallel quartz veins are the brittle expression of the $\sigma_1 - \sigma_2$ plane (cf. Secor 1965) and can be used as a palaeostress indicator. Macro- and microstructural analysis of the shear planes in the microfabric of the bedding-parallel veins and NW–SE-trending slickensides on the vein walls have been used to indicate that Variscan shortening occurred in an overall NW–SE-directed compression during bedding-parallel veining. In the compressional regime, veins follow the bedding anisotropy (lower tensile strength because of lower cohesion); in the extensional regime they form in intact rock (higher tensile strength).

Bedding-parallel veining is followed by the formation of NW-verging folds and a SE-dipping axial planar cleavage during Variscan fold-and-cleavage development in the North Eifel (Fielitz 1992; Van Noten *et al.* 2008). Similar to the North Eifel, the major part of the Rhenish Massif is characterized by a strong NW-directed vergence that reflects an overall NW–SE compression during the main phase of Variscan contraction (Weber 1981).

Basic assumptions as starting point for the reconstruction of 3D stress transitions

The orientations of the two vein sets in the North Eifel have been used as a tool to evaluate the stress-state and fluid-pressure evolution of a basin during incipient orogenic shortening. Although the state of stress at the time of veining is well understood, with the two vein sets as the end members reflecting the stress state in the extensional and compressional regime, the switch between two stress regimes has (to date) only been approached from a crust with isotropic stress properties (e.g. Sibson 2000, 2004). From the consistent vein orientation in the extensional regime and in the compressional regime in the High-Ardenne case study it is however already clear that mostly anisotropic stress conditions are present in the Earth's crust. To illustrate the complexity of 3D stress transitions in a triaxial stress state, we constructed three different scenarios in which a basin with a predefined sedimentary geometry experiences a compressional tectonic inversion induced by a consistently increasing horizontal tectonic stress. In each scenario, we investigate the stress-state evolution of the basin following two options by changing the orientation of σ_T, causing shortening from a NW–SE to a NE–SW direction. Although the main tectonic compression in the case study was NW–SE, it is still unclear whether the direction of tectonic compression remained constant through time or if it has rotated prior to the main deformation.

After presenting these scenarios, we discuss those stress states which are favourable to form the two successive vein sets that are observed in the High-Ardenne case study.

The starting point for constructing these three different scenarios is based on the following questions:

(1) How does a increasing horizontal tectonic stress σ_T influence the eventual orientation of extension veins?
(2) Does the initial basin geometry influence the orientation of extension veins (Fig. 5)?
(3) Do veins with a regionally consistent orientation only form in one specific stress regime, or can they occur in several different stress regimes?
(4) What is the effect of the intermediate principal stress σ_2 during inversion?

The different scenarios are exemplified by means of 3D stress-state graphics (Figs 6–8); the longest axis represents σ_1, the intermediate axis σ_2 and the shortest axis σ_3. Each specific stress state during the progressive stress-state changes is reconstructed in a Mohr–Coulomb diagram. Stress abbreviations and symbols used are listed in Table 1. In order to evaluate the different scenarios during inversion properly, the constraints described in the following sections are assumed.

Constant σ_V during tectonic inversion

The 3D stress-state reconstructions are constructed in such a way that σ_V is assumed to have a constant magnitude, corresponding to σ_1 in the extensional regime prior to tectonic inversion and σ_3 in the compressional regime after tectonic inversion. This constraint implies that the thickness of the overburden is not drastically changed during tectonic inversion. In the High-Ardenne case study, the assumption of a constant vertical principal stress is valid due to the limited compressibility of rocks both at maximum burial and during initial tectonic compression (i.e. the timing of bedding-normal and bedding-parallel veining, respectively). This constraint furthermore implies that the Poisson ratio (ν) of the fractured sandstones (in the case study $\nu = 0.25$) remains constant during inversion.

Horizontal σ_T

The tectonic inversion is induced by a horizontal tectonic stress σ_T that is consistent in orientation at the onset of orogeny. It is assumed to be parallel to one of the two horizontal principal stresses, σ_H or σ_h, and increases in magnitude from extension to compression. If the tectonic stress σ_T is parallel to σ_1, σ_2 or σ_3 it is indicated as $\sigma_{1,T}$, $\sigma_{2,T}$ or $\sigma_{3,T}$ respectively (see Table 1). If the increasing σ_T is parallel to, for example, σ_H, then σ_h will also indirectly increase due to the compressibility of the host rock (Mandl 2000) but at a lower rate than the increasing σ_H.

$\sigma_1 > \sigma_2 > \sigma_3$ and $\sigma_H > \sigma_h$

The basic assumption is that both veining events in the case study reflect Andersonian stress regimes with one principal stress vertical (σ_V) and the other two principal stresses in the horizontal plane, parallel to the Earth's surface, and with σ_h being smaller than σ_H (Anderson 1951). By definition, σ_1 is larger than σ_2 which in turn exceeds σ_3. The latter implies that if σ_3 increases because of a consistent σ_T that is oriented parallel to σ_3, then $\sigma_{3,T}$ can become so large that it passes σ_2 in such a way that $\sigma_{3,T}$ becomes $\sigma_{2,T}$

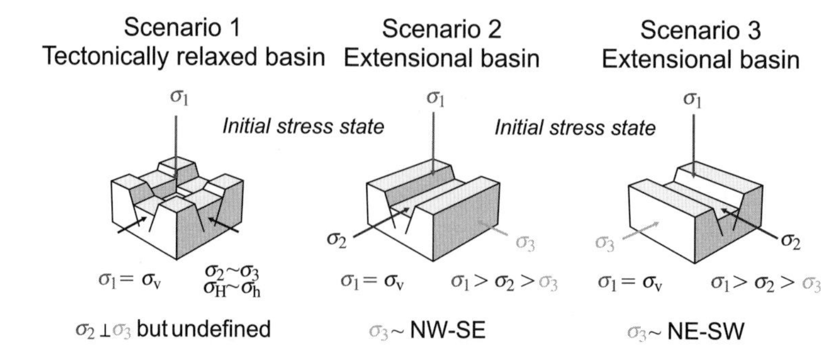

Fig. 5. Simplified representation of different basins as a starting point for the stress reconstructions. Scenario 1: tectonically relaxed basin that is able to extend in all directions without any predefined horizontal principal stress. Scenario 2: extensional basin that has a NE–SW-directed sedimentary basin elongation and extends perpendicular to σ_3. Scenario 3: extensional basin with a NE–SW-directed extension and a NW–SE elongation. Mirror image of scenario 2.

Table 1. *Abbreviations used in the stress-state reconstructions*

Notation	Definition
T	Tensile strength
P_f	Pore-fluid pressure
σ'	Effective stress
σ_V	Vertical principal stress
σ_T	Tectonic stress
σ_H	Maximum horizontal principal stress
σ_h	Minimum horizontal principal stress
σ_1	Maximum principal stress
σ_2	Intermediate principal stress
σ_3	Minimum principal stress
$\sigma_{1,T}$	Maximum principal stress increased by tectonic stress
$\sigma_{2,T}$	Intermediate principal stress increased by tectonic stress
$\sigma_{3,T}$	Minimum principal stress increased by tectonic stress

and σ_2 turns into σ_3. A similar transition can also occur between σ_2 and σ_1 if the tectonic principal stress σ_T influences σ_2. Only two kinds of tectonic stress transition can thus occur; one between σ_3 and σ_2 and one between σ_2 and σ_1. A direct transition between σ_3 and σ_1 is by definition not possible.

Principal stresses

All stresses tend to be compressional in the Earth's crust and true tensile stresses are uncommon at larger depths (Cosgrove 1995). In the different scenarios, principal stresses are used to illustrate the stress changes during tectonic inversion. During the whole evolution of inversion, it is assumed that hydrofracturing can take place continuously by reducing the effective stresses ($\sigma' = \sigma - P_f$) by an elevated fluid pressure (P_f) that overcomes σ_3 to allow Mode I fracturing. In the High-Ardenne case study, the presence of overpressured fluids during the positive tectonic inversion is demonstrated by microthermometric studies, making this assumption valid. Differential stresses are thereby inferred to be low enough to allow extension fracturing. This constraint allows us to evaluate vein formation during each different stage.

Vein rotation

Although a slight reorientation of the original vein orientation has been observed from the SW (NE–SW; Bertrix) to the NE (NNE–SSW; Rursee) in the High-Ardenne slate belt (see Fig. 2), the orientation of the veins is kept constant for the simplicity of the constructed stress models. If veins form, they are assumed to be NE–SW. This simplified model allows the prediction of the specific moment during which the NE–SW vein set is formed.

Mechanical models

Scenario 1: Tectonically relaxed basin

In a first scenario, the stress-state reconstruction starts with a tectonically relaxed sedimentary basin in the extensional regime (Figs 5 & 6). The boundary conditions of a tectonically relaxed basin are such that σ_1 corresponds to the vertical overburden stress ($\sigma_1 \sim \sigma_V$) and there is no distinction between the two horizontal principal stresses ($\sigma_2 = \sigma_3$ or $\sigma_H = \sigma_h$; see Cosgrove 1997, 2001). In a tectonically relaxed basin, σ_V and σ_H are related by common constraints of the rock in such way that $\sigma_H = \sigma_V/(m-1)$ in which m is the reciprocal of the Poisson ratio ν (Price 1966).

Stage 1. A tectonic stress σ_T is assumed to be absent during subsidence in the tectonically relaxed basin. If extension veins could form due to elevated fluid pressures in stage 1, there would be a tendency to form vertical veins but not in a certain direction. Theoretically, if the veins are close enough to interfere, polygonal vein arrays would develop at this specific stress state in which $\sigma_V > \sigma_H = \sigma_h$ at low differential stresses smaller than 4 T (Cosgrove 1995, 1997).

Stage 2. As soon as a positive σ_T starts to increase in the basin at the onset of orogeny, σ_H and σ_h are defined as result of σ_T influencing one of the horizontal principal stresses. It depends on the NW–SE- or NE–SW-directed orientation of σ_T (1a or 1b in Fig. 6, respectively) in which direction $\sigma_H \sim \sigma_{2,T}$ will develop. Concomitant with the increasing $\sigma_{2,T}$, $\sigma_3 \sim \sigma_h$ increases but at a lower rate depending on the compressibility of the host rock. Eventually, the initial increase of σ_T results in a decrease of the differential stress $\sigma_1 \sim \sigma_3$. Due to the protracted compression and increasing

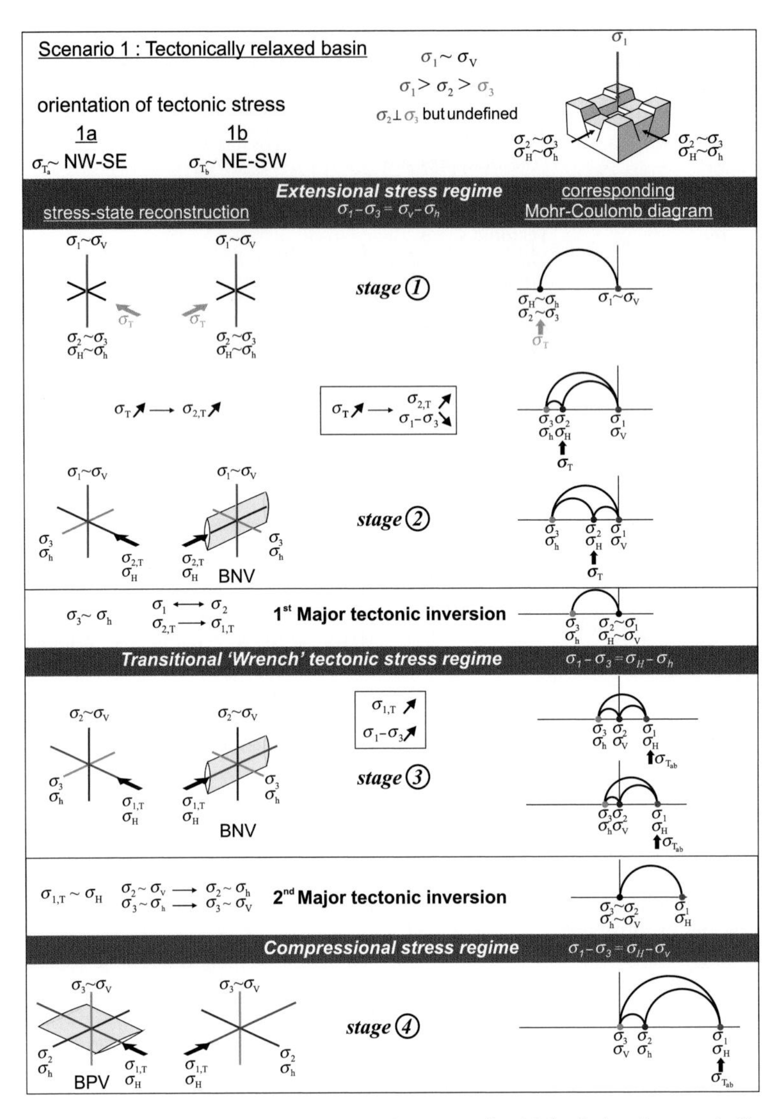

Fig. 6. Scenario 1 illustrating the 3D stress-state changes and corresponding Mohr circles of a tectonically relaxed basin that is shortened by a consistently oriented tectonic stress at the onset of orogeny and during compressional tectonic inversion. In the four stages, $\sigma_1 - \sigma_3 < 4\,T$. BNV, bedding-normal quartz veins; BPV, bedding-parallel quartz veins. The vertical line in the Mohr circles corresponds to σ_V.

σ_T, $\sigma_{2,T}$ approaches the magnitude of $\sigma_1 \sim \sigma_V$. When both stresses are subsequently equal in magnitude, a major tectonic inversion occurs during which $\sigma_{2,T}$ now becomes $\sigma_{1,T} \sim \sigma_H$ while σ_1 turns into $\sigma_2 \sim \sigma_V$. The tectonic transition at this point implies a transition between two stress regimes, changing from an extensional regime into a 'wrench' tectonic regime with σ_2 as σ_V. In the Mohr–Coulomb diagram (Fig. 6), this major tectonic inversion between the two stress regimes is illustrated by a *switch point* between σ_2 and σ_1 at the magnitude of σ_V. At this switch point, σ_3 remains substantially smaller than σ_2 and σ_1 so that a residual differential stress remains present during this first major inversion and no isotropic stress condition is reached.

Stage 3. After this major tectonic inversion from the extensional into the 'wrench' tectonic regime, $\sigma_{1,T} \sim \sigma_H$ passes the magnitude of σ_V and keeps increasing with increasing σ_T. Similar to stage 2, $\sigma_3 \sim \sigma_h$ increases concomitant with $\sigma_{1,T} \sim \sigma_H$ but at a lower rate and approaches the magnitude of σ_2. As a result of the compressibility effect, the differential stress ($\sigma_1 - \sigma_3$) increases again. This shows that a differential stress state of zero is never reached during progressive compression, but that a residual differential stress at a value substantially smaller than 4 T remains present. A second major tectonic inversion occurs when $\sigma_3 \sim \sigma_h$ eventually exceeds $\sigma_2 \sim \sigma_V$. This transition is illustrated as a switch point in which σ_2 and σ_3 change in such way that σ_V now becomes σ_3 and σ_h becomes σ_2. This second switch point implies a tectonic inversion from a 'wrench' tectonic regime into a compressional tectonic regime, which occurs during increasing differential stress.

Stage 4. Finally, in the compressional regime, differential stress increases as long as σ_T does not stagnate and $\sigma_{1,T}$ keeps increasing. Eventually, the tectonic inversion from extension to compression of a tectonically relaxed basin in this scenario results in a triaxial stress state with three well-defined principal stresses.

Scenario 2: Extensional basin with predefined σ_3 c. NW–SE

In a second scenario, we examine the effect of an increasing tectonic stress component applied on an extensional basin with a predefined structural orientation due to the preceding regional extension during sedimentary basin development. This scenario reflects a more common case in which the two horizontal stresses are different. The elongation of the basin is parallel to σ_2 and the basin opens perpendicular to σ_3 in a NW–SE direction (Figs 5b & 7).

Stage 1. We examine two options in which σ_T varies from orthogonal ($\sigma_{Ta} \sim$ NW–SE) to parallel ($\sigma_{Tb} \sim$ NE–SW) to the predefined structural elongation of the basin. The initial stress state prior to inversion is illustrated in the Mohr–Coulomb diagram by a universal triaxial stress state in which σ_1 corresponds to σ_V. The initial starting point is equal for the two options; however, as soon as the additional σ_T starts to increase, the resulting stress state is different for the two options. When σ_{Ta} increases in a NW–SE-directed contraction (Fig. 7; option 2a), $\sigma_{3,Ta} \sim \sigma_h$ and $\sigma_2 \sim \sigma_H$ will both also increase although with a different rate because of the compressibility effect of the host rock. Because of this different rate, $\sigma_{3,Ta} \sim \sigma_h$ first approaches σ_2 and will afterwards become equal in magnitude to $\sigma_2 \sim \sigma_H$ at a certain moment. This equality is exemplified by a switch point in the Mohr–Coulomb diagram in which $\sigma_{3,Ta}$ switches into $\sigma_{2,Ta}$ and σ_2 becomes $\sigma_3 \sim \sigma_h$. This minor horizontal switch has no influence on the overall extensional stress regimes because σ_1 remains vertical and σ_2 and σ_3 remain in the horizontal plane corresponding to the two different horizontal principal stresses. However, if an increasing σ_{Tb} in a NE–SW-direction is applied to the predefined basin (Fig. 7; option 2b), then the previously described minor tectonic switch between σ_3 and σ_2 will never occur between stage 1 and stage 2 because σ_{Tb} immediately influences the initial σ_2.

Stage 2. In option 2b (Fig. 7), $\sigma_{2,Tb}$ subsequently approaches σ_1 at a faster rate than the increase of σ_3 so that a minor horizontal switch between both principal stresses can never occur in the horizontal plane. Despite the difference between options 2a and 2b in the extensional regime, eventually a major tectonic inversion occurs in both options. At the switch point in the Mohr–Coulomb diagram $\sigma_{2,Tb}$ equals σ_1, and during subsequent progressive compression they mutually change into a stress state in which $\sigma_2 \sim \sigma_V$ and σ_1 become $\sigma_{1,Tab} \sim \sigma_H$. This specific stress state has also been recognized during the tectonic inversion of a tectonically relaxed basin in scenario 1 and corresponds to a 'wrench' tectonic stress regime. In both options 2a and 2b, a residual differential stress remains present during this first major tectonic inversion because of the difference between σ_1 and σ_3. Although the stress state at the inversion is illustrated in one specific Mohr–Coulomb diagram for both options, there could be a substantial difference in differential stress between both options, with a lower differential stress state when σ_{Ta} first starts to work on σ_3 (option 2a) than if σ_{Tb} works on σ_2 (option 2b).

Stage 3. After the first major tectonic inversion $\sigma_{1,T} \sim \sigma_H$ keeps increasing and, because of the compressibility effect of the host rock in the horizontal plane, $\sigma_3 \sim \sigma_h$ approaches $\sigma_2 \sim \sigma_V$. This leads to a second major tectonic inversion of the 'wrench' tectonic regime into the compressional tectonic regime in which σ_2 and σ_3 change so that σ_V becomes σ_3 and σ_h becomes σ_2.

Stage 4. σ_3 eventually turns towards the vertical axis after the second tectonic inversion and $\sigma_{1,Tab}$ and σ_2 are now both situated in the horizontal plane, as expected in the compressional stress regime. However, because of the orientation of the initial tectonic stress component σ_{Ta} or σ_{Tb}, the eventual structural grain of the deformed basin is oriented differently in the two options.

Scenario 3: Extensional basin with predefined $\sigma_3 \sim$ NE–SW

In a third scenario, we examine the effect of a stress component that is applied on an extensional

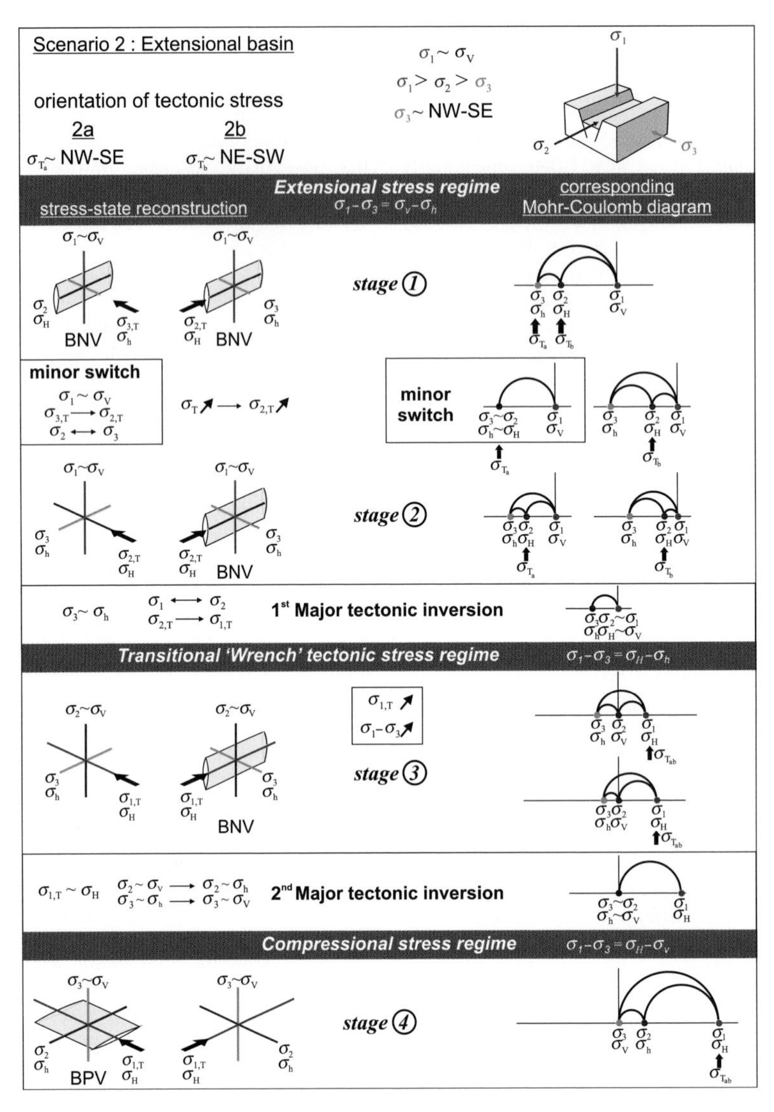

Fig. 7. Scenario 2 illustrating the 3D stress-state changes and corresponding Mohr circles of an extensional basin with a predefined σ_3 oriented NW–SE that is shortened by a consistently oriented tectonic stress at the onset of orogeny and during compressional tectonic inversion. Only scenario 2a reflects the successive veining conditions in the High-Ardenne slate belt (North Eifel, Germany). In the four stages, $\sigma_1 - \sigma_3 < 4\,T$. BNV, bedding-normal quartz veins; BPV, bedding-parallel quartz veins. The vertical line in the Mohr circles corresponds to σ_V.

basin with a predefined structural orientation in which the elongation of the basin (σ_2) is now oriented in a NE–SW direction and the basin opens perpendicular to σ_3 in a NW–SE-direction (Figs 5 & 8). Similar to the second scenario, two options are investigated in which a consistent positive tectonic stress is applied on the predefined basin in a NW–SE-direction (i.e. σ_{Ta}; Fig. 8, option 3a) and in a NE–SW-direction (i.e. σ_{Tb}; option 3b).

Stage 1. The initial starting point is again equal for the two options, with σ_1 oriented vertically and two predefined horizontal σ_h and σ_H.

Stage 2. As σ_T starts to increase, no minor tectonic switch occurs in option 3a if the applied tectonic stress $\sigma_{T,a}$ immediately works on $\sigma_{2,Ta}$ and subsequently approaches $\sigma_1 \sim \sigma_V$ at the end of stage 2. However, when a tectonic stress $\sigma_{T,b}$ increases parallel to σ_3 (option 3b), a minor tectonic switch

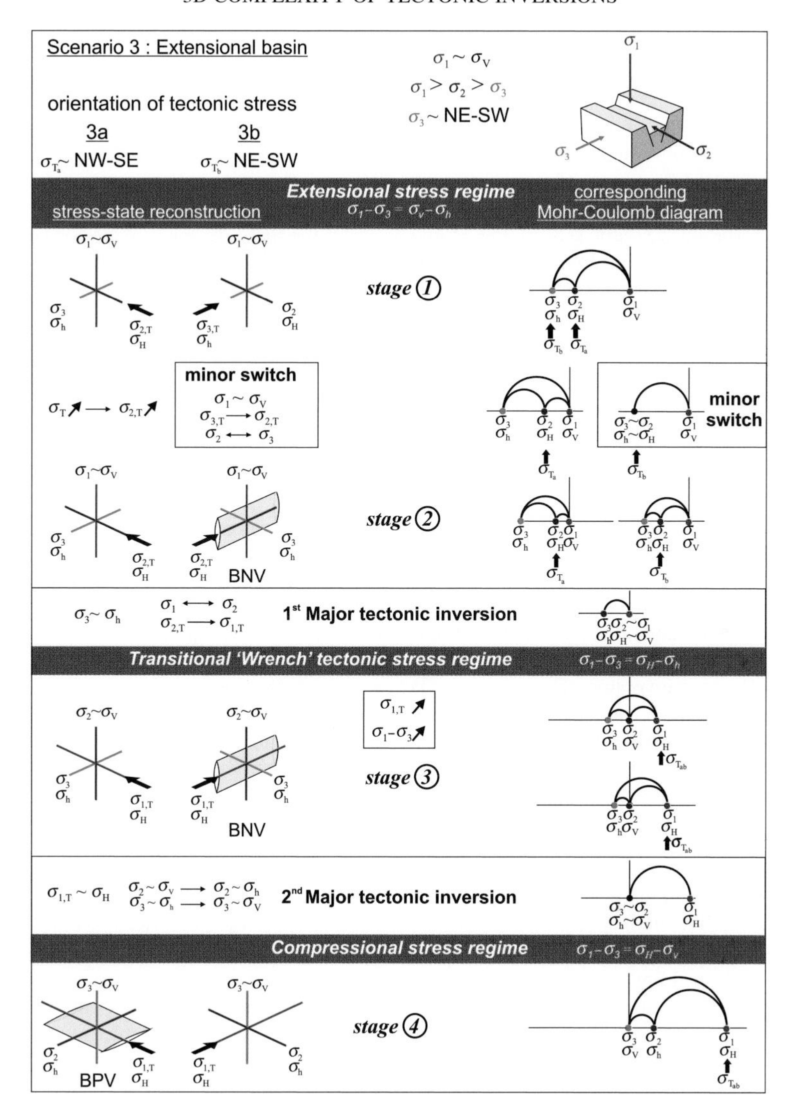

Fig. 8. Scenario 3 illustrating the 3D stress-state changes and corresponding Mohr circles of an extensional basin with a predefined σ_3 oriented NE–SW that is shortened by a consistently oriented tectonic stress at the onset of orogeny and during compressional tectonic inversion. In the four stages, $\sigma_1 - \sigma_3 < 4\,T$. BNV, bedding-normal quartz veins; BPV, bedding-parallel quartz veins. The vertical line in the Mohr circles corresponds to σ_V.

occurs because $\sigma_{3,\mathrm{Tb}} \sim \sigma_\mathrm{h}$ approaches $\sigma_2 - \sigma_\mathrm{H}$ faster than σ_2 approaches $\sigma_1 \sim \sigma_\mathrm{V}$. At the switch point of this minor tectonic inversion, $\sigma_{3,\mathrm{Tb}}$ changes into $\sigma_{2,\mathrm{Tb}}$ and σ_2 changes into $\sigma_3 \sim \sigma_\mathrm{h}$ at constant $\sigma_1 \sim \sigma_\mathrm{V}$. After this minor inversion in the horizontal plane, the differential stress further decreases because of increasing σ_{Ta} and σ_{Tb}, eventually leading to the first major tectonic inversion at the end of stage 2.

Stage 3. Similar to scenario 2 in both options 3a and 3b, $\sigma_{2,\mathrm{T}}$ equals σ_1 at the switch point and, during subsequent progressive compression, they mutually change at the point during which $\sigma_{2,\mathrm{T}}$ turns towards the vertical σ_V and σ_1 becomes $\sigma_{1,\mathrm{T}} \sim \sigma_\mathrm{H}$. The latter is indicative for a 'wrench' tectonic stress regime, in which extensional fracturing can occur at low differential stresses >0. With increasing differential stress at stage 3, σ_3 approaches $\sigma_2 \sim \sigma_\mathrm{V}$ which leads to the second major tectonic inversion.

Stage 4. After inversion σ_3 eventually switches towards the vertical σ_V, corresponding to the compressional regime.

Discussion on fracturing and vein formation

The stress-state evolution of a shortened sedimentary basin during compressional tectonic inversion, which is demonstrated by means of three different scenarios, yields several possibilities during which the two successive quartz vein sets of the High-Ardenne case study can be formed. Concerning the timing of bedding-normal veining, the different stages during which the NE–SW bedding-normal extension veins can be formed are illustrated by means of the veins drawn in the different stress-state reconstructions (Figs 6–8).

Starting from a relaxed basin (scenario 1), the only possibility of forming the proper NE–SW vertical extension veins is if a NE–SW-directed σ_T decreases the differential stress during progressive compression until extensional fracturing is allowed (Fig. 6; stage 2; option 1b). Vein formation can occur prior to the first major tectonic inversion in the extensional regime, as well as after the major tectonic inversion in the transitional 'wrench' tectonic regime, due to the fact that σ_3 remains constant during the first major inversion (Fig. 6; stage 3; option 1b). Subsequently, bedding-parallel veins with pronounced NW–SE fabric can only be formed in the compressional regime during consistent NW–SE-directed compression (Fig. 6; stage 4; option 1a).

In an extensional basin with predefined NW–SE-oriented σ_3 (scenario 2), bedding-normal veins can be formed in several phases. In stage 1 of both options 2a and 2b (Fig. 7), the NW–SE extension of the basin is initially the sole factor that determines if NE–SW bedding-normal veins can be formed under the circumstance that a NW–SE or NE–SW tectonic compression decreases the differential stress until extensional fracturing is allowed. After the predicted minor tectonic switch of σ_2 and σ_3 during progressive compression, the stress field is however misoriented to allow veining (Fig. 7; stage 2; option 2a). In option 2b, no minor switch occurs because the tectonic stress σ_{Tb} is oriented parallel to σ_2, resulting in a stress field that remains constant during stages 1 and 2. For similar reasons as in scenario 1, NE–SW veins can still be formed after the first major tectonic inversion in the 'wrench' transition regime due to the consistent σ_3 during inversion. Subsequently, the observed bedding-parallel veins are only formed in option 2a, reflecting NW–SE oriented compression.

Compression of a sedimentary basin in which the predefined σ_3 is NE–SW oriented (scenario 3), NE–SW veins can only be formed in stage 2 after the minor horizontal tectonic switch during NE–SW-directed contraction of σ_{Tb} (Fig. 8; stage 3; option 3b). Similar to the other two scenarios, veins can be formed both prior to as well as after the major tectonic inversion to a 'wrench' transitional tectonic regime. Similar to the other scenarios, the subsequent bedding-parallel veins with a NW–SE fabric are formed in the compressional regime after the second tectonic inversion, due to a NW–SE-oriented σ_{Ta} (Fig. 8; stage 4; option 3a).

Of the three different scenarios considered only option 2a (Fig. 7) complies with the observations in the High-Ardenne case study, properly demonstrating how the NE–SW bedding-normal veins and subsequently the bedding-parallel veins with a NW–SE-directed internal fabric are formed. In this option, the model starts from an extensional basin that has a predefined structural NE–SW orientation (Fig. 5b) due to the geodynamic context of the Ardenne-Eifel basin on the passive margin of the Rhenohercynian Ocean. This particular basin configuration is corroborated by the identification of major basin-bounding NE–SW-trending normal fault systems indicated by stratigraphical mapping (see Mansy *et al.* 1999; Lacquement 2001). During extension of the basin, in which the load of the overburden corresponds to σ_1 in the basin, there is already a σ_T in a NW–SE direction parallel to σ_3 that causes an increase in the magnitude of σ_3 and consequently a decrease in differential stress, eventually leading to regionally consistent NE–SW extensional fracturing. This model furthermore implies that no veins are formed after the minor tectonic switch and in the transitional 'wrench' tectonic regime, because of misorientation of the stress regime.

The proper configuration in option 2a allows the development of a stress-state reconstruction by means of several Mohr–Coulomb diagrams that illustrate successive fracturing and vein formation during the compressional tectonic inversion. Microthermometry testifies that maximum fluid overpressures of *c.* 190 MPa and *c.* 205 MPa are reached during bedding-normal and bedding-parallel extension veining, respectively (Van Noten 2011; Van Noten *et al.* 2011). The magnitude of the constant burial-related σ_v that corresponds to the load of the overburden at a depth of *c.* 7 km reflects a lithostatic pressure of *c.* 185 MPa, which is kept constant in the Mohr circle reconstructions (Fig. 9). During bedding-normal veining, the value of σ_h relates to σ_V in such way that (Cosgrove 1995; Cox *et al.* 2001):

$$\sigma_h = \sigma_v\left(\frac{\nu}{1-\nu}\right) + \sigma_T \qquad (1)$$

with a common Poisson ratio of $\nu \sim 0.25$ for sandstone (Mandl 2000).

In a tectonically relaxed basin in which $\sigma_T = 0$, this equation corresponds to $\sigma_h \sim 62$ MPa. However, the basin had a predefined geometry with

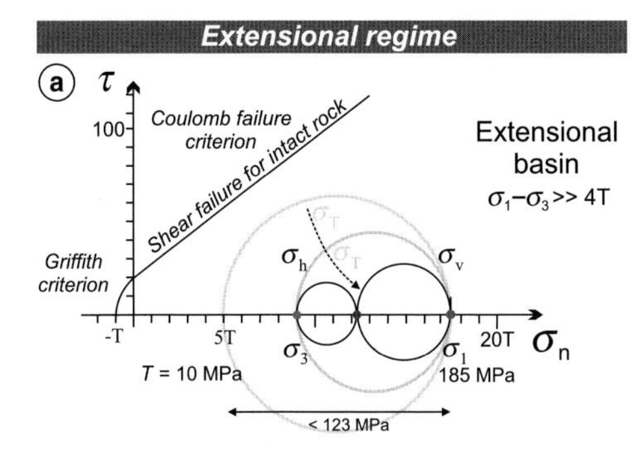

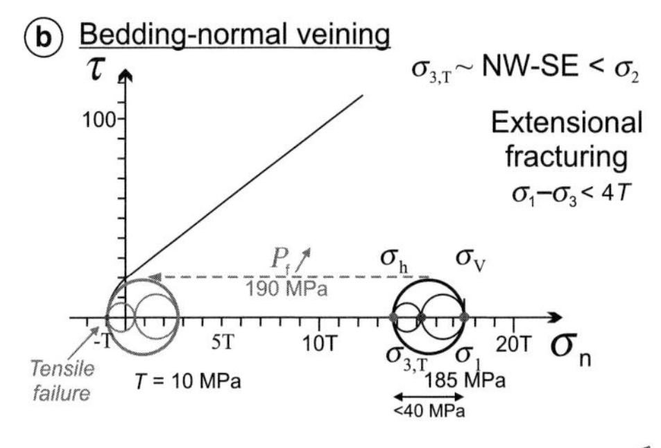

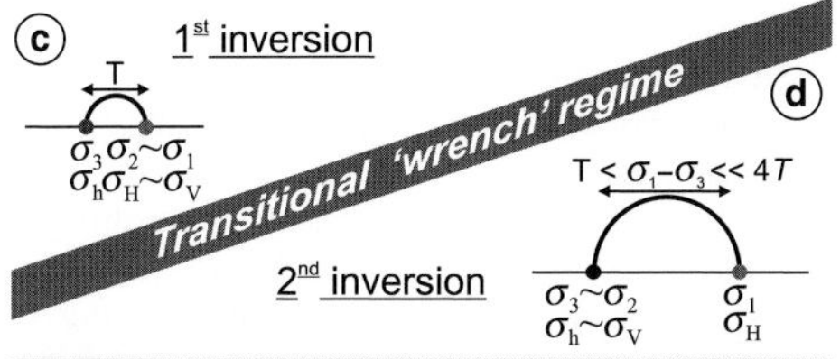

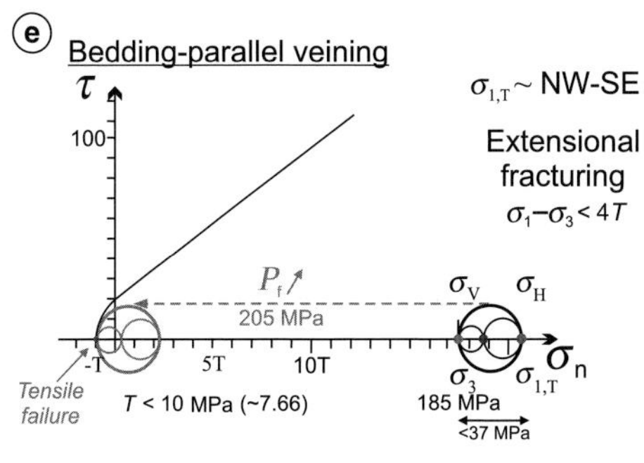

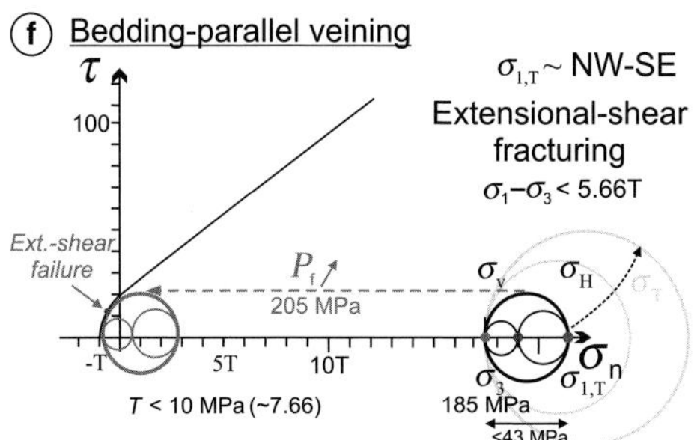

extension oriented NW–SE perpendicular to $\sigma_3 \sim \sigma_h$, indicating that the magnitude of $\sigma_3 \sim \sigma_h$ of the basin in extension must therefore have been below $\sigma_h = 62$ MPa. In such a configuration, the differential stress $\sigma_1 - \sigma_3$ (that corresponds to $\sigma_v - \sigma_h$) is *c.* 123 MPa (Fig. 9a). This value is too large to allow the formation of extension fractures developing at $\sigma_1 - \sigma_3 < 4\,T$ at *c.* 40 MPa in which $T = 10$ MPa is representative for intact sandstone. Therefore, σ_T acting on $\sigma_{3,T}$ (Fig. 7; option 2a; stage 1) must at least increase up to 83 MPa in order to decrease the differential stress until the stress configuration of $\sigma_1 - \sigma_{3,T} < 40$ MPa is fulfilled (Fig. 9b). At this stage, a P_f of *c.* 190 MPa is able to overcome *T*, allowing the formation of bedding-normal veins against the action of total pressure. This is illustrated by the fluid pressure that pushes the Mohr circle into the tensile domain (Fig. 9b). Note that the stress-state evolution from Figure 9b is illustrated as a static reconstruction; in reality, this is a dynamic process in which σ_T increases with P_f.

After the minor switch of $\sigma_{3,T}$ to $\sigma_{2,T}$, the first major tectonic inversion occurs (Fig. 9c). At this point, a residual differential stress remains present at an estimated value of approximately $\sigma_1 - \sigma_3 =$ *c. T*. After this stage, differential stress increases again in the transitional 'wrench' tectonic regime until the second tectonic inversion takes place at a differential stress above *T* but below a substantial value of 4 *T* (Fig. 9d). After the reorientation of the principal stresses in the compressional regime, fracturing and vein formation takes place at a value of $\sigma_1 - \sigma_3 < 30$ MPa for extensional failure (Fig. 9e) and slightly increases to $\sigma_1 - \sigma_3 < 43$ MPa during subsequent extensional-shear failure (Fig. 9f). Extensional failure therefore takes place at a lower differential stress in the compressional regime than compared to the extensional regime because of the lower tensile strength due to the bedding anisotropy ($T < 10$ MPa) during compression. Since $\sigma_1 - \sigma_3 = \sigma_H - \sigma_v$ and from Equation 1 modified for σ_H, σ_T is allowed to rise up to values of $\sigma_T = 155$ MPa during extensional failure and $\sigma_T = 167$ MPa during extensional-shear failure in the compressional regime.

Fig. 9. Stress-state Mohr circle reconstruction of option 2a which best fits with the observations of the formation of the successive vein sets in the High-Ardenne slate belt. Stages (**a**–**f**) reflect the different stages of extensional brittle failure from the extensional regime over the intermediate 'wrench' tectonic regime into the compressional regime. See Table 1 for definitions of notation. BNV, bedding-normal veins; BPV, bedding-parallel veins.

At this stage, the (supra-)lithostatic fluid pressure of P_f at *c.* 205 MPa is able to overcome *T*. This is illustrated by the shift of the Mohr circle to the tensile domain, allowing failure when the Mohr circle hits the failure envelope. Furthermore, note that in a Mohr–Coulomb diagram supra-lithostatic failure is not illustrated as the Mohr circle cannot pass the Griffith criterion. Subsequently, σ_T increases during further contraction in the compressional regime (grey circle in Fig. 9f). A further Mohr circle reconstruction is prohibited as folds and cleavage subsequently developed, defining a plastic deformation which cannot be illustrated in a brittle failure diagram.

General implications

Apart from the regional implications, the different scenarios imply that if a basin is subjected to a tectonic compression, two or three tectonic switches are always necessary to explain the complex 3D tectonic inversion from an extensional to a compressional tectonic regime. The orientation of σ_T with respect to σ_3 is the sole factor that determines whether or not a minor horizontal tectonic switch between σ_2 and σ_3 will occur prior to the first major tectonic inversion. Furthermore, no matter the orientation of the basin geometry or the orientation of the increasing tectonic shortening, the presence of a transitional 'wrench' tectonic regime with a vertical σ_2 is always predicted during compressional tectonic inversion in a progressive deformation history. Although strike-slip related brittle features should be expected in such a stress configuration (Anderson 1951), the latter has to date never been observed or these features are not preserved in a shortened basin that has been subjected by a compressional tectonic inversion. This is possibly due to the (very) low differential stresses that remain present during inversion or due to the presence of previously formed structures (e.g. veins) that may form a strong anisotropy and may hamper the formation of new strike-slip related structures. Additionally, the time aspect during which tectonic inversion from extension to compression occurs may also play an important role. If, depending on the geodynamic setting of an area, a (relatively slow) tectonic inversion could take place during several millions years, a very long intermediate stage could occur in a specific situation during which differential stresses remain very small. In this specific case, structures related to the intermediate stage should be found.

To conclude, despite the fact that Sibson (1998, 2004) already mentioned that 'wrench' regimes have intermediate values bounded by the compressional and extensional regime criteria (i.e. at the zero differential stress line on Fig. 1), the presence of a transitional 'wrench' tectonic regime in between the extensional and compressional regime has to date not been visualized in a brittle failure mode plot. This study emphasizes that, depending on the value of σ_2, a whole range of possibilities are present during the tectonic inversion. Both the presence of an intermediate 'wrench' tectonic regime and the 3D complexity of a tectonic switch should therefore be taken into account in the construction of brittle failure mode plots. Tectonic inversions between stress regimes should not be considered as a single switch point, but should be subdivided in a double tectonic switch with a short-lived intermediate 'wrench' tectonic regime.

Conclusions

Although brittle failure mode plots are very useful to visualize the fluid pressure and stress-state evolution during tectonic inversion, they oversimplify the true stress switches in the Earth's crust during tectonic inversion at low differential stresses. In this study a 3D stress-state reconstruction of a compressional tectonic inversion has been carried out, based on two successive extensional quartz vein sets that are oriented normal and parallel to bedding and which occur in siliciclastic multilayers of the High-Ardenne slate belt (Belgium, Germany). The two vein occurrences are formed during the extensional and compressional regime and thus reflect Early Variscan tectonic inversion affecting the Ardenne-Eifel basin. In order to determine the stress-field orientation during compressional tectonic inversion, several 3D stress-state evolutions are reconstructed by applying a consistent tectonic stress component to a basin with a predefined structural orientation. By changing both the initial sedimentary basin geometry and the direction of the tectonic compression, several scenarios are developed; only one could explain the vein occurrences observed in the High-Ardenne slate belt. In the model which best fits the observations, the NE–SW bedding-normal extension veins can only be formed during a stress state in which a NW–SE-directed tectonic stress σ_T, oriented parallel to the opening direction (σ_3) of the extensional Ardenne-Eifel basin, reduces the differential stress substantially in order to allow extensional veining. Bedding-parallel veins with the observed NW–SE internal fabric can subsequently only be formed under NW–SE-directed compression.

More generally, these stress-state reconstructions predict that if an increasing positive tectonic stress starts to compress a sedimentary basin at constant vertical principal stress (= overburden pressure), two or three tectonic stress switches are always necessary to explain the inversion from extension to compression depending on the orientation of σ_T with respect to the extension direction of the basin. If σ_T is oriented parallel to σ_3, a first minor stress switch occurs in the horizontal plane between σ_3 and σ_2. This minor switch is however absent if a tectonic stress increases in a tectonically relaxed basin or if it increases parallel to σ_2 (i.e. the elongation of the basin). During further progressive compression however, no matter the orientation of the basin geometry or the direction of the tectonic stress, in all the different scenarios an intermediate transitional 'wrench' tectonic stress regime is predicted in between the extensional and compressional regime. No evidence of this intermediate 'wrench' tectonic regime has yet been found in particular brittle features in a shortened basin that has been subjected to a tectonic inversion at low differential stress. Eventually, these reconstructions illustrate that if a basin experiences a tectonic inversion under triaxial stress conditions, stress transitions are more complex than classically represented. Ideally, the predicted intermediate 'wrench' tectonic regime should be implemented in the brittle failure mode plots, either in a third dimension or in separate brittle failure mode plots that illustrate the differences of σ_1 and σ_3 with σ_2.

This study has benefited from many discussions at the Anderson Stress Conference, September 2010, Glasgow and we would like to thank Z. Shipton, R. Sibson, D. Healy, R. Butler and H. Moir for organising and for editing this special publication. A. Fagereng and F. Trippetta are acknowledged for their interesting reviews which improved the quality of the paper. Additional discussion with S. Cox was very fruitful. This research was conducted within the project KAN 1.5.128.05 of the Fonds voor Wetenschappelijk Onderzoek–Vlaanderen and results from the PhD research of KVN.

References

ANDERSON, E. M. 1951. *The Dynamics of Faulting and Dyke Formation with Application to Britain.* Oliver and Boyd, Edinburgh.

BLUNDELL, D. J. 2002. The timing and location of major ore deposits in an evolving orogen: the geodynamic context. *In*: BLUNDELL, D. J., NEUBAUER, E. & VON QUADT, A. (eds) *The Timing and Location of Major Ore Deposits in an Evolving Orogen.* Geological Society, London, Special Publications, **204**, 1–12.

BOULLIER, A.-M. & ROBERT, F. 1992. Palaeoseismic events recorded in Archaean gold-quartz vein networks, Val d'Or, Abitibi, Quebec, Canada. *Journal of Structural Geology*, **14**, 161–179.

COLLETTINI, C., DE PAOLA, N. & GOULTY, N. R. 2006. Switches in the minimum compressive stress direction induced by overpressure beneath a low-permeability fault zone. *Terra Nova*, **18**, 224–231.

COSGROVE, J. W. 1995. The expression of hydraulic fracturing in rocks and sediments. *In*: AMEEN, M. S. (ed.) *Fractography: Fracture Topography as a Tool in Fracture Mechanics and Stress Analysis.* Geological Society, London, Special Publications, **92**, 187–196.

COSGROVE, J. W. 1997. Hydraulic fractures and their implications regarding the state of stress in a sedimentary sequence during burial. *In*: SENGUPTA, S. (ed.) *Evolution of Geological Structures in Micro- to Macro-scales.* Chapman & Hall, London, 11–25.

COSGROVE, J. W. 2001. Hydraulic fracturing during the formation and deformation of a basin: a factor in the dewatering of low-permeability sediments. *AAPG Bulletin*, **85**, 737–748.

COX, S. F. 1987. Antitaxial crack-seal vein microstructures and their relationship to displacement paths. *Journal of Structural Geology*, **9**, 779–787.

COX, S. F. 1995. Faulting processes at high fluid pressures: an example of fault valve behavior from the Wattle Gully Fault, Victoria, Australia. *Journal of Geophysical Research*, **100**, 12 841–12 859.

COX, S. F. 2010. The application of failure mode diagrams for exploring the roles of fluid pressure and stress states in controlling styles of fracture-controlled permeability enhancement in fault and shear zones. *Geofluids*, **10**, 217–233.

COX, S. F., KNACKSTEDT, M. A. & BRAUN, J. 2001. *Principles of Structural Control on Permeability and Fluid Flow in Hydrothermal Systems Structural Controls on Ore Genesis Reviews in Economic Geology.* Society of Economic Geology, Boulder, **14**, 1–24.

CRESPI, J. M. & CHAN, Y.-C. 1996. Vein reactivation and complex vein intersection geometries in the Taconic Slate Belt. *Journal of Structural Geology*, **18**, 933–939.

DEBACKER, T. N., DUMON, M. & MATTHYS, A. 2009. Interpreting fold and fault geometries from within the lateral to oblique parts of slumps: a case study from the Anglo-Brabant Deformation Belt (Belgium). *Journal of Structural Geology*, **31**, 1525–1539.

FAGERENG, A., REMITTI, F. & SIBSON, R. H. 2010. Shear veins observed within anisotropic fabric at high angles to the maximum compressive stress. *Nature Geoscience*, **3**, 482–485.

FIELITZ, W. 1992. Variscan transpressive inversion in the northwestern central Rhenohercynian belt of western Germany. *Journal of Structural Geology*, **14**, 547–563.

FIELITZ, W. & MANSY, J.-L. 1999. Pre- and synorogenic burial metamorphism in the Ardenne and neighbouring areas (Rhenohercynian zone, central European Variscides). *Tectonophysics*, **309**, 227–256.

GILLESPIE, P. A., WALSH, J. J., WATTERSON, J., BONSON, C. G. & MANZOCCHI, T. 2001. Scaling relationships of joint and vein arrays from The Burren, Co. Clare, Ireland. *Journal of Structural Geology*, **23**, 183–201.

HEALY, D. 2009. Anisotropy, pore fluid pressure and low angle normal faults. *Journal of Structural Geology*, **31**, 561–574.

HILGERS, C., BÜCKER, C. & URAI, J. L. 2006*a*. Fossil overpressures compartments? a case study from the Eifel area and some general aspects. *In*: PHILLIPP, S., LEISS, B., VOLLBRECHT, A., TANNER, D. & GUDMUNDSSON, A.

(eds) *Symposium 'Tektonik, Struktur- und Kristallingeologie'*. Universitätsdrucke Göttingen, Göttingen, Germany, **11**, 87–89.

Hilgers, C., Kirschner, D. L., Breton, J.-P. & Urai, J. L. 2006*b*. Fracture sealing and fluid overpressures in limestones of the Jabal Akhdar dome, Oman mountains. *Geofluids*, **6**, 168–184.

Jackson, R. R. 1991. Vein arrays and their relationship to transpression during fold development in the Culm Basin, central south-west England. *Proceedings of the Ussher Society*, **7**, 356–362.

Kenis, I. & Sintubin, M. 2007. About boudins and mullions in the Ardenne-Eifel area (Belgium, Germany). *Geologica Belgica*, **10**, 79–91.

Kenis, I., Sintubin, M., Muchez, Ph. & Burke, E. A. J. 2002. The 'boudinage' question in the High-Ardenne Slate Belt (Belgium): a combined structural and fluid-inclusion approach. *Tectonophysics*, **348**, 93–110.

Kenis, I., Muchez, P., Verhaert, G., Boyce, A. J. & Sintubin, M. 2005. Fluid evolution during burial and Variscan deformation in the Lower Devonian rocks of the High-Ardenne slate belt (Belgium): sources and causes of high-salinity and C–O–H–N fluids. *Contributions to Mineralogy and Petrology*, **150**, 102–118.

Lacquement, F. 2001. *L'Ardenne Varisque. Déformation progressive d'un prisme sédimentaire pré-structuré, de l'affleurement au modèle de chaîne*. Société Géologique du Nord, Lille, **29**.

Laubach, S. E. & Diaz-Tushman, K. 2009. Laurentian palaeostress trajectories and ephemeral fracture permeability, Cambrian Eriboll Formation sandstones west of the Moine Thrust Zone, NW Scotland. *Journal of the Geological Society, London*, **166**, 349–362.

Laubach, S. E., Olson, J. E. & Gale, J. F. W. 2004. Are open fractures necessarily aligned with maximum horizontal stresses? *Earth and Planetary Science Letters*, **222**, 191–195.

Mandl, G. 2000. *Faulting in Brittle Rocks. An Introduction to the Mechanics of Tectonic Faults*. Springer-Verlag Berlin, Heidelberg.

Manning, C. E. & Bird, D. K. 1991. Porosity evolution and fluid flow in the basalts of the Skaergaard magma-hydrothermal system, East Greenland. *American Journal of Science*, **291**, 201–257.

Mansy, J. L., Everaerts, M. & De Vos, W. 1999. Structural analysis of the adjacent Acadian and Variscan fold belts in Belgium and northern France from geophysical and geological evidence. *Tectonophysics*, **309**, 99–116.

Mazzarini, F., Isola, I., Ruggieri, G. & Boschi, C. 2010. Fluid circulation in the upper brittle crust: thickness distribution, hydraulic transmissivity fluid inclusion and isotopic data of veins hosted in the Oligocene sandstones of the Macigno Formation in southern Tuscany, Italy. *Tectonophysics*, **493**, 118–138.

Meilliez, F. & Mansy, J.-L. 1990. Déformation pelliculaire différenciée dans une série lithologique hétérogène: le Dévono-Carbonifère de l'Ardenne. *Bulletin de la Société géologique de France*, **6**, 177–188.

Muchez, Ph., Sintubin, M. & Swennen, R. 2000. Origin and migration pattern of palaeofluids during orogeny: discussion on the Variscides of Belgium and northern France. *Journal of Geochemical Exploration*, **69–70**, 47–51.

Nielsen, P., Swennen, R., Muchez, Ph. & Keppens, E. 1998. Origin of Dinatian zebra dolomites south of the Brabant-Wales Massif, Belgium. *Sedimentology*, **45**, 727–743.

Price, N. J. 1966. *Fault and Joint Development in Brittle and Semi-Brittle Rock*. Pergamon Press, Oxford.

Ramsay, J. G. 1980. The crack-seal mechanism of rock deformation. *Nature*, **284**, 135–139.

Secor, D. T. 1965. Role of fluid pressure in jointing. *American Journal of Sciences*, **263**, 633–646.

Sibson, R. H. 1990. Conditions for fault–valve behaviour. *In*: Knipe, R. J. & Rutter, E. H. (eds) *Deformation Mechanisms, Rheology and Tectonics*. Geological Society, London, Special Publications, **54**, 15–28.

Sibson, R. H. 1995. Selective fault reactivation during basin inversion: potential for fluid redistribution through fault–valve action. *In*: Buchanan, J. G. & Buchanan, P. G. (eds) *Basin Inversion*. Geological Society, London, Special Publications, **88**, 3–19.

Sibson, R. H. 1998. Brittle failure mode plots for compressional and extensional tectonic regimes. *Journal of Structural Geology*, **20**, 655–660.

Sibson, R. H. 2000. Tectonic controls on maximum sustainable overpressure: fluid redistribution from stress transition. *Journal of Geochemical Exploration*, **69–70**, 471–475.

Sibson, R. H. 2004. Controls on maximum fluid overpressure definitions conditions for mesozonal mineralisation. *Journal of Structural Geology*, **26**, 1127–1136.

Stets, J. & Schafer, A. 2009. The Siegenian delta: land–sea transitions at the northern margin of the Rhenohercynian Basin. *In*: Koniggshof, P. (ed.) *Devonian Change: Case Studies in Palaeogeography and Palaeoecology*. Geological Society, London, Special Publications, **314**, 37–72.

Stowell, J. F. W., Watson, A. P. & Hudson, N. F. C. 1999. Geometry and populations systematics of a quartz vein set, Holy Island, Anglesey, North Wales. *In*: McCaffrey, K. J. W., Lonergan, L. & Wilkinson, J. J. (eds) *Fractures, Fluid Flow and Mineralisation*. Geological Society, London, Special Publications, **155**, 17–33.

Teixell, A., Durney, D. W. & Arboleya, M.-L. 2000. Stress and fluid control on décollement within competent limestone. *Journal of Structural Geology*, **22**, 349–371.

Tunks, A. J., Selley, D., Rogers, J. R. & Brabham, G. 2004. Vein mineralization at the Damang Gold Mine, Ghana: controls on mineralization. *Journal of Structural Geology*, **26**, 1257–1273.

Van Baelen, H. 2010. *Dynamics of a Progressive vein Development During the Late-Orogenic Mixed Brittle-Ductile Destabilisation of a Slate Belt. Examples of the High-Ardenne Slate Belt (Herbeumont, Belgium)* Aardkundige mededelingen series, **24**, 1–221.

Van Baelen, H. & Sintubin, M. 2008. Kinematic consequences of an angular unconformity in simple shear: an example from the southern border of the Lower Palaeozoic Rocroi inlier (Naux, France). *Bulletin de la Société géologique de France*, **179**, 73–87.

Van Noten, K. 2011. *Stress-State Evolution of the Brittle Upper Crust during Early Variscan Tectonic Inversion as Defined by Successive Quartz Vein-Types in the High-Ardenne Slate Belt, Germany*. Aardkundige mededelingen series, **28**, 1–241.

Van Noten, K. & Sintubin, M. 2010. Linear to nonlinear relationship between vein spacing and layer thickness in centimetre- to decimetre-scale siliciclastic multilayers from the High-Ardenne slate belt (Belgium, Germany). *Journal of Structural Geology*, **32**, 377–391.

Van Noten, K., Hilgers, C., Urai, J. L. & Sintubin, M. 2008. Late burial to early tectonic quartz veins in the periphery of the High-Ardenne slate belt (Rursee, North Eifel, Germany). *Geologica Belgica*, **11**, 179–198.

Van Noten, K., Berwouts, I., Muchez, Ph. & Sintubin, M. 2009. Evidence of pressure fluctuations recorded in crack-seal veins in low-grade metamorphic siliciclastic metasediments, Late Palaeozoic Rhenohercynian fold-and-thrust belt (Germany). *Journal of Geochemical Exploration*, **101**, 106.

Van Noten, K., Muchez, Ph. & Sintubin, M. 2011. Stress-state evolution of the brittle upper crust during compressional tectonic inversion as defined by successive quartz vein types (High-Ardenne slate belt, Germany). *Journal of the Geological Society, London*, **168**, 407–422.

Von Winterfeld, C.-H. 1994. *Variszische Deckentektonik und devonische Beckengeometrie der Nord-Eifel - Ein quantitatives Modell (Profilbilanzierung und Strain-Analyse im Linksrheinischen Schiefergebirge)*. Aachener Geowissenschaftliche Beiträge, RWTH Aachen, **2**, 1–319.

Weber, K. 1981. The structural development of the Rheinische Schiefergebirge. *Geologie en Mijnbouw*, **60**, 149–159.

Wiltschko, D. V., Lambert, G. R. & Lamb, W. 2009. Conditions during syntectonic vein formation in the footwall of the Absaroka Thrust Fault, Idaho–Wyoming–Utah fold and thrust belt. *Journal of Structural Geology*, **31**, 1039–1057.

Geomechanical modelling of fault reactivation in the Ceduna Sub-basin, Bight Basin, Australia

J. MACDONALD[1]*, G. BACKÉ[1], R. KING[2], S. HOLFORD[1] & R. HILLIS[3]

[1]*Australian School of Petroleum, University of Adelaide, Santos Petroleum Engineering Building, Adelaide, 5005, Australia*

[2]*School of Earth and Environmental Sciences, University of Adelaide, Mawson Laboratories, Adelaide, 5005, Australia*

[3]*Deep Exploration Technologies Cooperative Research Centre, University of Adelaide, Mawson Laboratories, Adelaide, 5005, Australia*

**Corresponding author (e-mail: jmacdonald@asp.adelaide.edu.au)*

Abstract: The Ceduna Sub-basin is located within the Bight Basin on the Australian southern margin. Recent structural analysis using newly acquired two-dimensional (2D) and three-dimensional (3D) seismic data demonstrates two Late Cretaceous delta–deepwater fold–thrust belts (DDWFTBs), which are overlain by Cenozoic sediments. The present-day normal fault stress regime identified in the Bight Basin indicates that the maximum horizontal stress (S_{Hmax}) is margin parallel; Andersonain faulting theory therefore suggests the delta-top extensional faults are oriented favourably for reactivation. A breached hydrocarbon trap encountered in the Jerboa-1 well demonstrates this fault reactivation. Faults interpreted from 3D seismic data were modelled using the Poly3D© geomechanical code to determine the risk of reactivation. Results indicate delta-top extensional faults that dip 40–70° are at moderate–high risk of reactivation, while variations in the orientation of the fault planes results in an increased risk of reactivation. Two pulses of inversion are identified in the Ceduna Sub-basin and correlate with the onset of rifting and fault reactivation in the Santonian. We propose a ridge-push mechanism for this stress which selectively reactivates extensional faults on the delta-top, forming inversion anticlines that are prospective for hydrocarbon exploration.

Indications of past hydrocarbon accumulations have been found in the Jerboa-1 well of the Eyre Sub-basin on the southern margin of Australia (Totterdell *et al.* 2000; Ruble *et al.* 2001; Somerville, 2001; Struckmeyer *et al.* 2001; Fig. 1). However, exploration success has been lacking in the Ceduna Sub-basin due to geographic constraints such as water depth and accessibility, as well as failure to identify a significant hydrocarbon system in only two exploration wells (Potoroo-1 and Gnarlyknots-1; Figs 1 & 2).

The Ceduna Sub-basin is located in the Bight Basin at the centre of the southern margin of Australia, which is defined by the bathymetry of the Ceduna Terrace (Fig. 2). The Bight Basin is significantly underexplored yet is a prospective hydrocarbon province that extends along the coast between Port Lincoln (South Australia) and Cape Leeuwin (Western Australia), a distance in excess of 3000 km, and offshore for hundreds of kilometres (Fig. 1). Only 12 exploration wells have been drilled to date in the Bight Basin, two of which were drilled in the Ceduna Sub-basin. The Ceduna Sub-basin covers an area of approximately 90 000 km^2 with water depths ranging from 200 m to more than 5000 m.

The thickness of Middle Jurassic–Cenozoic sedimentary rocks in the Ceduna Sub-basin exceeds 12 km (Struckmeyer *et al.* 2001; Krassay & Totterdell 2003). The Ceduna Sub-basin contains two spatially and temporally separate Cretaceous delta–deepwater fold–thrust belt (DDWFTB) systems: the late Albian–Santonian White Pointer DDWFTB and the late Santonian–Maastrichtian Hammerhead DDWFTB. These systems detach above the mud-rich Blue Whale and Tiger supersequences, respectively (Fig. 3).

Previous work completed by Geoscience Australia (in 1999–2008) and Woodside (including drilling of the most recent well, Gnarlyknots-1) resulted in an improved understanding of the hydrocarbon prospectivity of the Bight Basin, particularly the Ceduna Sub-basin. This work focused on the sequence stratigraphy, tectonics and petroleum systems of the Bight Basin to identify organic-rich supersequences, understand the regional maturity of these rocks and identify suitable traps and seals for hydrocarbon accumulation (Blevin *et al.* 2000;

From: HEALY, D., BUTLER, R. W. H., SHIPTON, Z. K. & SIBSON, R. H. (eds) 2012. *Faulting, Fracturing and Igneous Intrusion in the Earth's Crust*. Geological Society, London, Special Publications, **367**, 71–89. http://dx.doi.org/10.1144/SP367.6

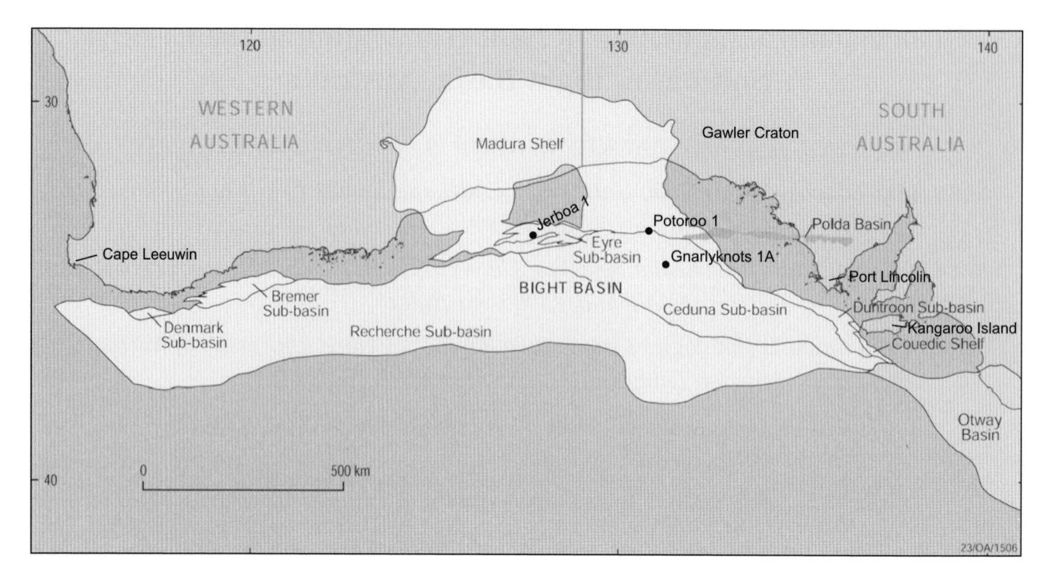

Fig. 1. Location map showing the position of the Ceduna Sub-basin in the Bight Basin on the southern Australian margin (from Totterdell & Bradshaw 2004).

Totterdell *et al.* 2000, 2008; Ruble *et al.* 2001; Sayers *et al.* 2001; Somerville 2001; Struckmeyer *et al.* 2001, 2002; Krassay & Totterdell 2003; Totterdell & Krassay 2003; King & Mee 2004; Totterdell & Bradshaw 2004; Tapley *et al.* 2005).

In this paper, we demonstrate that extensional faults on the delta top are optimally oriented for fault reactivation, based on the current understanding of the *in situ* stress regime in the Ceduna Sub-basin. Using 3D seismic data and Poly3D© (Maerten

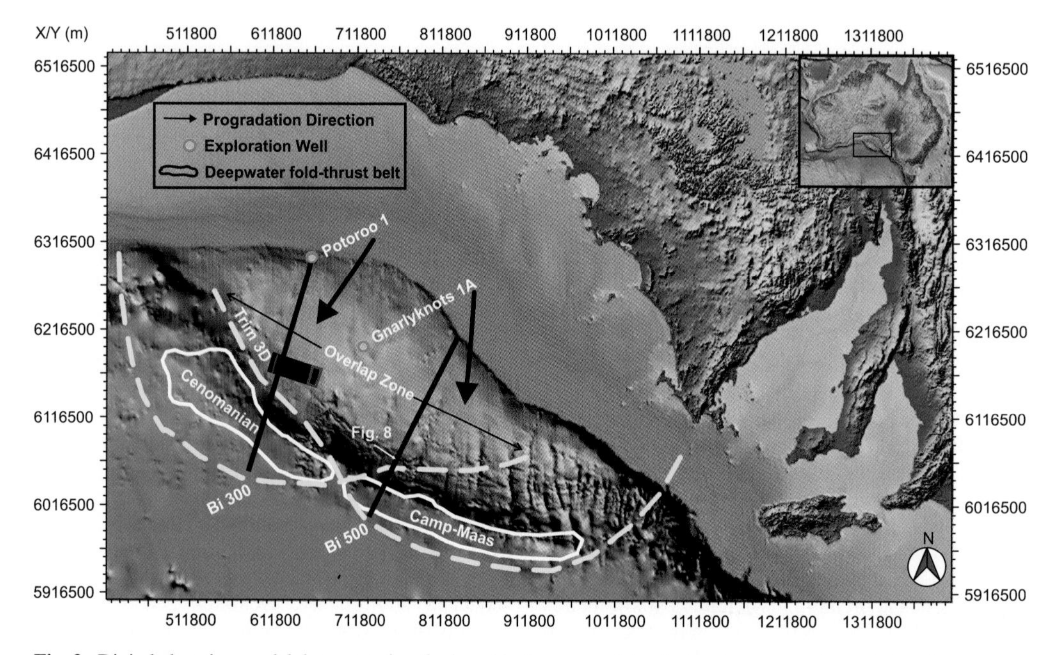

Fig. 2. Digital elevation model demonstrating the broad bathymetry of the Ceduna Terrace, outline of delta lobes and overlap zone (yellow dashed line) with approximate progradation direction, location of delta–deepwater fold–thrust belts (DDWFTBs), well locations and location of seismic surveys used in this study.

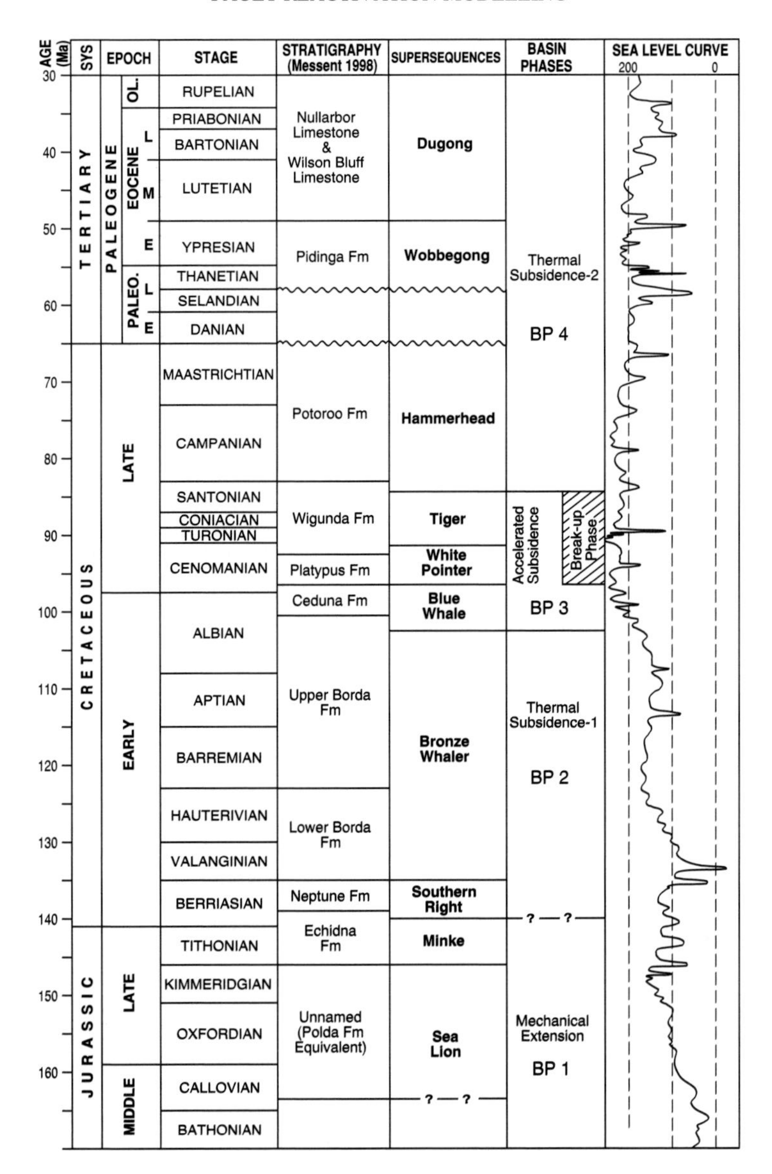

Fig. 3. Sequence stratigraphic framework (supersequences) for the Bight Basin with basin phases represented by numbers 1–4. Global sea level curve on right side from Haq *et al.* (1988). Modified from Krassay & Totterdell (2003).

et al. 2005, 2009), we demonstrate the geometry and reactivation potential of the extensional faults in the delta top. We apply this model as a proxy to further assess the reactivation risk of the extensional faults for the regional 2D seismic data.

Tectono-stratigraphic framework

The Bight Basin is located in the Great Australian Bight (Fig. 1) and demonstrates a broad bathymetric terrace (the Ceduna Terrace; Fig. 2) which forms the continental slope and rise of the southern Australian margin (Totterdell *et al.* 2000). The Bight Basin developed during Late Jurassic–Early Cretaceous rifting and continued to develop with the break-up of Gondwana, separating Australia and Antarctica during the Late Cretaceous (Totterdell *et al.* 2000; Sayers *et al.* 2001). Late Cretaceous break-up resulted in the separation of the basin into three structurally controlled sub-basins: the Eyre, Ceduna and Recherche (Fig. 1). The Ceduna Sub-basin is the largest depocentre and the focus of this study

(Totterdell *et al.* 2000; Krassay & Totterdell 2003; Fig. 1). The sub-basins are separated by NW-striking accommodation zones, which were formed by the rifting and crustal thinning of the Australian Plate (Stagg *et al.* 1990; Willcox & Stagg 1990; Totterdell *et al.* 2000).

The Ceduna Sub-basin is bound to the north by Proterozoic and older terranes and to the east by the Proterozoic of the Gawler Craton (Krassay & Totterdell 2003; Fig. 1). The Eyre Sub-basin defines the western limit of the Ceduna Sub-basin while the Duntroon Basin forms the SE boundary; to the south it tapers out onto the thin oceanic crust of the South Australian Abyssal Plain (Fraser & Tilbury 1979; Bein & Taylor 1981; Willcox & Stagg 1990; Hill 1995; Norvick & Smith 2001; Totterdell & Krassay 2003; Fig. 1).

The stratigraphy of the Ceduna Sub-basin comprises 10 supersequences which relate to the four Bight Basin phases first described by Totterdell *et al.* (2000; Fig. 3). The sub-basin evolution involved two successive periods of extension and thermal subsidence that commenced in the Late Jurassic (Totterdell *et al.* 2000; Totterdell & Krassay 2003). Basin Phase 1 marks the onset of sedimentation in the Bight Basin during the Middle–Late Jurassic extension, with the contemporaneous formation of a series of extensional and transtensional half-graben structures. Pre-existing basement trends appear to have focused the deformation during this phase (Totterdell *et al.* 2000). Basin Phase 2 records the slow thermal subsidence during the Early Cretaceous, which ended abruptly with the onset of rapid subsidence (Basin Phase 3) during the Late Albian (Totterdell *et al.* 2000). Seafloor spreading in the Late Santonian caused the break-up of Australia and Antarctica and marks the end of Basin Phase 3 in the Bight Basin. A period of thermal subsidence (Basin Phase 4) followed, which marks the initiation of the Late Santonian South Australian passive margin (Totterdell & Krassay 2003). These four Bight Basin phases are observed in the Ceduna Sub-basin (Fig. 3).

Of the ten supersequences defined in Totterdell *et al.* (2000), those of most importance to this study are the stratigraphic units that comprise the two DDWFTB systems and their respective detachments. At the northern margin of the Ceduna Sub-basin these include: the Albian Blue Whale supersequence; Cenomanian White Pointer supersequence; Turonian–Santonian Tiger supersequence; Campanian–Maastrichtian Hammerhead supersequence; and the Cenozoic rocks above the intra-Maastrichtian regional unconformity (Fig. 3). Basin modelling of the distal parts of the sub-basin indicates that the Blue Whale, White Pointer and Tiger supersequences are most prospective in terms of source rocks (Blevin *et al.* 2000).

The supersequence lithology descriptions below come from Totterdell *et al.* (2000), who used well data from the Bight, Eyre and Duntroon sub-basins. The oldest unit relevant to this study is the Middle Albian–Cenomanian Blue Whale supersequence, which records the first major marine flooding event with the deposition of restricted marine siltstones nearest the palaeoshelf and an inferred thick package of marine mudstones in the basin. This package of marine mudstone forms the detachment for the overlying Cenomanian White Pointer DDWFTB, comprising the White Pointer supersequence (Totterdell & Bradshaw 2004). Rapid (*c.* 5 Ma) deposition of the aggradational White Pointer supersequence likely produced overpressure in the underlying ductile marine mudstone, enabling gravity-driven deformation which resulted in the formation of growth faults and toe thrusts that sole out or are rooted in the Blue Whale mudstone and form the White Pointer DDWFTB (Totterdell & Krassay 2003). The White Pointer supersequence is composed primarily of fluvial to lagoonal siltstone and mudstone intercalated with minor sandstone and coal units. Dewatering of the underlying shale resulted in a loss of overpressure (Totterdell *et al.* 2000; Totterdell & Krassay 2003) and provided the likely mechanism for the cessation of all gravity-driven deformation.

Overlying the White Pointer DDWFTB is the Turonian–Santonian Tiger supersequence, containing an aggradational package of what is suggested to be marine shale based on seismic sequence stratigraphy and extrapolation of well data from Potoroo-1 in the northern Ceduna Sub-basin (Totterdell *et al.* 2000). The unit is heavily faulted in the Ceduna Sub-basin due to reactivation of older faults (Mulgara Fault Family), controlled by the underlying ductile shales of the Blue Whale supersequence. The timing of this fault reactivation is well constrained to the Late Santonian as this faulting had ceased before deposition of the overlying Hammerhead supersequence (Totterdell *et al.* 2000).

The Santonian–Maastrichtian Hammerhead supersequence has an overall progradational to aggradational character and is believed to be almost entirely composed of channel sandstones near the present-day shelf in the Ceduna Sub-basin (Totterdell *et al.* 2000). Three stratigraphic sequences have been interpreted in the Hammerhead supersequence from seismic data: the lowermost and middle sequences are progradational and the uppermost sequence is aggradational. The lowermost and middle sequences of the Hammerhead supersequence show seismic facies character that lack strong stratal geometries on the shelf and are homogeneous. This seismic character is consistent with a basinward change to a more shale-rich system. The 19 Ma period of progradation of the Hammerhead

DDWFTB across the palaeoshelf resulted in the initiation of gravity-driven deformation with the top of the underlying Tiger supersequence forming the detachment surface (Totterdell & Krassay 2003; King & Backé 2010). The gravitational tectonics initiated at the palaeoshelf margin, where the Hammerhead supersequence is thickest (*c.* 5000 m at this location; Totterdell & Krassay 2003).

The Hammerhead supersequence is overlain by the Paleocene–Early Eocene Wobbegong supersequence, which consists of marginal marine-to-deltaic sandstone and siltstone that was deposited on a hiatus which is believed to represent 5–7 Ma. Above this, the Middle Eocene–Pleistocene Dugong supersequence is present, consisting of a basal coarse sandstone and thick uniform cold water carbonate succession related to the development of a stable carbonate shelf.

Structural geometry of delta–deepwater fold–thrust belts

The structural geometry of delta–deepwater fold–thrust belts has been widely studied where they form prolific hydrocarbon provinces such as the Niger (e.g. Doust & Omatsola 1990; Morley & Guerin 1996; Briggs *et al.* 2006) and Baram DDWFTBs (e.g. Tingay *et al.* 2005, 2009; King *et al.* 2010; Morley *et al.* 2011). Delta–deepwater fold–thrust belts typically form at continental margins where rapid progradation of deltaic sediments over salt or water-prone mud results in overpressure development (in water-prone mud) and deformation under gravitational forces (Morley & Guerin 1997; Morley 2003; Rowan *et al.* 2004). The result is a broad segregation of the delta into extensional and compressional provinces, whereby margin-parallel gravitational extensional stresses on the delta top drive downdip margin-normal compressional stresses in the deepwater fold–thrust belt (Yassir & Zerwer 1997; Corredor *et al.* 2005; King *et al.* 2009).

The delta top is typically characterized by regional and counter-regional listric normal growth faults that sole out at the level of the underlying pro-delta sediments (salt or shale; Mandl & Crans 1981; Morley & Guerin 1996; Rowan *et al.* 2004). These extensional structures are balanced by the deepwater fold–thrust belt, which is composed of imbricate thrust sheets and associated fault-propagation folds rooted at the basal detachment (Morley & Guerin 1996; McClay *et al.* 2003; King *et al.* 2009). The extensional and compressional provinces are commonly separated by a transitional province that is characterized by detachment folds and reactive, active and passive diapirs of shale or salt (Jackson *et al.* 1994; Morley & Guerin 1996).

Shale detachment horizons are dependent on the development and maintenance of overpressure, with rapid progradation of delta sediments over the water-prone mud contributing the initial disequilibrium compaction overpressure of the muds (Morley *et al.* 2008, 2011). Overpressure is not required for salt as it is naturally weak and mobile, whereas shale requires overpressure to induce weakness and mobility (Davis & Engelder 1985; Morley & Guerin 1996; Costa & Vendeville 2002; Rowan *et al.* 2004; Bilotti & Shaw 2005). Subsequent inflationary overpressures can also develop, most often by dewatering during the smectite–illite transition and volume increase due to hydrocarbon generation; these serve to maintain overpressure (Morley & Guerin 1996; Osborne & Swarbrick 1997; Morley *et al.* 2011).Variations in sediment supply and detachment parameters [such as lithology and thickness of the detachment horizon(s)] and development of overpressure are some of the key factors that contribute to the varied structural styles observed in DDWFTBs worldwide.

Delta–deepwater fold–thrust belts often form excellent hydrocarbon provinces due to the generation of hydrocarbons from marine sediments and an abundance of trapping mechanisms in the deltaic sediments, as a result of the gravity-driven deformation (Morley *et al.* 2011). There are numerous structures formed in these settings, many of which have excellent hydrocarbon-trapping potential. For example, rollover antiforms and tilted fault blocks develop in the extensional province (Finkbeiner *et al.* 2001). Hanging-wall and footwall folds developed due to thrust fault propagation in a deepwater fold–thrust belt also form excellent fault-independent and -dependent traps (Ingram *et al.* 2004; Cobbold *et al.* 2009). The imbricate thrusts can provide fluid pathways from the often hydrocarbon-generating shale detachment horizon into the overlying traps (Ingram *et al.* 2004; Cobbold *et al.* 2009).

Regional structural style of the Ceduna Sub-basin

The Ceduna Sub-basin contains two Cretaceous DDWFTBs: the late Albian–Cenomanian White Pointer and the Late Santonian–Maastrichtian Hammerhead. The geometry of the Ceduna DDWFTBs was first recognized by Boeuf & Doust (1975) and Fraser & Tilbury (1979) and has since been investigated by several authors including Totterdell & Krassay (2003), Totterdell & Bradshaw (2004), Espurt *et al.* (2009), King & Backé (2010) and MacDonald *et al.* (2010).

The Ceduna Sub-basin provides a unique opportunity to study two separate progradational

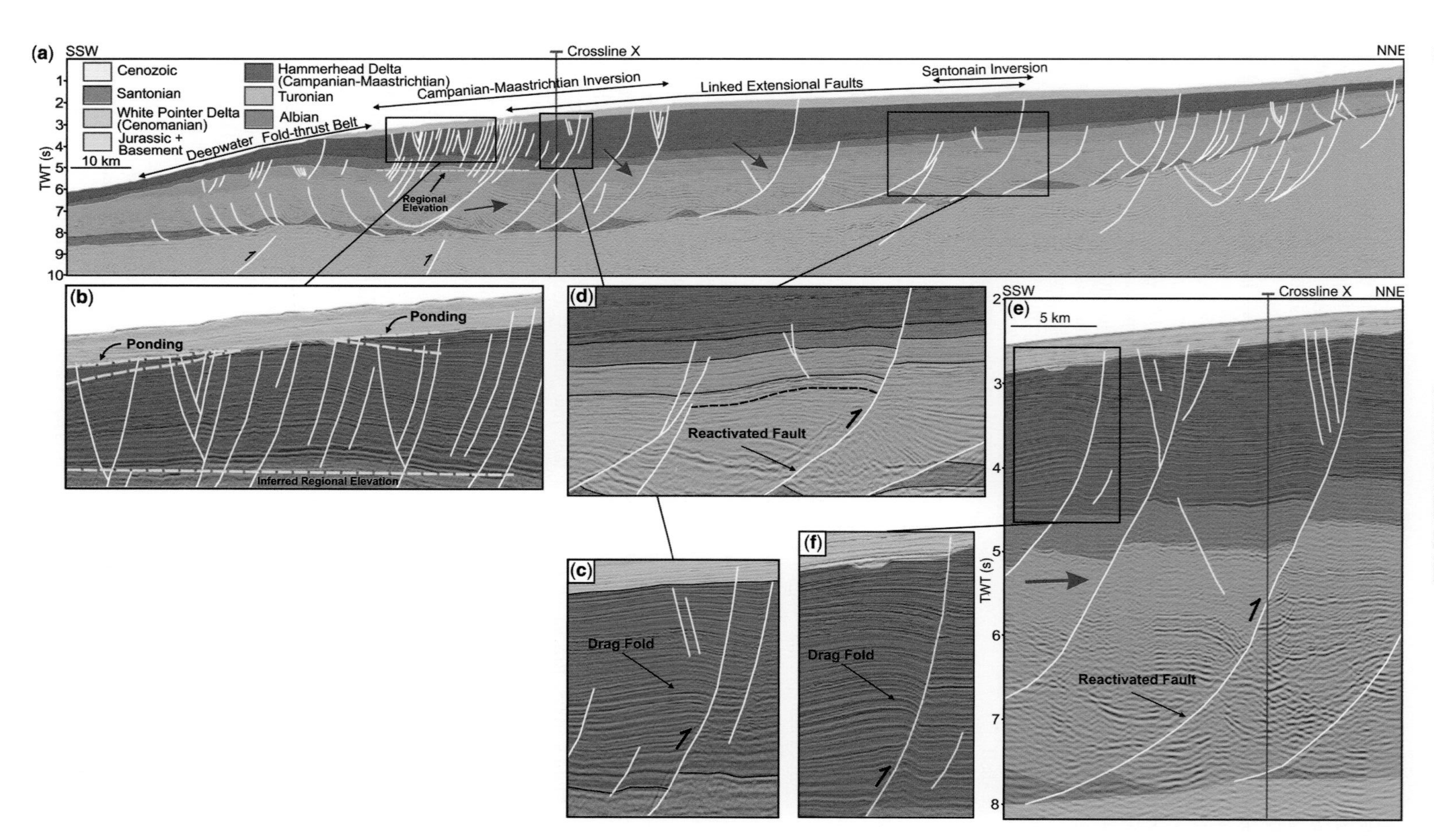

Fig. 4. Seismic line Bi 300. (**a**) Sequence stratigraphic framework for the western Ceduna Sub-basin including the White Pointer DDWFTB (blue) with extensional (delta top) and shortening (delta toe) provinces. Two phases of inversion are indicated: (**b**) the Campanian–Maastrichtian (purple) inversion resulting in broad uplift and erosion and (**c**) selected fault reactivation forming an inversion anticline. Inset (b) demonstrates that the boundary between the Santonian (orange) and Campanian (purple) is structurally elevated above the regional level (yellow dashed line) while there is minor erosion and ponding on the flanks of the inversion anticline at the top of the Campanian–Maastrichtian

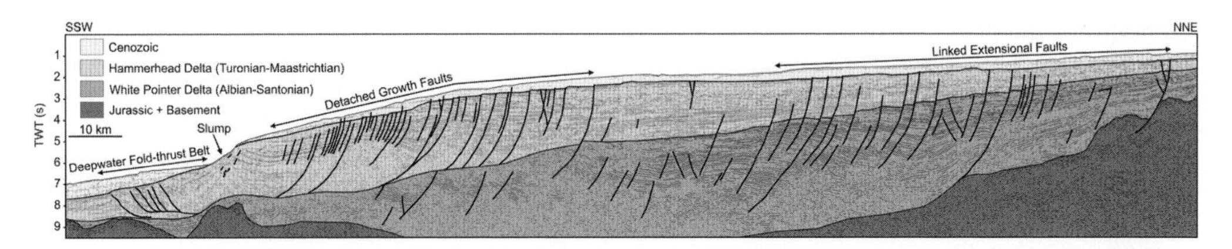

Fig. 5. Seismic line Bi 500 from the eastern Ceduna Sub-basin demonstrating growth of deeper White Pointer DDWFTB extensional faults upwards into the Hammerhead DDWFTB nearest the margin. Outboard of this, extensional listric growth faults of the Hammerhead DDWFTB detach above Turonian–Santonian strata and are linked downdip to a deepwater fold–thrust belt (delta toe). The White Pointer supersequence images well-developed extensional faults, but is missing a linked downdip delta toe.

DDWFTB systems that are independent in size, shape and structural geometry (MacDonald *et al.* 2010). Both DDWFTBs exhibit well-developed gravity-driven deformation structures, making them analogous to other delta systems around the world (Totterdell & Krassay 2003; MacDonald *et al.* 2010). The structural interpretation of the White Pointer and Hammerhead DDWFTBs is based on newly acquired 2D BightSPAN© seismic data and a public domain Trim3D seismic dataset. Both 2D and 3D seismic datasets were interpreted in the time domain while the fault model from the 3D dataset was depth converted. Depth conversion employed a simple velocity model due to the lack of well data available in the Ceduna Sub-basin. Water depth varies considerably along the regional 2D seismic lines from several hundred metres to over four kilometres, which can result in dramatic differences in dip of structures towards the basin due to the simple velocity model used in depth conversion. It is important to note that true dip and interlimb angles are apparent in the time domain. However, the overall structural style will not be modified by migration of the data in the depth domain. In this study the fault reactivation analysis is based on a depth-migrated fault model where true, rather than apparent, dips and geometries are displayed. This fault model has been used as a proxy for similar geometry faults elsewhere in the basin, which are imaged in the time domain on the regional 2D seismic lines.

In this study, we present two representative 2D seismic lines that demonstrate the regional variation in the structural style of the Late Cretaceous DDWFTB systems (Figs 4 & 5). In addition to this, we examine a 3D dataset on the delta top to constrain fault geometry in three dimensions. Results from previous interpretation of the public domain 2D seismic data covering the Ceduna Sub-basin were considered when picking the representative 2D regional seismic lines.

Seismic line Bi 300 (Fig. 4a) images the western side of the Ceduna Sub-basin and illustrates a linked system of extension and shortening within the Cenomanian White Pointer DDWFTB. This linked system is overlain by the Tiger supersequence (Turonian–Santonian) and the Hammerhead supersequence (Campanian–Maastrichtian). However, in the western Ceduna Sub-basin the Hammerhead supersequence did not develop into a DDWFTB system (MacDonald *et al.* 2010). In the eastern Ceduna Sub-basin, seismic line Bi 500 (Fig. 5) illustrates a linked system of extension and shortening within the Hammerhead supersequence, forming a DDWFTB system. In this section, the Cenomanian White Pointer supersequence is purely an extensional system that is lacking a downdip deepwater fold–thrust belt (Espurt *et al.* 2009; MacDonald *et al.* 2010). It therefore cannot be classified as a linked DDWFTB system at this location.

Examination of additional regional 2D seismic data indicates that the White Pointer and Hammerhead DDWFTBs prograded in slightly different directions based on the positions of their deepwater fold–thrust belts (Fig. 2). Overlap occurs between the lobes on the delta top (Fig. 2; yellow dashed line) due to proximity to the source and the deeper Mulgara Fault System, which developed in the White Pointer supersequence, is shared at this location (MacDonald *et al.* 2010). Linkage between the White Pointer, Tiger and Hammerhead supersequences on the delta top is achieved via the Mulgara Fault System, whereby optimally oriented faults were reactivated as the Hammerhead

Fig. 4. (*Continued*) (purple) strata. (**d**) The centre fault block in inset demonstrates a Santonian aged inversion anticline. (**e**) Along-strike seismic data demonstrates the continuation of the inversion anticline imaged in inset (c) (red vertical line where projected crossline X would intersect both). (**f**) A second inversion anticline with excellent fault-independent closure is shown basinwards of the structure imaged in (e). Scale and legend for inset images as per (a). Red arrows indicate examples of growth faults that show no evidence for inversion.

supersequence prograded over the existing White Pointer system. This selective reactivation is evidenced by the growth strata in the Tiger and Hammerhead supersequences and is localized to areas where the sedimentary wedge is thickest (Figs 4a, e & 5). In areas where the Mulgara Faults are observed to propagate through the Tiger and Hammerhead supersequences but do not demonstrate evidence for growth, these faults are interpreted as a result of ongoing compaction of the existing White Pointer DDWFTB below. To the east and west of the overlap zone between the DDWFTB systems, the linked faults become less prominent and the extensional faults that belong to each respective system become more abundant. This is observed in Figure 5 where the outboard growth faults within the Hammerhead DDWFTB detach at the level of the Tiger supersequence, rather than linking to the existing Mulgara Fault System below.

The limited 3D seismic survey in the Ceduna Sub-basin is located in the overlap zone (Fig. 2) where the Mulgara Fault System has been reactivated by the presence of the Tiger and Hammerhead supersequences above. This presents an excellent opportunity to map these linked faults and determine the reactivation risk by applying the present-day stress regime of the Bight Basin (after Reynolds *et al.* 2003) to the 3D fault models using Poly3D© (Maerten *et al.* 2005), a 3D boundary element code for heterogeneous, linear and elastic whole or half-space developed by Igeoss/Schlumberger.

Inversion in the Ceduna Sub-basin

Although the southern margin of Australia is a rifted margin, it has been subjected to episodic compressional forces since early rifting in the Late Jurassic (Teasdale *et al.* 2003; Holford *et al.* 2011). These forces have been attributed to ridge push, which imposes a compressional stress on a rifted margin upon initiation of seafloor spreading (cf. Bott 1991), as well as compressional forces transmitted through the Indo-Australian plate via collisions along its northern and eastern boundaries during the Cenozoic (Reynolds *et al.* 2002; Sandiford *et al.* 2004). Inversion structures in the Ceduna Sub-basin were first identified by Totterdell & Bradshaw (2004), who attributed the reverse reactivation of some normal faults to minor compressional forces in the basin, possibly coupled with Cenozoic magmatic activity, resulting in the formation of slight hanging-wall anticlines. The subtle inversion structures described here are interpreted as Late Cretaceous in age, and thus most likely resulted from ridge-push forces generated following progressive separation between the Australian and Antarctic plates. Of the four examples of inverted normal faults described here, three still retain net-normal displacement with all marker horizons exhibiting net-normal offset but appearing to have been uplifted above their original regional elevation (cf. Williams *et al.* 1989; Cooper & Warren 2010). Although this inversion is relatively minor, it indicates that some faults in the Ceduna Sub-basin have undergone reactivation (with opposing sense of slip) in response to changing stress field conditions, subsequent to their formation. Detailed mapping of the newly acquired 2D BightSPAN© seismic data and existing 3D seismic data has enabled identification and relative timing of two pulses of Late Cretaceous inversion in the Ceduna Sub-basin.

Santonian inversion

The oldest phase of inversion in the Ceduna Sub-basin is displayed in seismic line Bi 300 (Fig. 4a, d) whereby a Cenomanian growth fault (Mulgara Fault System) within the White Pointer DDWFTB is reactivated in the inboard part of the basin. Here, uplift and erosion of the Santonian Tiger supersequence resulted in a thinning of the Santonian above the core of the inversion anticline (Fig. 4d). The displacement on the fault is net normal, but the structure has been subjected to minor inversion as the geometry of the hanging-wall anticline is such that the south side of the fault block is structurally lower than the north side (within a south-facing growth fault). The geometry of the hanging-wall anticline in Figure 4d is unique in that whereas most of the hanging-wall rollover anticlines in the basin (Fig. 4a, red arrows) have a gentle south limb and steeper north limb, this anticline displays nearly equal dips on both limbs. These three relationships (the thinning of Santonian strata, structurally elevated boundaries and near-equal limb dips) suggest this first identifiable phase of inversion occurred within the Santonian, which directly corresponds to the onset of Late Cretaceous seafloor spreading (Sayers *et al.* 2001). Due to the timing of the inversion we propose a ridge-push mechanism for the compressional stress required to form this structure. This example provides a unique view into the history of the Ceduna Sub-basin in that the Santonian inversion, albeit minimal, is preserved. Subsequent to this phase of inversion, many of the linked Mulgara Faults in the overlap zone (Fig. 2 & 4e) would have accommodated sediment input from the Hammerhead DDWFTB. At this stage, the stress field due to external ridge-push forcing was likely overwhelmed due to the internal forces resulting from sediment loading. This may indicate that inversion in the Santonian was more widespread than what we observe today; however, the deposition of the Hammerhead supersequence would have effectively destroyed much of this evidence through subsequent 'negative inversion' or restoration of hanging-wall strata.

Campanian–Maastrichtian inversion

The most recent episode of inversion evidenced in the Ceduna Sub-basin, also attributed to be the result of a ridge-push mechanism due to contemporaneous seafloor spreading, is demonstrated in Figure 4a, b. Here, a Cenomanian growth fault (Mulgara Fault System) of the White Pointer DDWFTB has been reactivated resulting in broad uplift of the Campanian–Maastrichtian strata in the hanging wall of the fault. In Figure 4a the boundary between the Santonian Tiger supersequence and Campanian–Maastrichtian Hammerhead supersequence is structurally elevated above the regional level to the north by *c.* 0.5 s (two-way time), suggesting fairly mild inversion compared to more severe examples from the St George's Channel Basin, offshore UK (Williams *et al.* 2005) or the East Java Sea Basin (Turner & Williams 2004). The contact between the Campanian–Maastrichtian and the Tertiary strata exhibits minor erosion of the inversion anticline with slight ponding of growth strata on both flanks of the structure (Fig. 4b). Further to the north along seismic line Bi 300, there is an additional inversion anticline (Fig. 4c) which again shows net-normal displacement and is controlled by a Cenomanian growth fault (Mulgara Fault System). Here, the displacement occurs primarily near the top of the reactivated growth fault, resulting in a characteristic drag-fold hanging-wall anticline (Fig. 4c). There therefore exists evidence of two faults in seismic line Bi 300 (Fig. 4a), with Campanian–Maastrichtian-aged inversion anticlines at a more distal location to the Santonian-aged inversion anticline described above.

Along-strike to the west of Figure 4a, a crossline from the Trim 3D seismic dataset (Fig. 2 red line; Fig. 4e) also images the same fault and associated inversion anticline interpreted in Figure 4c. The anticline has undergone more displacement here, although still retaining net-normal offset, and has steeper limb dips (Fig. 4g). Drag-fold geometry is observed at the Cenomanian–Turonian, Turonian–Santonian and Santonian–Campanian boundaries along the fault in this location, indicating increased reverse displacement along the fault to the west. In addition to this structure, Figure 4e, f demonstrates that the adjacent basinwards fault has undergone slight reactivation resulting in an inversion anticline. This anticline also has a net-normal displacement which occurs nearest the top of the fault, indicating the continuity of these inversion anticlines along-strike for *c.* 12 km. The inverted growth faults described above have a very different geometry to the other growth faults in the basin, which are indicated in Figure 4a by the red arrows.

The inversion structure imaged in Figure 4c, f displays potential to trap hydrocarbons for a number of reasons: it displays four-way closure along-strike with a significant fault-independent volume (from 3D seismic); it is linked to potential source rocks of the Albian Blue Whale, Cenomanian White Pointer and Turonian–Santonian Tiger supersequences; and formed in the upper Campanian or Maastrichtian. This age is based on the thickness of the Hammerhead supersequence and position of the inverted strata within the upper portion of this sedimentary package (Fig. 4c). Some structures with similar geometries in the Otway Basin (to the east of the Ceduna Sub-basin) contain significant volumes of methane gas, although these structures formed during younger Mid–Late Cenozoic inversion episodes (Holford *et al.* 2010).

From a regional standpoint, seismic line Bi 500 (Fig. 5) does not reveal any evidence for inversion which may be attributed to the geometry of the Cenomanian faults in the eastern part of the Ceduna Sub-basin. Here, the Mulgara Fault System is not commonly linked to the above fault system within the Hammerhead DDWFTB. On the eastern side of the sub-basin, the Hammerhead DDWFTB is composed of growth faults that sole out or detach within the formerly mobile shales of the Turonian–Santonian Tiger supersequence. It is plausible that any minor pulse of Santonian-aged reactivation and inversion in this area may be obscured by the thickness and ductility of the Tiger supersequence, given that there is a well-established gravity-sliding DDWFTB actively deforming above it. Any inversion structure that began to propagate upwards into the Tiger or Hammerhead supersequences may subsequently have been effectively accommodated by remobilization of the overpressured shale. Other possible explanations for the lack of inversion of the Cenomanian White Pointer faults (Mulgara Fault System) in the east may be the slight change in strikes of the fault system there, rotating azimuths slightly from their position in the overlap zone between the two DDWFTB systems (i.e. Fig. 4a) or the occurrence of sub-seismic scale inversion that has not been imaged (Holford *et al.* 2009).

Fault orientation analysis

The Trim3D seismic survey (Fig. 2) was used to constrain the orientation of the extensional faults (Fig. 6a) in the overlap zone between the two DDWFTB systems. The major faults (Fig. 6b, c) within the eastern side of the Trim3D survey were mapped and a 3D fault model was constructed for both orientation and fault reactivation analysis. The faults within the 3D seismic survey were simplified such that only the large synthetic (Faults 2 and 4) and antithetic (Faults 1 and 3) faults were mapped (Fig. 6b, c). The fault model was then depth converted to show true rather than apparent orientations, and imported into 3D MOVE© for the structural analysis (Fig. 6a).

The analysis of the depth-converted faults shows an excellent correlation between the orientations of both the synthetic and antithetic faults (Fig. 6a). Faults 2 and 4, the synthetic growth faults, show a wider dip-azimuth distribution on the rose diagrams yet very similar averages (Fault 2: 228°; Fault 4: 224°; Fig. 6a) and a larger point cluster (Fig. 6a) due to the larger size and strike length of these faults. Fault 4 displays the largest point cluster, due to the slightly deeper detachment level (Fig. 6b). The mean principal poles for Faults 2 and 4 are also slightly different, again due to the larger surface area and increased variability in dip of Fault 4.

The antithetic faults are very similar in orientation (dip-azimuth) in both the rose diagrams with 052° (Fault 1) and 048° (Fault 3; Fig. 6a) and in terms of point cluster size (Fig. 6a). The mean principal pole for Faults 1 and 3 are almost exactly the same, indicating very similar geometries along both strike and dip of the faults (Fig. 6b, c). The similarity of these two structures, with nearly identical orientations, is observed elsewhere in the Ceduna Sub-basin in regional 2D seismic data (Figs 4a & 5).

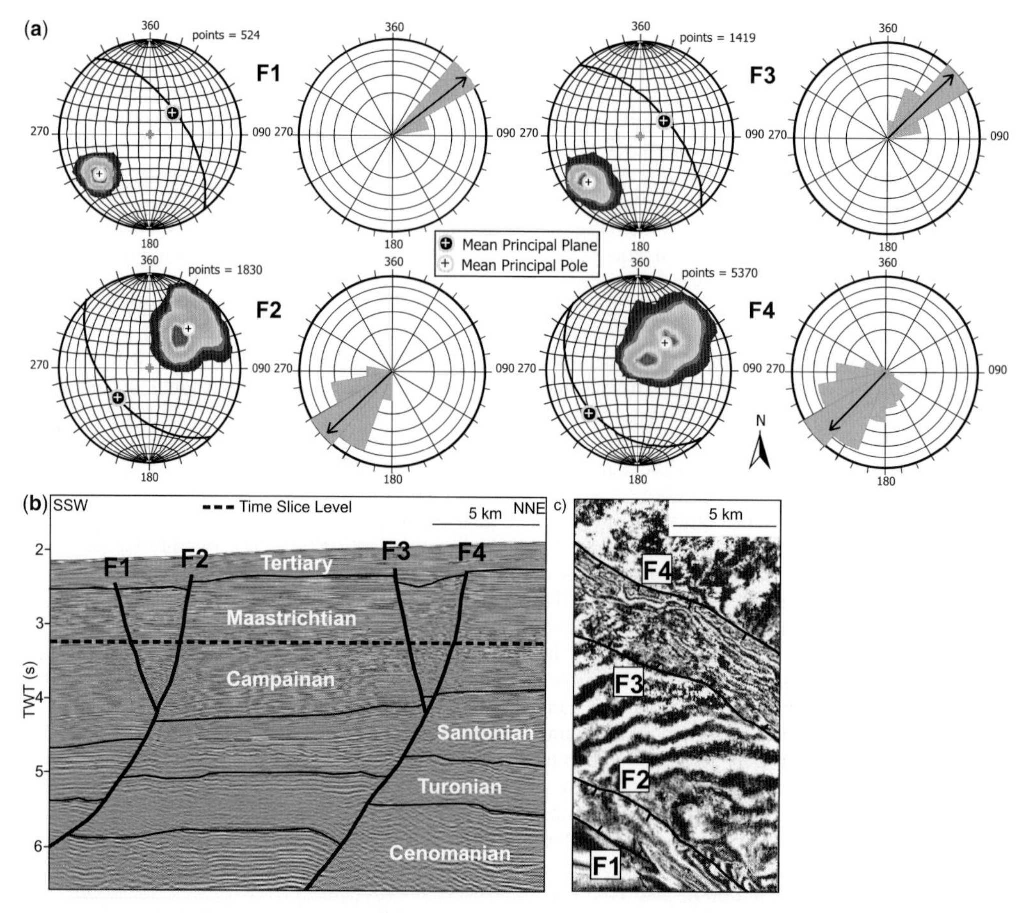

Fig. 6. (**a**) Fault analysis data plotted on equal-area stereographic projections for 3D fault model shows similarity between antithetic (Faults 1 and 3) and synthetic (Faults 2 and 4) faults with rose diagram (dip-azimuth) and contoured poles to plane (dip-azimuth). (**b**) Interpreted seismic section from Trim3D showing extensional Faults 1–4 in 2D with black dashed line indicating observation grid level at 1000 m below seabed. (**c**) Time slice from Trim3D survey at −1000 m observation level, showing extensional faults used for 3D fault model.

Regional stress regime of the Bight Basin

The first investigation of the *in situ* stress regime of the Bight Basin was conducted by Reynolds *et al.* (2003), whereby the stress tensor was calculated from the available well data in the Bight Basin and a fault reactivation analysis was completed. The absence of additional well data in the basin since this work was completed limits our ability to further refine the *in situ* stress analysis. However, with newly acquired 2D and 3D seismic data allowing for detailed investigation of the 3D fault geometries and new techniques for modelling fault reactivation using Poly3D©, we can more accurately predict the risk of reactivation in the basin. The reactivation risk for a well-constrained fault model can then be used as a proxy for other faults in data-poor areas of the basin. The following fault reactivation risk analysis is therefore based entirely on the *in situ* stress regime as it was calculated in Reynolds *et al.* (2003).

The *in situ* stress regime was calculated using the drilling and logging data from ten open file exploration wells within the Bight Basin, particularly the Duntroon and eastern Ceduna sub-basins and the Polda Trough to the north (Reynolds *et al.* 2003; Figs 1 & 7). The variation in water depth across the Bight Basin from several hundred metres to over four kilometres posed a significant problem for calculating the total stress in the basin; the analyses are therefore based on effective stress (total stress minus pore pressure; Reynolds *et al.* 2003).

Reynolds *et al.* (2003) determined the maximum horizontal stress (S_{Hmax}) orientation in the Bight Basin by interpreting borehole breakout directions in high-resolution 4-arm dipmeter (HDT) logs from four wells (Fig. 7)–Echidna-1 (Duntroon Sub-basin), Duntroon-1, Platypus-1 (Ceduna Sub-basin) and Columbia-1 (Polda Basin)–and image log data (Formation Microscanner, FMS) from two wells (Greenly-1 and Borda-1; eastern Ceduna Sub-basin).

To determine the S_{Hmax} value for the Bight Basin, Reynolds *et al.* (2003) used a total of 78 breakouts with a combined length of 1208 m from the six wells in the basin covering a depth range 1460–4791 m below rotary table. The average S_{Hmax} orientation calculated from the six wells was N130°E (130). The interpreted stress data from Reynolds *et al.* (2003) is plotted on Figure 7 along with the Australian stress map, demonstrating the sparse data coverage in the western Ceduna Sub-basin. The average S_{Hmax} orientation of N130°E (130) for the available wells in the Bight Basin is consistent with S_{Hmax} orientations in the Otway Basin to the east (Fig. 7).

The stress trajectories or regionally averaged stress orientations for the Australian stress field have been modelled by Hillis & Reynolds (2000) and plotted in Figure 7. These trajectories allow for a better understanding of the regional stress field over the entire Bight Basin rather than any one sub-basin. On the western side of the Bight Basin, the stress trajectories indicate a more east--west orientation which reflects the (data constrained) east–west S_{Hmax} orientation in the Perth Basin to the west (Hillis & Reynolds 2000; Reynolds & Hillis 2000; King *et al.* 2008). Due to the lack of available well data in the western Bight Basin, Reynolds *et al.* (2003) were unable to verify if the S_{Hmax} orientation rotates to an east–west orientation in the western part of the Bight Basin. An average S_{Hmax} orientation of N130°E (130) was therefore used for the entire Bight Basin.

Reynolds *et al.* (2003) determined the magnitudes of the vertical stress (S_v), minimum horizontal stress (S_{hmin}) and S_{Hmax} from the petroleum wells in the Bight Basin. They integrated density logs and check-shot data to determine S_v at 10.5 MPa km^{-1}. They used leak-off tests and formation integrity tests to calculate S_{hmin} at 6.0 MPa km^{-1} and used frictional limits calculations to determine the upper limit of S_{Hmax} at 18.7 MPa km^{-1}. However, Reynolds *et al.* (2003) were unable to define S_{Hmax} any further. They therefore concluded that the Bight Basin may demonstrate a normal fault stress regime ($S_v > S_{Hmax} > S_{hmin}$), a borderline normal fault to strike-slip fault stress regime ($S_v \geq S_{Hmax} > S_{hmin}$) or a strike-slip fault stress regime ($S_{Hmax} > S_v > S_{hmin}$; Fig. 8; Table 1).

For details of the calculation of individual stress tensor components and their magnitudes, including the vertical stress, minimum horizontal stress, maximum horizontal stress and pore pressure, we refer the reader to Reynolds *et al.* (2003).

Fault reactivation in the Ceduna Sub-basin

Previous work completed on fault reactivation in the Bight Basin utilized the FAST technique (Fault Analysis Seal Technology, Mildren *et al.* 2002; Reynolds *et al.* 2003). In the FAST technique, the risk of fault reactivation is determined using the stress tensor (Mohr circle) and fault-rock strength (failure envelope). This technique was applied to regional faults from the Bight Basin whereby strike and dip data were determined from regional 2D seismic mapping. The analysis completed in the Reynolds *et al.* (2003) work is therefore based on generalized fault orientations rather than a precise fault model from 3D seismic data, which includes combined mechanical interaction due to fault slip.

To evaluate fault reactivation, Reynolds *et al.* (2003) investigated three stress regimes: (1)

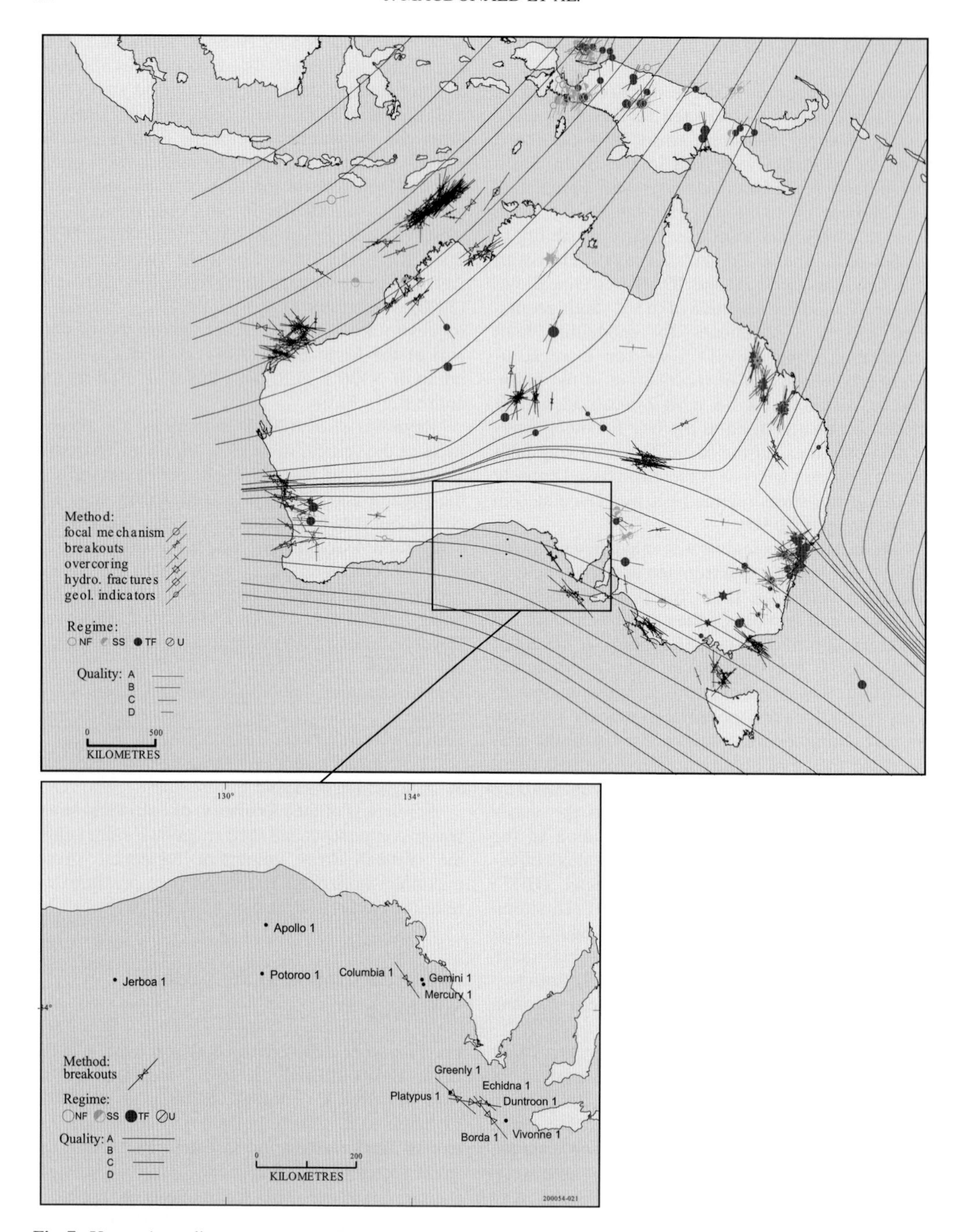

Fig. 7. Upper: Australian stress map (A–D quality) with the Bight Basin stress data with superimposed stress trajectory map from Hillis & Reynolds (2000) demonstrates the regional trends across the Australian continent. The S_{Hmax} orientation for the Bight Basin is reasonably consistent with the stress trajectories in the region. Lower: smaller-scale stress map of the Bight Basin showing A–D quality stress indicators and well names used in this study. Orientation and length of vector represents the data quality, while wells with no data or E-quality data are represented by a dot (after Reynolds *et al.* 2003).

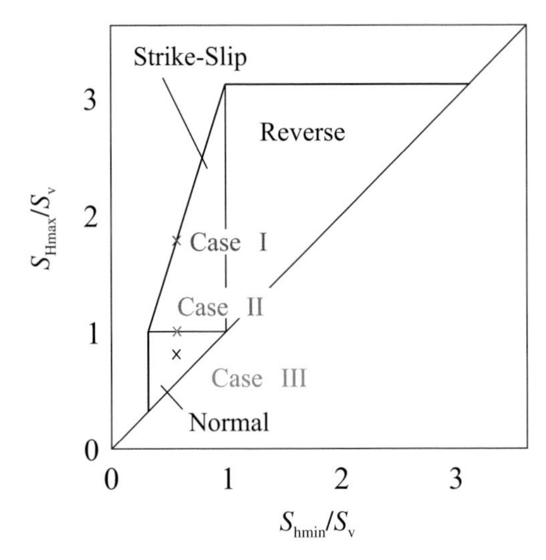

Fig. 8. Location in stress space of the three *in situ* stress cases evaluated (Table 1; after Reynolds *et al.* 2003).

strike-slip fault stress regime ($S_{hmin} < S_v < S_{Hmax}$); (2) normal fault stress regime ($S_{hmin} < S_{Hmax} < S_v$); and (3) borderline normal fault to strike-slip fault stress regime (Fig. 8; Table 1). The magnitude of the *in situ* stress field for the three cases was determined at a depth of 1000 m below seabed; an observation grid at this depth was therefore used for the Poly3D© analysis.

Geomechanical stress analysis

We used the boundary element code Poly3D to characterize the reactivation risk along the faults interpreted from the 3D seismic cube in the Ceduna Delta. Poly3D© utilizes polygon elements and linear elasticity theory (Thomas 1993; Maerten *et al.* 2000, 2002; Fiore *et al.* 2006). Models are simplified in the boundary element method (BEM) as only the discontinuities (such as faults) are gridded and therefore taken into account in the calculation. The faults were meshed using a 500 m triangular mesh size, which was a good compromise between precision and computation time. Specific boundary conditions were applied, which govern how the elements should respond to far-field stress or strain states. Different attributes can be used or studied to describe the propensity of a fault to slip under a given stress field (e.g. Morris *et al.* 1996; Ferrill & Morris 2003; Healy 2009; Morris & Ferrill 2009). Here, we used the total displacements (total, vertical or horizontal displacements) calculated along the fault planes when subjected to the known present-day stress field in order to estimate the reactivation potential of the listric faults imaged by the Trim3D 3D seismic cube. The displacements were computed for cohesionless fault surfaces, with boundary conditions set on the triangular boundary elements making up the faults as follows.

- The traction components parallel to the element plane in the dip direction and in the strike direction are both set to zero so the element surfaces may slip freely.
- The displacement discontinuity component normal to the element plane is set to zero to avoid opening or interpenetration of the fault surfaces.

After depth conversion using a simple velocity model, the faults surfaces were imported and remeshed in Poly3D©. We also created an observation grid where attributes such as total displacement, direction and magnitudes of stresses were also computed close to and away from the faults. The results of the models allow the reactivation potential of the main faults in the Trim3D 3D seismic cube to be investigated.

The fault reactivation analysis is based on two independent fault sets (Fig. 6b) from the overlap zone between the two DDWFTB systems (Fig. 2). These two fault sets were selected as a proxy as they are well imaged in the Trim3D seismic dataset (Fig. 2) and are representative of growth fault geometries on the delta top(s). To utilize the calculated stress magnitudes for each component of the stress tensor (Table 1), we have used the same observation level of 1000 m below seabed

Table 1. *Calculated stress magnitudes of Reynolds* et al. *(2003), used to model fault reactivation risk with the geomechanical modelling software Poly3D©*

Case	S_{Hmax} (MPa)	S_v (MPa)	S_{hmin} (MPa)	Fault regime	S_{Hmax} orientation (°N)
I	18.7	10.5	6.0	Strike-slip	130
II	10.5	10.5	6.0	Strike-slip/normal	130
III	8.5	10.5	6.0	Normal	130
I	18.7	10.5	6.0	Strike-slip	090
II	10.5	10.5	6.0	Strike-slip/normal	090
III	8.5	10.5	6.0	Normal	090

for this analysis as for the stress analysis (Fig. 6b) of Reynolds *et al.* (2003). Figure 6c represents a time slice through this observation level (1000 m below seabed), where the four faults that were analysed are shown (Faults 1–4). Faults 1 and 3 are antithetic faults while Faults 2 and 4 are synthetic growth faults (Fig. 6c).

The synthetic growth faults (Fault 2 and 4; Figs 6b & 9a) detach within the Albian Blue Whale supersequence and were reactivated, likely after the Santonian, allowing for growth to occur along the faults within the Campanian–Maastrichtian section (Fig. 4a, e). The antithetic faults formed as a result of increased offset on the synthetic faults and are relatively shallow and younger (Figs 6b & 9a). The fault model was depth converted, triangulated and imported into Poly3D© and three fault regime scenarios were run on the model (Figs 8 & 9a–f).

The orientation of S_{Hmax} was set at N130°E (130) for each fault regime scenario and the calculated stress magnitudes from Reynolds *et al.* (2003) were used (Table 1) starting with Case I strike-slip fault stress regime ($S_{hmin} < S_v < S_{Hmax}$; Fig. 9a). Case II normal fault stress regime ($S_{hmin} < S_{Hmax} < S_v$; Fig. 9b) was run next followed by Case III (Fig. 9c), the borderline normal fault to strike-slip fault stress regime. The results (Fig. 9a–c) demonstrate that variability in the stress magnitudes has only a minor effect on reactivation risk. All three cases demonstrate that the antithetic faults are optimally oriented for reactivation (Faults 1 and 3; Fig. 9a–c) while the larger synthetic growth faults are less optimally oriented for reactivation and demonstrate a moderate to low risk (Faults 2 and 4; Fig. 9a–c).

An important consideration that was not addressed in the work of Reynolds *et al.* (2003) was fault plane roughness, likely due to the lack of data and scope of the study. With the Poly3D© fault model constrained from 3D seismic data, we can demonstrate that variability in fault plane roughness affects reactivation along the strike of the fault (e.g. in Figure 9b where the east side of Fault 4 is at a high risk of reactivation in the normal fault stress regime but is only at moderate risk of reactivation in the other two scenarios; Fig. 9a, c). Depending on the orientation of the structure with respect to S_{Hmax}, the fault plane roughness can have a significant impact on reactivation risk

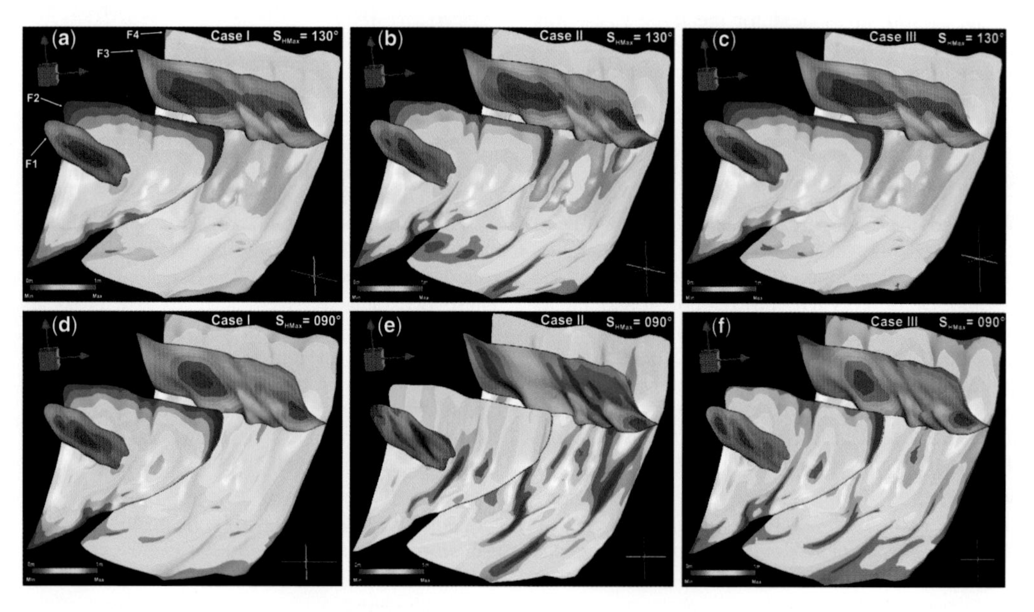

Fig. 9. Modelled 3D antithetic (Faults 1 and 3) and synthetic (Faults 2 and 4) faults from Trim3D seismic data (Figs 2 & 6). Modelled S_{Hmax} orientation of (**a–c**) N130°E (130) and (**d–f**) N090°E (090). Three cases were modelled for each S_{Hmax} orientation: (a, d) Case I strike-slip faulting stress regime ($S_{hmin} < S_v < S_{Hmax}$); (b, e) Case II normal faulting stress regime ($S_{hmin} < S_{Hmax} < S_v$); and (c, f) Case III strike-slip normal faulting boundary. Orientation in space is shown by grey cube in top left corner where blue arrow is z axis, green is y axis (north) and red is x axis. Orientation and relative magnitude of stress field are shown in bottom-right corner where blue represents maximum horizontal stress, red represents minimum horizontal stress and green represents vertical stress as per Table 1 calculated values. Contours define the modelled displacement in metres as a result of the stress applied to the boundary of the model and along the faults.

across the fault as demonstrated when the orientation of S_{Hmax} is rotated to east–west or N090°E (090; Fig. 9d–f).

Previous Finite Element Method (FEM) modelling of the Australian stress field incorporating all plate boundary forces that act on the Indo-Australian Plate indicates an *c.* east–west S_{Hmax} for the Bight Basin (Reynolds *et al.* 2003; Fig. 7). As the individual principal stress axis (averaged by black lines; Fig. 7) demonstrates a more east–west S_{Hmax} orientation in the FEM model, the Poly3D© fault model was also run with this orientation (N090°E) using the same three fault regimes and magnitudes as the initial model [with a S_{Hmax} orientation of N130°E (130)]. Here, Figure 9d–f demonstrates the same fault model, fault regimes and stress magnitudes with a rotated S_{Hmax} orientation to N090°E (090). There are two reasons for running a second model with a rotated S_{Hmax} orientation: (1) to investigate the difference in the modelled (FEM) versus calculated present-day stress field from Reynolds *et al.* (2003); and (2) to demonstrate that even a rotation of 40° from the interpreted plate scale stress trajectory (Fig. 7) would still infer that the Ceduna Sub-basin faults are at moderate to high risk of reactivation (Fig. 9d–f).

By changing the S_{Hmax} orientation to N090°E (090), we can now investigate the risk of reactivation for the three fault regimes (Case I–Case III; Fig. 9d–f). This model demonstrates that Faults 1 and 3 in Case I and Case III (Fig. 9d, f) remain at a moderate to high reactivation risk, albeit lower than for the S_{Hmax} orientation of N130°E (130) scenario (Fig. 9a, c). Case II faults demonstrate a much lower risk of reactivation in this scenario (Fig. 9e) with a low to moderate risk rather than a high risk as in the orientation of N130°E (130) scenario (Fig. 9b). The fault plane roughness has a more acute effect on reactivation risk; a modelled S_{Hmax} orientation of N090°E (090) with select areas of the fault plane demonstrates a high risk of reactivation (Fig. 9f) compared to moderate risk in the S_{Hmax} orientation of N130°E (130) scenario (Fig. 9c).

In both scenarios described so far, the antithetic faults (Faults 1 and 3) display a high risk of reactivation (aside from Fig. 9e) relative to the synthetic faults (Fig. 9). This is again a result of the fault plane roughness along Faults 1 and 3 as the 'bullseye' occurs in the same place each time (Fig. 9a–c). The uneven nature of the fault surface is represented clearly on Faults 1 and 3; Faults 2 and 4 display smoother fault planes and therefore have an overall lower risk of reactivation (Fig. 9).

The two examples of varying the S_{Hmax} orientation described above demonstrate that the orientation of the structure with respect to S_{Hmax} is more substantial for fault reactivation risk than changing the stress magnitude. This is justified by the absence of significant change between the three cases and corresponding stress magnitudes tested in the model (Fig. 9). This analysis also suggests that the moderately dipping (40–70°) segments of Faults 2 and 4 are at a higher risk of reactivation than the shallow ($<40°$) or steeply dipping ($>70°$) portions of the fault, which is in agreement with the results of Reynolds *et al.* (2003). This evidence suggests that dips of the fault surfaces do play a role in the reactivation risk. However, the variability of the fault plane roughness has a more substantial effect on risk of reactivation along any single fault, especially when the variability occurs in the moderately dipping section of the fault (Fig. 9).

The fault sets modelled above are orientated approximately N140°E (140; synthetic) and N320°E (320; antithetic). The S_{Hmax} orientation of N130°E (130) therefore produces a small angle between the modelled faults and the S_{Hmax} orientation (*c.* 10°), which is a generally favourable fault reactivation (Anderson 1951; Healy *et al.* 2006). The second orientation of S_{Hmax} at N090°E (090) results in a larger angle between the modelled faults and the S_{Hmax} orientation (*c.* 50°), which is reflected in the modelled decrease in risk of reactivation with this scenario (Fig. 9d–f).

Given that the majority of the faults in the Mulgara Fault System and Hammerhead DDWFTB system have a similar strike and dip (Figs 4a & 5) and a 40° rotation of the S_{Hmax} orientation only minimally reduces the reactivation risk (Fig. 9), this fault model allows us to extrapolate risk of reactivation to include other similarly oriented faults in the Ceduna Sub-basin.

Fault model implications

From the calculated and modelled stress orientation(s) and magnitude(s) from Reynolds *et al.* (2003) and the 3D fault geometries and fault reactivation analysis described above, we can now examine the implications for regional fault reactivation in the Ceduna Sub-basin. Because we have 3D seismic control on the delta-top extensional faults, emphasis will be placed on comparison to other extensional faults in the sub-basin given that these are likely targets for exploration (Totterdell *et al.* 2000).

The regional extensional faults in the Ceduna Sub-basin have an average strike of N140°E–N320°E (140–320) and are listric normal growth faults associated with two periods of delta-top extension (Fig. 6). The majority of the potential hydrocarbon trap geometries in the Ceduna Sub-basin are within the extensional delta top(s) and rely at least partially on fault seal for viability. Understanding the reactivation risk is therefore

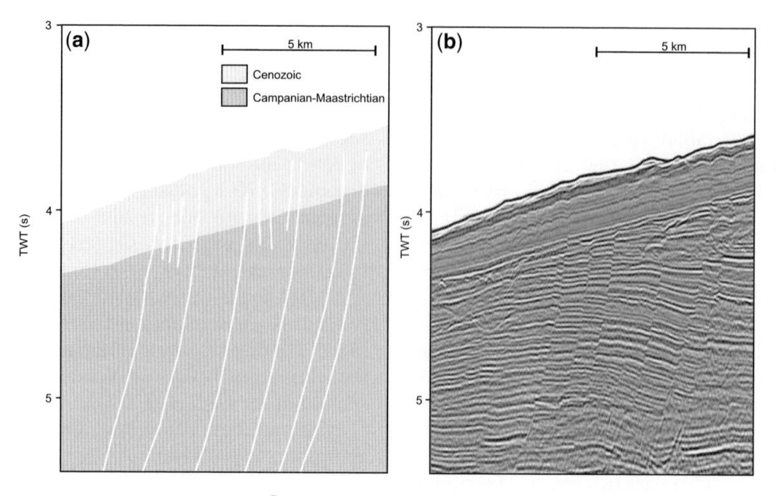

Fig. 10. (**a**) Interpreted and (**b**) un-interpreted 2D seismic image of Campanian–Maastrichtian faults intersecting the Cenozoic strata and reaching the seabed. This represents post-Maastrichtian fault reactivation in the Ceduna Sub-basin.

essential (Blevin *et al.* 2000; Totterdell *et al.* 2000, 2008; Ruble *et al.* 2001; Somerville 2001; Struckmeyer *et al.* 2001; 2002; King & Mee 2004; Tapley *et al.* 2005). Given that the modelled extensional faults from the 3D seismic data demonstrate a moderate to high risk of reactivation along the centre of the fault surfaces, where they are moderately dipping (>40° and <70°; Fig. 9) it is possible that these and other extensional faults on the delta top (Figs 4a & 5) are either at risk of reactivation or have already been reactivated. There is evidence for fault seal breach having occurred in the Bight Basin, such as the failure of Jerboa-1 well that demonstrated a charged and breached petroleum trap, likely due to Late Cretaceous reactivation (described above; Totterdell *et al.* 2000; Ruble *et al.* 2001; Somerville 2001; Struckmeyer *et al.* 2001). Strong evidence for fault reactivation is demonstrated where delta-top extensional faults are observed to disrupt the Cenozoic sediments and, in some cases, result in seafloor topography (Fig. 10).

The presence of Late Cretaceous inversion structures in the Ceduna Sub-basin (Fig. 4) indicates that far-field stresses affected isolated extensional structures, creating another potential reactivation mechanism in the area prior to establishment of the present-day stress regime. The exact orientation of this mechanism is unknown; however, if the stress was due to ridge push as proposed here, the extensional faults would have been optimally oriented for reactivation during the Late Cretaceous. In addition, it may be possible that this inversion pulse created fault-independent traps that could be prospective targets for hydrocarbon exploration in the Ceduna Sub-basin (Fig. 4c–f).

Conclusions

The Ceduna Sub-basin is composed of two temporally and spatially separate DDWFTBs, each demonstrating updip extension and downdip shortening (Figs 2, 4a & 5). From detailed mapping of two fault sets (in the extensional delta-top regions) from 3D seismic data and regional 2D seismic data we can now accurately model fault plane geometries at N320°E (320) and N140°E (140) orientations (Fig. 9). The detailed mapping of 3D and 2D seismic data has demonstrated the approximate timing of two newly identified phases of inversion in the Ceduna Sub-basin during the Santonian and Campanian–Maastrichtian. Both inversion events correspond to the initiation of sustained rifting of the Australian Southern Margin. These inversion structures may be responsible for the breach of the Jerboa-1 well in the Late Cretaceous and may also provide potential fault-independent trapping mechanisms via inversion anticlines (Fig. 4).

Using the fault plane geometries in combination with the previously calculated *in situ* stress regime and new geomechanical modelling, we can now determine the risk of fault reactivation in the Ceduna Sub-basin (Fig. 9). The results indicate that moderately dipping faults and/or areas of the fault planes are at the highest reactivation risk while the shallow (<40°) and steeply (>70°) dipping fault and/or areas of fault planes have a lower reactivation risk. The modelled antithetic faults have a higher risk of reactivation for both S_{Hmax} orientations [N130°E (130) and N090°E (090)] due to a combination of acute fault plane roughness, causing heterogeneities along the fault surface and the moderate dips between 40° and 70°. In addition, the modelling also demonstrates that for both S_{Hmax} orientations of N130°E (130) and N090°E (090) all three stress regime scenarios pose a moderate to high risk of fault reactivation; this indicates that fault reactivation is less sensitive to stress magnitude and most sensitive to the orientation of S_{Hmax} and fault plane roughness. Presence of fault reactivation is also identified in 2D seismic data (Fig. 10), indicating post-Maastrichtian fault reactivation and establishment of a normal or strike-slip normal faulting stress regime in the basin. Our results demonstrate reactivation is still likely to occur at present day, leading to potential seal breach and hydrocarbon migration in the Bight Basin.

Further exploration will ultimately lead to better data coverage, which will allow for more accurate depth conversions and additional well data for regional stress analysis.

The authors would like to thank the Australian Research Council and the University of Adelaide for their financial support of this study. The authors would also like to thank Igeoss/Schlumberger, Ion Geophysical, Midland Valley, SMT and PIRSA for the data and software used in this study. The authors would especially like to thank the reviewers of this paper for the constructive and thoughtful comments. This forms TRaX record 142.

References

ANDERSON, E. M. 1951. *The Dynamics of Faulting and Dyke Formation with Applications to Britain*. 2nd edn. Oliver and Boyd, Edinburgh.

BEIN, J. & TAYLOR, M. L. 1981. The Eyre Sub-basin: recent exploration results. *The Australian Petroleum Production and Exploration Association Journal*, **21**, 91–98.

BILOTTI, F. & SHAW, J. H. 2005. Deep-water Niger Delta fold and thrust belt modeled as a critical taper wedge: the influence of elevated basal fluid pressure on structural styles. *American Association of Petroleum Geologists Bulletin*, **89**, 1475–1491.

BLEVIN, J. E., TOTTERDELL, J. M., LOGAN, G. A., KENNARD, J. M., STRUCKMEYER, H. I. M. & COLWELL, J. B. 2000. Hydrocarbon prospectivity of the Bight Basin – petroleum systems analysis in a frontier basin. *In*: *2nd Spring Symposium – Frontier Basins, Frontier Ideas, Adelaide, June 29–30th, 2000*. Geological Society of Australia, Abstracts, **60**, 24–29.

BOTT, M. H. P. 1991. Ridge push and associated plate interior stress in normal and hot spot regions. *Tectonophysics*, **200**, 17–32.

BOEUF, M. G. & DOUST, H. 1975. Structure and development of the southern margin of Australia. *The Australian Petroleum Production and Exploration Association Journal*, **15**, 33–43.

BRIGGS, S. E., DAVIES, R. J., CARTWRIGHT, J. A. & MORGAN, R. 2006. Multiple detachment levels and their control on fold styles in the compressional domain of the deepwater west Niger Delta. *Basin Research*, **18**, 435–450, doi: 10.1111/j.1365-2117. 2006.00300.x.

COBBOLD, P. R., CLARKE, B. J. & LØSETH, H. 2009. Structural consequences of fluid overpressure and seepage forces in the outer thrust belt of the Niger Delta. *Petroleum Geoscience*, **15**, 3–15, doi: 10.1144/1354-079309-784.

COOPER, M. & WARREN, M. J. 2010. The geometric characteristics, genesis and petroleum significance of inversion structures. *In*: LAW, R. D., BUTLER, R. W. H., HOLDSWORTH, R. E., KRABBENDAM, M. & STRACHAN, R. A. (eds) *Continental Tectonics and Mountain Building: The Legacy of Peach and Horne*. Geological Society, London, Special Publications, **335**, 827–846.

CORREDOR, F., SHAW, J. H. & BILOTTI, F. 2005. Structural styles in the deep-water fold and thrust belts of the Niger Delta. *American Association of Petroleum Geologists Bulletin*, **89**, 753–780.

COSTA, E. & VENDEVILLE, B. C. 2002. Experimental insights on the geometry and kinematics of fold-and-thrust belts above weak, viscous evaporitic décollement. *Journal of Structural Geology*, **24**, 1729–1739.

DAVIS, D. M. & ENGELDER, T. 1985. The role of salt in fold-and-thrust belts. *Tectonophysics*, **199**, 67–88.

DOUST, H. & OMATSOLA, E. 1990. Niger delta. *In*: EDWARDS, J. D. & SANTOGROSSI, P. A. (eds) *Divergent/Passive Margin Basins*. American Association of Petroleum Geologists, Boulder, Memoir, **48**, 201–238.

ESPURT, N., CALLOT, J.-P., TOTTERDELL, J., STRUCKMEYER, H. & VIALLY, R. 2009. Interactions between continental breakup dynamics and large-scale delta system evolution: Insights from the Cretaceous Ceduna delta system, Bight Basin, Southern Australia margin. *Tectonics*, **28**, TC6002, doi: 10.1029/2009TC002447.

FERRILL, D. A. & MORRIS, A. P. 2003. Dilational normal faults. *Journal of Structural Geology*, **25**, 183–196.

FINKBEINER, T., ZOBACK, M., FLEMINGS, P. & STUMP, B. 2001. Stress, pore pressure, and dynamically constrained hydrocarbon columns in the South Eugene Island 330 field, northern Gulf of Mexico. *American Association of Petroleum Geologists Bulletin*, **85**, 1007–1031.

FIORE, P. E., POLLARD, D. D., CURRIN, W. R. & MINER, D. M. 2006. Mechanical and stratigraphic constraints on the evolution of faulting at Elk hills, California. *American Association of Petroleum Geologists Bulletin*, **91**, 321–341.

FRASER, A. R. & TILBURY, L. A. 1979. Structure and stratigraphy of the Ceduna Terrace region, Great Australian Bight. *The Australian Petroleum Production and Exploration Association Journal*, **19**, 53–65.

HAQ, B. U., HARDENBOL, J. & VAIL, P. R. 1988. Mesozoic and Cenozoic chronostratigraphy and eustatic cycles. *In*: WILGUS, C. K., HASTINGS, B. S., KENDALL, C. G. St. C., POSAMENTIER, H. W., ROSS, C. A. & VAN WAGONER, J. C. (eds) *Sea-Level Changes: An Integrated Approach*. Society of Economic Paleontologists and Mineralogists, Tulsa, Special Publication, **42**, 71–108.

HEALY, D. 2009. Anisotropy, pore fluid pressure and low angle normal faults. *Journal of Structural Geology*, **31**, 561–574.

HEALY, D., JONES, R. R. & HOLDSWORTH, R. E. 2006. Three-dimensional brittle shear fracturing by tensile crack interaction. *Nature*, **439**, 64–67.

HILLIS, R. R. & REYNOLDS, S. D. 2000. The Australian stress map. *Journal of the Geological Society, London*, **157**, 915–921.

HOLFORD, S. P., TURNER, J. P., GREEN, P. F. & HILLIS, R. R. 2009. Signature of cryptic sedimentary basin inversion revealed by shale compaction data in the Irish Sea, western British Isles. *Tectonics*, **28**, TC4011, doi: 10.1029/2008TC002359.

HOLFORD, S. P., HILLIS, R. R., DUDDY, I. R., GREEN, P. F., TUITT, A. K. & STOKER, M. S. 2010. Impacts of neogene-recent compressional deformation and uplift on hydrocarbon prospectivity of the 'passive' southern Australian margin. *The Australian Petroleum Production and Exploration Association Journal*, **50**, 267–284.

HOLFORD, S. P., HILLIS, R. R. *ET AL.* 2011. Cenozoic post-breakup compressional deformation and exhumation of the southern Australian margin. *The Australian Petroleum Production and Exploration Association Journal*, **51**, 613–638.

INGRAM, G. M., CHISHOLM, T. J., GRANT, C. J., HEDLUND, C. A., STUART-SMITH, P. & TEASDALE, J. 2004. Deep-water North West Borneo: hydrocarbon accumulation in an active fold thrust belt. *Marine and Petroleum Geology*, **21**, 879–887.

JACKSON, M. P. A., VENDEVILLE, B. C. & SCHULZ-ELA, D. D. 1994. Salt-related structures in the Gulf of Mexico: a field guide for geophysicists. *The Leading Edge*, **13**, 837–842.

KING, S. J. & MEE, B. C. 2004. The seismic stratigraphy and petroleum potential of the Late Cretaceous Ceduna Delta, Ceduna Sub-basin, Great Australian Bight. *In*: BOULT, P. J., JOHNS, D. R. & LANG, S. C. (eds) *Eastern Australasian Basins Symposium II*. Petroleum Exploration Society of Australia, Special Publication, 63–73.

KING, R. C. & BACKÉ, G. 2010. A balanced 2D structural model of Hammerhead Delta – deepwater fold–thrust belt, Bight Basin, Australia. *Australian Journal of Earth Sciences*, **57**, 1005–1012.

KING, R. C., HILLIS, R. R. & REYNOLDS, S. D. 2008. In situ stresses and natural fractures in the Northern Perth Basin, Australia. *Australian Journal of Earth Sciences*, **55**, 685–701.

KING, R. C., HILLIS, R. R. & TINGAY, M. R. P. 2009. Present-day stress and neotectonic provinces of the Baram Delta and deepwater fold–thrust belt. *Journal of the Geological Society, London*, **166**, 197–200.

KING, R. C., BACKÉ, G., MORLEY, C. K., HILLIS, R. R. & TINGAY, M. R. P. 2010. Balancing deformation in NW Borneo: quantifying plate-scale v. gravitational tectonics in a delta and deepwater fold–thrust belt systems. *Marine and Petroleum Geology*, **27**, 238–246.

KRASSAY, A. A. & TOTTERDELL, J. M. 2003. Seismic stratigraphy of a large, Cretaceous shelf-margin delta complex, offshore southern Australia. *American Association of Petroleum Geologists Bulletin*, **87**, 935–963.

MACDONALD, J. D., KING, R., HILLIS, R. R. & BACKÉ, G. 2010. Structural style of the white pointer and Hammerhead Delta – deepwater fold–thrust belt, Bight Basin, Australia. *The Australian Petroleum Production and Exploration Association Journal*, **50**, 487–510.

MAERTEN, F. 2010. Adaptive cross-approximation applied to the solution of system of equations and post-processing for 3D elastostatic problems using the boundary element method. *Engineering Analysis with Boundary Elements*, **34**, 483–491.

MAERTEN, L., GILLEPIE, P. & POLLARD, D. D. 2002. Effects of local stress perturbation on secondary fault development. *Journal of Structural Geology*, **24**, 145–153.

MAERTEN, F., RESOR, P. G., POLLARD, D. D. & MAERTEN, L. 2005. Inverting for slip on three dimensional fault surfaces using angular dislocations. *Bulletin of the Seismological Society of America*, **95**, 1654–1665.

MAERTEN, F., MAERTEN, L. & COOKE, M. 2009. Solving 3d boundary element problems using constrained iterative approach. *Computational Geosciences*, **14**, 551–564.

MANDL, G. & CRANS, W. 1981. Gravitational gliding in deltas. *In*: PRICE, N. J. & MCCLAY, K. R. (eds) *Nappe and Thrust Tectonics*. The Geological Society, London, Special Publications, **9**, 41–54.

MCCLAY, K., DOOLEY, T. & ZAMORA, G. 2003. Analogue models of delta systems above ductile substrates. *In*: VAN RENSBERGEN, P., HILLIS, R. R., MALTMAN, A. J. & MORLEY, C. K. (eds) *Subsurface Sediment Mobilization*. Geological Society, London, Special Publications, **216**, 411–428.

MILDREN, S. D., HILLIS, R. R. & KALDI, J. 2002. Calibrating predictions of fault seal reactivation in the Timor Sea. *The Australian Petroleum Production and Exploration Association Journal*, **42**, 187–202.

MORLEY, C. K. 2003. Mobile shale related deformation in large deltas developed on passive and active margins. *In*: VAN RENSBERGEN, P., HILLIS, R. R., MALTMAN, A. J. & MORLEY, C. K. (eds) *Subsurface Sediment Mobilization*. Geological Society, London, Special Publications, **216**, 335–357.

MORLEY, C. K. & GUERIN, G. 1996. Comparison of gravity-driven deformation styles and behaviour associated with mobile shales and salt. *Tectonics*, **15**, 1154–1170.

MORLEY, C. K., TINGAY, M. R. P., HILLIS, R. R. & KING, R. C. 2008. Relationship between structural style, overpressures and modern stress, Baram Delta province, NW Borneo. *Journal of Geophysical Research*, **113**, BO9410, doi: 10.1029/2007JB005324.

MORLEY, C. K., KING, R. C., HILLIS, R. R., TINGAY, M. R. P. & BACKÉ, G. 2011. Deepwater fold and thrust belt classification and hydrocarbon prospectivity. *Earth Science Reviews*, **104**, 41–91.

MORRIS, A. & FERRILL, D. A. 2009. The importance of the effective intermediate principal stress (s_2') to fault slip patterns. *Journal of Structural Geology*, **31**, 950–959.

MORRIS, A. P., FERRILL, D. A. & HENDERSON, D. B. 1996. Slip tendency and fault reactivation. *Geology*, **24**, 275–278.

NORVICK, M. S. & SMITH, M. A. 2001. Mapping the plate tectonic reconstruction of southern and southeastern Australia and implications for petroleum systems. *The Australian Petroleum Production and Exploration Association Journal*, **41**, 15–35.

OSBORNE, M. J. & SWARBRICK, R. E. 1997. Mechanisms for generating overpressure in Sedimentary Basins: a reevaluation. *American Association of Petroleum Geologists Bulletin*, **81**, 1023–1041.

REYNOLDS, S. D. & HILLIS, R. R. 2000. The in situ stress field of the Perth Basin, Australia. *Geophysical Research Letters*, **27**, 3421–3424.

REYNOLDS, S. D., COBLENTZ, D. D. & HILLIS, R. R. 2002. Tectonic forces controlling the regional intraplate stress field in continental Australia: results from new finite element modelling. *Journal of Geophysical Research*, **107B**, 21–31.

REYNOLDS, S. D., HILLIS, R. R. & PARASCHIVOIU, E. 2003. The in situ stress field, fault reactivation and seal integrity in the Bight Basin, South Australia, Australia. *Exploration Geophysics*, **34**, 174–181.

ROWAN, M. G., PEEL, F. J. & VENDEVILLE, B. C. 2004. Gravity-driven fold belts on passive margins. *In*: MCCLAY, K. R. (ed.) *Thrust Tectonics and Hydrocarbon Systems*. American Association of Petroleum Geologists, Boulder, Memoir, **82**, 157–182.

RUBLE, T. E., LOGAN, G. A. *ET AL*. 2001. Geochemistry and charge history of a paleo-oil column: Jerboa-1, Eyre Sub-basin, Great Australian Bight. *In*: HILL, K. C. & BERNECKER, T. (eds) *Eastern Australasian Basins Symposium. A Refocused Energy Perspective for the Future*. Petroleum Exploration Society of Australia, Special Publication, 521–530.

SANDIFORD, M., WALLACE, M. & COBLENTZ, D. 2004. Origin of the in situ stress field in south-eastern Australia. *Basin Research*, **16**, 325–338.

SAYERS, J., SYSMONDS, P., DIREEN, N. G. & BERNARDEL, G. 2001. Nature of the continent–ocean transition on the non-volcanic rifted margin of the central Great Australian Bight. *In*: WILSON, R. C. L., WHITMARSH, R. B., TAYLOR, B. & FROITZHEIM, N. (eds) *Non-Volcanic Rifting of Continental Margins: A Comparison of Evidence from Land and Sea*. Geological Society, London, Special Publications, **187**, 51–76.

SOMERVILLE, R. 2001. The Ceduna Sub-basin – a snapshot of prospectivity. *The Australian Petroleum Production and Exploration Association Journal*, **41**, 321–346.

STAGG, H. M. V., COCKSHELL, C. D. *ET AL*. 1990. Basins of the Great Australian Bight region: geology and petroleum potential. *Bureau of Mineral Resources, Australia, Continental Margins Program, Folio 5*.

STRUCKMEYER, H. I. M., TOTTERDELL, J. M. *ET AL*. 2001. Character, maturity and distribution of potential Cretaceous oil source rocks in the Ceduna Sub-basin, Bight Basin, Great Australian Bight. *In*: HILL, K. C. & BERNECKER, T. (eds) *Eastern Australasian Basin Symposium, a Refocused Energy Perspective for the Future*. Petroleum Exploration Society of Australia, Special Publication, 543–552.

STRUCKMEYER, H. I. M., WILLIAMS, A. K., COWLEY, R., TOTTERDELL, J. M., LAWRENCE, G. & O'BRIEN, G. W. 2002. Evaluation of hydrocarbon seepage in the Great Australian Bight. *The Australian Petroleum Production and Exploration Association Journal*, **42**, 371–385.

TAPLEY, D., MEE, B. C., KING, S. J., DAVIS, R. C. & LEISCHNER, K. R. 2005. Petroleum potential of the Ceduna Sub-basin: impact of Gnarlyknots-1A. *The Australian Petroleum Production and Exploration Association Journal*, **45**, 365–380.

TEASDALE, J. P., PRYER, L. L., STUART-SMITH, P. G., ROMINE, K. K., ETHERIDGE, M. A., LOUTIT, T. S. & KYAN, D. M. 2003. Structural framework and basin evolution of Australia's southern margin. *The Australian Petroleum Production and Exploration Association Journal*, **43**, 13–37.

THOMAS, A. L. 1993. Poly3D: *A three-dimensional, polygonal element, displacement discontinuity boundary element computer program with applications to fractures, faults, and cavities in the Earth's crust*. MSc thesis, Stanford University.

TINGAY, M. R. P, HILLIS, R. R., MORLEY, C. K., SWARBRICK, R. E. & DRAKE, S. J. 2005. Present-day stress orientation in Brunei: a snapshot of 'prograding tectonics' in a Tertiary delta. *Journal of the Geological Society, London*, **162**, 39–49.

TINGAY, M., HILLIS, R., MORLEY, C., KING, R., SWARBRICK, R. & DAMIT, A. 2009. Present-day stress and neotectonics of Brunei: implications for petroleum exploration and production. *American Association of Petroleum Geologists Bulletin*, **93**, 75–100.

TOTTERDELL, J. M. & KRASSAY, A. A. 2003. The role of shale deformation and growth faulting in the Late Cretaceous evolution of the Bight Basin, offshore southern Australia. *In*: VAN RENSBERGEN, P., HILLIS, R. R., MALTMAN, A. J. & MORLEY, C. K. (eds) *Subsurface Sediment Mobilisation*. Geological Society, London, Special Publications, **216**, 429–442.

TOTTERDELL, J. M. & BRADSHAW, B. E. 2004. The structural framework and tectonic evolution of the Bight Basin. *In*: BOULT, P. J., JOHNS, D. R. & LANG, S. C. (eds) *Eastern Australasian Basins Symposium II*. Petroleum Exploration Society of Australia, Special Publication, 41–61.

TOTTERDELL, J. M., BLEVIN, J. E., STRUCKMEYER, H. I. M., BRADSHAW, B. E., COLWELL, J. B. & KENNARD, J. M. 2000. A new sequence framework for the Great Australian Bight: starting with a clean slate. *The Australian Petroleum Production and Exploration Association Journal*, **40**, 95–117.

TOTTERDELL, J. M., STRUCKMEYER, H. I. M., BOREHAM, C. J., MITCHELL, C. H., MONTEIL, E. & BRADSHAW, B. E. 2008. Mid–Late Cretaceous organic-rich rocks from the eastern Bight Basin: implications for prospectivity. *PESA Eastern Australasian Basins Symposium III*, 137–158.

TURNER, J. P. & WILLIAMS, G. D. 2004. Sedimentary basin inversion and intra-plate shortening. *Earth-Science Reviews*, **65**, 277–304.

WILLCOX, J. B. & STAGG, H. M. J. 1990. Australia's southern margin: a product of oblique extension. *Tectonophysics*, **173**, 269–281.

WILLIAMS, G. A., TURNER, J. P. & HOLFORD, S. P. 2005. Inversion and exhumation of the St Georges Channel Basin, offshore Wales, UK. *Journal of the Geological Society, London*, **162**, 97–110.

WILLIAMS, G. D., POWELL, C. M. & COOPER, M. A. 1989. Geometry and kinematics of inversion tectonics. *In*: COOPER, M. A. & WILLIAMS, G. D. (eds) *Inversion Tectonics*. Geological Society, London, Special Publications, **44**, 3–15.

YASSIR, N. A. & ZERWER, A. 1997. Stress regimes in the Gulf Coast, offshore Louisiana: data from well-bore breakout analysis. *American Association of Petroleum Geologists Bulletin*, **81**, 293–307.

Quantifying Neogene plate-boundary controlled uplift and deformation of the southern Australian margin

DAVID R. TASSONE*[1], SIMON P. HOLFORD[1], RICHARD R. HILLIS[2] & ADRIAN K. TUITT[1]

[1]*Australian School of Petroleum, Centre for Tectonics, Resources and Exploration (TRaX), University of Adelaide, North Terrace, Adelaide, SA 5005, Australia*

[2]*Deep Exploration Technologies CRC, 26 Butler Boulevard, Burbridge Business Park, Adelaide Airport, SA 5950, Australia*

**Corresponding author (e-mail: david.tassone@adelaide.edu.au)*

Abstract: Parts of the Australian continent, including the Otway Basin of the southern Australian margin, exhibit unusually high levels of neotectonic deformation for a so-called stable continental region. The onset of deformation in the Otway Basin is marked by a regional Miocene–Pliocene unconformity and inversion and exhumation of the Cretaceous–Cenozoic basin fill by up to *c.* 1 km. While it is generally agreed that this deformation is controlled by a mildly compressional intraplate stress field generated by the interaction of distant plate-boundary forces, it is less clear whether the present-day record of deformation manifested by seismicity is representative of the longer-term geological record of deformation. We present estimates of strain rates in the eastern Otway Basin since 10 Ma based on seismic moment release, geological observations, exhumation measurements and structural restorations. Our results demonstrate significant temporal variation in bulk crustal strain rates, from a peak of *c.* $2 \times 10^{-16}\ s^{-1}$ in the Miocene–Pliocene to *c.* $1.09 \times 10^{-17}\ s^{-1}$ at the present day, and indicate that the observed exhumation can be accounted for solely by crustal shortening. The Miocene–Pliocene peak in tectonic activity, along with the orthogonal alignment of inverted post-Miocene structures to measured and predicted maximum horizontal stress orientations, validates the notion that plate-boundary forces are capable of generating mild but appreciable deformation and uplift within continental interiors.

It is becoming increasingly recognized that many passive-margin sedimentary basins contain evidence of episodes of uplift and deformation that have interrupted their otherwise continuous post-break-up subsidence histories (e.g. Brodie & White 1995; Praeg *et al.* 2005; Stoker *et al.* 2005, 2010; Holford *et al.* 2009*a*). Understanding the origins of such episodes is important because they can have significant negative impacts on the hydrocarbon systems of passive-margin basins (Doré *et al.* 2002). The most commonly invoked driving mechanisms for such episodes include dynamic uplift resulting from mantle convection either associated with or independent of mantle plumes (e.g. Sandiford 2007; Champion *et al.* 2008), igneous underplating (Brodie & White 1995; White & Lovell 1997) and crustal shortening induced by plate-boundary controlled intraplate stresses (Ziegler *et al.* 1995; Hillis *et al.* 2008). However, some workers have regarded the latter process as an inefficient means of generating regional uplift of passive-margin basins. For example, Cenozoic compressional shortening and exhumation is well documented across the British Isles (Hillis *et al.* 2008), although Brodie & White (1995) have argued that the lithosphere is too weak for shortening to generate regional uplift and that observed shortening magnitudes across this region are too small to account for observed levels of uplift and exhumation.

Here we focus on the Neogene record of uplift and deformation within the Otway Basin of the southern Australian margin (see Fig. 1). This basin initially formed as a result of lithospheric extension associated with Australian–Antarctica continental break-up that began during the Late Jurassic–Early Cretaceous and was largely accomplished by the Early Paleogene (Norvick & Smith 2001; Krassay *et al.* 2004). However, the post-rift subsidence history of the Otway Basin has been interrupted by several periods of uplift, exhumation and deformation, including the Mid-Cretaceous and Mid-Eocene, with the most recent phase beginning during the Late Miocene–Early Pliocene and continuing to the present day as evidenced by elevated levels of seismicity for a passive margin adjoining a stable continental region (Hill *et al.* 1995; Dickinson *et al.* 2002; Sandiford 2003*b*; Sandiford *et al.* 2004; Hillis *et al.* 2008; Holford *et al.*

From: Healy, D., Butler, R. W. H., Shipton, Z. K. & Sibson, R. H. (eds) 2012. *Faulting, Fracturing and Igneous Intrusion in the Earth's Crust*. Geological Society, London, Special Publications, **367**, 91–110. http://dx.doi.org/10.1144/SP367.7

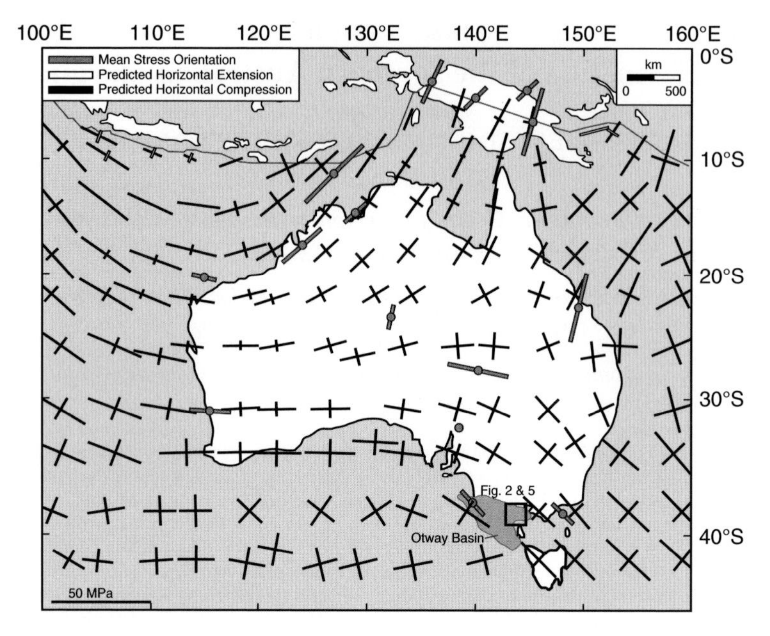

Fig. 1. Predicted maximum horizontal stress orientations across Australia based on finite element modelling compared with average maximum horizontal stress orientations for a number of tectonic provinces (Reynolds *et al.* 2003). Area outlined by black box is shown in Figures 2 and 5, while the grey shading displays the extent of the Otway Basin.

2011*b*). The onset of this most recent 'neotectonic' episode is marked by a regional tectonic unconformity that can be traced for *c.* 1500 km along the basins of SE Australia, attesting to widespread uplift along the margin (Dickinson *et al.* 2002). There is evidence for reverse faulting and folding of Miocene and older sediments within the Otway Basin, implying an important role for compressive tectonics in Neogene deformation (Holford *et al.* 2011*a*). These structures generally trend NE–SW, orthogonal to the present-day NW–SE maximum horizontal stress (S_{Hmax}) orientation, which is itself consistent with regional stress trends predicted by plate-scale stress modelling (Hillis & Reynolds 2000; Reynolds *et al.* 2003).

Although Neogene deformation in south-eastern Australia is largely thought to be controlled by an intraplate stress regime generated by the interactions of forces generated at the distant boundaries of the Indo-Australian plate (e.g. Hillis *et al.* 2008), it has also been suggested that the southern Australian margin has experienced long-wavelength (of the order 10^3 km) dynamic uplift in response to the northwards motion of Australia towards the subduction zones along the northern and eastern boundaries of the Indo-Australian plate (Sandiford 2007). Deformation in the eastern Otway Basin (the area comprising the Colac Trough, Port Campbell Embayment, Otway Ranges and Torquay Sub-basin) is accompanied by substantial localized exhumation with magnitudes estimated as high as *c.* 1 km (Holford *et al.* 2010) over recognized Late Miocene–Early Pliocene compressional structures, but there is also up to *c.* 400 m of exhumation in parts of the basin where there are few obvious signs of shortening (Holford *et al.* 2011*a*).

The primary aims of this study are to assess the extent to which observed magnitudes of post-10 Ma exhumation in the Otway Basin can be accounted for by crustal shortening, and to determine the temporal variation of intraplate strain rates over this time period. We estimated the present-day (seismogenic) strain rates from the seismic moment release rate for earthquakes between 1970 and 2007, and compared this with Quaternary geological evidence and longer-term bulk crustal or total volumetric strain rates constrained from published line-length restorations and exhumation measurements. Our results indicate that observed degrees of crustal shortening (*c.* 5%; Cooper & Hill 1997) can account for the observed estimates of Late Miocene–Pliocene exhumation. Furthermore, our results demonstrate that strain rates have likely declined from peak values of *c.* 2×10^{-16} s^{-1} at the onset of the Late Miocene to *c.* 1.09×10^{-17} s^{-1} at the present day, indicating that plate-boundary-controlled intraplate stress fields and deformation can vary significantly over timescales of *c.* 10^6–10^7 a. The observed decline in bulk crustal strain rates may be explained by present-day deformation having migrated and localized elsewhere along the southern margin (e.g. the Flinders and Mt Lofty Ranges of South Australia).

In situ stress field of the southern Australian margin

Both present-day stress orientations and neotectonic palaeostress orientations exhibit marked variations across the Australian continent, with continent-scale curvatures in S_{Hmax} orientation (Hillis & Reynolds 2000; Fig. 1). Results from plate-scale stress modelling indicate that present-day stress orientations across Australia are consistent with first-order controls by the net torques of the forces that drive and resist motion of the Indo-Australian plate (Hillis & Reynolds 2000; Reynolds *et al.* 2003). The plate-scale stress modelling incorporates the complex nature of the convergent and divergent Indo-Australian plate boundaries, and stress measurements are determined by methods such as borehole breakouts (BO) and drilling-induced tensile fractures (DITF).

BO and DITF data collected from exploration wells in the Otway Basin show that S_{Hmax} rotates from *c.* 125°N in western parts of the basin (i.e. South Australia; Hillis *et al.* 1995) to *c.* 137°N in the eastern Otway Basin (i.e. Victoria; Nelson *et al.* 2006; Vidal-Gilbert *et al.* 2010; Fig. 2). Detailed studies of the origin of the *in situ* stress field along the southern Australian margin (e.g. Sandiford *et al.* 2004; Dyksterhuis & Müller 2008) have indicated that the broadly NW–SE oriented and mildly compressional S_{Hmax} observed in the Otway Basin primarily reflects the increased coupling of the Australian and Pacific plate boundary, associated with the formation of the southern Alps in New Zealand since the Late Miocene–Early Pliocene (Sandiford *et al.* 2004).

While there is general agreement regarding stress orientations in the Otway Basin, in recent years there has been debate regarding the state of stress particularly in the eastern part of the basin. Using an extensive dataset of stress measurements, based mainly on wells from the Port Campbell Embayment and Shipwreck Trough, Nelson *et al.* (2006) concluded that the present-day stress regime is strike-slip (approaching reverse). Leak-off pressure test and extended leak-off pressure test data reported by Nelson *et al.* (2006) indicate an increase in the magnitude of S_{Hmin} from the western to eastern Otway Basin, indicating that the eastern part of the basin is approaching a reverse stress regime (Nelson *et al.* 2006). However, in a study of the stress regime at a CO_2 storage site in the Port Campbell Embayment, Bérard *et al.* (2008) constrained a normal stress regime; this is less consistent with the geological evidence for neotectonic reverse faulting and compressional deformation in this region (Nelson *et al.* 2006). The disagreement in interpreted stress regime between these studies may be due to the different techniques used to determine the S_{Hmax} magnitude (Vidal-Gilbert *et al.* 2010), with Nelson *et al.* (2006) using DITF occurrences and frictional limits and Bérard *et al.* (2008) basing their analysis on dipole sonic log data. Independent support for a present-day stress regime in the Otway Basin that is more consistent with the neotectonic record of deformation is provided by a focal mechanism solution determined for the $M_{\mathrm{L}} = 4.4$ 1977 Balliang earthquake located *c.* 35 km north of Geelong (Denham *et al.* 1981). The focal parameters for this earthquake indicated thrust (i.e. reverse) faulting due to compressive forces acting *c.* NW–SE (Denham *et al.* 1981). This focal mechanism solution is consistent with the *in situ* S_{Hmax} orientation derived from BO and DITF data, and is approximately orthogonal to the strikes of most Neogene structures that have been identified in the Otway Basin (Fig. 2; Nelson *et al.* 2006; Hillis *et al.* 2008).

Couzens-Schultz & Chan (2010) recently presented a new interpretation of leak-off pressure test results, whereby the magnitude of the minimum principal stress is constrained by assuming that the leak-off test causes shear failure along pre-existing planes of weaknesses rather than the traditional assumption of failure via tensile openings. Preliminary applications of this new interpretation to leak-off test data from Otway Basin wells increases S_{Hmin} to values greater than the vertical stress such that the faulting regime becomes reverse (i.e. $S_{\mathrm{Hmax}} > S_{\mathrm{Hmin}} > S_{\mathrm{v}}$), consistent with geological evidence (King *et al.* 2011).

Neogene deformation, exhumation and uplift

There is extensive evidence for Neogene deformation in the Otway Basin (Hillis *et al.* 2008). A seismic reflection profile oriented subparallel to the S_{Hmax} orientation (Fig. 3a) reveals *c.* NE–SW-striking inversion anticlines (Fig. 2) that have folded Miocene and older rocks, such as the anticline penetrated by the Nerita-1 well. These anticlines probably formed as a result of reverse reactivation of optimally oriented (i.e. strike) normal faults that formed during Cretaceous–Paleogene rifting (Holford *et al.* 2011*a*). NE–SW-trending Miocene–Recent inversion anticlines are widespread throughout the onshore and offshore Otway Basin (Fig. 2), and several of these structures host important hydrocarbon accumulations

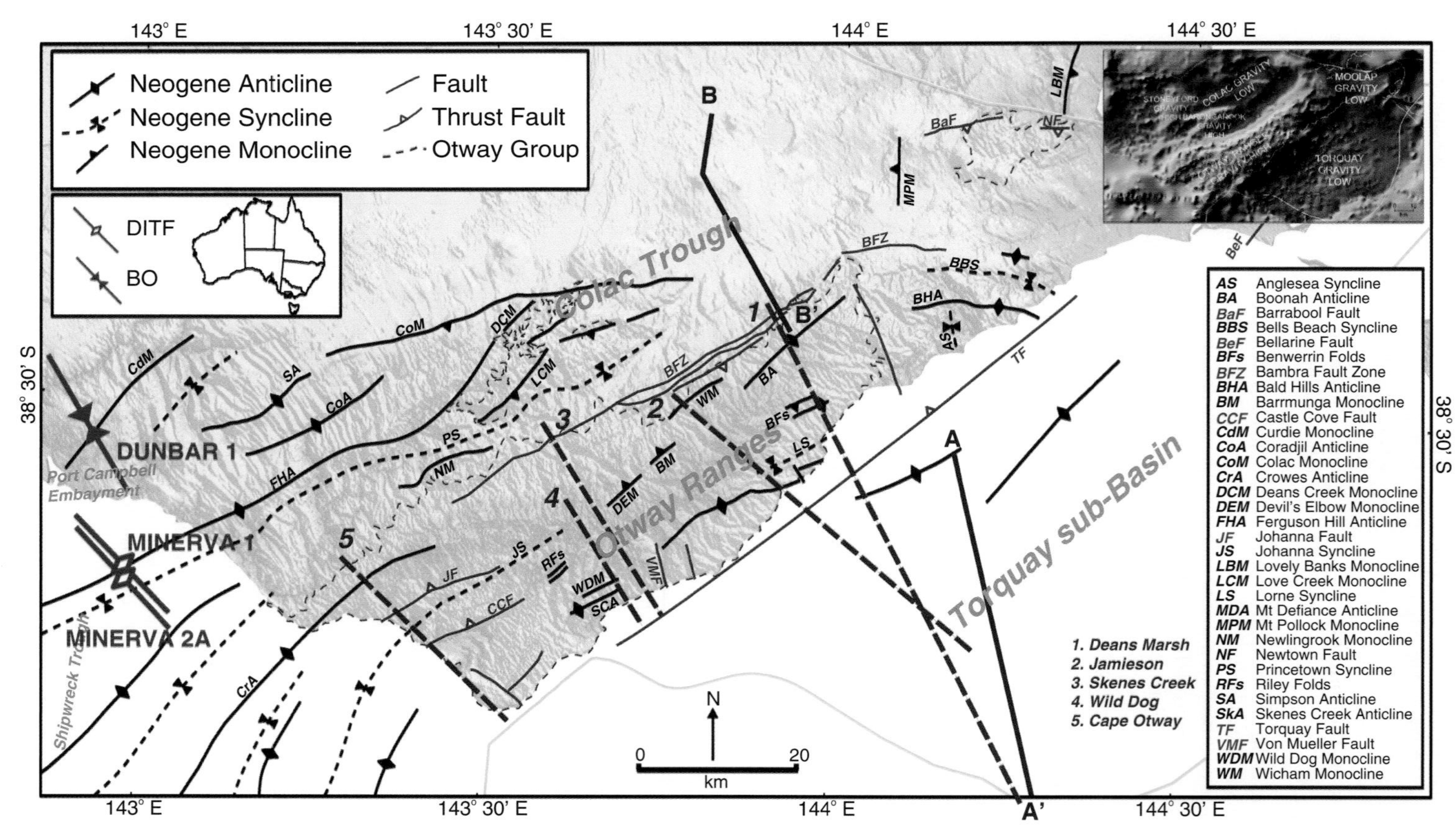

Fig. 2. Topographic map of the Colac Trough and Otway Ranges (Constantine & Liberman 2001) showing known Neogene anticlines, synclines, monoclines and faults (Holford *et al.* 2011*a*) and maximum horizontal stress data orientations determined from BO and DITF data; Dunbar-1 *c.* 151°N; Minerva-1 *c.* 137°N; Minerva-2A *c.* 138°N (Nelson *et al.* 2006). Not shown on map: Eric the Red-1 *c.* 137°N (Nelson *et al.* 2006) and CRC-1 *c.* 142°N (Vidal-Gilbert *et al.* 2010). Seismic profiles A–A′ and B–B′ are shown in Figure 3. Sections 1–5 refer to the seismic lines which Cooper (1995) and Cooper & Hill (1997) restored in order to estimate Miocene–Pliocene shortening. Inset map in the top-right corner is a Bouguer gravity map of the eastern Otway Basin (Constantine & Liberman 2001) with purple-blue colours representing gravity lows and red-brown colours representing gravity highs.

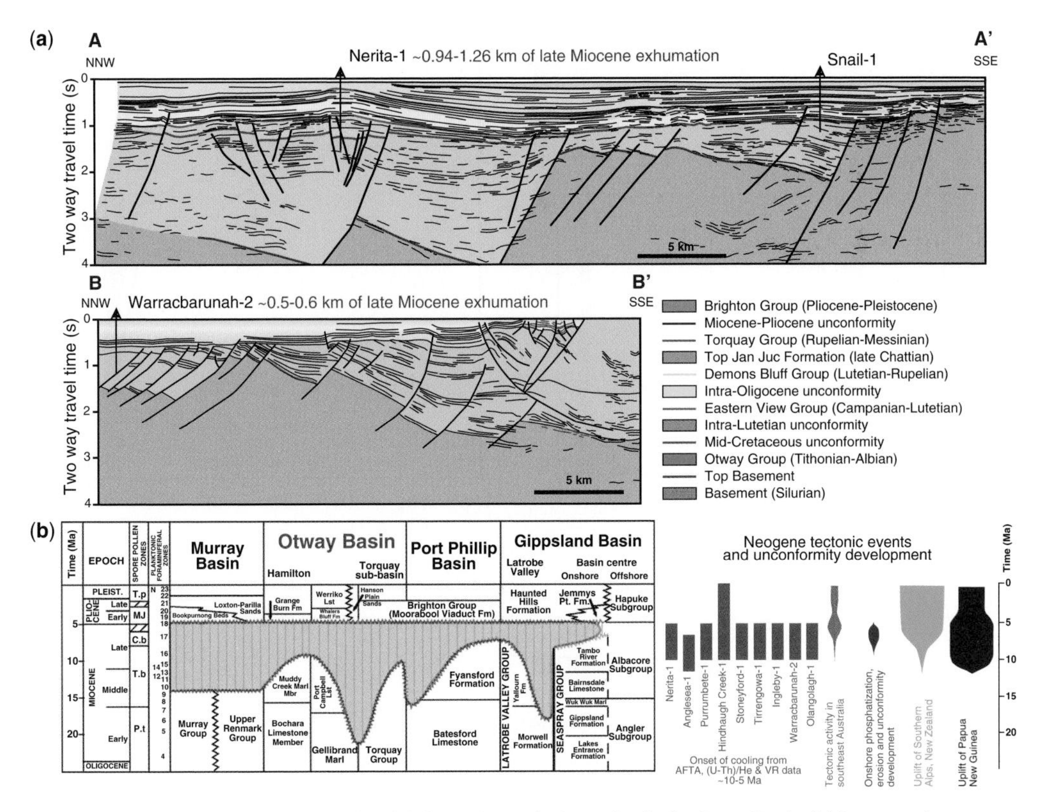

Fig. 3. Evidence for Neogene compressional deformation and exhumation in the Otway Basin. (**a**) Interpreted two-way time seismic sections showing evidence for Neogene compressional deformation in the Torquay Sub-basin and Colac Trough (Holford *et al.* 2011*a*). (**b**) Comparative Late Oligocene–Holocene stratigraphy of basins along the SE 'passive' Australian continental margin indicating the regional tectonic Late Miocene–Early Pliocene unconformity (Dickinson *et al.* 2002). The timing of this regional unconformity correlates well with the onset of exhumation of Miocene and older rocks from apatite fission track, apatite (U–Th)/He and vitrinite reflectance data (Green *et al.* 2004; Holford *et al.* 2011*a*; unpublished Geotrack International reports). It also coincides with the change in relative plate motion between the Pacific and Indo-Australian plates due to arc collision of Australia with SE Asia to the north coupled with ridge push to the south and compressional forces along the New Zealand and south of New Zealand plate boundaries (Hill *et al.* 1995; Dickinson *et al.* 2002; Sandiford 2003*a*). Modified from Holford *et al.* (2011*a*).

such as the Minerva gas field in the Shipwreck Trough, offshore Port Campbell (Hillis *et al.* 2008).

A valuable constraint on the timing of this deformation is provided by a regional tectonic, low-angle unconformity of Late Miocene–Early Pliocene age that can be traced for a distance of *c.* 1500 km throughout the basins of SE Australia (Dickinson *et al.* 2002; Fig. 3b). Biostratigraphic and isotopic data constrain the ages of the youngest underlying Miocene successions that subcrop this unconformity to *c.* 10–11.5 Ma, while the ages of the oldest overlying Pliocene section are *c.* 5.5–6 Ma (Dickinson *et al.* 2002; Sandiford *et al.* 2004).

Both the regional Late Miocene–Early Pliocene unconformity and some of the Neogene compressional structures in the Otway Basin are locally associated with substantial exhumation. The timing, magnitude and distribution of this exhumation have been independently constrained by analysing the thermal histories of Miocene and older sediments using apatite fission-track data from hydrocarbon exploration wells (Figs 4 & 5; Cooper & Hill 1997; Green *et al.* 2004; Holford *et al.* 2010, 2011*a*; unpublished Geotrack International reports). Apatite fission-track data from the Anglesea-1 and Nerita-1 wells indicate that cooling and exhumation of Miocene and older rocks in the eastern Otway Basin commenced between 10 and 5 Ma (Fig. 3c; Green *et al.* 2004; Holford *et al.* 2010), which compares well with the age of the Late Miocene–Early Pliocene unconformity (Fig. 3b).

Exhumation magnitudes attain values as high as *c.* 1 km over recognized Late Miocene–Early Pliocene compressional structures in the Torquay

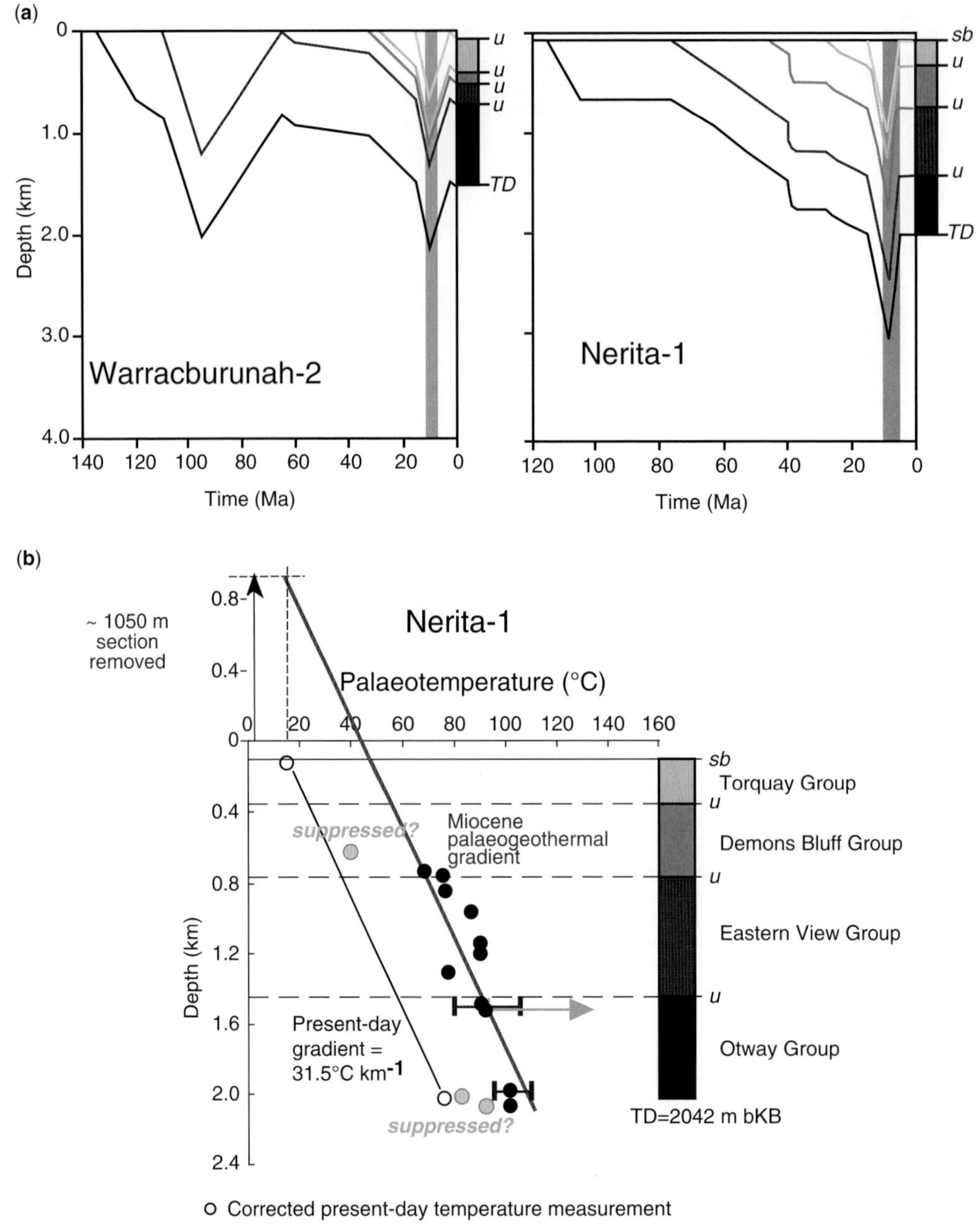

Fig. 4. Timing and magnitude of Neogene exhumation in the eastern Otway Basin. (**a**) Reconstructed burial histories derived from palaeothermal data (apatite fission track and vitrinite reflectance data) from Warracbarunah-2 (unpublished Geotrack International reports) in the Colac Trough and Nerita-1 (Holford *et al.* 2011*a*) in the Torquay Sub-basin highlighting the onset of Late Miocene cooling and exhumation. (**b**) Palaeothermal data (i.e. apatite fission track, apatite (U–Th)/He and vitrinite reflectance data) from Nerita-1 indicating *c.* 1 km of Late Miocene–Early Pliocene section exhumation (Holford *et al.* 2011*a*).

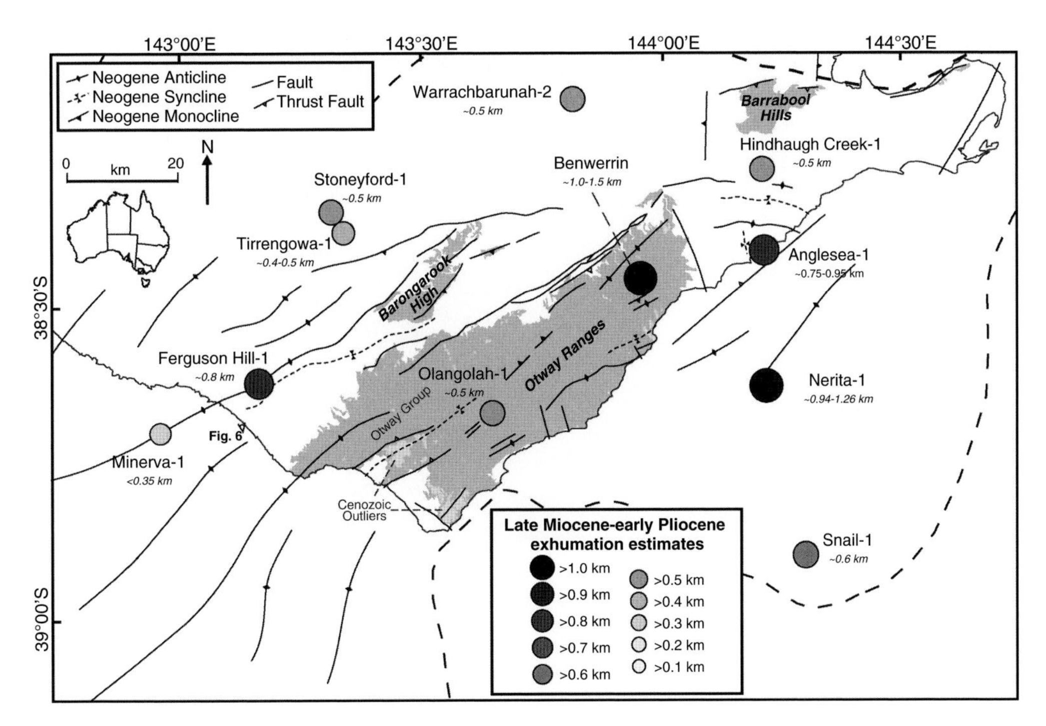

Fig. 5. Distribution of Late Miocene–Early Pliocene exhumation magnitudes in the eastern Otway Basin based on palaeothermal (i.e. apatite fission track and vitrinite reflectance) data. Exhumation estimates from Cooper (1995), Cooper & Hill (1997), Green *et al.* (2004), Holford *et al.* (2010) and unpublished Geotrack International reports. Modified from Holford *et al.* (2011*a*).

Sub-basin such as the Nerita Anticline (Figs 4 & 5; Holford *et al.* 2010, 2011*a*). Similar Late Neogene exhumation magnitudes are inferred from apatite fission track data from the onshore Anglesea-1 well in the Torquay Sub-basin (*c.* 0.85 km; Green *et al.* 2004) and from the Olangolah-1 well that was drilled into the Otway Ranges (*c.* 1 km; Cooper & Hill 1997; Fig. 5). Further to the north, Late Neogene exhumation of up to *c.* 0.6 km is inferred from apatite fission track data from exploration wells in the Colac Trough (e.g. Warracbarunah-2; Figs 4a & 5), where seismic data reveal fewer obvious signs of compressional deformation (Fig. 3a). These results attest to the occurrence of widespread Late Miocene–Early Pliocene uplift across the Otway Basin (cf. Dickinson *et al.* 2002), possibly in response to crustal shortening controlled by plate-boundary-derived stresses.

It is clear that deformation within the eastern Otway Basin continued after the Late Miocene–Early Pliocene peak in Neogene deformation (Sandiford *et al.* 2004). This is evidenced by faulting and folding of onshore (Fig. 6) and offshore Pliocene sediments (Dickinson *et al.* 2002), and incision of valleys into *c.* 1–2 Ma basaltic lava flows in the vicinity of the onshore Simpson and Ferguson Hill anticlines, suggesting continuing growth of these structures (Sandiford 2003*a*). Sandiford (2003*b*) estimated that up to *c.* 200 m of fault-related topography has been generated adjacent to the Otway Ranges since the Late Miocene–Early Pliocene deformation peak based on the present-day elevations of Late Pliocene palaeoshorelines (strandlines).

To date there have been few attempts to estimate shortening values associated with Late Neogene compressional deformation in the Otway Basin. Cooper (1995) and Cooper & Hill (1997) constructed five regional cross-sections from depth-converted seismic profiles and field data through the onshore Otway Ranges and Torquay Sub-basin that were oriented subparallel to the present-day S_{Hmax} orientation. They employed a forward modelling technique that backstripped and geometrically restored key horizons to a regional (undeformed) elevation using a fault-slip-fold mechanism, enabling both the compressional and extensional strains associated with basin formation and subsequent deformation to be estimated (Cooper & Hill 1997). Fault blocks were conserved during the restoration stage; Late Miocene–Early Pliocene denudation of *c.* 1 km was estimated based on thermal data and replaced

Fig. 6. Evidence for post-break-up reverse faulting near the Otway Ranges. Small low-angle (*c.* 40°) reverse fault in cliffs of Mid–Late Miocene Port Campbell Limestone at Twelve Apostles (see Fig. 5 for location), suggesting fault must have formed in the past *c.* 10 Ma. The fault dips towards *c.* E–SE and reverse motion is inferred from small offsets visible at the top of the cliff and in thin grey clayey layers in the cliff face. Cliff face is approximately *c.* 40 m high and people at the top of the cliff can be used for scale. Photographer facing *c.* N–NNE. Modified from Holford *et al.* (2011*c*).

to allow for accurate decompaction (Cooper & Hill 1997). Restoration of the Deans Marsh (over 81.3 km), Skenes Creek Road (over 36.7 km) and Cape Otway (over 37.4 km) sections yielded Late Miocene–Recent shortening estimates of 1.13, 6.07 and 7.3% respectively (Cooper 1995). Cooper & Hill (1997) suggested that an average of *c.* 5% shortening across the eastern Otway Basin is more realistic over this time interval.

We are interested in constraining the variation of strain rates in the Otway Basin over the past *c.* 10 Ma. We approach this problem by estimating present-day strain rates based on seismic moment release rate, longer-term strain rates derived from both measured regional shortening estimates and inferred shortening magnitudes that are required to explain the Neogene exhumation recorded by palaeothermal data. We then assess whether or not these strain rates are consistent with geological evidence and whether plate-boundary forces can account for the Neogene deformation, uplift and exhumation observed in this part of the southern Australian margin.

Seismicity and present-day strain rates

Distribution of seismicity

During the past 167 years, 5 earthquakes of magnitude 5 or greater have been recorded within the Otway Basin. The largest instrumentally recorded earthquake in this study area is *M c.* 5.7 (based on Geoscience Australia's earthquake database), while in south-eastern Australia the largest recorded earthquake is *M c.* 6.4 (Célérier *et al.* 2005).

A number of recent studies (e.g. Sandiford *et al.* 2003, 2004; Célérier *et al.* 2005; Leonard 2008; Braun *et al.* 2009; Sandiford & Quigley 2009) have provided quantitative constraints on neotectonic deformation and strain rates across different parts of Australia using the seismic moment tensor of earthquakes (Kostrov 1974). We have employed a similar approach to these studies in order to estimate the present-day seismic strain rate in the eastern Otway Basin (specifically, within the area defined by latitudes 37.5–39.5°S and longitudes 142.5 to 144.5°E). Following Braun *et al.* (2009), we have utilized Geoscience Australia's earthquake database. This database comprises earthquakes complied from a number of organizations including Geoscience Australia, the Department of Primary Industries South Australia, the Seismological Research Centre (part of Environment Systems and Services), Queensland University, the South East Queensland Water Corporation Ltd, the Research School of Earth Sciences at the Australian National University and the International Seismological Centre. The compilation includes both historical and instrumental recordings of earthquake magnitudes (not tensors) and extends from March 1840 until April 2007. This database does not differentiate between the various magnitude scales used (e.g. moment magnitude, surface wave magnitude, local magnitude, duration magnitude, body wave magnitude) by each organization over different time periods. The magnitudes of earthquakes are therefore only an approximate estimate of the size of the earthquake (Braun *et al.* 2009).

An important consideration when assessing the spatial variation in seismic activity for strain rate calculations is the impact of catalogue completeness. Catalogue completeness for Australian earthquakes has recently been reviewed by Leonard (2008) who suggests completeness levels of $M_L = 3.5–4$ for 1980+, $M_L = 4–5$ for 1970+, $M_L = 5–5.5$ for 1960+ and $M_L = 5.5$ for 1910+. For the $2° \times 2°$ (*c.* 173.5 × 222 km) area that we have investigated, seismic monitoring is considered more or less complete for earthquakes down to a local magnitude $M_L = 2$ for the period between January 1970 and April 2007. Any duplicate earthquake events within this area (i.e. those recorded by more than one organization) were removed, along with aftershocks, using declustering algorithms provided by Leonard (2008).

Seismic strain rate

The elevated levels of shallow seismicity observed within the Otway Basin (Fig. 7) provide evidence for ongoing tectonic deformation (in this case compressive) of the brittle upper crust (cf. Braun *et al.* 2009). The total crustal deformation resulting from an earthquake is represented by the seismic moment tensor, M_{ij} (Johnston 1994). If all the crustal deformation that occurs within a volume V

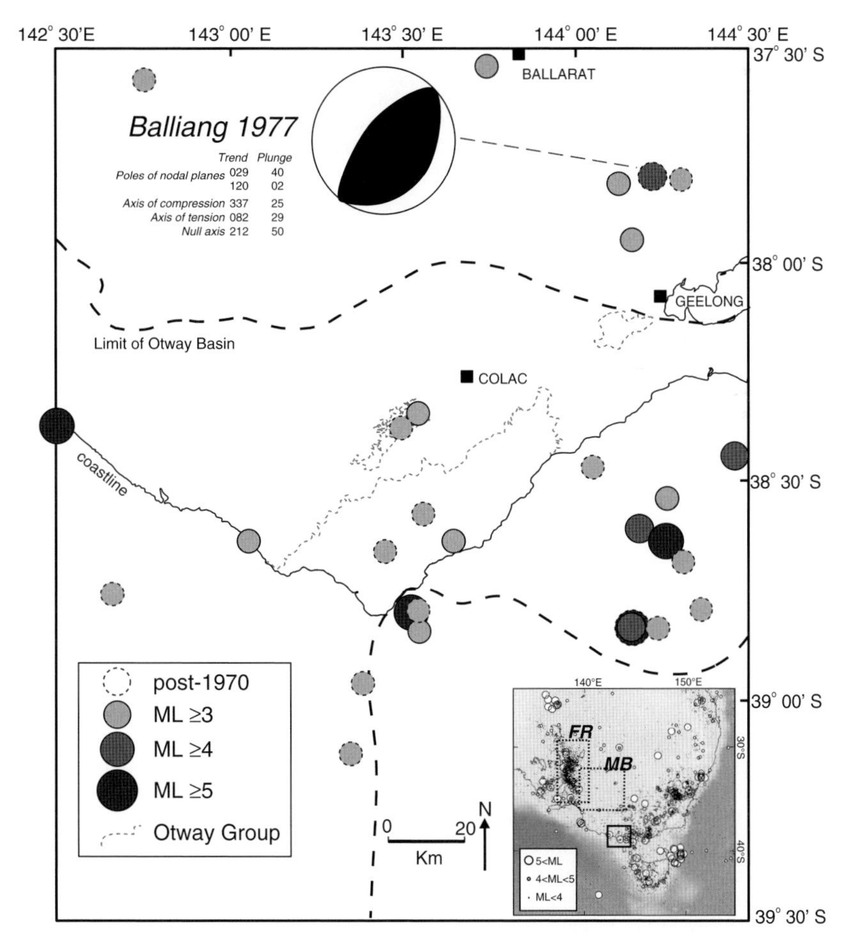

Fig. 7. Recorded earthquakes within the eastern Otway Basin. The focal mechanism solutions of the Balliang 1977 earthquake is shown, indicating a pure reverse faulting regime with S_{Hmax} oriented *c.* NW–SE (Denham *et al.* 1981). Inset shows the distribution of earthquakes in south-eastern Australia with the Flinders Ranges (FR) and Murray Basin (MB) highlighted for their high and low levels of seismicity, respectively (modified from Sandiford 2003*b*).

is seismic, Kostrov (1974) showed that there is a relationship between the average irrotational seismic strain rate ε'_{ij} of the volume and the sum of the seismic moment tensors of all the earthquakes N within the volume during a time interval T (Jackson & McKenzie 1988; Johnston 1994):

$$\varepsilon'_{ij} = \frac{1}{2\mu VT}\left(\sum_{n=1}^{N} M_{ij}^{(n)}\right) \quad (1)$$

where μ is the elastic shear modulus. The ε'_{ij} is expressed as a tensor; the seismic strain rate value therefore not only represents a magnitude, but also represents movement (e.g. extension, contraction, shear) and implies that there is no net rotation in the crustal volume (Johnston 1994). However, given that the *in situ* stress field in the Otway Basin is well constrained (S_{Hmax} *c.* 140°–320° *c.* NW–SE; Fig. 2) and the 1977 Balliang earthquake focal mechanism (Fig. 7) and reinterpretations of leak-off pressure data (i.e. S_{Hmin}) indicate a purely NW–SE reverse/compressive tectonic regime, the sum of the moment tensor elements may be expressed in terms of the cumulative moment scalar ΣM_0 when the x-direction is oriented parallel to S_{Hmax} in a stable continental region and the z-direction is vertical (Johnston 1994). That is, $\Sigma M_{xx} = \Sigma M_0$ and $\Sigma M_{zz} = -\Sigma M_0$ when the tectonic regime is characterized by contraction in the x-direction and crustal thickening in the z-direction. The other seven tensor elements (M_{xy}, M_{xz}, M_{yx}, M_{yy}, M_{yz}, M_{zx}, M_{zy}) all equal zero. We can therefore use the distribution and magnitude of earthquakes in the Otway Basin measured since 1970 to calculate an estimate of the average horizontal (parallel to S_{Hmax}) present-day seismic (i.e. brittle) strain rate using (Kostrov 1974):

$$\varepsilon'_{H} = \frac{1}{2\mu(A \times z_{sc})}\Sigma M'_0 \quad (2)$$

where A is the defined area of interest (i.e. 2° × 2° or 173.5 × 222 km), z_{sc} is the thickness of the seismogenic zone of the crust (which we estimate as *c.* 15 ± 5 km; Fig. 8) and $\Sigma M'_0$ is the average annual rate of seismic moment release. Although seismologists generally use 3–4 × 10^{10} N m^{-2} for the elastic shear modulus μ, we have assumed a value of 8 × 10^{10} N m^{-2} as used in similar studies for south-eastern Australia and the Flinders Ranges in South Australia (e.g. Sandiford *et al.* 2003, 2004; Célérier *et al.* 2005). The average annual rate of seismic moment release $\Sigma M'_0$ is calculated from the seismicity of a specific region, which may be difficult to estimate in seismically quieter areas and in areas where spatial coverage of seismic stations is poor (Braun *et al.* 2009). Regardless,

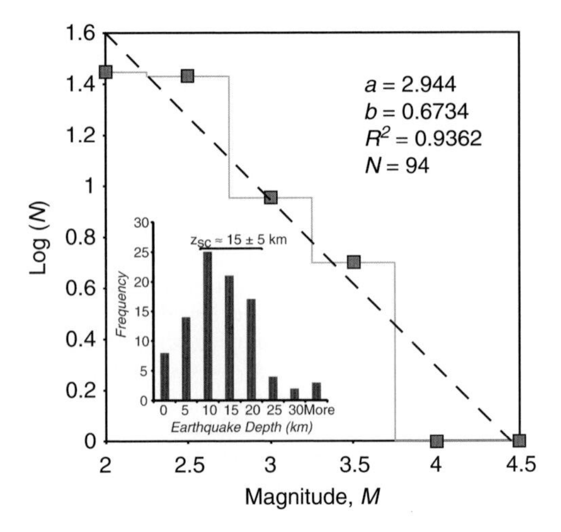

Fig. 8. Guttenberg–Richter relations for earthquakes in the eastern Otway Basin recorded between January 1970 and April 2007 (within the area 37.5–39.5°S and 142.5–144.5°E; 2° × 2° or 173.5 × 222 km = 3.852 × 10^{10} m^2). Inset is a histogram showing the frequency of earthquakes at different depths, from which we estimate the thickness of the seismogenic zone to be *c.* 15 ± 5 km.

the seismicity of a region may be quantitatively described by:

$$\log_{10}(N) = a - bM \quad (3)$$

$$\log_{10}(M_0) = cM + d \quad (4)$$

where Equation (3) quantifies the intensity of seismic activity within the region using the Gutenberg–Richter relationship (Gutenberg & Richter 1944) and Equation (4) converts the magnitude scale of earthquakes to the magnitude of seismic moment release (Hanks & Kanamori 1979). In Equation (3), N is the number of earthquakes with magnitude M or greater (Gaul *et al.* 1990) and a and b are linear regression parameters that vary spatially, depend on the magnitude scale and magnitude range (Sandiford *et al.* 2003) and are calibrated from the observed magnitude frequency distribution for an area of interest over a period of time where the earthquake catalogue is considered complete (Braun *et al.* 2009). As described previously, the catalogue is considered complete for the Otway Basin down to local magnitude $M_L = 2$ over the 37.28 year period between January 1970 and April 2007. Using the Gutenberg–Richter relationship (Gutenberg & Richter 1944), we estimate $a = 2.944$ and $b = 0.6734$ (Fig. 8). In Equation (4), M_0 is the magnitude of the total seismic moment release, M is the earthquake magnitude and c and d are linear regression constants that relate M to M_0 and are equal to 1.5

and 9.1 Nm, respectively (Hanks & Kanamori 1979). Combining and integrating Equations (3) and (4), Johnston (1994) showed that $\Sigma M'_0$ could be calculated using:

$$\Sigma M'_0 \approx \frac{1}{T}\left[\frac{b(10^{a+d})}{c-b}\right] \times \left[10^{(c-b)M_{\max}} - 10^{(c-b)M_{\min}}\right] \quad (5)$$

where M_{max} is the maximum expected magnitude for earthquakes in the study area, T is the time span of the seismic record (i.e. the period between January 1970 and April 2007, which equals *c*. 37.28 a or *c*. 1 176 356 876 s) and a, b, c and d are the known calibrated constants described above. M_{min} is the minimum expected magnitude for earthquakes in the study area; this term can be neglected when $b \leq 1.35$ (Johnston 1994), as is the case in the eastern Otway Basin. As can be deduced from Equation (5), $\Sigma M'_0$ is highly sensitive to the value of b. Since this value is relatively low for the eastern Otway Basin, it therefore concentrates a higher proportion of the moment release at the earthquakes of larger magnitudes (Figs 8 & 9; Johnston 1994). Given that the majority of the annual seismic moment release rate is carried out by the largest, most infrequent earthquakes rather than the numerous smaller ones (Johnston 1994; Célérier *et al.* 2005), another main uncertainty in calculating seismic strain rates using this approach is the value of M_{max} (Fig. 9). It is unlikely that the historic record of seismicity, which dates back only *c*. 200 a, includes the maximum earthquake magnitude for an intraplate region such as Australia; the historic record therefore only provides a lower bound on M_{max}. The largest instrumentally recorded earthquake in Australia was M_w *c*. 6.8 in 1941 at Meeberrie, Western Australia. However, palaeoseismic studies have suggested that M_L events >7 have likely occurred within the Flinders Ranges in South Australia (Quigley *et al.* 2006). By assuming an M_{max} value of 7 and $z_{sc} = 15$ km, the maximum horizontal present-day brittle strain rate calculated for the crustal volume within the eastern Otway Basin is therefore *c*. 5.07×10^{-18} s^{-1}.

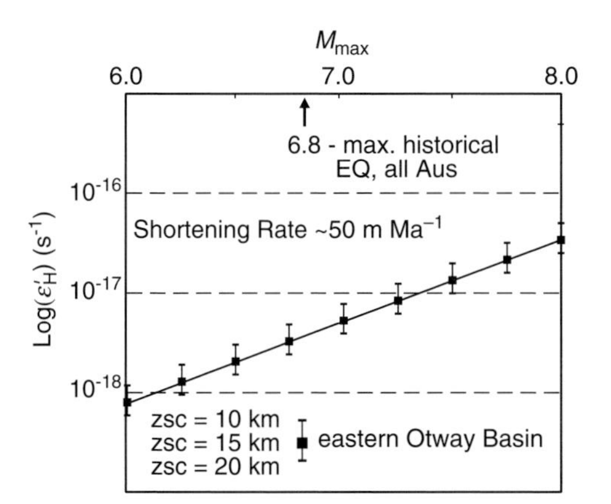

Fig. 9. Seismic strain rate estimates for the eastern Otway Basin as a function of maximum expected earthquake magnitude (M_{max}). The seismogenic crustal zone is estimated to be *c*. 15 ± 5 km (see Fig. 8). The largest instrumentally recorded earthquake in Australia (1941 Meeberrie earthquake) is also plotted.

Strain rates derived from shortening and exhumation magnitudes

The onset of Miocene–Pliocene compressional deformation, shortening and exhumation in the Otway Basin at *c*. 10 Ma is well constrained from seismic, stratigraphic and palaeothermal data (Figs 3 & 4). Several studies have suggested that the acme of deformation was reached by *c*. 5 Ma (Dickinson *et al.* 2002; Sandiford *et al.* 2004), although stress data and modelling (e.g. Nelson *et al.* 2006; Dyksterhuis & Muller 2008) and geological observations (e.g. Sandiford 2003*a*, *b*) confirm that this region continues to experience compressional deformation. Therefore, in addition to calculating strain rates based on present-day seismicity, we can also calculate Neogene strain rates using known shortening (Cooper 1995; Cooper & Hill 1997) and exhumation (Green *et al.* 2004; Holford *et al.* 2011*a*) constraints. This approach permits the evolution of total volumetric or bulk crustal strain rates over the past *c*. 10 Ma to be assessed. Two scenarios are considered when estimating longer-term bulk crustal strain rates. The first scenario assumes that compressional shortening has remained constant since the onset of exhumation and neotectonic deformation at *c*. 10 Ma until the present day. The second scenario assumes that the compressional shortening revealed by structural restorations was accomplished at a constant rate between 10 and 5 Ma, consistent with the notion of a Late Miocene–Early Pliocene deformation acme (e.g. Dickinson *et al.* 2002).

We utilize two different approaches to estimate longer-term bulk crustal strain rates. The first method involves conversion of forward-modelled Neogene shortening estimates derived from structurally restored seismic profiles (Cooper 1995; Cooper & Hill 1997) to bulk crustal strain rates. The second method involves conversion of denudation estimates (e.g. Green *et al.* 2004; Holford *et al.* 2011*b*) to bulk crustal strain rates using known relationships between uplift, denudation and shortening (e.g. Brodie & White 1995).

Converting shortening estimates to bulk crustal strain rates ε_s' is simply calculated using (Davis & Reynolds 1984):

$$\varepsilon_s' = \frac{e}{T} \tag{6}$$

where e is the shortening percentage ($e = 1.13$–6.07%; Cooper 1995; Cooper & Hill 1997), T is time period in which deformation has occurred (i.e. 10 Ma $= 3.15576 \times 10^{14}$ s or 5 Ma $= 1.57788 \times 10^{14}$ seconds) and ε_s' and e are positive when in compression. The bulk crustal strain rates calculated using this approach yield estimates of *c.* 3.58–$19.23 \times 10^{-17}\ \text{s}^{-1}$ for deformation occurring continuously since the onset of exhumation at *c.* 10 Ma and *c.* 7.16 to $38.47 \times 10^{-17}\ \text{s}^{-1}$, for deformation occurring over the first 5 Ma after the onset of exhumation.

A theoretical approach using known denudation and shortening relations to calculate bulk crustal strain rates is also considered to assess whether or not the magnitudes of Late Neogene exhumation recorded by palaeothermal data can be satisfactorily accounted for by crustal and/or lithospheric strain rates, or whether another tectonic process must be invoked to explain these observations. Permanent, isostatically compensated uplift can only be generated by thickening the crust or by adding material which is less dense than the asthenosphere to the lithospheric column (Brodie & White 1995). If a standard column of continental lithosphere is shortened uniformly by a factor f, then the amount of uplift U can therefore be quantified using either (Brodie & White 1995):

$$U_{\text{mt}} = \frac{a\left\{(\rho_m - \rho_c) \times \frac{t_c}{a}\left[1 - \left(\frac{\alpha T_1}{2} \times \frac{t_c}{a}\right)\right] - \frac{\alpha T_1}{2} \times \rho_m\right\}}{\rho_m \times (1 - \alpha T_1)} \times (f - 1) \tag{7}$$

or

$$U_{\text{n\_mt}} = t_c\left\{1 - \left[\left(\frac{\rho_c}{\rho_m}\right) \times \left(\frac{2\alpha - \alpha T_1 t_c}{2\alpha(1 - \alpha T_1)}\right)\right]\right\} \times (f - 1) \tag{8}$$

where $f = 1/(1-e)$ when e is positive (i.e. in compression) and depending on whether or not mantle thickening has occurred (U_{mt}: mantle thickening has occurred; $U_{\text{n\_mt}}$: no mantle thickening has occurred). The relationship between denudation (i.e. exhumation) D and uplift U is expressed in Equation (8), which takes into consideration the amplification effect caused by erosion (Brodie & White 1995):

$$D = \frac{\rho_a}{(\rho_a - \rho_s)} U. \tag{9}$$

The notation used in Equations (7)–(9) is defined in Table 1. Crustal thickness beneath the Otway Basin is estimated at *c.* 32.5 km based on seismic-refraction, seismic-reflection and receiver-function data (Collins *et al.* 2003). In order to calculate the required strain rates to account for the observed magnitudes of Late Miocene–Recent exhumation, measured values of D derived from palaeothermal data were input into Equation (8) to calculate U. The resultant values of U were then respectively input into Equations (7) and (8) to calculate f for the scenarios of both mantle thickening and no mantle thickening. This parameter was consequently converted to a shortening percentage e and strain rate ε_d' for the two scenarios (Table 2). The difference in strain rates calculated for estimates of D that vary between *c.* 200 to 1000 m for the two scenarios and different shortening criteria remain constant in the log ε_d' scale, which can be seen in Figure 10 and Table 2. For *c.* 1 km of Neogene exhumation (i.e. $D = 1$ km and $U = 0.25$ km), the

Table 1. *Symbols and values of parameters used to calculate shortening and consequently strain rates given known estimates of denudation recorded by thermal data (after Brodie & White 1995)*

Symbol	Parameter	Input value
a	Lithospheric thickness	125 km
t_c	Crustal thickness	32.5 km
ρ_w	Sea water density	*c.* 1.0 g cm^{-3}
ρ_c	Crustal density (at 0 °C)	*c.* 2.8 g cm^{-3}
ρ_m	Mantle density (at 0 °C)	*c.* 3.33 g cm^{-3}
ρ_a	Asthenosphere density (at 1333 °C)	*c.* 3.2 g cm^{-3}
ρ_s	Sediment density	*c.* 2.4 g cm^{-3}
A	Thermal expansion coefficient	*c.* 3.28×10^{-5} °C^{-1}
T_1	Asthenospheric temperature	*c.* 1333 °C

Table 2. *The required shortening and strain rate over a 10 and 5 Ma period using Equation (7) (lower limit: no mantle thickening) and Equation (6) (upper limit: mantle thickening) for measured amounts of denudation and uplift*

Denudation D (m)	Uplift U (m)	Shortening e (%)	Strain rate ε'_d	
			Required over a 10 Ma period ($\times 10^{-17}$ s^{-1})	Required over a 5 Ma period ($\times 10^{-17}$ s^{-1})
1000	250	5.8–9.0	18.27–28.59	36.54–57.18
800	200	4.7–7.4	14.79–23.29	29.58–46.59
600	150	3.5–5.6	11.22–17.80	22.44–35.59
400	100	2.4–3.8	7.575–12.09	15.14–24.18
200	50	1.2–1.9	3.83–6.16	7.66–12.33

shortening percentage is *c*. 5.8–9.0% and the strain rate is *c*. 18.27–28.59 $\times 10^{-17}$ s^{-1} over 10 Ma and *c*. 36.54–57.18 $\times 10^{-17}$ s^{-1} over 5 Ma. The lower limit of these estimates is the shortening percentage and strain rate assuming lithospheric shortening involved no mantle thickening, while the upper limit takes mantle thickening into consideration. The values of parameters used in these equations are the same as those used by Brodie & White (1995) (except for crustal thickness). The effect to which varying these parameters affects the calculation of strain rates is illustrated in Figure 10. As an example, for the maximum denudation case that we consider (i.e. $D = 1000$ m and hence $U = 250$ m), varying the ρ_c between 2.5 and 3.0 g cm^{-3} results in a variation in strain rate by nearly an order of magnitude for the shortening scenario where mantle thickening occurs (Fig. 10).

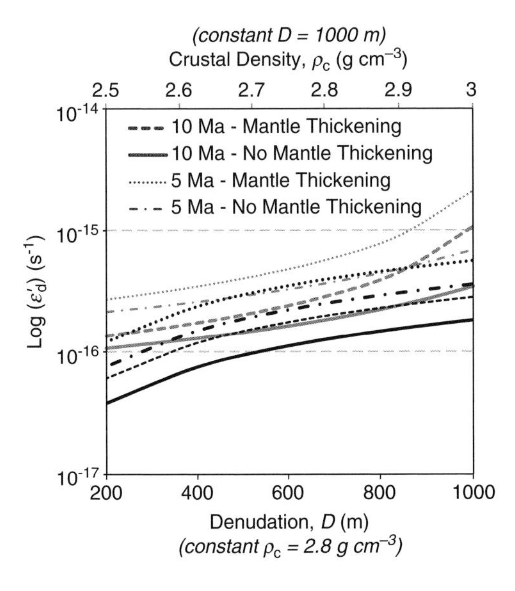

Fig. 10. Strain rates derived from Late Miocene–Early Pliocene denudation (i.e. exhumation) estimates, based on known relationships between uplift, denudation and shortening (see Equations (7) and (8); Brodie & White 1995). The values of parameters used are provided in Table 1. The variation of strain rate with carrying crustal densities when denudation is constant (1 km) is highlighted.

Discussion

Reconciling the present-day seismic strain rate with Pliocene–Quaternary geological evidence

Because of the ambiguities associated with determining parameters such as M_{max}, seismogenic crustal zone thickness, elastic shear modulus and the calibrated parameters *a*, *b*, *c* and *d*, our estimate of the present-day seismic brittle strain rate parallel to S_{Hmax} is subject to large uncertainties (i.e. at least 50%; Sandiford *et al.* 2004). For example, increasing M_{max} from 6 to 8 results in a difference in calculated strain rates of *c*. 3.32 $\times 10^{-17}$ s^{-1} (Fig. 8). Further uncertainties arise when translating strain rates determined using seismicity to bulk crustal strain rates (Johnston 1994). Our approach assumes that all ideally oriented fault planes (i.e. orthogonal to S_{Hmax}) within the specified volume of lithosphere are contributing to the calculated seismic strain rate, but when faulting does not occur on such ideally oriented fault planes Equation (2) will overestimate ε'_H (Johnston 1994). Furthermore, Equation (2) is strictly only true if the strain is elastic and released abruptly in earthquakes that generate the seismic waves from which the annual moment release rate is obtained (Johnston 1994). Thus, ε'_H does not represent the 'total volumetric' or bulk crustal strain rate, but rather a maximum 'seismic' strain rate which can be used as a lower bound estimate for the regional strain rate (Johnston 1994). The discrepancy between 'seismic' and 'total' strain rates is due to the possible contribution of significant (but unknown) 'aseismic' contributions to strain (Johnston 1994), and this issue is examined in more depth later in the 'Causes of aseismic deformation?' section.

Despite the uncertainties outline above, estimates of present-day seismic strain rate remain a useful constraint that can provide insight into minimum fault-slip rates of structures identified from palaeoseismic and geological investigations if the present-day seismicity is representative of longer-term geological deformation and seismic strain rates (Johnston 1994; Célérier *et al.* 2005; Sandiford & Quigley 2009). For example, if we assume that the neotectonic record of reverse faulting in the eastern Otway Basin is indicative of essentially uniaxial NW–SE compression uniformly across a *c.* 290 km wide zone within the 2° × 2° area considered in this study, then a present-day seismic strain rate of *c.* $5.07 \times 10^{-18}\,s^{-1}$ would equate to a minimum shortening rate of *c.* 50 m Ma^{-1}. A shortening rate of this order could be accounted for by a minimum of 1–5 faults, each slipping between *c.* 55 and 12 m Ma^{-1} assuming the faults are new non-inverted reverse faults dipping 30° (compare with Sandiford *et al.* 2004 for a similar analysis for the Flinders Ranges in South Australia). We note that it is likely that a large proportion of the shortening strain in the Otway Basin is accommodated by slip on normal faults that have been reactivated under a compressional stress regime. It is difficult to quantify shortening on inverted faults since their fault heaves are apparent (Holford *et al.* 2009*b*).

However, faulting in Pliocene sediments on the south-eastern side of the Otway Ranges record displacements indicative of shortening rates of at least *c.* 100 m Ma^{-1} (Sandiford 2003*b*), and a number of small Late Pliocene–Quaternary (i.e. *c.* 1–2 Ma) faults and monoclinal flexures with displacements (i.e. fault slips) of the order *c.* 10–50 m are known to occur adjacent to the northern flank of the Otway Ranges around Ferguson Hill (Sandiford 2003*b*). Geological evidence therefore implies that the present-day seismic strain rate calculated in this study does not adequately represent the total bulk crustal strain rate; it therefore represents a minimum estimate. If a shortening rate of *c.* 100 m Ma^{-1} adequately accounts for the observed Late Pliocene–Quaternary deformation in the eastern Otway Basin, then over a *c.* 290 km zone (consistent with the length used to calculate the seismic strain rate) the bulk crustal strain rate over this Late Pliocene–Quaternary period would be ε'_g *c.* $1.09 \times 10^{-17}\,s^{-1}$. It is therefore reasonable to assume the present-day bulk crustal strain rate is similar (i.e. *c.* $1.09 \times 10^{-17}\,s^{-1}$) and that the difference of at least *c.* $3.33 \times 10^{-18}\,s^{-1}$ between the present-day bulk crustal strain rate (based on Late Pliocene–Quaternary geological evidence) and present-day seismic strain rate (i.e. $\varepsilon'_g - \varepsilon'_H$), or *c.* 30% of the total present-day strain rate, is attributed to aseismic deformation. It should be noted that this bulk crustal strain rate would increase if the length of the transect decreases, although we use the length defined by the seismicity analysis for consistency.

By normalizing this seismic strain rate of *c.* $5.07 \times 10^{-18}\,s^{-1}$ to an area of 10 000 km^2 and a time period of one year, it is possible to compare our results with similar studies focused on the Flinders Ranges and Murray Basin (Sandiford *et al.* 2004; Célérier *et al.* 2005). This normalization yields a strain rate of *c.* $1.95 \times 10^{-17}\,s^{-1}$, which accounts for *c.* 3% of the total Australian seismic strain (i.e. *c.* $8.9 \times 10^{-18}\,s^{-1}$ over *c.* 5300 km; Célérier *et al.* 2005). This normalized estimate lies in between strain rates that have been estimated for the Flinders Ranges (*c.* $1.0 \times 10^{-16}\,s^{-1}$; Célérier *et al.* 2005) and Murray Basin (*c.* $1.0 \times 10^{-17}\,s^{-1}$; Sandiford *et al.* 2004). The estimates of strain rates for these areas are to be expected given that the Flinders Ranges are among the most seismically active parts of the Australian continent with a number of large Pliocene–Quaternary faults, while the Murray Basin is seismically quiet and relatively undeformed (Sandiford 2003*a*; Holford *et al.* 2011*b*; Fig. 8). It is apparent that eastern Otway Basin seismic strain rate derived here is only slightly greater than the seismic strain rate for the Murray Basin, even though significantly more neotectonic deformation and exhumation has occurred in the eastern Otway Basin. This may be because the total and seismic strain rates in the Murray Basin are similar with little to no aseismic deformation, whereas up to *c.* 30% of the total strain rate in the eastern Otway Basin may be accommodated by aseismic deformation.

Causes of aseismic deformation?

We have estimated that *c.* 30% of the present-day total volumetric strain in the eastern Otway Basin is potentially attributable to aseismic deformation. There are a number of ways where compressional deformation and shortening can be accommodated without releasing significant seismic energy. Aseismic volumetric strains in the upper crust could be accommodated by small sub-seismogenic structures (i.e. small faults, joints, cleavage) and/or ductile deformation mechanisms such as folding (Holford *et al.* 2009*b*). This may be important in parts of the Otway Basin stratigraphy dominated by less-competent fine-grained rocks such as the shales and mudstones of the Lower Cretaceous Eumeralla Formation.

Earthquake data in the eastern Otway Basin indicates that the frequency of earthquakes below 20 km has been relatively low since 1970, with the deepest recorded earthquake at 32 km (i.e. approximately at Moho depths). It is likely that the low levels of seismic moment release by deep earthquakes is due to strain in the hotter middle and lower crust (i.e. at sub-seismogenic depths) being primarily accommodated by creeping and crystal plastic failure rather than brittle faulting (McKenzie *et al.* 2000).

Another possible mechanism whereby aseismic shortening strain may be accommodated is horizontal compaction of porous sedimentary rocks (Turner & Williams 2004; Holford *et al.* 2009*b*). The eastern Otway Basin has likely been in a reverse faulting regime since the onset of exhumation in the Late Miocene–Early Pliocene (Nelson *et al.* 2006); S_{Hmax} has therefore been the maximum principal effective stress reducing porosity (Giles *et al.* 1998). Horizontal compaction may be an efficient mechanism of accommodating aseismic strain in the siliciclastic and calcareous Cenozoic sedimentary rocks in the Otway Basin, although it is likely that much of the porosity reduction in older sedimentary rocks (such as those of the Lower Cretaceous Eumeralla Formation which crop out across much of the Otway Ranges) occurred during burial and/or shortening during the Mid-Cretaceous (Cooper & Hill 1997; Green *et al.* 2004).

Have strain rates remained constant since the onset of Miocene–Pliocene exhumation?

Long-term bulk crustal strain rates (i.e. over the intervals of 10–5 and 10–0 Ma) in the Otway Basin have been constrained from structural restorations and inferred from the shortening required to explain Late Miocene–Early Pliocene exhumation estimates. These results are generally in good agreement, especially for exhumation-derived estimates for the case where no mantle thickening has occurred

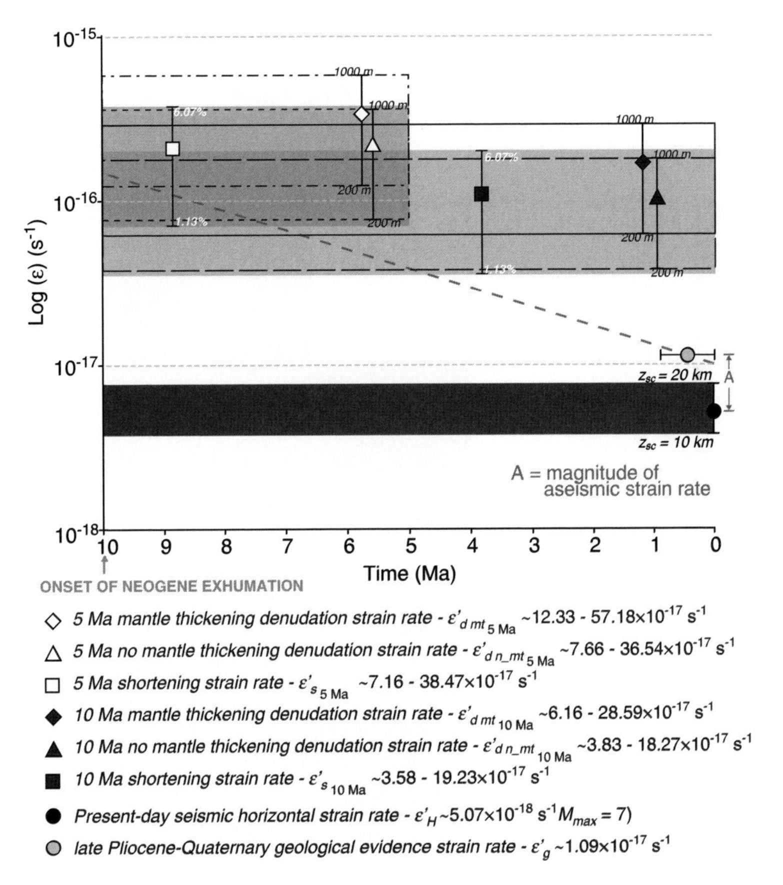

Fig. 11. Comparison of horizontal strain rates calculated from Quaternary geological evidence (Sandiford 2003*b*), seismicity, shortening estimates (Cooper 1995; Cooper & Hill 1997) over 10 and 5 Ma time periods, and observed denudation estimates (Holford *et al.* 2011*a*) over 10 and 5 Ma time periods. If we assume shortening and denudation occurred at the onset of exhumation (*c.* 10 Ma), then this suggests that the horizontal strain rates have decreased since the onset of deformation.

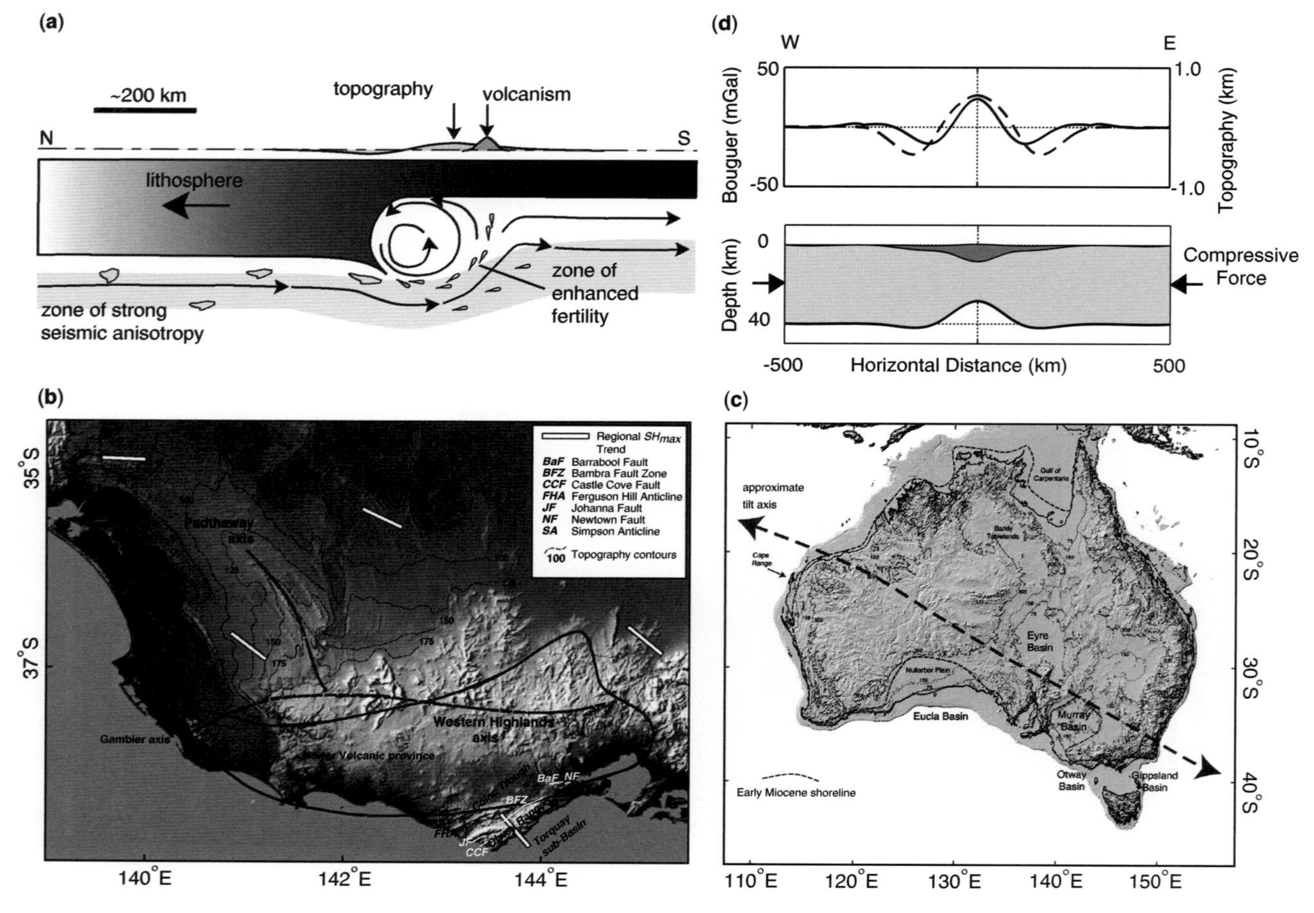

Fig. 12. Potential mechanisms for generating long-wavelength Neogene uplift along the southern Australian margin. (**a**) Edge-driven mantle convection responsible for Quaternary volcanism and uplift of the Padthaway Ridge (after Demidjuk *et al.* 2007). (**b**) The extent of Quaternary volcanism (i.e. Newer Volcanic Province) and the Padthaway Ridge topographic high and the locations of Quaternary faults in the eastern Otway Basin (Demidjuk *et al.* 2007). (**c**) Model for regional Neogene tilting of Australia with dynamic topographic uplift of the southern Australian margin due to the Australian continent moving towards the Indonesian subduction zone since the Late Miocene (Sandiford 2007). (**d**) Elastic thin-plate model of the Flinders Ranges with coinciding gravity and topographic highs as a result of lithospheric flexure (Célérier *et al.* 2005).

(Fig. 11). However, the long-term bulk crustal strain rates are between one and one-and-a-half orders of magnitude higher than the present-day bulk crustal strain rate of *c.* $1.09 \times 10^{-17}\ s^{-1}$ (Fig. 11). This result suggests that strain rates in an intraplate setting such as the Otway Basin can vary significantly over timescales of *c.* 10^6-10^7 a. Our results provide further support for the notion that Late Neogene deformation in south-eastern Australian peaked during the Late Miocene–Early Pliocene, coincident with the initial formation of the Southern Alps in New Zealand (cf. Sandiford *et al.* 2004). It is generally agreed that far-field plate-boundary forces had a significant role in the generation of the *in situ c.* NW–SE-oriented stress field in south-eastern Australia (Sandiford 2003*a*; Sandiford *et al.* 2004; Nelson *et al.* 2006). Given that the majority of Neogene compressional structures strike approximately orthogonally to the maximum horizontal stress orientation, there is strong evidence to conclude that the intraplate deformation of the Otway Basin is controlled by plate-boundary forces and that the stress fields generated by distant plate-boundary interactions are capable of generating mild but appreciable deformation, uplift and exhumation within plate interiors (cf. Hillis *et al.* 2008; Holford *et al.* 2009*a*, 2011*a*).

A remaining question is why bulk crustal strain rates in the Otway Basin appear to have declined by at least an order of magnitude from between 10 and 5 Ma to the present day. Although there are a number of important assumptions and uncertainties associated with both the long-term and short-term strain rates that we have determined (how representative is the *c.* 37 year earthquake record of deformation over longer timescales, for example $>10^2-10^3$ a?), geological evidence from the Otway Basin provides independent corroboration for a decrease in the intensity of deformation to the present day from an acme around the Late Miocene–Early Pliocene boundary (e.g. Dickinson *et al.* 2002; Sandford 2003*a*, *b*; Holford *et al.* 2011*a*). Seismogenic strain rates in the Flinders Ranges in South Australia are estimated at *c.* $10^{-16}\ s^{-1}$ (Célérier *et al.* 2005), suggesting that the locus of deformation along the southern margin of Australia may have migrated over the past *c.* 10 Ma, reflecting either changing mechanical properties of the crust and lithosphere or variations in the magnitude of the stress field.

Can shortening solely account for Neogene uplift and exhumation?

Although much of the Neogene uplift and exhumation in the eastern Otway Basin can be ascribed to crustal and/or lithospheric shortening in response to a compressional intraplate stress field generated by plate-boundary interactions, several additional mechanisms have been proposed to account for Late Cenozoic uplift along parts of the southern Australian margin. In addition to uplift caused by reverse faulting, Célérier *et al.* (2005) have suggested that the topography of the Flinders Ranges in South Australia reflects a component of low-amplitude (*c.* 200–500 m) long-wavelength (*c.* 200 km) lithospheric flexure (Fig. 12d). Based on isotope geochemistry and surface wave tomography data, Demidjuk *et al.* (2007) suggest that the topographically elevated Padthaway Ridge, which bounds the north-western margin of the Otway Basin, developed in response to mild upwelling generated by edge-driven convection in the upper mantle (Fig. 12a, b). Mantle structuring is also invoked by Sandiford (2007) to explain up to *c.* 300 m of Neogene uplift along the southern Australian margin, due to the continent-wide tilting of Australia as it has moved towards the Indonesian subduction zone (Fig. 12c).

Of these processes, it is probable that a certain component of the Neogene uplift and exhumation of the eastern Otway Basin can be ascribed to dynamic topographic uplift of the southern Australian margin (cf. Sandiford 2007). The role of edge-driven convection as proposed to explain the uplift of the Padthaway Ridge is less certain. This feature is associated with Late Pliocene–Quaternary volcanism but little or no faulting, whereas evidence for Late Pliocene–Quaternary faulting increases to the east towards the Otway Ranges (Fig. 12b; Sandiford 2003*b*), implying that uplift and exhumation in the eastern Otway Basin is more likely associated with fault-related deformation. The observation of similar patterns of Bouguer gravity anomalies in the Flinders Ranges and Otway Ranges, that is, gravity highs (the Ranges themselves) bordered by gravity lows (marginal depressions such as the Colac Trough and Torquay Sub-basin; Fig. 2), suggests that the topography of the Otway Ranges may have been accentuated by lithospheric flexure in response to the loading induced by crustal and/or lithospheric shortening (cf. Célérier *et al.* 2005).

Conclusion

In an attempt to constrain the impacts of a stress field generated by distant plate-boundary interactions on intraplate tectonic processes, we have investigated the neotectonic record of deformation in the Otway Basin, southern Australian margin, where there is abundant evidence for shortening, uplift and exhumation of Late Miocene–Pliocene and older sedimentary rocks. We have used structural restorations and seismicity data to assess the variation in intraplate total volumetric or bulk crustal and seismic strain rates, respectively, over the past *c.* 10 Ma. We have also considered whether observed horizontal bulk crustal strain rates can explain observed magnitudes of upper crustal exhumation over this time interval. Present-day (seismogenic) strain rates were calculated from the seismic moment release rate for earthquakes between 1970 and 2007 and were compared with Quaternary bulk crustal strain rates based on geological evidence to indicate that *c.* 30% of Quaternary deformation is aseismic. The present-day bulk crustal strain rates were then compared with longer-term (i.e. 5–10 Ma) strain rates and shortening estimates determined from published line-length restorations (e.g. Cooper & Hill 1997). Our results indicate that observed degrees of crustal shortening (*c.* 5%; Cooper & Hill 1997) yield strain rates sufficient to generate *c.* 1 km of Late Miocene–Pliocene exhumation. Furthermore, our results demonstrate that bulk crustal strain rates have likely declined from peak values of *c.* $2 \times 10^{-16}\ s^{-1}$ at the onset of the Late Miocene to *c.* $1.09 \times 10^{-17}\ s^{-1}$ at the present day, demonstrating that plate-boundary-controlled intraplate stress fields and deformation can vary significantly over timescales of *c.* 10^6–10^7 a. The observed decline in strain rates may be explained by present-day deformation and intraplate stresses having migrated and localized elsewhere along the southern margin (e.g. the Flinders and Mt Lofty Ranges of South Australia), or a decrease in plate-boundary forcing. Nevertheless, the stress fields generated by distant plate-boundary interactions are capable of generating mild but appreciable deformation, uplift and exhumation within plate interiors (cf. Hillis *et al.* 2008; Holford *et al.* 2009*a*, 2011*a*).

This work forms part of ARC Discovery Project DP0879612, ASEG Research Foundation Project RF09P04 and represents TRaX contribution #146. We thank W. Jones, A. Smith, I. Brown and H. Edwards of PGS for provision of SAMDA, and gratefully acknowledge Geoscience Australia, PIRSA and DPI Victoria for access to seismic and well data. P. Green and I. Duddy of Geotrack International are thanked for access to their non-exclusive and propriety apatite fission-track datasets, and for valuable discussions regarding the geology of the Otway Basin.

References

Bérard, T., Sinha, B. K., Van Ruth, P., Dance, T., John, Z. & Tan, C. 2008. Stress estimation at the Otway CO_2 storage site, Australia. *In*: Society of Petroleum Engineers (ed.) SPE Asia Pacific Oil & Gas Conference and Exhibition, Perth, 20–22 October 2008, SPE 116422.

Braun, J., Burbridge, D. R., Gesto, F. N., Sandiford, M., Gleadow, A. J. W., Kohn, B. P. & Cummings, P. R. 2009. Constraints on the current rate of deformation and surface uplift of the Australian continent from a new seismic database and low-T thermochronological data. *Australian Journal of Earth Sciences*, **56**, 99–110.

Brodie, J. & White, N. 1995. The link between sedimentary basin inversion and igneous underplating. *In*: Buchanan, J. G. & Buchanan, P. G. (eds) *Basin Inversion*. Geological Society, London, Special Publications, **88**, 21–38.

Célérier, J., Sandiford, M., Lundbek Hansen, D. & Quigley, M. 2005. Modes of active intraplate deformation, Flinders Ranges, Australia. *Tectonics*, **24**, TC6006, doi: 10.1029/2004TC001679.

Champion, M. E. S., White, N. J., Jones, S. M. & Lovell, J. P. B. 2008. Quantifying transient mantle convective uplift: an example from the Faroe-Shetland basin. *Tectonics*, **27**, TC1002, doi: 10.1029/2007TC002106.

Collins, C. D. N., Drummond, B. J. & Nicoll, M. G. 2003. Crustal thickness patterns in the Australian continent. *In*: Hillis, R. R. & Muller, R. D. (eds) *Evolution and Dynamics of the Australian Plate*. Geological Society of Australia, Sydney, Special Publication, **22**, 121–128.

Constantine, A. E. & Liberman, N. 2001. Hydrocarbon prospectivity package for VIC/O-01(1), VIC/O-01(2) and VIC/O-01(3), Eastern Onshore Otway Basin, Victoria, Australia: 2001 Acreage release. *Victorian Initiative for Minerals & Petroleum*, VIMP Report 70.

Cooper, G. T. 1995. *Structural geology, thermochronology and tectonic evolution of the Torquay Embayment, eastern Otway Basin*. PhD thesis, Monash University, Australia.

Cooper, G. T. & Hill, K. C. 1997. Cross-section balancing and thermochronological analysis of the Mesozoic development of the eastern Otway Basin. *The APPEA Journal*, **37**, 390–414.

Couzens-Schultz, B. A. & Chan, A. W. 2010. Stress determination in active thrust belts: an alternative leak-off pressure interpretation. *Journal of Structural Geology*, **32**, 1061–1069.

Davis, G. H. & Reynolds, S. J. 1984. *Structural Geology of Rocks and Regions*. 2nd edn. John Wiley & Sons, New York, 128.

Demidjuk, Z., Turner, S., Sandiford, M., George, R., Foden, J. & Ethridge, M. 2007. U-series isotope and geodynamic constraints on mantle melting processes beneath the Newer Volcanic Province in South Australia. *Earth and Planetary Science Letters*, **261**, 517–533.

Denham, D., Weekes, J. & Krayshek, C. 1981. Earthquake evidence for compressive stress in the southeast Australian crust. *Journal of the Geological Society of Australia*, **28**, 323–332.

Dickinson, J. A., Wallace, M. W., Holdgate, G. R., Gallagher, S. J. & Thomas, L. 2002. Origin and timing of the Miocene–Pliocene unconformity in Southeast Australia. *Journal of Sedimentary Research*, **72**, 288–303.

Doré, A. G., Corcoran, D. V. & Scotchman, I. C. 2002. Prediction of the hydrocarbon system in exhumed basins, and application to the NW European margin. *In*: Doré, A. G., Cartwright, J. A., Stoker, M. S., Turner, J. P. & White, N. J. (eds) *Exhumation of the North Atlantic Margin: Timing, Mechanisms and*

Implications for Petroleum Exploration. Geological Society, London, Special Publications, **196**, 401–429.

Dyksterhuis, S. & Muller, R. D. 2008. Cause and evolution of intraplate orogeny in Australia. *Geology*, **36**, 495–498.

Gaul, B. A., Michael-Leiba, M. O. & Rynn, J. M. W. 1990. Probabilistic earthquake risk maps of Australia. *Australian Journal of Earth Science*, **37**, 169–187.

Giles, M. R., Indrelid, S. L. & James, D. M. D. 1998. Compaction – the great unknown in basin modelling. *In*: Duppenbecker, S. J. & Iliffe, J. E. (eds) *Basin Modelling: Practice and Progression*. Geological Society, London, Special Publications, **141**, 15–43.

Green, P. F., Crowhurst, P. V. & Duddy, I. R. 2004. Integration of AFTA and (U–Th)/He thermochronology to enhance the resolution and precision of thermal history reconstruction in the Anglesea-1 well, Otway Basin, SE Australia. *In*: Boult, P. J., Johns, D. R. & Lang, S. C. (eds) *Eastern Australasian Basins Symposium II*. Petroleum Exploration Society of Australia, West Perth, Special Publication, 117–131.

Gutenberg, B. & Richter, C. F. 1944. Frequency of earthquakes in California. *Bulletin of the Seismological Society of America*, **34**, 185–188.

Hanks, T. C. & Kanamori, H. 1979. A moment magnitude scale. *Journal of Geophysical Research*, **84**, 2348–2350.

Hill, K. C., Hill, K. A., Cooper, G. T., O'Sullivan, A. J., O'Sullivan, P. B. & Richardson, M. J. 1995. Inversion around the Bass Basin, SE Australia. *In*: Buchanan, J. G. & Buchanan, P. G. (eds) *Basin Inversion*. Geological Society, London, Special Publications, **88**, 525–547.

Hillis, R. R. & Reynolds, S. D. 2000. The Australian stress map. *Journal of the Geological Society, London*, **157**, 915–921.

Hillis, R. R., Monte, S. A., Tan, C. P. & Willoughby, D. R. 1995. The contemporary stress field of the Otway Basin, South Australia: implications for hydrocarbon exploration and production. *The APPEA Journal*, **35**, 494–506.

Hillis, R. R., Sandiford, M., Reynolds, S. D. & Quigley, M. C. 2008. Present-day stresses, seismicity and Neogene-to-Recent tectonics of Australia's 'passive' margin: intraplate deformation controlled by plate boundary forces. *In*: Johnson, H., Doré, A. G., Gatliff, R. W., Holdsworth, R., Lundin, E. R. & Ritchie, J. D. (eds) *The Nature and Origin of Compression in Passive Margins*. Geological Society, London, Special Publications, **306**, 71–90.

Holford, S. P., Green, P. F., Duddy, I. R., Turner, J. P., Hillis, R. R. & Stoker, M. S. 2009*a*. Regional intraplate exhumation episodes related to plate-boundary deformation. *Geological Society of America Bulletin*, **121**, 1611–1628, doi: 10.1130/B26481.1.

Holford, S. P., Turner, J. P., Green, P. F. & Hillis, R. R. 2009*b*. Signature of cryptic sedimentary basin inversion revealed by shale compaction data in the Irish Sea, western British Isles. *Tectonics*, **28**, TC4011, doi: 10.29/2008TC002359.

Holford, S. P., Hillis, R. R., Duddy, I. R., Green, P. F., Tuitt, A. K. & Stoker, M. S. 2010. Impacts of Neogene–Recent compressional deformation and uplift on hydrocarbon prospectivity of the passive southern Australian margin. *The APPEA Journal*, **50**, 267–286.

Holford, S. P., Hillis, R. R., Duddy, I. R., Green, P. F., Tassone, D. R. & Stoker, M. S. 2011*a*. Palaeothermal and seismic constraints on late Miocene–Pliocene uplift and deformation in the Torquay sub-basin, southern Australian margin. *Australian Journal of Earth Sciences*, **58**, 543–562.

Holford, S. P., Hillis, R. R., Hand, M. & Sandiford, M. 2011*b*. Thermal weakening localizes intraplate deformation along the southern Australian continental margin. *Earth and Planetary Science Letters*, **305**, 207–214, doi: 10.1016/j.epsl.2011.02.056.

Holford, S. P., Hillis, R. R. *et al*. 2011*c*. Cenozoic post-breakup compressional deformation and exhumation of the southern Australian margin. *The APPEA Journal*, **51**, 613–638.

Jackson, J. & McKenzie, D. 1988. The relationship between plate motions and seismic moment tensors, and the rates of active deformation in the Mediterranean and Middle East. *Geophysical Journal*, **93**, 45–73.

Johnston, A. C. 1994. Seismotectonic interpretations and conclusions from the stable continental region seismicity database. *In*: Johnston, A. C., Coppersmith, K. J., Kanter, L. R. & Cornell, C. A. (eds) *The Earthquakes of Stable Continental Region – Volume 1: Assessment of Large Earthquake Potential*. Electric Power Research Institute, Palo Alto, California, Report TR-102261-V1.

King, R. C., Holford, S. P., Hillis, R. R., Backé, G., Tingay, M. R. P. & Tuitt, A. 2011. Reassessing the in-situ stress regimes of Australia's petroleum basins. AAPG 2011 Annual Convention and Exhibition, 10–13 April, Houston.

Kostrov, V. V. 1974. Seismic moment and energy of earthquakes, and seismic flow of rocks. *Earth Physics*, **1**, 23–40.

Krassay, A. A., Cathro, D. L. & Ryan, D. J. 2004. A regional tectonostratigraphic framework for the Otway Basin. *In*: Boult, P. J., Johns, D. R. & Land, S. C. (eds) *Eastern Australasian Basins Symposium II*. Petroleum Exploration Society of Australia, West Perth, Special Publication, 97–116.

Leonard, M. 2008. One hundred years of earthquake recording in Australia. *Bulletin of the Seismological Society of America*, **98**, 1458–1470.

McKenzie, D., Nimmo, F., Jackson, J., Gans, P. & Miller, E. 2000. Characteristics and consequences of flow in the lower crust. *Journal of Geophysical Research*, **105**, 11 029–11 046.

Nelson, E., Hillis, R. R., Sandiford, M., Reynolds, S. D. & Mildren, S. 2006. Present-day state-of-stress of southeast Australia. *The APPEA Journal*, **46**, 283–305.

Norvick, M. S. & Smith, M. A. 2001. Mapping the plate tectonic reconstruction of southern and southeastern Australia and implications for petroleum systems. *The APPEA Journal*, **41**, 15–35.

Praeg, D., Stoker, M. S., Shannon, P. M., Ceramicola, S., Hjelstuen, B. O., Laberg, J. S. & Mathiesen, A. 2005. Episodic Cenozoic tectonism and the development of the NW European 'passive' continental margin. *Marine and Petroleum Geology*, **22**, 1007–1030.

Quigley, M., Cupper, M. & Sandiford, M. 2006. Quaternary faults of south-central Australia: palaeoseismicity, slip rates and origin. *Australian Journal of Earth Sciences*, **53**, 285–301.

Reynolds, S. D., Coblentz, D. & Hillis, R. R. 2003. Influences of plate-boundary forces on the regional intraplate stress field of continental Australia. *In*: Hillis, R. R. & Muller, R. D. (eds) *Evolution and Dynamics of the Australian Plate*. Geological Society of Australia, Sydney, Special Publication, **22**, 59–70.

Sandiford, M. 2003*a*. Geomorphic constraints on the late Neogene tectonics of the Otway Ranges. *Australian Journal of Earth Sciences*, **50**, 69–80.

Sandiford, M. 2003*b*. Neotectonics of south-eastern Australia: linking the Quaternary faulting record with seismicity and in situ stress. *In*: Hillis, R. R. & Muller, R. D. (eds) *Evolution and Dynamics of the Australian Plate*. Geological Society of Australia, Sydney, Special Publication, **22**, 107–119.

Sandiford, M. 2007. The tilting continent: a new constraint on the dynamic topographic field from Australia. *Earth and Planetary Science Letters*, **261**, 152–163.

Sandiford, M. & Quigley, M. 2009. TOPO-OZ: insights into the various modes of intraplate deformation in the Australian continent. *Tectonphysics*, **474**, 405–416.

Sandiford, M., Leonard, M. & Coblentz, D. 2003. Geological constraints on active seismicity in southeast Australia. *In*: Wilson, N. J. L., Nam, N. K. & Gibson, G. (eds) *Earthquake Risk Mitigation*. Australian Earthquake Engineering Society, Victoria, 1–10.

Sandiford, M., Wallace, M. & Coblentz, D. 2004. Origin of the in situ stress field in south-eastern Australia. *Basin Research*, **16**, 325–338.

Stoker, M. S., Hoult, R. J. *et al.* 2005. Sedimentary and oceanographic responses to early Neogene compression on the NW European margin. *Marine and Petroleum Geology*, **22**, 1031–1044.

Stoker, M. S., Holford, S. P., Hillis, R. R., Green, P. F. & Duddy, I. R. 2010. Cenozoic post-rift sedimentation off northwest Britain: recording the detritus of episodic uplift on a passive continental margin. *Geology*, **38**, 595–598.

Turner, J. P. & Williams, G. A. 2004. Sedimentary basin inversion and intra-plate shortening. *Earth Science Reviews*, **65**, 277–304.

Vidal-Gilbert, S., Tenthorey, E., Dewhurst, D., Ennis-King, J., Van Ruth, P. & Hillis, R. R. 2010. Geomechanical analysis of the Naylor Field, Otway Basin, Australia: implications for CO_2 injection and storage. *International Journal of Greenhouse Gas Control*, **4**, 827–839.

White, N. & Lovell, B. 1997. Measuring the pulse of a plume with the sedimentary record. *Nature*, **387**, 888–891, doi: 10.1038/43151.

Ziegler, P. A., Cloetingh, S. & Van Wees, J. D. 1995. Dynamics of intra-plate compressional deformation: the Alpine foreland and other examples. *Tectonophysics*, **252**, 7–59.

Pressure conditions for shear and tensile failure around a circular magma chamber; insight from elasto-plastic modelling

MURIEL GERBAULT

Université de Nice Sophia-Antipolis, Institut de Recherche pour le Développement (UR 082), Observatoire de la Côte d'Azur, Géoazur, 250 av Einstein 06560 Valbonne, France (e-mail: gerbault@geoazur.unice.fr)

Abstract: Overpressure within a circular magmatic chamber embedded in an elastic half space is a widely used model in volcanology. However, this overpressure is generally assumed to be bounded by the bedrock tensile strength since gravity is neglected. Critical overpressure for wall failure is thus greater. It is shown analytically and numerically that wall failure occurs in shear rather than in tension, because the Mohr–Coulomb yield stress is less than the tensile yield stress. Numerical modelling of progressively increasing overpressure shows that bedrock failure develops in three stages: (1) tensile failure at the ground surface; (2) shear failure at the chamber wall; and (3) fault connection from the chamber wall to the ground surface. Predictions of surface deformation and stress with the theory of elasticity break down at stage 3. For wall tensile failure to occur at small overpressure, a state of lithostatic pore-fluid pressure is required in the bedrock which cancels the effect of gravity. Modelled eccentric shear band geometries are consistent with theoretical solutions from engineering plasticity and compare well with shear structures bordering exhumed intrusions. This study shows that the measured ground surface deformation may be misinterpreted when neither plasticity nor pore-fluid pressure is accounted for.

Supplementary material: The numerical benchmark data are available at: http://www.geolsoc.org.uk/SUP18517.

An unsolved question in volcanology is that of the conditions for and progression of country rock failure around near-surface pressurized magma bodies. Anderson (1936) was the first to apply mathematical solutions derived from the theory of elasticity to explain the formation of cone-sheets and ring-dykes around circular igneous intrusions: 'In explaining central intrusions it is necessary to involve both types of rupture, the tensile type to produce cone-sheets, and the shearing-type to account for the production of ring-dykes, giving rise to nearly vertical ring-fractures'. Despite several decades of analytical and numerical models of deformation around magmatic chambers (e.g. reviews by Acocella 2007; Marti *et al.* 2008; Geyer & Marti 2009) and a considerably longer history of field studies examining exposed structures (e.g. Gudmundsson 2006), many questions remain. The crude state of our knowledge with respect to magma chamber deformation is also revealed with the application of elastic models to geodetic monitoring of active volcanoes (e.g. Mogi 1958; Masterlark 2007; Segall 2009). As the non-unique fit between measured and modelled deformation spans a wide range of parameters (chamber geometry, country rock rheology, pore pressure and magmatic overpressure, etc.) it is clear that more constraints are desirable. One fundamental approach is to simulate the failure patterns that arise from an idealistic volcanic chamber with a minimum number of parameters; this is the objective of this work.

In this paper it is shown, analytically and numerically, how commonly used solutions of critical overpressure for bedrock failure can be biased when neglecting the role of gravity. Whereas Grosfils (2007) raised this problem of yield stress conditions for tensile failure around a circular magmatic chamber, here it is demonstrated how in fact shear failure should theoretically occur rather than tensile failure. Previous models that use self-consistent elasto-plasticity or visco-elasto-plasticity have not directly addressed the critical pressure condition for bedrock failure. As a matter of fact, the precise state of internal pressure associated with the initiation and propagation of faulting around an idealistic circular magmatic chamber remains poorly constrained.

First, the conditions for failure around a circular pressurized inclusion are developed. The reasoning is then supported with numerical models that incorporate self-consistent elasto-plasticity with a Mohr–Coulomb failure criterion. The progression of plastic yielding is explored, displaying peculiar geometrical patterns that are produced with increasing internal overpressure, and their similarity to solutions from engineering plasticity are demonstrated.

From: Healy, D., Butler, R. W. H., Shipton, Z. K. & Sibson, R. H. (eds) 2012. *Faulting, Fracturing and Igneous Intrusion in the Earth's Crust*. Geological Society, London, Special Publications, **367**, 111–130. http://dx.doi.org/10.1144/SP367.8

The role of pore-fluid pressure is also illustrated, as it nullifies the gravity component in the shear failure yield criterion and allows for tensile failure to occur instead.

In discussing conditions for shear failure in the real world, a few examples of shearing structures exposed at the borders of magmatic bodies are displayed. The main point of this paper is to show that it is necessary to consider a state of near-lithostatic fluid overpressure in the bedrock in order to explain field observations of tensile failure, predicted to occur for less than or an average of 10 MPa of internal overpressure around an ideal spherical magmatic chamber.

Analytical pressure solutions for tensile and shear failure

Classical solution for tensile failure

Simple solutions of surface displacement over a spherical source, with depth much greater than radius, were approximated by Mogi (1958) using a dilational point source. This solution remains widely used for fitting geodetic measurements above active volcanoes (Bonafede *et al.* 1986; Masterlark & Lu 2004; Pritchard & Simons 2004; Trasatti *et al.* 2005; Ellis *et al.* 2007; Masterlark 2007; Bonafede & Ferrari 2009; Foroozan *et al.* 2010). Crustal deformation in the idealistic elastic crust is assumed to be the result of a pressure change ΔP in the source with respect to 'hydrostatic' pressure. Assuming Poisson's ratio ν is equal to 0.25 and G is the shear modulus, calculated horizontal and vertical displacements at the surface are:

$$u_x = \frac{3\Delta P}{4G}\frac{R^3 x}{(x^2+H^2)^{3/2}}, \quad u_y = \frac{3\Delta P}{4G}\frac{R^3 H}{(x^2+H^2)^{3/2}} \tag{1}$$

where R and H are the radius and the depth of the chamber, respectively (Fig. 1). In plane-strain, the magma chamber becomes an infinitely long cylinder and the analytical solution differs by a factor of about 4 depending on R and H; stress functions for a generic class of problems with complex variables are provided by Verruijt (1998).

The distribution of the stress field produced by a pressurized circular cavity in an elastic half-space was solved by Jeffery (1920) using curvilinear coordinates and plane strain. At the free surface, he showed that the horizontal component of the stress field (σ_{xx}) is:

$$\sigma_{xx} = \frac{-4\Delta P R^2(x^2 - H^2 + R^2)}{(x^2+H^2-R^2)^2}. \tag{2}$$

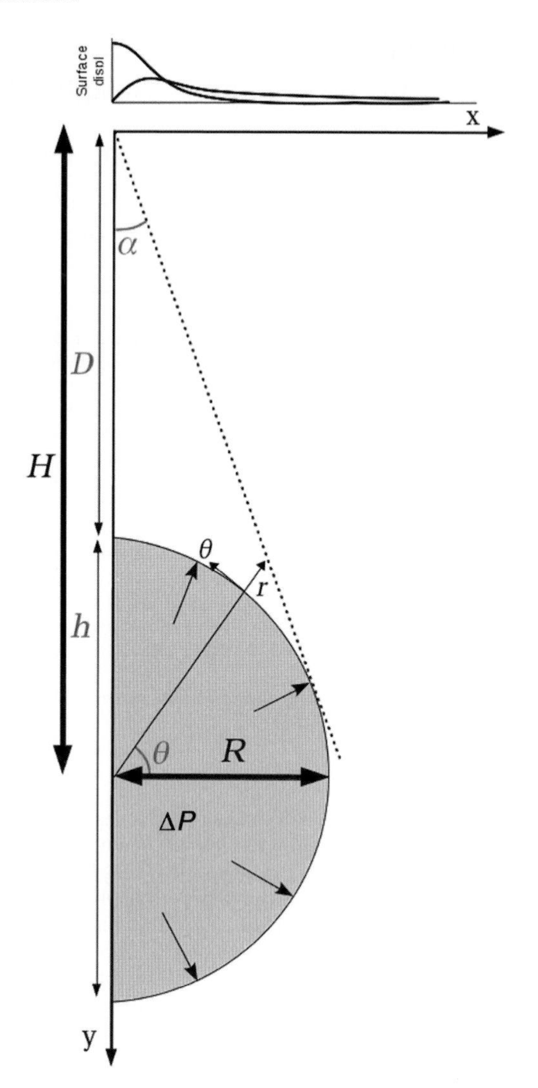

Fig. 1. Parametric definition of the problem. Grey values are those defined by Grosfils (2007).

Its maximum is located above the chamber at X_o, with $\sigma_{xx} = 4\Delta P R^2/(H^2 - R^2)$. At the chamber wall, the hoop stress is expressed according to either the angle α between the vertical axis of symmetry and the line joining the surface origin X_o to a point on the chamber wall (Jeffery 1920) or to the polar angle θ taken from the horizontal at the centre of the chamber to that same point (Grosfils 2007):

$$\sigma_{\theta\theta} = \Delta P(1 + 2\tan^2\alpha). \tag{3}$$

Jeffery (1920) defined the free surface correction factor $C = 1 + 2\tan^2\alpha$, and showed that the hoop stress at the wall is maximum when

$\alpha_m = \arcsin(R/H)$ (Fig. 1). Then, $\sigma_m = -\Delta P(H^2 + R^2)/(H^2 - R^2)$.

Jeffery (1920) also discussed the conditions for failure: 'if H is greater than $R\sqrt{2}$ and less than $R\sqrt{3}$, and internal pressure increased until failure occurs, the crack will begin on the surface according to the greatest tension theory or on the edge of the hole if the greatest stress-difference theory holds'.

The majority of authors, such as Gudmundsson (1988), Tait *et al.* (1989), Gudmundsson *et al.* (1997) and Pinel & Jaupart (2003, 2005), assume that the same tensile failure criterion applies to the Earth's surface and the chamber walls, that is, when the deviatoric stress $(\sigma_1 - \sigma_3)/2$ exceeds the rock's tensile strength T.

- At the surface where the gravity component vanishes, failure occurs when $(\sigma_{xx} - \sigma_{zz})/2 = T$. By combining with Equation (3), the internal overpressure reaches the critical value

$$\Delta P_S = T\left(\frac{H^2 - R^2}{2R^2}\right). \tag{4}$$

- At the chamber wall, the same tensile failure criterion gives $(\sigma_{\theta\theta} - \Delta P)/2 = T$. By combining with Equation (2), the critical internal overpressure is

$$\Delta P_T = T\left(\frac{H^2 - R^2}{H^2}\right). \tag{5}$$

Tensile failure is therefore traditionally predicted to initiate at the previously given location α_m along the chamber wall for magmatic overpressures $\Delta P_T \leq T$, which ranges between *c.* 6 MPa (Gudmundsson *et al.* 2002) and 20 MPa (Pinel & Jaupart 2003).

Grosfils' solution for tensile failure

Grosfils (2007) demonstrated analytically and numerically that a higher critical overpressure was required to initiate wall failure, invoking the necessity to account for the gravity body force. Considering a spherical chamber of equal magma and rock densities, his analytical reasoning is summarized as follows.

(a) The total magma pressure is first defined as the sum of an internal overpressure ΔP and the lithostatic stress, $P_t = \Delta P - \rho g y$.

(b) Grosfils (2007) then proceeded as in previous studies (e.g. Gudmundsson 1988) by assuming that the tangential stress $\sigma_{\theta\theta}$ equals half the normal stress balance across the wall modified by the free surface factor C, so that:

$$\sigma_{\theta\theta} = C\Delta P/2. \tag{6}$$

(c) Grosfils (2007) expressed C as a function of relative depth h taken from the crest of the chamber, itself located at depth $D = H - R$ (Fig. 1):

$$C = 1 + 2\tan^2\alpha = \frac{2h(2R + D) + D^2 - h^2}{(D + h)^2}. \tag{7}$$

C has a parabolic shape >1, and equals 1 when h equals 0 or $2R$.

(d) Differently from previous authors, Grosfils (2007) then argued that tensile failure at the chamber's wall occurs when the tangential stress $\sigma_{\theta\theta}$ exceeds not only the tensile strength of rocks T but also the wall-parallel component of the lithostatic stress P_l, that is

$$\sigma_{\theta\theta} = T + P_l = T + \rho g(D + h). \tag{8}$$

(e) Combining Equations (6) and (8), Grosfils' overpressure for tensile failure $\Delta P = \Delta P_{TG}$ is defined:

$$\Delta P_{TG} = \frac{2T}{C} + 2\rho g\frac{D + h}{C}. \tag{9}$$

In summary, 'the total pressure P_t required for tensile failure approaches the limit of 3 times the lithostatic stress σ_z at great depth $(D \gg R)$, whereas it becomes less than σ_z if $R > 0.6D$' (Grosfils 2007). The associated critical overpressure ΔP_{TG} contrasts with the prediction of ΔP_T of failure for only a few MPa above lithostatic pressure (Equation (5), e.g. Tait *et al.* 1989; Gudmundsson *et al.* 1997, 2002; Pinel & Jaupart 2003, 2005; Gudmundsson 2006). Grosfils (2007) discussed in detail how this difference stems from the absence of gravity in the formulation of most analytical models and contrary to the common belief: 'a self-consistent numerical approach that incorporates explicit boundary stresses and body loads becomes a necessary reference against which analytical models should be improved'.

Conditions for shear failure

Now we shall evaluate the internal overpressure (ΔP_{MC}) necessary to produce shear failure in the bedrock surrounding the chamber, and compare it with the above-mentioned overpressure prediction required for tensile failure.

The classical Mohr–Coulomb criterion of failure relates the tangential and normal stresses τ_s and σ_n along any given plane of a medium with internal friction angle φ and cohesion S_o. This criterion can also be written in terms of the pressure P and the deviatoric shear stress σ_{II}:

$$\tau_s = S_o - \sigma_n \tan\varphi, \quad \tau_s = \sigma_{II}\cos\varphi, \quad \sigma_n = -P + \sigma_{II}\sin\varphi. \tag{10}$$

We first neglect gravity, and assume that the minimum and maximum principal stress components at the chamber wall are of opposite sign so that $P = 0$ and $\sigma_{II} = \Delta P_{MC}$ (ΔP_{MC} is the applied internal overpressure; note that in 3D, $\sigma_{II} = \Delta P_{MC}/2$; Timoshenko & Goodier 1970). When adding gravity, the isotropic lithostatic component appears in the total pressure $P = -\rho g y$ (at negative depth y) which, when inserted into Equation (10), provides:

$$\Delta P_{MC} = S_o \cos\varphi - \rho g y \sin\varphi. \tag{11}$$

In order to account for the free surface (e.g. Gudmundsson 1988; Parfitt *et al.* 1993), the factor C appears in the relationship between the principal stresses at the wall: $\sigma_3 = -C\sigma_1$. The stress invariants P and σ_{II} are then expressed:

$$P = -\rho g y + \Delta P \frac{C-1}{2}, \quad \sigma_{II} = \Delta P \frac{C+1}{2}.$$

Inserting these expressions into the Mohr–Coulomb criterion (Equation (10)) and defining a theoretical tensile strength deduced from the cohesion $T_o = S_o/\tan\varphi$ provides the critical overpressure required for shear failure ΔP_{MCS} :

$$\Delta P_{MCS} = 2\sin\varphi \frac{-\rho g y + T_o}{1 + \sin\varphi + C(1 - \sin\varphi)}. \tag{12}$$

At the vertical axis of symmetry where $C = 1$, this expression is equivalent to Equation (11).

Figure 2 displays all three analytical predictions of the internal pressures required for (a) tensile failure at the chamber wall according to classical predictions (Equation (5)), (b) Grosfils' prediction (Equation (9)), and (c) shear failure according to Equation (12). Friction and cohesion are defined as

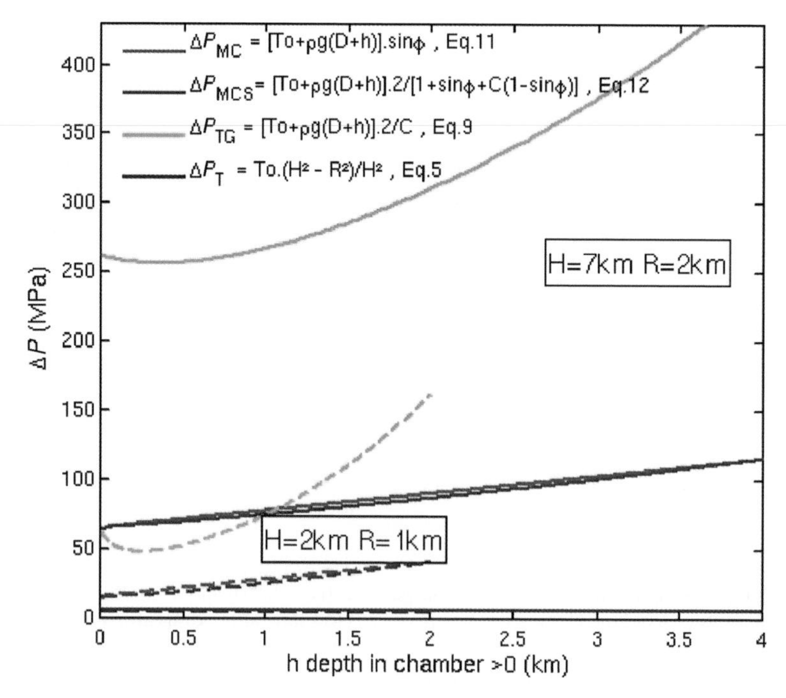

Fig. 2. Pressure difference ΔP required to initiate failure at the chamber wall according to different assumptions: tensile failure accounting for free surface without gravity (ΔP_T, Equation (5)) and with gravity (Grosfils 2007; ΔP_{TG}, Equation (9)) and Mohr–Coulomb shear failure with and without a free surface (ΔP_{MC}, Equation (11); ΔP_{MCS}, Equation (12)). Here both depth and overpressure are set positive. Failure onset is predicted where the critical overpressure is minimal: it initiates at the crest when $H = 7$ km and $R = 2$ km (depth $h = 0$ and angle $\theta = 90°$) and at h *c.* 0.2 km when $H = 2$ km and $R = 1$ km.

$\varphi = 30°$, $S_o = 10$ MPa. Two cases with different depth and radius are displayed, one at depth $H = 7$ km and radius $R = 2$ km ($R/D = 0.4$) and another at depth $H = 2$ km and $R = 1$ km ($R/D = 1$). For both cases, the critical overpressure for shear failure (ΔP_{MC}) is smaller than that for tensile failure when gravity is accounted for. The effect of the free-surface modifies only the internal slope between the crest and the base of the chamber, which indicates that shear failure can initiate over a broad domain extending laterally from the chamber roof.

To summarize, it has been shown that a chamber wall should yield by shear failure (mode II) rather than by tensile failure (mode I).

Predictions from other studies

Many studies have developed analogue and numerical models of failure around a magmatic chamber (e.g. Roche & Druitt 2001; Acocella *et al.* 2004; Acocella 2007; Marti *et al.* 2008) but surprisingly, apart from Grosfils (2007), few directly address the association between the mode of failure, the geometry of the failure domain and the internal overpressure. There are several methodological reasons for this.

First, a majority of studies model only elastic behaviour of the bedrock, and contour the domains that exceed a failure stress threshold that is chosen a priori (e.g. Sartoris *et al.* 1990; Gudmundsson 1988, 2006; Gudmundsson *et al.* 2002; Masterlark 2007). This method has been used since the 1950s (Hafner 1951) but, while it generally yields good results to a first order, it does not account for self-consistent plasticity. This method therefore cannot address when exactly failure initiates with respect to the level of internal magmatic overpressure.

Other models incorporate self-consistent elasto-plasticity, but a dilational deformation is applied instead of an internal overpressure (Chery *et al.* 1991; Kusumoto & Takemura 2003; Gray & Monaghan 2004; Hardy 2008). Only Chery *et al.* (1991) evaluate chamber overpressures of 50–60 MPa around a magmatic chamber 10 km deep embedded in a temperature-dependent elasto-plastic-ductile crust; however, a state of hydrostatic pore-fluid pressure was also assigned in the bedrock, and the precise shape of the associated failure domain was not given (the mesh resolution was poor with a chamber wall composed of 12 elements).

Models that define a magmatic chamber with sharp edges produce failure much ‘more easily’ than when an ideal circular body is used. For example, Burov & Guillou-Frottier (1999) and Guillou-Frottier *et al.* (2000) modelled a cycle of inflation and collapse with only 10 MPa overpressure, applied in a middle-crust rectangular chamber, with visco-elasto-plastic rheology. In these models, shear band structures develop between the corners of the chamber and the surface.

Finally, Trasatti *et al.* (2005) tested a number of parameters to model surface uplift at Campi Flegrei volcano. Whereas Mogi-type elastic models without gravity reproduced the measured uplift for an internal overpressure of at least 80 MPa (the lithostatic pressure at the chamber’s crest D *c.* 3200 m, recognized as unrealistic), elasto-plastic models including gravity and a Von Mises failure threshold set to 15 MPa reproduced the uplift for an internal overpressure of only 45–50 MPa. This study illustrates how greater surface uplift is obtained when accounting for shear failure, and for a ‘reasonable’ overpressure greater than the tensile strength. (Note that this overpressure fits Equation (12) at the crest; sin(30) ($\rho g D + T$) *c.* 48 MPa). Unfortunately, the geometrical pattern of bedrock failure was not provided in this study.

From an engineering plasticity viewpoint

In engineering mechanics, a problem similar to a magmatic chamber is that of a pressurized cavity, applied to metal indentation and tunnelling. A technique known as slip-line field theory is used. For example, Nadai (1950) displayed the slip-lines solution associated with the indentation of an infinite plastic medium (Fig. 3a). Defining the radial and tangential normal principal stresses as σ_r and σ_θ and k as Tresca’s yield stress, the plasticity condition reduces to $\sigma_r - \sigma_\theta = \pm 2k$ and solutions take the form:

$$\sigma_r = \pm 2k \ln\left(\frac{R}{r}\right), \qquad \sigma_\theta = \pm 2k\left[1 + \ln\left(\frac{R}{r}\right)\right].$$

The resulting slip-lines are two orthogonal families of logarithmic spirals, commonly observed in steel plates pressed against a cylindrical stamp.

Prediction of plastic flow patterns for the even ‘simpler’ problem of flat indentation remains difficult because there are many possible slip systems. Closed-form kinematical solutions for flat indentation were calculated by Salençon (1969, Fig. 3b) and will be used further at the end of this paper. A lower bound for the internal pressure associated with failure around a cylindrical cavity was provided by Salençon (1966), whereas Caquot & Kerisel (1956) and d’Escatha & Mandel (1974) analysed the static admissible stress field for an upper bound using various friction and cohesion values. d’Escatha and Mandel presented graphical solutions based on the slip-line characteristics method (shown in Fig. 3c), and Caquot & Kerisel (1956) provided closed-form solutions. Fairhurst & Carranza-Tores (2002) used the numerical code Fast Lagrangian Analysis of Continuum (FLAC) to simulate a progressively decreasing support pressure on the wall

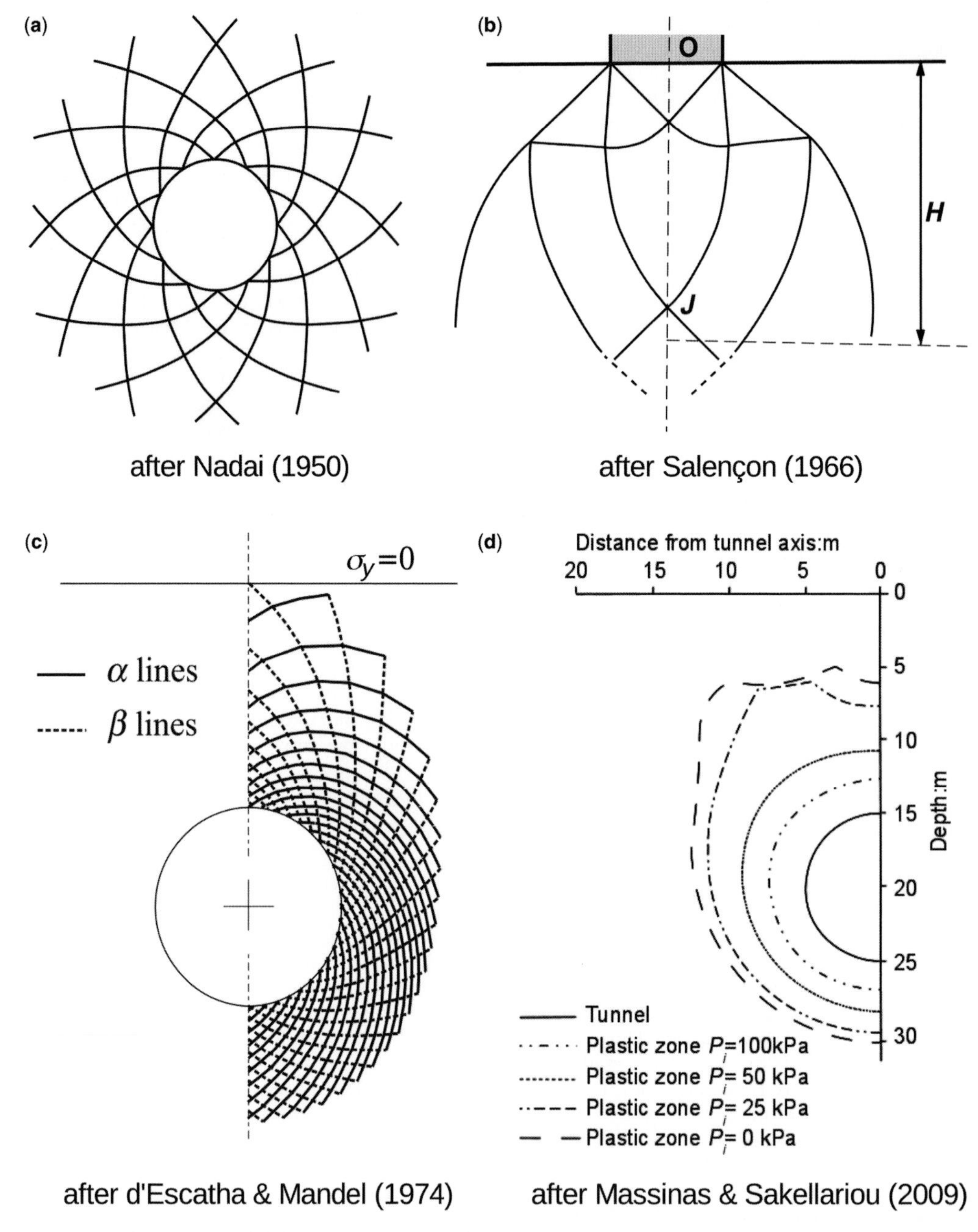

Fig. 3. Solutions from engineering mechanics. (**a**) Slip-line solution for circular indentation in a perfectly plastic infinite medium (after Nadai 1950, in which a photograph of steel indentation shows striking similarity). (**b**) Closed-form solution for flat indentation, displaying slip-line intersection at the axis of symmetry of flat indentation for $OJ/a > 5.298$ (after Salençon 1969). (**c**) Slip-line characteristics calculated by d'Escatha & Mandel (1974) for a friction angle $\varphi = 20°$. (**d**) Plasticized domain calculated by Massinas & Sakellariou (2009) for different internal support pressures (P_i) in tunnels (note the ear shape appearing at some stage).

of a circular tunnel, and observed that failure did not start until the pressure was reduced to 80% of Caquot & Kerisel's (1956) upper bound value.

More recently, Massinas & Sakellariou (2009) provided another closed-form solution of the critical support pressure P_{cr} for wall failure around a tunnel.

Bipolar coordinates α and β were used similarly to that of Jeffery (1920), in which $\alpha_1 = \cosh^{-1}(H/R)$. Here, P_o is defined as the uniform external pressure (set as a proxy to the gravity load at the cavity's centre) and the Mohr–Coulomb friction angle φ is defined as before. Failure occurs at the wall for the minimum value of internal support pressure P_{cr}, which occurs when coordinate β equals 0 at the tunnel crest:

$$P_{cr} = \frac{P_o(1 - \sin\varphi)(1 + \sin^2\beta/\sinh^2\alpha_1) - T_o \sin\varphi}{1 + (1 - \sin\varphi)\sin^2\beta/\sinh^2\alpha_1}. \quad (13)$$

For $\varphi = 30^\circ$ and $\beta = 0$,

$$P_{cr} = \frac{P_o - T_o}{2}.$$

Note how this critical overpressure compares with that calculated in Equation (11): it is equal when considering (1) that signs are opposite for a depressed cavity and for an inflating chamber and (2) that the integrated average of a linearly increasing pressure with depth corresponds to half the maximum pressure P_o.

According to Massinas & Sakellariou (2009) and references therein, for deep tunnels with $H/R \geq 7$, the geometry of the plasticized domain is circular around the tunnel as the effect of the free surface is less than 10%. For shallow tunnels with $H/R < 7$, the geometry of the plasticized domain becomes eccentric (ear-shaped; Fig. 3d).

The role of pore-fluid pressure

The role of pore-fluid pressure in volcanic systems has long been thought to contribute to complex mechanisms such as heating of confined pore water by intrusions, degassing of intrusions, discharges of highly pressurized fluids from depth or deformation by faulting (e.g. Day 1996). Here, we will use a very simple assumption about the role of pore-fluid pressure in the bedrock, and argue how it can play a fundamental role in the mode of fracturing around magmatic chambers. Hubbert & Rubey (1959) proposed that fluids in pores produce an effective normal stress $\sigma_{eff} = \sigma_n - p_f$ that is represented in the Mohr–Coulomb yield criterion as:

$$\tau = S_o - (\sigma_n - p_f)\tan(\phi). \quad (14)$$

Hubbert & Rubey (1959) expressed p_f in terms of the vertical lithostatic stress and the pore-fluid pressure ratio λ, so that $p_f = -\lambda \rho g y$. In a rock of density 2500 kg m^{-3}, $\lambda = 0.4$ for hydrostatic pore-fluid pressure and $\lambda = 1$ for lithostatic pore-fluid pressure. While Townend & Zoback (2000) argued convincingly that hydrostatic pore-fluid pressure appropriately describes an equilibrium state within the crust, lithostatic pore-fluid pressure may also be relatively common in nature since many well data indicate values of $\lambda = 0.9$ (Engelder & Leftwich 1997; Hillis 2003). If we consider this extreme case of lithostatic pore-fluid pressure in the bedrock surrounding a magmatic chamber, then the component of gravity is cancelled out in the failure criterion (Equation (14)) and wall failure at the chamber wall can then be evaluated without accounting for this body force (Equation (5) becomes valid again).

If the development of up to lithostatic pore-fluid pressures in the neighbourhood of a magmatic chamber can be justified, then shear and tensile failure are predicted to occur under relatively small internal overpressures of the order of the tensile strength of rocks (10 MPa).

A state of lithostatic pore-fluid pressure surrounding magmatic chambers should therefore be envisaged when field observations report a majority of mode I opening dykes (e.g. Gudmundsson 2006). This point will be further discussed in the Conclusion.

Numerical modelling of inflating magma chambers

Numerical method and set-up

The finite-differences code Parovoz is used (Poliakov & Podladchikov 1992; Podladchikov *et al.* 1993), which is based on the FLAC method (Cundall & Board 1988). In this method the equations of motion are solved explicitly in time and in large-strain mode, retaining a locally small strain formulation commonly used in continuum mechanics. This method is well known for being able to reproduce the initiation and propagation of non-predefined faults (treated as shear bands, e.g. Poliakov & Podladchikov 1992), and it has been used in a number of geodynamical settings (e.g. Burov & Guillou-Frottier 1999; Gerbault *et al.* 1998; Burov *et al.* 2003; Lavier *et al.* 2000). The program executes the following procedure in one time-step. Velocities are first calculated from the equation of motion with density ρ, time t, velocity vector V, stress tensor σ and gravitational acceleration g:

$$\rho\frac{dV_i}{dt} + \frac{d\sigma_{ij}}{dx_i} = \rho g \quad (15)$$

where d/dt and d/dx_i are the time and space derivatives. The deformation rate is:

$$\varepsilon_{ij} = \frac{1}{2}\left(\frac{dV_i}{dx_j} + \frac{dV_j}{dx_i}\right) \quad (16)$$

and is used to calculate the new stress distribution from the elasto-plastic constitutive law. From these new stresses, nodal forces and displacements are evaluated and inserted in the next time-step. Elasticity relates stress and strain using the Lamé parameters λ and G (δ_{ij} is the Kroenecker symbol):

$$\sigma_{ij} = \lambda\varepsilon_{ij} + 2G\varepsilon_{ij}\delta_{ij}. \quad (17)$$

Non-associated plastic flow is modelled (e.g. Cundall 1989; Gerbault *et al.* 1998; Kaus 2010) with a frictional Mohr–Coulomb stress criteria given in Equation (10) and a dilatancy set to 0 (Vermeer & de Borst 1984). Common failure parameters are chosen with friction $\varphi = 30°$, cohesion $S_0 = 10$ MPa and tensile cut-off strength $T = 5$ MPa. Modelled shear bands are commonly assimilated to shear 'faults'. Tensile failure is detected when one or more stress components exceeds the tensile cut-off T. Despite Parovoz being a two-dimensional plane-strain code, the condition for failure is evaluated with the three stress components. Although Parovoz can identify domains of tensile failure, the mesh continuum cannot split such as in real mode I crack opening. However, White *et al.* (2004) successively compared domains of tensile failure identified by FLAC^{2D}, with a Particle Flow Code that can track individual microfractures via breakable bonds.

Our problem is modelled in plane-strain. The left border ($X = 0$) represents the vertical axis of symmetry passing through the centre of a circular magma chamber. Domain dimensions are 100 km in length and 80 km in depth, far enough from the zone of interest to minimize border effects. The chamber has radius $R = 2$ km and is located at $H = 7$ km depth.

The mesh is defined with quadrilateral elements of height and width equal to 25 m over the first 12 km of the model domain. Grid resolution progressively reduces in both directions to *c.* 1 km at the bottom-right corner. The total number of mesh elements is 275,000.

In the model, all borders apart from the free ground surface have free-slip boundary conditions. A uniform rock density is set at $\rho = 2500$ kg m^{-3} (Table 1 summarizes all analytical and numerical

Table 1. *List of parameters and specific values used for the reference model*

Symbol	Description	Value
R	Radius of chamber	2 km
H	Depth to centre of chamber	7 km
D	Depth to crest of chamber, used by Grosfils (2007) $H-R$	5 km
G	Shear modulus: bedrock, chamber	20 GPa, 2 GPa
ν	Poisson's ratio	0.25
ρ	Density	2500 kg m^{-3}
g	Gravity	9.81 m^2 s^{-1}
α	Angle between vertical axis and line at surface origin to any point at wall, used by Jefferys (1920)	
θ	Angle between horizontal axis at centre of chamber and any point at wall, used by Grosfils (2007)	
σ_{xx}	Horizontal stress	
$\sigma_{\theta\theta}$	Hoop stress at the chamber wall	
C	Free surface factor : $1 + 2\tan\alpha''$, used by Grosfils (2007)	
φ	Bedrock friction angle	30°
S_o	Bedrock cohesion	10 MPa
T_o	Bedrock tension deduced from S_o	17.3 MPa
T	Bedrock cut-off tensile strength	5 MPa
P, σ_{II}	First and second stress invariants	
τ, σ_n	Tangential and normal stress along any given plane	
ΔP	Internal overpressure	
ΔP_S	Critical overpressure for tensile failure at the surface	Equation (4)
ΔP_T	Critical overpressure for tensile failure predicted from common studies, neglecting gravity	Equation (5)
ΔP_{TG}	Critical overpressure for tensile failure predicted by Grosfils (2007)	Equation (7)
ΔP_{MC}, ΔP_{MCS}	Critical overpressure for shear failure predicted in this study	Equations (11), (12)
P_o, P_{cr}	Uniform external pressure, and critical internal overpressure defined by Massinas & Sakellariou (2009)	Equation (13)
ΔP_1, ΔP_2, ΔP_3	Modelled internal overpressure at different stages of failure	
λ	Pore pressure ratio	0/1
p_f	Pore-fluid pressure	

variables). The model is initially set up with isotropic lithostatic components (weight of overburden rocks) so that a strain of only 1‰ develops during readjustment to the plane-strain conditions (e.g. Turcotte & Schubert 1982).

Numerical benchmarks of the results that are described in the following sections are included in the Supplementary Material. The formation and development of precise shear bands is conditioned by a sufficiently high mesh resolution, which requires computationally expensive runs. The major numerical concern is mesh-locking effects as shear bands form and propagate from the chamber wall. The finite element code Adeli (Hassani *et al.* 1997; Chery *et al.* 2001) is used to benchmark our solutions. The benchmark shows the effects of meshed versus unmeshed chamber and of coarse mesh resolution, especially when the domain becomes highly plasticized. Localizing plastic deformation is known to be complicated to model numerically (e.g. Yarushina *et al.* 2010); a research objective could therefore be to improve this benchmark with high-resolution models, similar perhaps to those that have been conducted for the problem of fault propagation in accretionary prisms (Buiter *et al.* 2008).

Assumptions about the rheological behaviour of the chamber

Many models of an inflating magmatic chamber do not include the internal domain of the magmatic chamber in their mesh. However, this has been included here in order to achieve high mesh resolution with the quadrilateral elements formulation of Parovoz. The rheology of the chamber must therefore be defined using Lamé parameters. While a magmatic chamber filled with low-viscosity fluid should be assumed incompressible (Poisson's ratio = 0.5), many studies point to the important proportion of volatile phases which reduce both its elastic rigidity and its incompressibility (Bower & Woods 1997; Huppert & Wood 2002; Rivalta & Segall 2008). The effects of the elastic properties of the chamber were therefore verified. The greater the internal Young's Modulus the more the chamber dilates and 'absorbs' its internal pressure, thus transferring less pressure to the outer walls and bedrock domain. The pressure felt out of the chamber walls is best measured via the second invariant of the deviatoric shear stress, similar to the approach used by Chery *et al.* (1991).

If the Young's Modulus is reduced to 10 times that of the external bedrock, then more than 90% of the applied internal pressure is transferred to the outer domain. In this case, the value of Poisson's ratio does not act significantly. The tests show that modelled stress and deformation become indistinguishable for models with $\nu = 0.25$ and $\nu = 0.45$. The presented reference model assumes Lamé parameters equal to $1/20$ those outside the chamber (corresponding to $\nu = 0.25$ and $E = 2.5$ GPa, whereas in the bedrock $E = 50$ GPa).

Numerical experiments follow a procedure in which the internal overpressure (ΔP) increases progressively. This pressure increase occurs proportionally to the time-step of the model, according to $\Delta P = A \times$ time where A is a coefficient fixed so that the pressure increases (1) fast enough for the total computing time of the run to be reasonable and (2) slow enough so that deformation resulting from a specific ΔP detected during our sampling frequency provides a quasi-static solution. Application of an internal overpressure that overshoots the yield strength of elements can lead to numerical inconsistencies. This justifies the application of a continuous radial deformation to model magmatic inflation (e.g. Chery *et al.* 1991). However, for the specific purposes of our study, application of a radial overpressure is more appropriate (as in tunnelling engineering, see above).

In the real world, the onset of an eruption or dyke injection releases confined magmatic fluids from the chamber, inhibiting further increase in internal overpressure. However, Wegler *et al.* (2006) interpreted continuously increasing shear wave velocities below the Merapi volcano as an indicator of increasing magmatic pressure in between two consecutive eruptions in 1998. The application of an elevated internal overpressure in our models should therefore be considered within the context of rapid arrival of over-pressurized magma and as a preliminary stage of regional microcracking and damaging of the host rock, prior to the onset of dyke injection or magmatic eruption (see Discussion section).

Different stages of deformation with increasing pressure

Sequential results of a reference model (M1 in Table 2) are presented (Fig. 4) at increasing internal overpressure. Zones of failure, shear strain, deviatoric and horizontal stresses are described below for three different stages of evolution.

Stage 1: Development of surface tensile failure (ΔP_1)

Internal overpressure increases progressively from an initial uniform lithostatic state. Tensile failure is first reached at the origin X_o on the ground surface, and expands progressively both in width and depth. Since this surface is defined as stress free, failure is only limited by the equality of σ_{xx} with the tensile strength $T = 5$ MPa. Locally, the

Table 2. *Numerical models. Initial conditions are distinguished according to the magmatic chamber being included or not in the mesh, pore-fluid pressure (hydrostatic or lithostatic), tensile strength* T *(elastic means the entire bedrock is elastic as opposed to elasto-plastic) and applied overpressure* ΔP

Model name	Process name	Fig.	Chamber (GPa)	Fluid pressure	T (MPa)	Pressure ΔP (MPa)	Fracturing stage	Surface uplift (m)
M1	M3dpten5b	4a	$\lambda = \mu = 1$	–	5	54	1-surface	1.5
		4b				114	2-wall	4.1
		4c				127	3-connection	5.5
		4d				135	3-connection	6.7
M2	M3dpten5L1	6a	$\lambda = \mu = 1$	Hydrostatic	5	55	2-wall	1.5
M3	M3dpcten5L3	6b	$\lambda = \mu = 1$	Lithostatic	5	20	2-wall	0.5
M4	M3dpR3L3a	6c	$\lambda = \mu = 1$	Lithostatic	17.3	20	3-connection	1.0
A1a	M3f50r3	A1	$\lambda = \mu = 1$	–	17.3	50	1-surface	2.06
A1b	M3cp50elas	A1	$\lambda = \mu = 1$	–	Elastic	50	Elastic	2.05
A1c	bull9	A1	Empty	–	17.3	50	1-surface	2.13
A1d	bull9elas	A1	Empty	–	Elastic	50	Elastic	2.13
A2a	M3cp120	A2	$\lambda = \mu = 1$	–	17.3	120	2-wall	4.9
A2b	M3cp120elas	A2	$\lambda = \mu = 1$	–	Elastic	120	2-wall	5.11
A2c	bull9	A2	Empty	–	17.3	120	2-wall	5.12
A2d	bull9elas	A2	Empty	–	Elastic	120	2-wall	5.25
A3a	M3cp20nog	A3	$\lambda = \mu = 1$	–	17.3	15.2	2-wall	1.08
A3b	bull9nog	A3	Empty	–	17.3	15.2	2-wall	1.18
A3c	M3dpcL3	A3	$\lambda = \mu = 1$	Lithostatic	17.3	18.1	2-wall	1.15
A3d	bull8nog	A3	Empty	–	17.3	18.8	2-wall	2.5

Results are summarized in terms of stage of fracturing and amount of surface uplift. Models A1a–A3d are displayed in the Supplementary Material.

horizontal stress is slightly higher with $\sigma_{xx} = 6$ MPa because of values taken at the centre of 25 m-thick mesh elements.

The analytical prediction (Equation (4)) without gravity provides failure at the surface for $\Delta P_1 = T(H^2 - R^2)/2R^2 = 28.13$ MPa. The numerical model produces surface rupture for $27.4 < \Delta P_1 < 28.4$ MPa (numerical time-step sampling is automatic and cannot correspond to exact critical analytical pressures). The consistency of the numerical value of ΔP_1 with the analytical solution of Equation (4) indicates that failure at the surface is less dependent on the gravitational load acting on the system than on the local state of stress at the very top surface where gravity stresses vanish.

Stage 2: Development of shear failure along the chamber wall (ΔP_2)

As internal overpressure continues to increase, tensile failure ceases to propagate and is replaced by normal shear faults that also propagate downwards. Shear failure then initiates around the chamber (Fig. 4b). The numerical model indicates that this occurs at an overpressure of 70 MPa, in excellent agreement with the Mohr–Coulomb yield prediction evaluated in Equation (11): $\Delta P_{MC} = (-\rho g y + T_o) \sin \varphi = 71$ MPa.

Shear failure first initiates in the upper quarter of the chamber wall, consistent with analytical predictions and Figure 2. With increasing overpressure, shear bands initiating at the chamber wall are oriented at an angle $45 \pm \varphi/2 = 30^\circ$ to the most compressive radial direction, in agreement with expectations for a non-associative Mohr–Coulomb material (e.g. Vermeer & de Borst 1984; Kaus 2010). These shear bands develop eccentrically from the chamber, and cross each other at an angle of 60° consistent with a friction angle $\phi = 30^\circ$ (e.g. d'Escatha & Mandel 1974; Vermeer & de Borst 1984; Gerbault *et al.* 1998). Note that the difference in orientation between these modelled shear bands and those calculated by d'Escatha &

Fig. 4. Reference model M1 for different stages of increasing internal pressure (ΔP), tensile strength $T = 5$ MPa. (**a**) Stage 1: tensile failure initiates at the surface when $\Delta P_1 = 28$ MPa, later forming inward-dipping normal faults. (**b**) Stage 2: shear failure initiates around the chamber walls from $\Delta P_2 = 70$ MPa, and forms an eccentric fault pattern here shown at $\Delta P = 100$ MPa. (**c**) Stage 3: Both fault systems connect at $\Delta P_3 = 130$ MPa. (**d**) The fault system expands, forming a vertical fault pattern from the chamber to the surface. Figures show (from left to right) rupture zones in blue, the 2nd invariant of the cumulated deviatoric shear strain and the 2nd invariant of the stress tensor.

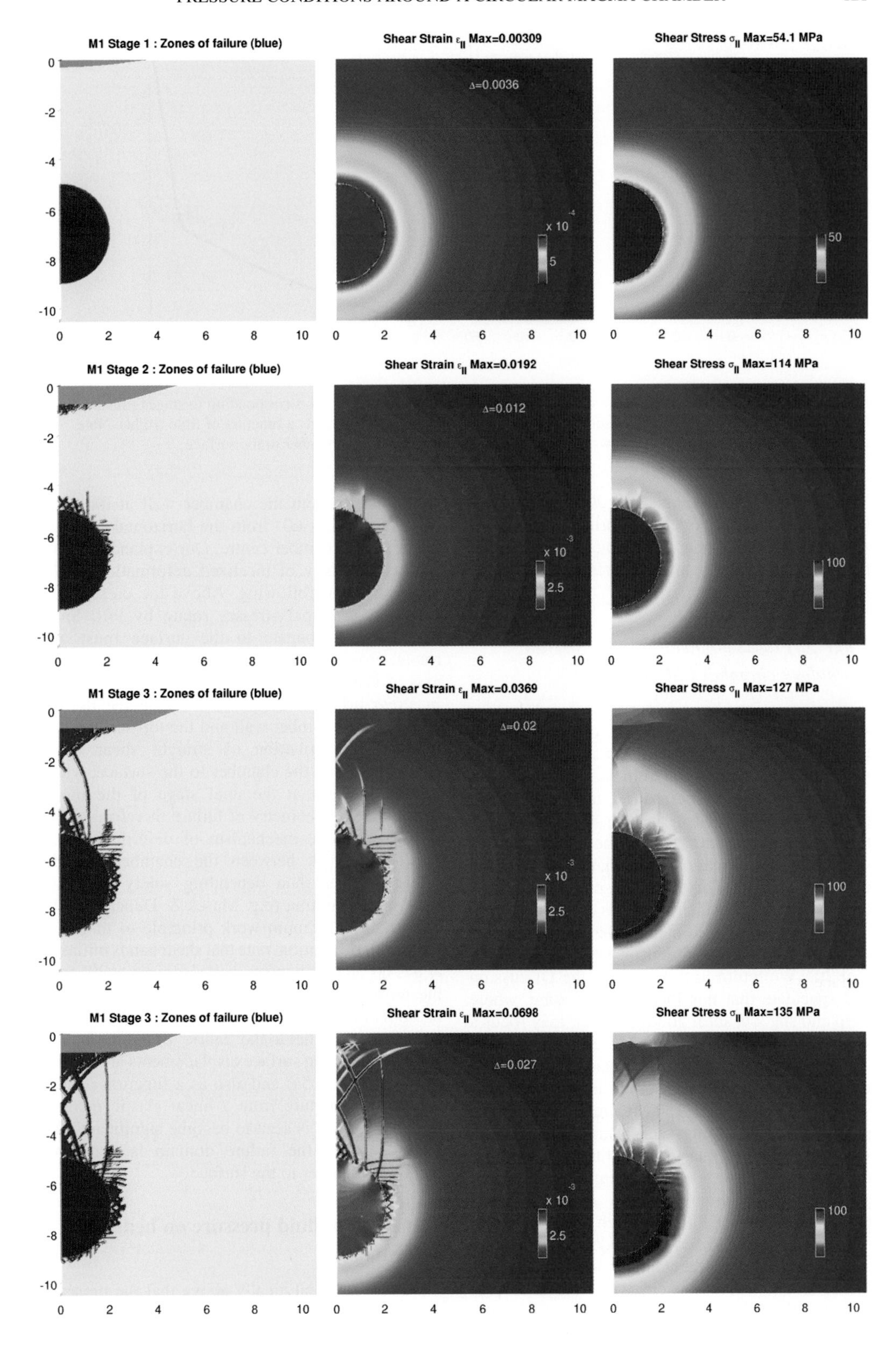
M1 Stage 1 : Zones of failure (blue)
Shear Strain ε_II Max=0.00309
Shear Stress σ_II Max=54.1 MPa
Δ=0.0036
M1 Stage 2 : Zones of failure (blue)
Shear Strain ε_II Max=0.0192
Shear Stress σ_II Max=114 MPa
Δ=0.012
M1 Stage 3 : Zones of failure (blue)
Shear Strain ε_II Max=0.0369
Shear Stress σ_II Max=127 MPa
Δ=0.02
M1 Stage 3 : Zones of failure (blue)
Shear Strain ε_II Max=0.0698
Shear Stress σ_II Max=135 MPa
Δ=0.027

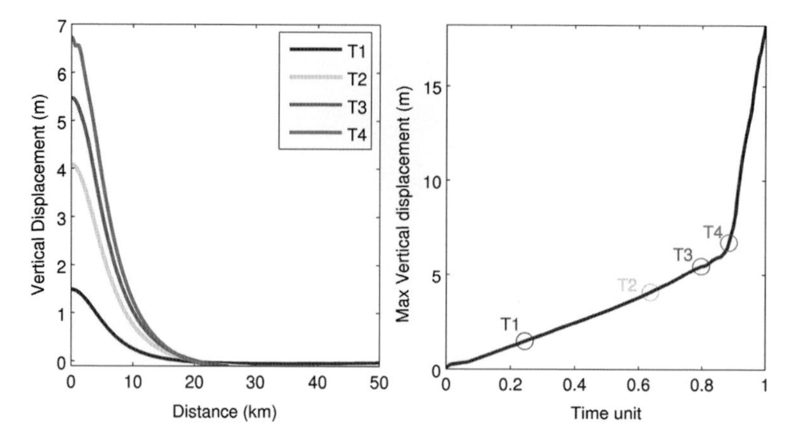

Fig. 5. Surface displacements from model M1 at different times T1, T2 and T3 corresponding to stages illustrated in Figure 4. Vertical displacement is shown as a function of distance (left) and as a function of time (right). Note departure from linear increase at T3, when the failure pattern connects the chamber to the surface.

Mandel (1974, friction angle $\varphi = 20^\circ$, Fig. 2) stems from perpendicularly orientated principal compressive stresses due to the application of either an overpressure (this study) or an underpressure (d'Escatha & Mandel 1974).

Stage 3: Faults connection and possible secondary chamber (ΔP_3)

With increasing overpressure, outward-dipping shear and reverse faults expand from the chamber upwards simultaneously with inward-dipping normal faults that propagate from the surface downwards. These two plasticized (i.e. faulted) domains eventually connect and merge (Fig. 4c). The connection occurs at a depth d_m along the vertical axis of symmetry as a function of the chamber width and depth (R and H). This depth d_m compares with a typical distance at which slip-lines cross each other in analytical solutions for flat indentation of a perfectly plastic material (Salençon 1966, distance OJ, Fig. 2b). We return to this point in the Discussion and speculate that this local dilation zone, where deep and shallow shear zones connect, may focus a secondary magmatic reservoir close to the surface.

This third stage where the domains of plasticized material connect the chamber to the surface occurs when ΔP reaches about 130 MPa in the model. The possibility of achieving such a high overpressure seems extreme in a quasi-static situation; such high overpressures may however occur as a transient phenomenon if extremely rapid arrival of overpressurized magmatic fluids, which have no time to relax within the chamber, is considered.

Whereas failure at the chamber wall initiates over the crest domain, consistent with analytical predictions illustrated in Figure 2, most active shear bands depart from the chamber wall at the angle interval θ *c.* 45–60° from the horizontal direction taken at the chamber centre. Our explanation for a greatest intensity of localized deformation at this location is the following. Above the crest of the chamber, principal stresses rotate by 90°. Shear bands that propagate to the surface must also rotate, cross each other and reflect on the vertical axis of symmetry. Instead, at θ *c.* 45–60° changes in stress orientations are minimal in the domain between the chamber wall and the top surface; this favours the formation of straight shear bands expanding from the chamber to the surface, which eventually form at the final stage of the model (Fig. 3d). The geometry of failure therefore results from a dynamic mechanism of deformation that minimizes work between the chamber and the surface, rather than depending solely upon the local stress minima (e.g. Masek & Duncan (1998) applied the minimum work principle to mountain building). In addition, note that shear bands initiating at θ *c.* 60° at the chamber wall also form at 30° from the most compressive stress (radial), and are thus naturally oriented at $60 + 30 = 90^\circ$ (i.e. the vertical) and closely parallel to the 'fabric' of the mesh.

When plotting surface displacements at different time-steps (Fig. 5a) and also as a function of time (Fig. 5b), departure from a linear elastic increase in surface uplift is seen to become significant after Stage 3 when the failure domain is connected from the chamber to the surface.

The effect of fluid pressure on bedrock deformation

It was argued analytically above that the presence of pore-fluid pressure p_f would enable failure for significantly lower internal overpressures because it annihilates the gravity component in the yield stress criterion. This effect is now illustrated with numerical models by taking into account pore-fluid pressure in the yield stress criterion.

Two sets of models are performed using Parovoz in which the yield criterion is expressed according to Equation (14), with a constant bedrock pore-fluid pressure set first to hydrostatic ($\lambda = 0.4$) and then to lithostatic ($\lambda = 1$) pressure.

When accounting for a hydrostatic pore-fluid pressure ($p_f = \rho_w g y$; $\lambda = 0.4$) in the bedrock, the progressive increase in internal overpressure leads to a first stage of tensile and normal faulting at the surface followed by the initiation of shear failure at the chamber wall (Stage 2) at an overpressure consistent with the analytical prediction $\Delta P_{MC} = \sin \varphi (T + (\rho_r - \rho_w) g D) = 43$ MPa. Figure 6a displays a snapshot of this model at 54 MPa, where the initiation of shear failure at the chamber wall can be observed (blue dots).

When accounting for a lithostatic fluid pressure ($p_f = \rho_r g y$, $\lambda = 1$) in the bedrock, the onset of wall shear failure is predicted when $\Delta P_{MC} = \sin \varphi \ T_o$ according to Equation (11). Thus, if tensile strength $T = T_o = S_o/\tan \varphi = 17.3$ MPa (there is no cut-off), then $\Delta P_{MC} = 8.6$ MPa. The onset of wall tensile failure is predicted from Equation (5) when overpressure $\Delta P_T = T(H^2 - R^2)/H^2$. In the case where the tensile strength $T = 17.3$ MPa, tensile failure therefore cannot occur since $\Delta P_T = 15.9$ MPa which is greater than ΔP_{MC}. However, if tensile strength $T = 5$ MPa, tensile failure is predicted at the wall for $\Delta P_T = 4.6$ MPa which is lower than ΔP_{MC}.

The numerical results are consistent with these predictions. With a tensile strength cut-off $T = 5$ MPa, the numerical model produces tensile failure at the wall and before any failure occurs at the ground surface (Fig. 6b). The plasticized domain has an almost circular shape; a subsequent increase in internal pressure increases its radius, with connection with the surface taking the shape of an amphora (not shown). ΔP needs to reach 21 MPa so that the plastic domain branches to the surface.

Another model is displayed in Figure 6c, in which the bedrock is again assumed to be at lithostatic pore-fluid pressure but for which there is no prescribed tensile strength cut-off (therefore $T = T_o = S_o/\tan \varphi = 17.3$ MPa). In this case, shear failure is shown to develop throughout the bedrock with a precise onset at the wall when $\Delta P = 8.6$ MPa (as predicted above).

Table 2 lists the fracturing stages and values of surface uplift for these models with hydrostatic and lithostatic pore-fluid pressure and shows that, for internal pressures equivalent to model M1, the fracturing pattern occurs systematically a stage ahead and is associated with greater deformation than in model M1.

Discussion of the models with respect to geological observations

Shear fracturing around exposed intrusions, open faults and fluids

Many field observations report mode I opening structures (mainly dyke intrusions) as a dominant process of deformation outside magmatic bodies (e.g. Gudmundsson 2006; Holohan *et al.* 2009). However, geological studies also report shear faulting prior to dyke intrusion. Figure 7 illustrates such observed shear structures.

The Solitario Laccolith, Trans-Pecos Texas, is a 16 km diameter dome that displays a complex sequence of doming with sill and dyke intrusions that were mapped and dated by Henry *et al.* (1997). They interpreted quartz-phyric rhyolite dykes (their Tir4) to have intruded along radial and concentric shear fractures during doming (Fig. 7a).

The island of Arran, Scotland, is a typical reference for preserved structures formed during magmatic intrusion (Fig. 7b). Describing these, Woodcock & Underhill (1987) note how 'late in the intrusion history of the Paleocene Northern granite, the major faults (in Permian–Triassic New Red Sandstones) on the southeast side of the granite formed a conjugate strike-slip system that accommodated radial expansion of the pluton'.

A 30 cm-scale rock is displayed in Figure 7c, which was sampled along the track of the Glen Rosa valley in Arran. The grey part of this rock (the 'bedrock') displays a regular pattern of conjugate shear bands crossing each other at *c.* 60°, adjacent to another unfaulted domain of whiter colour (presumably the 'fluid'). One of these shear bands (A) has been infiltrated by that fluid, and is about 2 mm thick. About 20 cm to the right, another much thicker (*c.* 2 cm) limb of fluid expands through the bedrock and overprints the conjugate system of faults (B). This observation may be interpreted as follows. First, shear fracturing has developed pervasively in the bedrock surrounding an over-pressurized medium (which did not fail). The fluid then infiltrated the shear bands, prior to or simultaneous with a more massive event in which tensile opening expelled greater amounts of fluid throughout the bedrock.

Other independent observations argue in favour of the occurrence of shear fracturing around magmatic chambers. First, in a general context of extension excluding fluid involvement, it is well known that tensile stress does not systematically lead to mode I opening. For example, Ramsey & Chester

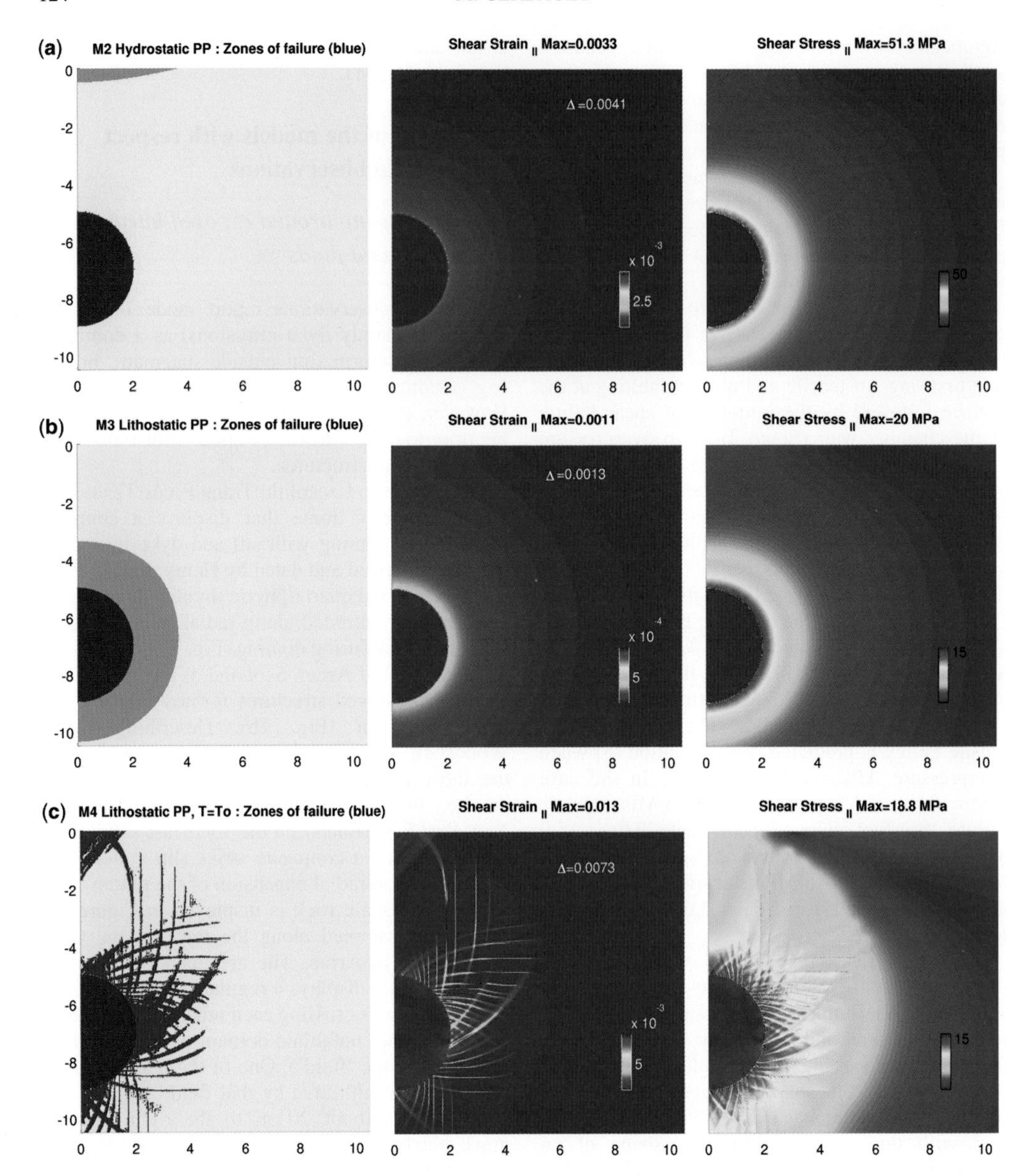

Fig. 6. Models accounting for an effective pore-fluid pressure in the host rock. (**a**) Hydrostatic pore pressure, tensile strength $T = 5$ MPa for $\Delta P = 54$ MPa. Tensile failure develops at the top surface and shear failure occurs at the wall (dark blue dots). (**b**) Lithostatic pore pressure, tensile strength $T = 5$ MPa for $\Delta P = 20$ MPa. Tensile failure develops all around the chamber wall. (**c**) Lithostatic pore pressure with no tensile strength cut-off ($T = T_o = 17.3$ MPa) at $\Delta P = 20$ MPa. Shear failure develops everywhere in the bedrock.

(2004) showed from laboratory experiments under triaxial tension, that rock failure can evolve continuously from mode I opening cracks to mode II shear fractures as the confining pressure increases. Second, the observation of seismicity under volcanic edifices indicates double-couple earthquake focal mechanisms, associated with or prior to an eruption (e.g. Waite & Smith 2002; Kumagai *et al.* 2011). Double-couple focal mechanisms are an indicator for failure occurring in mode II, independently of the recognized presence of large amounts of fluids in active volcanic zones.

The present analytical and numerical study indicates that the occurrence of mode I fracturing

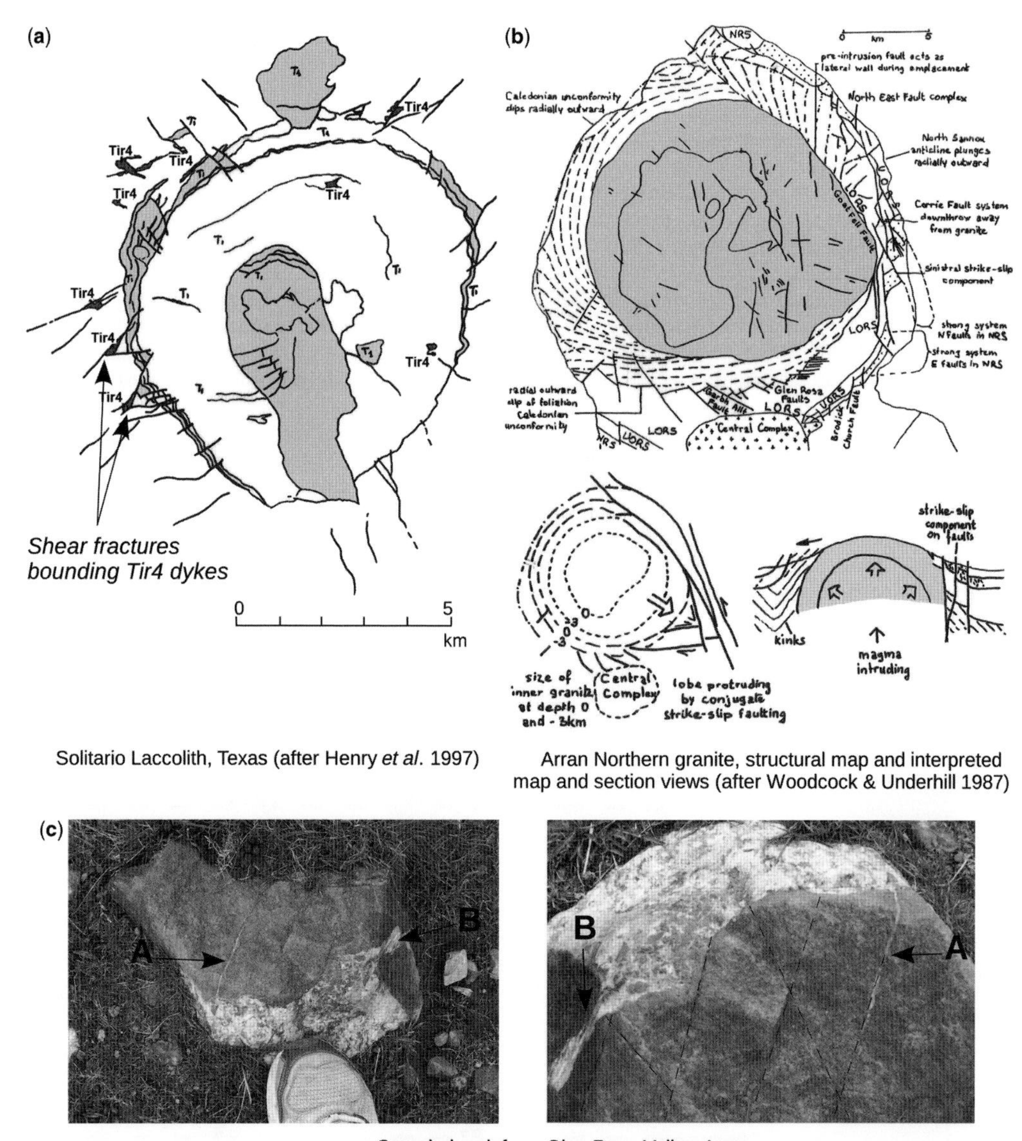

Fig. 7. Field examples of shear fractures around magmatic bodies. (**a**) The Solitario lacccolith, Texas, showing dyke intrusions bounded by pre-existing shear faults (simplified after Henry *et al.* 1997). (**b**) Geological structural map of Isle of Arran, Scotland, with interpreted map and plane views of structures forming during pluton ascent, after Woodcock & Underhill (1987). (**c**) Rock sampled at Isle of Arran in the Glen Rosa valley, displaying a conjugate fracture pattern along which fluid penetration occurs at location A and a non-oriented more massive intrusion at location B.

around an inflating magmatic chamber requires that the bedrock medium contains sufficient pore-fluid pressures close to a lithostatic state. It suggests that once the bedrock around an inflating chamber is sufficiently weakened by either processes of shear fracturing, microcracking or transport of fluids in pores associated with the inflation, the subsequent massive stage of mode I fracturing can occur by means of dyke or sill emplacement and eruption. A more complex dynamic model accounting for double-phase hydro-mechanical processes is required to demonstrate the validity of such mechanisms.

Insights from the modelled failure geometry

The sophisticated faulting geometries obtained in the numerical experiments (stage 2 to stage 3) result from the complex stress pattern produced by the circular overpressure, and was achieved due to an uncommonly high numerical mesh resolution; no previously published numerical studies demonstrate such shear band geometries (to the author's knowledge).

When gravity is neglected (or when lithostatic fluid pressures are present), the geometry of the modelled shear bands is similar to slip-line solutions predicted by circular plane-strain indentation and displays logarithmic eccentricity (Nadai 1950; Fig. 3a). When gravity is accounted for, numerical results show strong similarities to slip-line characteristics around a collapsing tunnel by d'Escatha & Mandel (1974, Fig. 3c).

An interesting comparison with flat indentation can also be made. Salençon (1969) showed that the maximum distance $d_m = OJ$ from the indenter (at which slip-lines would cross each other along the central axis of symmetry before reaching the opposite free surface) satisfies a maximum height to width ratio $d_m/a = 5.298$ (where a is the half-width of the indenter and J is slip-line intersection point beyond which triangular block motion links with the opposite surface; Fig. 3b). Such a local zone of dilation due to crossing shear zones has previously been applied in the field of Earth Sciences in order to explain the development of tensile structures in compressional tectonic regimes. For instance, Lake Baikal results from the indentation of Asia by India (e.g. Tapponnier & Molnar 1976) and the Eifel volcanic zone is a result of the Alpine–European indentation (Regenauer & Petit 1997). Salençon's relationship ($H > 5.3\,R$ approximately) may be invoked to a first order to assess whether a magmatic chamber is deep enough to generate a secondary magmatic reservoir very close to the surface. This speculative process requires a systematic investigation of double sources below volcanoes worldwide (e.g. below Askja, Sturkell

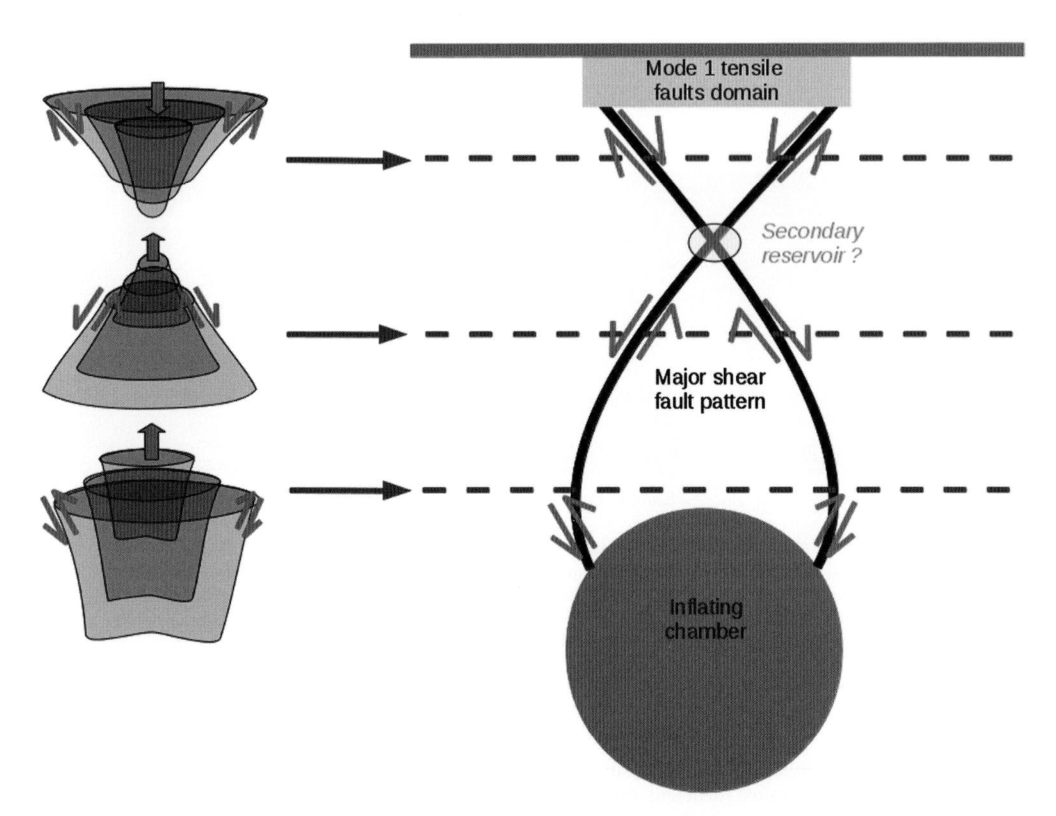

Fig. 8. Diagram of failure patterns from the surface to the top of a circular pluton, displaying the imbrications of individual faulted blocks for different depths. A zone of dilation forms locally at the crossing point between upwards- and downwards-propagating shear zones, suggesting the formation of a secondary magmatic reservoir (compare to Fig. 3b).

et al. 2006; below la Soufrière, Foroozan *et al.* 2010; and below Tungurahua, Ruiz *et al.* 2010).

Structurally, the present modelled shear band structures dip inwards when initiating from the ground surface and dip outwards when initiating from the chamber wall. Despite the similarity, these geometrical results remain difficult to compare to numerical and analogue studies that consider deflation instead of inflation (e.g. Roche & Druitt 2001; Kusumoto & Takemura 2003; Acocella *et al.* 2004; Acocella 2007). Figure 8 depicts our modelled geometries and relates them to structures likely observable in the field, with different depths corresponding to different levels of surface erosion and exhumation. Different imbricated structures appear. At the surface, and possibly coeval with tensile cracks, normal faults form cones imbricated one into the other downwards; close to the chamber, reverse faults develop like imbricated flower pots. At intermediate depths, upwards-open cones and previous imbricated cones may coexist simultaneously and overlap as deformation progresses. Such geometries are best found in the field in relatively old and well-preserved plutonic and annular intrusive complexes, for which relatively slow ascent and cooling would maintain the structures deforming around them (e.g. Woodcock & Underhill 1987; Maza *et al.* 1998; Lafrance & John 2001). Clearly, a thorough investigation of natural analogues is needed to demonstrate the validity of these structural geometries.

Conclusion

In studying the stress conditions for failure around a spherical magma chamber subjected to an internal overpressure, assuming an elasto-plastic bedrock and a simple approximation of pore-fluid pressure, the following points have been demonstrated.

Failure occurs in shear mode along the chamber walls and for an internal overpressure (in excess of the state of lithostatic equilibrium) about half an order of magnitude (depending on the radius-to-depth ratio) greater than the usually inferred limit given by tensile strength of the rocks. The important role of the wall-parallel component of the lithostatic stress on failure conditions (Grosfils 2007) is therefore confirmed. It has also been shown that a chamber wall should fail in shear mode rather than in tensile mode, a common behaviour also assumed in tunnelling engineering mechanics (e.g. Massinas & Sakellariou 2009).

Only in the specific case where the bedrock is at a state of lithostatic pore-fluid pressure does tensile failure occur around the chamber for an internal overpressure close to the tensile strength (5–10 MPa). Consequently, if the bedrock surrounding an active volcano contains less than lithostatic pore-fluid pressures, then researchers investigating geodetic studies that use the Mogi solutions to infer chamber depth and radius should not be surprised to achieve their data fit with associated ‘high’ overpressures. In addition, ground surface deformation rapidly exceeds elastic solutions as soon as the failure domain connects the chamber to the surface.

The exceptionally precise geometries of the modelled shear bands were obtained due to a high mesh resolution, but these results should be confirmed with future benchmarks (Gerbault *et al.* 2012). These high-resolution shear bands are comparable to slip-line plasticity solutions in displaying eccentric structures and suggest that an initially deep magmatic chamber may generate a secondary reservoir closer to the surface, at the locus of intersection of major shear zones along the vertical axis of symmetry. Further systematic documentation of double reservoirs in magmatic systems is needed to confirm this scenario.

Naturally, all these results are valid in the context of an ideally circular magmatic chamber in a homogeneous, isotropic, elasto-plastic medium. A finger- or dyke-shaped intrusion would be better modelled with the crack theory, sensitive to the host rock tensile strength. As mentioned by Marti *et al.* (2008) or modelled by McLeod & Tait (1999), dyke intrusion may significantly affect the conditions for fault nucleation and propagation; the applicability of our model results is obviously limited to the circular geometry of the chamber. In addition, deformation is also governed by the rheology of the magma, temperature and viscosity of the bedrock, the driving pressure, stresses resulting from the intrusion itself, dynamic fluid transport through the chamber and the bedrock and many other temporally and spatially variable factors discussed elsewhere in the literature (e.g. Anderson 1936; Rubin & Pollard 1987; Lister & Kerr 1991; Rubin 1995; Day 1996; Fialko *et al.* 2001; Hurwitz *et al.* 2009; Karlstrom *et al.* 2010; Gressier *et al.* 2010; Galgana *et al.* 2011; de Saint Blanquat *et al.* 2011). Future research goals include the improvement of coupled hydromechanics modelling (Gerbault *et al.* 2012), the study of deflating conditions and the adaptation of these models to real volcanoes.

Thank are due to R. Hassani, J. Salençon and M. Sakellariou for sharing their knowledge on mechanics and O. Roche, V. Cayol, D. Healy, S. Holford, B. Mercier Lepinay, R. Plateaux and C. Ganino for sharing their knowledge on magmatic chambers. J. Giannetti provided priceless library support and J. Trevisan helped with the drawings. N. Croiset co-explored preliminary models and her work is acknowledged (Masters report, ENS Lyon, 2007). A detailed and enriching review by E. Grosfils improved the manuscript significantly.

References

ACOCELLA, V. 2007. Understanding caldera structure and development: an overview of analogue models compared to natural calderas. *Earth Sciences Reviews*, **85**, 125–160.

ACOCELLA, V., FUNICIELLO, R., MAROTTA, E., ORSI, G. & DE VITA, S. 2004. The role of extensional structures on experimental calderas and resurgence. *Journal of Volcanology and Geothermal Research*, **129**, 199–217.

ANDERSON, E. M. 1936. The dynamics of formation of cone-sheets, ring-dykes and cauldron-subsidences. *Proceedings of the Royal Society, Edinburgh*, **56**, 128–163.

BONAFEDE, M. & FERRARI, C. 2009. Analytical models of deformation and residual gravity changes due to a Mogi source in a viscoelastic medium. *Tectonophysics*, **471**, 4–13.

BONAFEDE, M., DRAGONI, M. & QUARENI, F. 1986. Displacement and stress fields produced by a centre of dilation and by a pressure source in a viscoelastic half-space: application to the study of ground deforma tion and seismic activity at Campi Flegrei, Italy. *Geophysical Journal of the Royal Astronomy Society*, **87**, 455–485.

BOWER, S. M. & WOODS, A. W. 1997. Control of magma volatile content and chamber depth on the mass erupted during explosive volcanic eruptions. *Journal of Geophysical Research*, **102**, 10 273–10 290.

BUITER, S. J. H., YU, A. ET AL. 2006. *The Numerical Sandbox: Comparison of Model Results for a Shortening and an Extension Experiment, Analogue and Numerical Modelling of Crustal-Scale Processes*. Geological Society, London, Special Publications, **253**, 29–64.

BUROV, E. & GUILLOU-FROTTIER, L. 1999. Thermomechanical behavior of large ash flow calderas. *Journal of Geophysical Research*, **104**, 23 081–23 109.

BUROV, E., JAUPART, C. & GUILLOU-FROTTIER, L. 2003. Ascent and emplacement of buoyant magma bodies in brittle–ductile upper crutst. *Journal of Geophysical Research*, **108**, 2177.

CAQUOT, A. & KERISEL, J. 1956. *Traité de Mécanique des Sols*. Gauthier-Villars, Paris.

CHERY, J., BONNEVILLE, A., VILLOTE, J. P. & YUEN, D. 1991. Numerical modeling of caldera dynamical behaviour. *Geophysical Journal International*, **105**, 365–379.

CHERY, J., ZOBACK, M. D. & HASSANI, R. 2001. An integrated mechanical model of the San Andreas Fault in central and northern California. *Journal of Geophysical Research*, **106**, 22 051–22 066.

CUNDALL, P. A. 1989. Numerical experiments on localization in frictional materials. *Ingenior Archives*, **59**, 148–159.

CUNDALL, P. & BOARD, M. 1988. A microcomputer program for modeling large-strain plasticity problems. *Numerical Methods in Geomechanics*, **6**, 2101–2108.

DAY, S. J. 1996. Hydrothermal pore fluid pressure and the stability of porous, permeable volcanoes. *In*: MCGUIRE, W. J., JONES, A. P. & NEUBER, J. (eds) *Volcano Stability on the Earth and Other Planets*. Geolocal Society, London, Special Publications, **110**, 77–93.

D'ESCATHA, Y. & MANDEL, J. 1974. Stabilité d'une galerie peu profonde en terrain meuble. *Industrie Minerale*, Numero Special, 45–53.

ELLIS, S. M., WILSON, C. J. N., BANNISTER, S., BIBBY, H. M., HEISE, W., WALLACE, L. & PATTERSON, N. 2007. A future magma inflation event under the rhyolitic Taupo volcano, New Zealand: numerical models based on constraints from geochemical, geological and geophysical data. *Journal of Volcanology and Geothermal Research*, **168**, 1–27.

ENGELDER, T. & LEFTWICH, J. T. 1997. A pore-pressure limit in overpressured South Texas oil and gas field. *In*: SURDAM, R. C. (ed.) *Seals, Traps, and the Petroleum System*. American Association of Petroleum Geologists, Tulsa, Memoir 67, Chapter 15, 255–267.

FAIRHURST, C. & CARRANZA-TORRES, C. 2002. Closing the circle, some comments on design procedures for tunnel supports in rock. *In*: LABUZ, J. F. & BENTLER, J. G. (eds) *Proceedings, University of Minnesota 50th Annual Geotechnical Conference,* Minneapolis University, Minnesota, 21–84.

FIALKO, Y., KHAZAN, Y. & SIMONS, M. 2001. Deformation due to a pressurized horizontal circular crack in an elastic half-space, with applications to volcano geodesy. *Geophysical Journal International*, **146**, 181–190.

FOROOZAN, R., ELSWORTH, D., VOIGHT, B. & MATTIOLI, G. S. 2010. Dual reservoir structure at Soufrière Hills volcano inferred from continuous GPS observations and heterogeneous elastic modeling. *Geophysical Research Letters*, **37**, doi: 10.1029/2010GL042511.

GALGANA, G. A., MCGOVERN, P. J. & GROSFILS, E. B. 2011. Evolution of large Venusian volcanoes: insights from coupled models of lithospheric flexure and magma reservoir pressurization. *Journal of Geophysical Research*, **116**, E03009, doi: 10.1029/2010JE 003654.

GERBAULT, M., POLIAKOV, A. & DAIGNIÈRES, M. 1998. Prediction of faulting from the theories of elasticity and plasticity: what are the limits? *Journal of Structural Geology*, **20**, 301–320.

GERBAULT, M., CAPPA, F. & HASSANI, R. 2012. Elasto-plastic and hydromechanical models of failure around an infinitely long magma chamber. *Geochemistry Geophysics Geosystems*, **13**, Q03009, doi: 10.1029/ 2011GC003917.

GEYER, A. & MARTI, J. 2009. Stress fields controlling the formation of nested and overlapping calderas: implications for the understanding of caldera unrest. *Journal of Volcanology and Geothermal Research*, **181**, 185–195.

GRAY, J. P. & MONAGHAN, J. J. 2004. Numerical modeling of stress fields and fracture around magma chambers. *Journal of Volcanology and Geothermal Research*, **135**, 259–283.

GRESSIER, J.-B., MOURGUES, R., BODET, L., MATTHIEU, J.-Y., GALLAND, O. & COBBOLD, P. 2010. Control of pore fluid pressure on depth of emplacement of magmatic sills: an experimental approach. *Tectonophysics*, **489**, 1–12.

GROSFILS, E. B. 2007. Magma reservoir failure on the terrestrial planets: assessing the importance of gravitational loading in simple elastic models. *Journal of Volcanology and Geothermal Research*, **166**, 47–75.

GUDMUNDSSON, A. 1988. Effect of tensile stress concentration around magma chambers on intrusion and extrusion frequencies. *Journal of Volcanology and Geothermal Research*, **35**, 179–194.

GUDMUNDSSON, A. 2006. How local stresses control magma-chamber ruptures, dyke injections, and eruptions in composite volcanoes. *Earth-Science Reviews*, **79**, 1–31.

GUDMUNDSSON, A., MARIF, J. & TURON, E. 1997. Stress field generating ring faults in volcanoes. *Geophysical Research Letters*, **24**, 1559–1562.

GUDMUNDSSON, A., FJELDSKAAR, I. & BRENNER, S. L. 2002. Propagation pathways and fluid transport of hydrofractures in jointed and layered rocks in geothermal fields. *Journal of Volcanology and Geothermal Research*, **116**, 257–278.

GUILLOU-FROTTIER, L., BUROV, E. B. & MILESI, J. P. 2000. Genetic links between ash flow calderas and associated ore deposits as revealed by large-scale thermo-mechanical modeling. *Journal of Volcanology and Geothermal Research*, **102**, 339–361.

HAFNER, W. 1951. Stress distribution and faulting. *Bulletin of the Geological Society of America*, **62**, 373–398.

HARDY, S. 2008. Structural evolution of calderas: insights from two-dimensional discrete element formulations? *Geology*, **36**, 927–930.

HASSANI, R., JONGMANS, D. & CHÉRY, J. 1997. Study of plate deformation and stress in subduction processes using two-dimensional numerical models. *Journal of Geophysical Research* **102**, 17 951–17 965.

HENRY, C. D., KUNK, M. J., MUEHLBERGER, W. R. & MCINTOSH, W. C. 1997. Igneous evolution of a complex laccolith-caldera, the Solitario, Trans-Pecos Texas: implications for calderas and subjacent plutons. *Geological Society of America Bulletin*, **10**, 1036–1054.

HILLIS, R. R. (ed.) 2003. *Pore Pressure/Stress Coupling and its Implications for Rock Failure*. Geological Society, London, Special Publications, **216**, 359–368, doi: 10.1144/GSL.SP.2003.216.01.23

HOLOHAN, P., TROLL, V. R., ERRINGTON, M., DONALDSON, C. H., NICOLL, G. R. & EMELEUS, C. H. 2009. The Southern Mountains Zone, Isle of Rum, Scotland: volcano-sedimentary processes upon an uplifted and subsided magma chamber roof. *Geological Magazine*, **146**, 400–418.

HUBBERT, M. K. & RUBEY, W. W. 1959. Role of fluid pressure in mechanics of overthrust faulting. *Geological Society of America Bulletin*, **70**, 115–166.

HUPPERT, H. & WOODS, A. W. 2002. The role of volatiles in magma chamber dynamics. *Nature*, **420**, 493–495.

HURWITZ, D. H., LONG, S. M. & GROSFILS, E. B. 2009. The characteristics of magma reservoir failure beneath a volcanic edifice. *Journal of Volcanology and Geothermal Research*, **188**, 379–394, doi: 10.1016/j.jvolgeores.2009.10.004.

JEFFERY, J. B. 1920. Plane stress and plane strain in bi-polar coordinates. *Transactions of the Royal Society, London*, **221**, 265–293.

KARLSTROM, L., DUFEK, J. & MANGA, M. 2010. Magma chamber stability in arc and continental crust. *Journal of Volcanology and Geothermal Research*, **190**, 249–270; doi: 10.1016/j.jvolgeores.2009.10.003.

KAUS, B. 2010. Factors that control the angle of shear bands in geodynamic numerical models of brittle deformation. *Tectonophysics*, **484**, 36–47.

KUMAGAI, H., PLACIOS, P., RUIZ, M., YEPES, H. & KOZONO, T. 2011. Ascending seismic source during an explosive eruption at Tunguruhua volcano, Ecuador. *Geophysical Research Letters*, **38**, L01306, doi: 10.1029/2010GL045944.

KUSUMOTO, S. & TAKEMURA, K. 2003. Numerical simulation of caldera formation due to collapse of a magma chamber. *Geophysical Research Letters*, **30**, doi: 10.1029/2003GL018380.

LAFRANCE, B. & JOHN, B. E. 2001. Sheeting and dyking emplacement of the Gunnison annular complex, SW Colorado. *Journal of Structural Geology*, **23**, 1141–1150.

LAVIER, L., BUCK, W. & POLIAKOV, A. 2000. Factors controlling normal fault offset in an ideal brittle layer. *Journal of Geophysical Research*, **105**, 23 431–23 442.

LISTER, J. R. & KERR, R. C. 1991. Fluid-mechanical models of crack propagation and their application to magma transport in dykes. *Journal of Geophysical Research*, **96**, 10 049–10 077.

MARTI, J., GEYER, A., FOLCH, A. & GOTTSMANN, J. 2008. A review of collapse caldera modeling. *In*: *Developments in Volcanology*, **10**, doi: 10.1016/S1871-644X(07)00006-X, 233–283.

MASEK, J. G. & DUNCAN, C. C. 1998. Minimum-work mountain building. *Journal of Geophysical Research*, **103**, 907–917.

MASSINAS, S. A. & SAKELLARIOU, M. G. 2009. Closed-form solution for a plastic zone formation around a circular tunnel in half-space obeying Mohr–Coulomb criterion. *Géotechnique*, **59**, 691–701, doi: 101680/geot8.069.

MASTERLARK, T. 2007. Magma intrusion and deformation predictions: sensitivities to the Mogi assumptions. *Journal of Geophysical Research*, **112**, doi: 10.1029/2006JB004860.

MASTERLARK, T. & LU, Z. 2004. Transient volcano deformation sources imaged with interferometric synthetic aperture radar: application to Seguam Island, Alaska. *Journal of Geophysical Research*, **109**, doi: 10.1029/2003JB002568.

MAZA, M., BRIQUEU, L., DAUTRIA, J.-M. & BOSCH, D. 1998. Le complexe annulaire d'age Oligocene de l'Achkal (Hogggar Central, Sud Algerie): temoin de la transition au Cenozoique entre les magmastismes tholeitique et alcalin. Evidences par les isotopes du Sr, Nd, Pb. *Comptes Rendus de l'Académie des Sciences – Series IIA*, **327**, 167–172, doi: 10.1016/S1251-8050(98)80004-9.

MCLEOD, P. & TAIT, S. 1999. The growth of dyke from magma chambers. *Journal of volcanology and Geothermal Research*, **92**, 231–245.

MOGI, K. 1958. Relations between the eruption of various volcanoes and the deformation of the ground surfaces around them. *Bulletin of the Earthquake Research Institute*, **36**, 99–134.

NADAI, A. 1950. *The Theory of Flow and Fracture of Solids*. McGraw-Hill, New York.

PARFITT, E. A., WILSON, L. & HEAD, J. W. 1993. Basaltic magma reservoirs: Factors controlling their rupture characteristics and evolution. *Journal of Volcanology and Geothermal Research*, **55**, 1–14, doi: 10.1016/0377-0273(93)90086-7.

PINEL, V. & JAUPART, C. 2003. Magma chamber behavior beneath a volcanic edifice. *Journal of Geophysical Research*, **108**, doi: 10.1029/2002JB001751.

PINEL, V. & JAUPART, C. 2005. Caldera formation by magma withdrawal from a reservoir beneath a volcanic edifice. *Earth and Planetary Science Letters*, **230**, 273–287.

PODLADCHIKOV, Y., POLIAKOV, A. & CUNDALL, P. 1993. An explicit inertial method for the simulation of viscoelastic flow: an evaluation of elastic effects on diapiric flow. *Flow and Creep in the Solar System: Observation, Modeling and Theory*, **391**, 175–195.

POLIAKOV, A. & PODLADCHIKOV, Y. 1992. Diapirism and topography. *Geophysical Journal International*, **109**, 553–564.

PRITCHARD, M. E. & SIMONS, M. 2004. An insar-based survey of volcanic deformation in central Andes. *Geochemistry Geophysics Geosystems*, **5**, doi: 10.1029/2003GC000610.

RAMSEY, J. M. & CHESTER, F. M. 2004. Hybrid fracture and the transition from extension fracture to shear fracture. *Nature*, **428**, 63–66.

REGENAUER-LIEB, K. & PETIT, J. P. 1997. Cutting of the European continental lithosphere: plasticity theory applied to the Alpine collision. *Journal of Geophysical Research*, **102**, 7731–7746.

RIVALTA, E. & SEGALL, P. 2008. Magma compressibility and the missing source for some dike intrusions. *Geophysical Research Letters*, **35**, L04306, doi: 10.1029/2007GL032521.

ROCHE, O. & DRUITT, T. H. 2001. Onset of caldera collapse during ignimbrite eruptions. *Earth and Planetary Science Letters*, **191**, 191–202.

RUBIN, A. M. 1995. Propagation of magma-filled cracks. *Annual Review, Earth and Planetary Sciences*, **23**, 287–336.

RUBIN, A. M. & POLLARD, D. D. 1987. Origins of blade-like dikes in volcanic rift zones. *US Geological Survey Professional Paper*, **1350**, 1449–1470.

DE SAINT BLANQUAT, M., HORSMAN, E., HABERT, G., MORGAN, S., VANDERHAEGHE, O., LAW, R. & TIKOFF, B. 2011. Multiscale magmatic cyclicity, duration of pluton construction, and the paradoxical relationship between tectonism and plutonism in continental arcs. *Tectonophysics*, **500**, 20–33.

SALENÇON, J. 1966. Expansion quasi-statique d'une cavité à symétrie sphérique ou cylindrique dans un milieu élastoplastique. *Annales des Ponts et Chaussées*, **III**, 175–187.

SALENÇON, J. 1969. *La théorie des charges limites dans la résolution des problèmes de plasticité en déformation plane*. PhD thesis, Laboratoire de Mécanique des Solides de l'École Polytechnique, Paris.

SARTORIS, G., POZZI, J. P., PHILIPPE, C. & LE MOUEL, J. L. 1990. Mechanical instability of shallow magma chambers. *Journal of Geophysical Research*, **95**, 5141–5151.

SEGALL, P. 2009. *Earthquake and Volcano Deformation*. Princeton University Press, Princeton.

STURKELL, E., SIGMUNDSSON, F. & SLUNGA, R. 2006. 1983–2003 decaying rate of deflation at Askja caldera: pressure decrease in an extensive magma plumbing system at a spreading plate boundary. *Bulletin of Volcanology*, **68**, 727–735, doi: 10.1007/s00445-005-0046-1.

TAIT, S., JAUPART, C. & VERGNIOLLE, S. 1989. Pressure, gas content and eruption periodicity of a shallow, crystallizing magma chamber. *Earth and Planetary Science Letters*, **92**, 107–123.

TAPPONNIER, P. & MOLNAR, P. 1976. Slip-line field theory and large-scale continental tectonics. *Nature*, **264**, 319–324.

TIMOSHENKO, S. & GOODIER, J. N. 1970. *Theory of Elasticity*. 2nd edn. McGraw-Hill Higher Education, New York.

TOWNEND, J. & ZOBACK, M. D. 2000. How faulting keeps the crust strong. *Geology*, **28**, 399–402.

TRASATTI, E., GIUNCHI, C. & BONAFEDE, M. 2005. Structural and rheological constraints on source depth and overpressure estimates at the Campi flegrei caldera, Italy. *Journal of Volcanology and Geothermal Research*, **144**, 105–118.

TURCOTTE, D. & SCHUBERT, G. 1982. *Geodynamics: Applications of Continuum Physics to Geological Problems.* Wiley, New York.

VERMEER, P. A. & DE BORST, R. 1984. Non-associated plasticity for soils, concrete and rocks. *Heron*, **29**, 1–75.

VERRUIJT, A. 1998. Deformations of an elastic half plane with a circular cavity. *International Journal of Solids Structures*, **35**, 2795–2804.

WAITE, G. P. & SMITH, R. B. 2002. Seismic evidence for fluid migration accompanying susbsidence of the Yellowstone caldera. *Journal of Geophysical Research*, **107**, doi: 10.1029/2001JB000586.

WEGLER, U., LÜHR, B.-G., SNIEDER, R. & RATDOMOPURBO, A. 2006. Increase of shear wave velocity before the 1998 eruption of Merapi volcano (Indonesia). *Geophysical Research Letters*, **333**, doi: 10.1029/2006GL025928,2006.

WHITE, B. G., LARSON, M. K. & IVERSON, S. R. 2004. Origin of mining-induced fractures through macro-scale distortion. *In*: *Gulf Rocks 2004. Rock Mechanics across Borders and Disciplines.* Proceedings of the Sixth North American Rock Mechanics Conference. Houston, Texas. Report ARMA/NARMS 04-569.

WOODCOCK, N. H. & UNDERHILL, J. R. 1987. Emplacement-related fault patterns around the Northern granite, Arran, Scotland. *Geological Society of America Bulletin*, **98**, 515–527.

YARUSHINA, V. M., DABROWSKI, M. & PODLADCHIKOV, Y. Y. 2010. An analytical benchmark with combined pressure and shear loading for elastoplastic numerical models. *Geochemistry Geophysics Geosystems*, **11**, doi: 10.1029/2010GC003130.

Stress fluctuation during thrust-related folding: Boltaña anticline (Pyrenees, Spain)

S. TAVANI[1,2]*, O. FERNANDEZ[1,3] & J. A. MUÑOZ[1]

[1]*Departament de Geodinamica i Geofisica, Geomodels Research Institute, Faculty of Geology, University of Barcelona, Spain*

[2]*Dipartimento di Scienze della Terra, Università Federico II, Napoli, Italy*

[3]*Repsol Exploración, Paseo de la Castellana 280, Madrid, Spain*

**Corresponding author (email: stefano.tavani@unina.it)*

Abstract: A common feature of thrust-related anticlines developing in thrust wedges is the presence of extensional structures paralleling the fold axial trend (i.e. longitudinal structures). These form in response to hinge-perpendicular stretching and indicate that the minimum stress component orients parallel to the regional shortening direction, that is, the maximum and minimum stress components locally invert. Under the assumption that the regional stress component paralleling the shortening direction is almost constant during folding, such an inversion requires a large local stress drop. In the thrust-related Boltaña anticline, folding was both predated and accompanied by the development of longitudinal extensional faults. Meso-scale contractional structures are only rarely found. Where contractional structures are present, cross-cutting relationships indicate that these structures episodically developed within a mainly extensional framework which was established during flexural bending of the foredeep and continued during thrust-related folding. We conclude that the coexistence of extensional and compressional structures relates to a stress fluctuation, with normal faulting and discontinuous fold growth occurring during 'extensional' and 'compressional' stages, respectively.

Understanding the genetic link between folding and fracturing is of primary importance for both academic and industrial purposes. Since the pioneering work of Stearns (1968), a large number of studies have focused on the description and interpretation of fracture patterns in reservoir-scale thrust-related anticlines (e.g. McQuillan 1974; Marshak & Engelder 1985; Cooper 1992; Tavani *et al.* 2006; Ahmadhadi *et al.* 2008). In particular, a very common feature of thrust-related anticlines is the presence of extensional structures such as joints and normal faults, striking parallel to the fold axial trend, which form both before (e.g. Billi & Salvini 2003) and during folding (e.g. Srivastava & Engelder 1990; Lemiszki *et al.* 1994; Fischer & Wilkerson 2000; Engelder & Peacock 2001; Stephenson *et al.* 2007; Tavani *et al.* 2008). In the second case, these structures develop due to a stress field characterized by a σ_3 (minimum stress) striking perpendicular to the fold axis and almost parallel to the regional σ_1 (maximum compressional stress), responsible for the development of the hosting thrust-related anticline. Assuming that the regional σ_1 is horizontal and perpendicular to the axis of the compressional folds, and almost constant in magnitude during the entire folding process, a significant decrease in regional σ_1 magnitude is required locally to account for axis-parallel extensional fracturing.

In this work we present meso-structural data from the Boltaña fault-propagation anticline (Spanish Pyrenees). Analysis of across- and along-strike deformation pattern variability and cross-cutting relationships between extensional and contractional deformation structures indicate that contractional structures form in response to episodic events within a coarse extensional framework. Implications for stress field distribution before and during thrust-related folding are discussed.

Geological framework

The north–south-trending and westwards-verging Eocene Boltaña fault-propagation anticline is located in the Spanish Pyrenees (Fig. 1) and involves Upper Cretaceous–Paleocene limestones and Ypresian–Middle Lutetian marls and siliciclastic sediments, detached above a thin level of Triassic evaporites (e.g. Seguret 1970; Cámara & Klimowitz 1985; Holl & Anastasio 1995; Fernandez 2004). The Boltaña anticline is part of an oblique set of anticlines and synclines along the eastern margin of the Gavarnie thrust sheet in the footwall of the

From: Healy, D., Butler, R. W. H., Shipton, Z. K. & Sibson, R. H. (eds) 2012. *Faulting, Fracturing and Igneous Intrusion in the Earth's Crust*. Geological Society, London, Special Publications, **367**, 131–140. http://dx.doi.org/10.1144/SP367.9

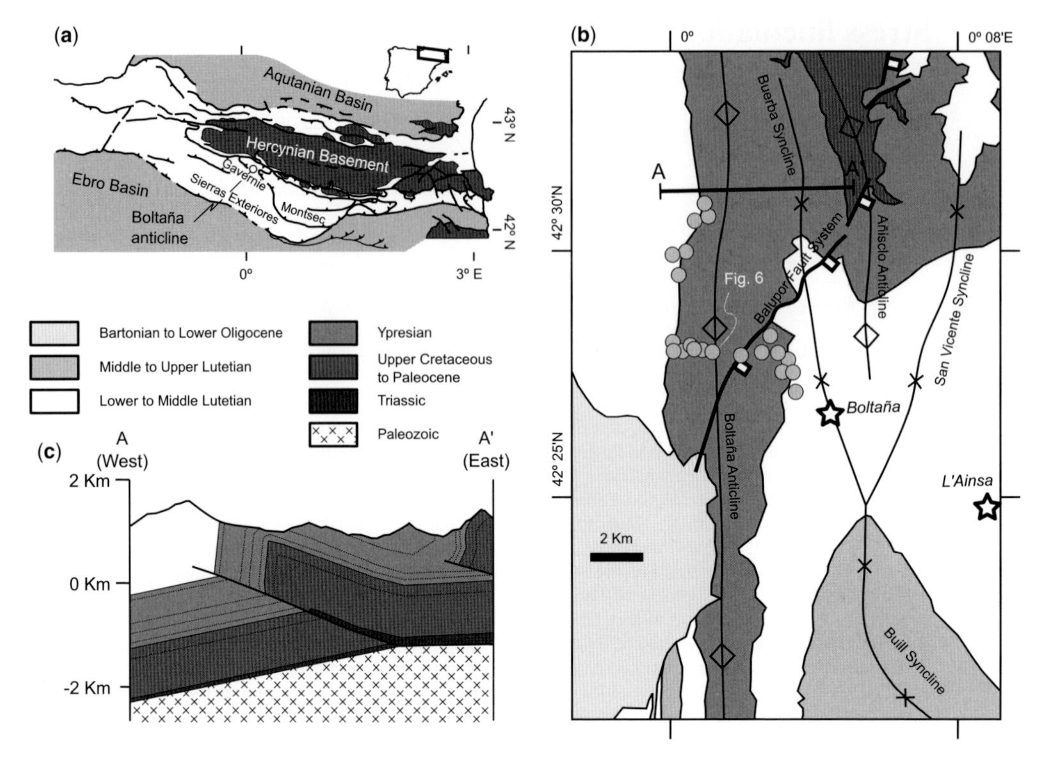

Fig. 1. (**a**) Schematic geological map of the Pyrenees; (**b**) geological map of the Boltaña anticlines with field sites (grey circles); and (**c**) cross-section along the trace A–A′ of Figure 1b.

Montsec thrust (Fig. 1). Palaeomagnetic data indicate that the Boltaña anticline and the neighbouring structures have rotated up to 80° in a clockwise direction during and after their growth (Dinares Turell 1992; Pares & Dinares Turell 1993; Pueyo *et al.* 2004). They originally formed as frontal structures with a roughly east–west strike, parallel to the dominant Pyrenean trend, and were progressively rotated to their current orientation during growth (Fernandez 2004). Despite this rotation, the compression direction registered in the Boltaña anticline (Mochales *et al.* 2010) and in the surrounding structures (e.g. Holl & Anastasio 1995; Tavani *et al.* 2006) is at all times perpendicular to the trend of anticlines (i.e. east–west shortening in their present-day orientation).

A westwards decrease of the contractional strain (Holl & Anastasio 1995) is documented in the area. Consistently, in the Añisclo anticline (located to the east of the Boltaña anticline) the meso-structural pattern mostly includes contractional structures (such as pressure-solution cleavage and conjugated strike-slip faults; Tavani *et al.* 2006). In the Boltaña anticline, the meso-structural pattern contains scarce contractional structures; east–west-oriented compression is mostly observed in the macro-scale folding and at the micro-scale in Anisotropy of Magnetic Susceptibility (AMS) data (Mochales *et al.* 2010).

Deformation pattern

A total of 208 orientation measurements of faults, joints, veins and pressure-solution cleavage were collected in 15 field sites located on both limbs and in the crest of the fold, along an east–west-oriented transect across the central portion of the anticline (Fig. 1b). Additional 97 measurements were collected to the north of the central transect, only in the forelimb.

The acquisition of orientation measurements was complemented with detailed field observations of outcrops to define the sense of displacement on faults, patterns of faults and other structures relative to each other and to bedding and cross-cutting relationships between different generations of structures.

Orientation measurements collected in the field have been analysed to group structures into families by orientation and to determine the timing of structures relative to folding. To this purpose, data have been analysed in both present-day orientation and in restored state (i.e. removing bedding dip). It has been found that many of the structures observed in the field pre-date folding as indicated by observations of the angle between joints and bedding (roughly perpendicular throughout the study area), the clustering of fault and fracture orientations once unfolded and detailed field observations, all of which are discussed below.

Orientation analysis

In the central transect, the meso-structural pattern mostly includes faults, joints and veins and (only occasionally) pressure-solution cleavage. In their present orientation, faults are clustered around different maxima: two main clusters of north–south-trending faults with dips of 50–60° to the east and west and a number of scattered clusters (Fig. 2a).

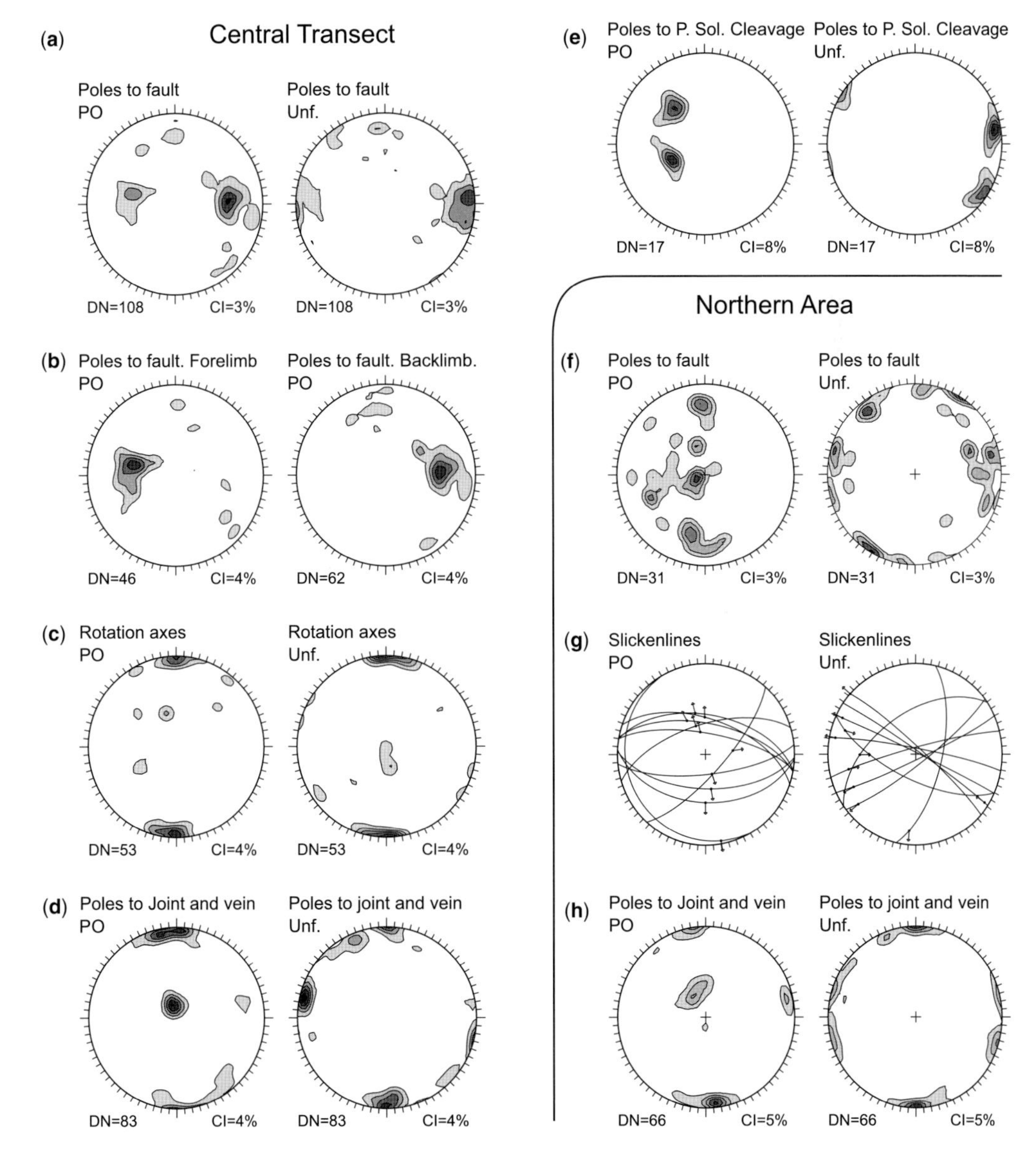

Fig. 2. Structural data. Central transect: (**a**) contouring of poles to faults in their present orientation (PO) and after bedding dip removal (Unf); (**b**) contouring of poles to faults in their present orientation, in the forelimb and backlimb; (**c**) contouring of rotation axes of faults and layer-parallel décollements; (**d**) contouring of poles to joints and veins; (**e**) contouring of poles to pressure-solution cleavages. Northern area: (**f**) contouring of poles to faults; (**g**) stereoplot of slickenlines; (**h**) contouring of poles to joints and veins. DN, data number; CI, contouring interval.

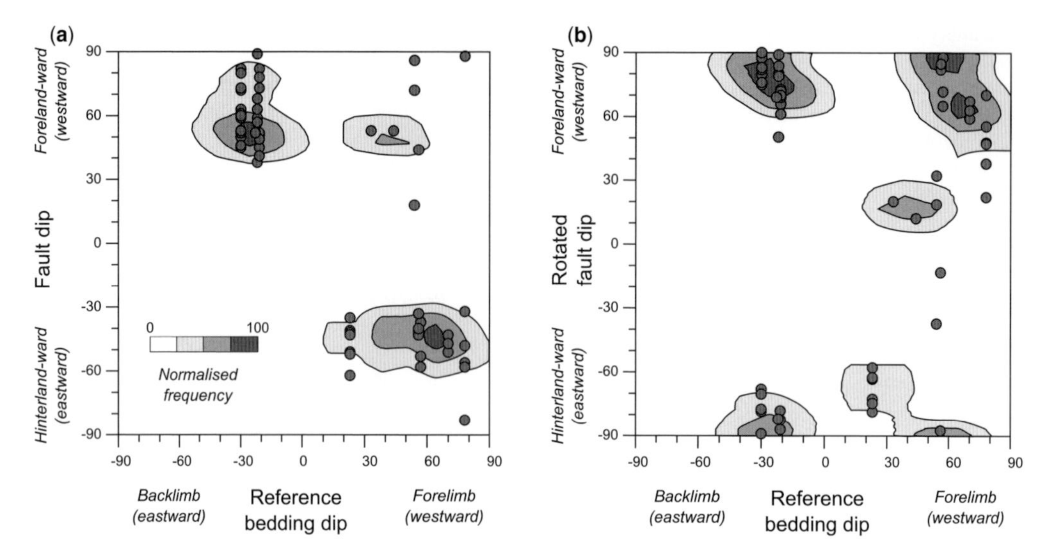

Fig. 3. Scatterplot and data density contours of longitudinal fault dip versus bedding dip: (**a**) in the present orientation and (**b**) after bedding dip removal. Data from central transect.

Longitudinal (north–south) faults in the backlimb have normal displacement and dip about 50–60° towards the west (Figs 2b, 3 & 4a). In the forelimb they mostly dip 50–60° toward the east, with a second less-abundant set of elements dipping 50–60° towards the west (Figs 2b, 3 & 4b–d). Longitudinal faults in the forelimb have both reverse and normal displacements. When bedding dip is removed, the longitudinal sets of faults on both limbs resolve to a dominant cluster of subvertical north–south-trending faults (Figs 2a & 3) with dominantly normal displacement and a subsidiary set of faults dipping 50–60° to the west present only in the forelimb. In addition to these two sets of longitudinal faults, unfolded orientations reveal a set of near-vertical faults: one striking roughly NE–SW and the other east–west (Fig. 2a). The NE–SW-striking system is parallel to the Balupor fault system, which completely cross-cuts the structure of the Boltaña anticline (Fernandez 2004; Fig. 1b).

The rotaxes (rotation axis; i.e. the axis lying on the fault plane and perpendicular to the slip direction) of faults and bedding surfaces reworked as faults are clustered along a north–south direction in both the present-day and bedding-corrected orientations (Fig. 2c). This is consistent with an east–west direction of transport (whether for extension or contraction), which is in turn parallel to the regional σ_1 responsible for the growth of the Boltaña anticline.

In their unfolded attitude joints and veins cluster into three sets, all of them vertical and roughly parallel to the three main fault orientations (N–S, NE–SW and E–W; Fig. 2a, d). Finally, pressure-solution cleavage is found only in the western forelimb. Cleavage is near-perpendicular to bedding and, when bedding dip is restored to the horizontal, strikes roughly north–south (Fig. 2e).

Poles to faults collected in the northern area are characterized by three orientation maxima (Fig. 2f). Once bedding dip is removed (as is the case in the central transect) most faults are subvertical and are grouped in three sets including a north–south-striking set, similar to that observed in the central transect. No slickenlines have been found along north–south-striking faults. However, displaced layers indicate that these are pre-tilting extensional faults. The east–west fault set observed to the south is replaced in this sector by two sets of WNW–ESE- and WSW–ENE-striking faults. These two sets form a conjugate strike-slip system that has been tilted with folding, consistent with an east–west-oriented shortening prior to folding (Fig. 2g). In the same area, joints and veins have a distribution similar to that observed in the central transect. These elements are at a high angle to bedding and are clustered in two broad maxima corresponding to north–south- and east–west-striking elements (Fig. 2h).

Structural interpretation

Observations reveal a rather simple deformation pattern in the backlimb, mostly composed of westwards-dipping longitudinal normal faults that are roughly perpendicular to bedding (Fig. 4a). On the contrary, the deformation pattern observed in the forelimb is more complex due the presence of different extensional stages. In the simplest cases,

Fig. 4. (**a**) Array of westward-dipping joints reworked as normal faults in the slightly eastward-dipping backlimb (eastern limb). (**b**) Eastward-dipping normal fault system in the steeply dipping forelimb (western limb). (**c**) Array of eastward-dipping faults in the near-vertical forelimb (indicated by arrows), possibly being pre-tilting normal faults. (**d**) The three faults in the photo (indicated by arrows) geometrically match a rotated conjugate normal fault system. However, all display normal displacement in their present orientation (as illustrated in the inset in the lower-right corner, showing slickenlines on the shallow-dipping fault), suggesting that they have been reworked after (or during) bed tilting.

Fig. 5. Frontal (down on bedding) view of near-vertical layers in the forelimb showing a tilted left-lateral fault offsetting two longitudinal veins.

longitudinal faults include pre-folding normal faults that have been rotated together with beds preserving their original offset (Fig. 4c). In other (frequent) cases, tilted conjugate-normal fault systems have been reworked during syn- to late-folding stages (Fig. 4d).

This variability in the deformation pattern across the central sector of the fold is key to evaluating the relationships between folding and deformation pattern development (e.g. Harris & Vand Der Plujim 1998; Tavani *et al.* 2006). The observation that most of the longitudinal-normal faults have opposing dip in the two limbs strongly supports a fold-related control and thus a syn-folding activity of faults. Nonetheless, the fact that many faults are almost perfectly perpendicular to bedding indicates that many of them probably developed as bedding-normal joints which were reworked as normal faults during bed tilting. Coupled with the observation that normal faults rotated with beds are also observed (Fig. 4c) and that these are mostly westwards-dipping regardless of their structural position, this is consistent with the onset of extensional conditions before fold growth.

In addition to extensional structures, contractional structures have been observed in both the central and northern sector, represented by pressure-solution cleavages, reverse faults and conjugate strike-slip faults. All these contractional structures relate to a σ_1 oriented about parallel to bedding and perpendicular to the strike of the Boltaña anticline.

Cross-cutting relationships

In both studied sectors of the Boltaña anticline, east–west-striking joints and veins abut on north–south striking elements. East–west-striking elements are therefore interpreted as cross-joints (e.g. Gross 1993). In the northern area, conjugate strike-slip faults displace longitudinal (north–south) veins, indicating the pre-contractional development of these veins (Fig. 5).

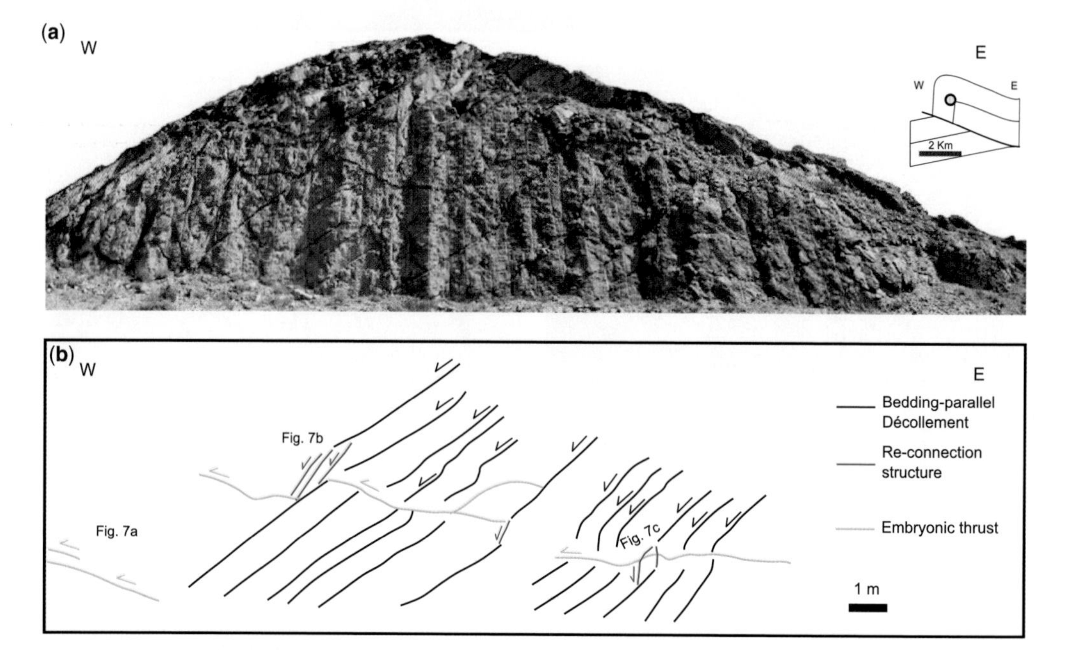

Fig. 6. Cross-cutting relationships between normal faults and compressional structures. (**a**) Photomosaic of the outcrop with (**b**) line drawing. Site location: 0.016617°E; 42.46728°N.

In the central transect, clear cross-cutting relationships between extensional and compressional structures are best observed at a site located at the transition between the crest and the forelimb (Fig. 6). Extensional structures are represented by tilted normal faults (including both westwards- and, subordinately, eastwards-dipping elements). Faults detach along thin marly layers which acted as extensional décollement horizons when bedding dip was subhorizontal. Data collected at this site along both faults and décollements indicate a general westwards hanging-wall motion during pre-folding extension. Compressional activity is recorded by the inversion of many of these extensional faults (Fig. 7a), the folding of the multilayer and the development of embryonic thrusts (Fig. 6b). After the development of compressional structures, bedding-parallel décollements reconnected across thrusts by developing complex accommodation structures (Fig. 7b). This occurred again in an extensional framework, as proved by the facts that: (1) slip direction continued to be westwards in the 're-connection' structures; and (2) many décollements did not reconnect but instead propagated downwards as normal faults (Fig. 7c).

Fig. 7. Details of the site in Figure 6. (**a**) Inverted westward-dipping normal fault; (**b**) duplexes (red arrows) reconnecting a layer-parallel décollement (solid black arrows); and (**c**) displaced layer-parallel décollement propagating downwards as a normal fault.

Discussion and conclusions

Based on the observations presented above, four different assemblages of structures can be defined in the Boltaña anticline. A first extensional assemblage is conformed by longitudinal (north–south) joints and veins at an high angle to bedding, east–west-striking cross-joints (e.g. Gross 1993) and north–south-striking extensional faults dipping toward the west when bedding dip is removed (Fig. 8a). These elements indicate an east–west direction of extension.

The second assemblage is compressional and contains conjugate strike-slip faults to the north and reverse faults and pressure-solution cleavage to the south (Figs 6 & 7). This assemblage post-dates extensional elements of the first assemblage and provides an east–west-oriented shortening direction (Fig. 8b). Cleavage and strike-slip conjugate faults are also present a few kilometres to the east in the Añisclo anticline (Fig. 1), where they have mostly developed before significant folding (Tavani *et al.* 2006). A pre- to early folding timing

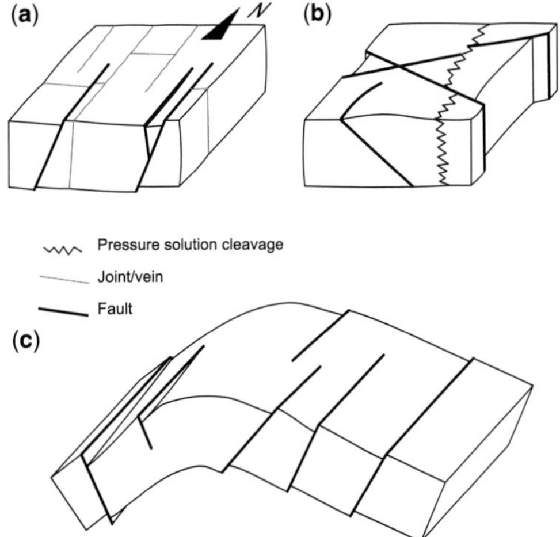

Fig. 8. Block diagrams illustrating the different assemblages found in the Boltaña anticline: (**a**) assemblages 1 and 3; (**b**) assemblage 2; and (**c**) assemblage 4. See text for details.

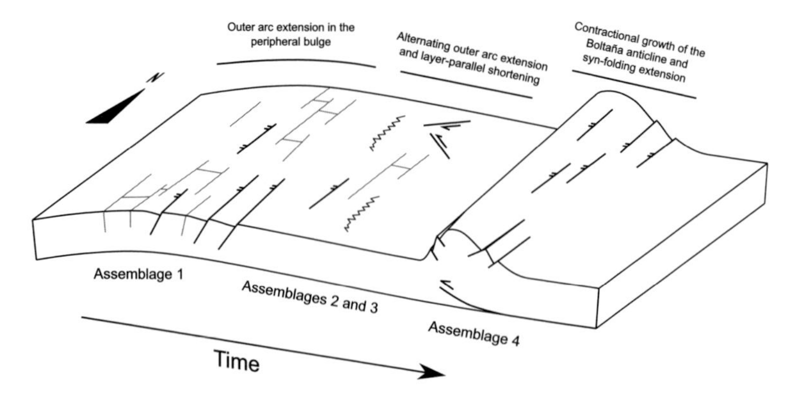

Fig. 9. Schematic evolution of the deformation pattern observed in the Boltaña anticline.

was also inferred for reverse faults in Figures 6 and 7, which are post-dated by extensional faults developed before significant folding. This contractional assemblage is therefore interpreted to have formed prior to or during early folding.

The third assemblage is extensional and is represented by post-contractional longitudinal extensional faults (as in Figs 6 & 7), which developed before or in the very early stages of folding. This assemblage provides the same extensional direction of the first extensional assemblage.

Finally, the fourth assemblage includes extensional faults developed during folding (Fig. 8c), which are characterized by a top to the hinge hanging-wall motion and reactivated previously developed joints or faults (Fig. 4b, d).

Widespread pre-contractional joints, veins and longitudinal extensional faults belonging to the first and third assemblages can be interpreted as developing due to outer arc extension when the study area formed part of the foreland basin undergoing flexural bending (e.g. Billi & Salvini 2003; Fig. 9). Combining the origin of joints and veins and the regional vertical axis rotation described in this area (Dinares Turell 1992; Pares & Dinares Turell 1993; Pueyo *et al.* 2004), the present-day north–south strike of joints and veins of the first assemblage is consistent with an east–west to NW–SE trend of the foredeep.

On the other hand, the second assemblage represents the classical pre- to early folding layer-parallel shortening pattern. Ambiguous cross-cutting relationships between pre- to early folding extensional (first and third assemblages) and contractional (second assemblage) structures indicate stress fluctuation in the foreland area.

Despite clearly being syn-folding extensional elements, the fourth assemblage cannot be associated with outer-arc extension (e.g. Ramsay 1967) or with flexural folding (e.g. Engelder & Peacock 2001). Outer-arc extension fails to explain the observed reactivation of joints in the constant-dip backlimb, where curvature is almost zero. On the other hand, extension associated with top to hinge flexural folding predicts a σ_3 oriented at about 45° to bedding. In the forelimb, this would imply a near-vertical σ_3. On the contrary, reactivated faults in the forelimb (Fig. 4d) mostly indicate a near-horizontal σ_3. Paradoxically, the nature of the fourth assemblage required the remote maximum stress to be near-vertical during fold growth. A possible explanation is that stress fluctuations inferred in the foreland area have also occurred during fold growth, with cyclic alternation between fold growth periods and extensional collapse stages.

As pointed out by Suppe *et al.* (1997) the rate of fold growth depends on the scale of observation. A steady-state behaviour can be assumed when considering an entire kilometric-scale fold at a 1–2 Ma timescale. Observations at more detailed scales however indicate that the rate of growth varies through time (e.g. Butler & Bowler 1995; Hubert-Ferrari *et al.* 2007; Charreau *et al.* 2008) and episodic co-seismic growth events have been documented (e.g. Chen *et al.* 2007). In summary, the folding rate fluctuates at different timescales. The amount and magnitude of these fluctuations increases with decreasing timescale.

In this work we present a well-documented field case revealing the fluctuation of the subhorizontal stress field. The combination of a contractional assemblage with pre- and syn-folding extensional assemblages is the expression of an episodic change of the tectonic framework. Accordingly, we interpret that extensional structures hosted in the Boltaña contractional anticline are syn-contractional when considering a long timescale of observation. In fact, they developed during extensional periods that occurred during the broader compressional folding event.

The fluctuation of stress field magnitude and folding rate can be interpreted as interconnected processes. Fast rates of fold growth can be associated with meso-scale contractional deformation and cause fold amplification (at the macro-scale). Slow to null growing rates produce (or are produced by) a drop of the horizontal stress magnitude that, where local and regional stress conditions are favourable, cause the onset of extensional deformation. In summary, variations of the regional σ_1 magnitude permit the onset of tectonic frameworks and related meso-structures that, when analysed, provide a record of a stress field apparently inconsistent with thrusting and folding.

The fluctuations in σ_1 magnitude explain not only longitudinal extensional structures, but also the typical assemblage found in reservoir-scale thrust-related anticlines. This assemblage includes one or more of the following elements, which are mostly oriented perpendicular to bedding: (1) longitudinal pressure-solution cleavage; (2) joints and veins striking parallel to cleavage; and (3) conjugate strike-slip faults bisecting both cleavage and joint/vein (e.g. Stearns 1968; Mitra *et al.* 1984; Marshak & Engelder 1985; Casas & Muñoz 1987; Price & Cosgrove 1990; Averbuch *et al.* 1992; Tavarnelli 1997; Tavani *et al.* 2006, 2008; Ahmadhadi *et al.* 2008). This pattern requires, paradoxically, a vertical/bedding-perpendicular σ_2, which is in disagreement with the presence of the thrust coring the anticline that requires a near vertical/bedding-perpendicular σ_3. In conclusion, the fluctuation of the stress field during thrust-related folding is an effective mechanism capable of explaining the vast presence of longitudinal syn-folding extensional structures in thrust-related anticlines but also, more generally, the presence of deformation structures developing within stress fields (apparently inconsistent with thrusting and folding).

D. Iacopini and R. Butler are thanked for their very helpful reviews. This work was carried out under the financial support of Cotiella (FBG306151) and INTECTOSAL (CGL2010-21968-C02-01/BTE) projects and Repsol YPF.

References

Ahmadhadi, F., Daniel, J.-M., Azzizadeh, M. & Lacombe, O. 2008. Evidence for pre-folding vein development in the Oligo-Miocene Asmari Formation in the Central Zagros Fold Belt, Iran. *Tectonics*, **27**, TC1016.

Averbuch, O., Frizon de Lamotte, D. & Kissel, C. 1992. Magnetic fabric as a structural indicator of the deformation path within a fold-thrust structure: a test case from the Corbières (NE Pyrenees, France). *Journal of Structural Geology*, **14**, 461–474.

Billi, A. & Salvini, F. 2003. Development of systematic joints in response to flexure-related fibre stress in flexed foreland plates: the Apulian forebulge case history, Italy. *Journal of Geodynamics*, **36**, 523–536.

Butler, R. W. H. & Bowler, S. 1995. Local displacement rate cycles in the life of a fold-thrust belt. *Terra Nova*, **7**, 408–416.

Cámara, P. & Klimowitz, J. 1985. Interpretación geodinámica de la vertiente centro-occidental surpirenaica (Cuencas de Jaca – Tremp). *Estudios geologicos*, **41**, 391–404.

Casas, J. M. & Muñoz, J. A. 1987. Sequences of mesostructures related to the development of Alpine thrusts in the Eastern Pyrenees. *Tectonophysics*, **135**, 67–75.

Charreau, J., Avouac, J-P., Chen, Y., Dominguez, S. & Gilder, S. 2008. Miocene to present kinematics of fault-bend folding across the Huerguosi anticline, northern Tianshan (China), derived from structural, seismic, and magnetostratigraphic data. *Geology*, **36**, 871–874.

Chen, Y.-G., Lai, K.-Y. *et al.* 2007. Coseismic fold scarps and their kinematic behavior in the 1999 Chi-Chi earthquake Taiwan. *Journal of Geophysical Research B: Solid Earth*, **112**, B03S02.

Cooper, M. 1992. The analysis of fracture systems in subsurface thrust structures from the foothills of the Canadian Rockies. *In*: McClay, K. R. (ed.) *Thrust Tectonics*. Chapman and Hall, London, 391–405.

Dinares Turell, J. 1992. *Paleomagnetisme a les unitats sudpirinenques superiors. Implicacions estructurals.* PhD thesis, University of Barcelona, Spain.

Engelder, T. & Peacock, D. C. P. 2001. Joint development normal to regional compression during flexural flow folding: the Lilstock buttress anticline, Somerset, England. *Journal of Structural Geology*, **23**, 259–277.

Fernandez, O. 2004. *Reconstruction of geological structures in 3D. An example from the southern Pyrenees.* PhD thesis, University of Barcelona.

Fischer, M. P. & Wilkerson, M. S. 2000. Predicting the orientation of joints from fold shape: results of pseudo–three-dimensional modeling and curvature analysis. *Geology*, **28**, 15–18.

Gross, M. R. 1993. The origin and spacing of cross joints: examples from the Monterey Formation, Santa Barbara Coastline, California. *Journal of Structural Geology*, **15**, 737–751.

Harris, J. H. & Van Der Pluijm, B. A. 1998. Relative timing of calcite twinning strain and fold–thrust belt development; Hudson Valley fold–thrust belt, New York, U.S.A. *Journal of Structural Geology*, **20**, 21–31.

Holl, J. E. & Anastasio, D. J. 1995. Cleavage development within a foreland fold and thrust belt, southern Pyrenees, Spain. *Journal of Structural Geology*, **17**, 357–369.

Hubert-Ferrari, A., Suppe, J., Gonzalez-Mieres, R. & Wang, X. 2007. Mechanisms of active folding of the landscape (southern Tian Shan, China). *Journal of Geophysical Research B: Solid Earth*, **112**, B03S09.

Lemiszki, P. J., Landes, J. D. & Hatcher, R. D., Jr. 1994. Controls on hinge-parallel extension fracturing in single-layer tangential-longitudinal strain folds. *Journal of Geophysical Research*, **99**, 22027–22042.

Marshak, S. & Engelder, T. 1985. Development of cleavage in limestones of a fold–thrust belt in eastern New York. *Journal of Structural Geology*, **7**, 345–359.

McQuillan, H. 1974. Fracture patterns on Kuh-e Asmari anticline, Southwest Iran. *AAPG Bulletin*, **58**, 236–246.

Mitra, G., Yonkee, W. A. & Gentry, D. J. 1984. Solution cleavages and its relationship to major structures in the Idaho–Utah–Wyoming thrust belt. *Geology*, **12**, 354–358.

Mochales, T., Pueyo, E. L., Casas, A. M., Barnolas, A. & Oliva-Urcia, B. 2010. Anisotropic magnetic susceptibility record of the kinematics of the Boltaña Anticline (Southern Pyrenees). *Geological Journal*, **45**, 544–561.

Pares, J. M. & Dinares Turell, J. 1993. Magnetic fabric in two sedimentary rock-types from the Southern Pyrenees. *Journal of Geomagnetism and Geoelectricity*, **45**, 193–205.

Price, N. J. & Cosgrove, J. W. 1990. *Analysis of Geological Structures*. Cambridge University Press, Cambridge.

Pueyo, E. L., Pocovi, A., Millan, H. & Sussman, A. J. 2004. Map-view models for correcting and calculating shortening estimates in rotated thrust fronts using paleomagnetic data. *In*: Sussman, A. J. & Weil, A. B. (eds) *Orogenic Curvature: Integrating Paleomagnetic and Structural Analyses*. Geological Society of America, Special Publication, **383**, 57–71.

Ramsay, J. G. 1967. *Folding and Fracturing of Rocks*. McGraw-Hill, New York.

Seguret, M. 1970. *Etude tectonique des nappes et series decollees de la partie centrale du versant sud des Pyrenees*. PhD thesis, University of Montpellier.

Srivastava, D. C. & Engelder, T. 1990. Crack-propagation sequence and pore-fluid conditions during fault-bend folding in the Appalachian Valley and Ridge, central Pennsylvania. *Geological Society of America, Bulletin*, **102**, 116–128.

Stearns, D. W. 1968. Certain aspect of fracture in naturally deformed rocks. *In*: Rieker, R. E. (ed.) *National Science Foundation Advanced Science Seminar in Rock Mechanics*. Special Report, Air Force Cambridge Research Laboratories, Bedford, Massachusetts, 97–118.

Stephenson, B. J., Koopman, A., Hillgartner, H., McQuillan, H., Bourne, S., Noad, J. J. & Rawnsley, K. 2007. Structural and stratigraphic controls on fold-related fracturing in the Zagros Mountains, Iran: implications for reservoir development. *In*: Lonergan, L., Jolly, R. J. H., Rawnsley, K. & Sanderson, D. J. (eds) *Fractured Reservoirs*. Geological Society, London, Special Publications, **270**, 1–21.

Suppe, J., Sabat, F., Munoz, J. A., Poblet, J., Roca, E. & Verges, J. 1997. Bed-by-bed fold growth by kink-band migration: Sant llorenc de Morunys, eastern Pyrenees. *Journal of Structural Geology*, **19**, 443–461.

Tavani, S., Storti, F., Fernández, O., Muñoz, J. A. & Salvini, F. 2006. 3-D deformation pattern analysis and evolution of the Añisclo anticline, southern Pyrenees. *Journal of Structural Geology*, **28**, 695–712.

Tavani, S., Storti, F., Salvini, F. & Toscano, C. 2008. Stratigraphic v. structural control on the deformation pattern associated with the evolution of the Mt. Catria anticline, Italy. *Journal of Structural Geology*, **30**, 664–681.

Tavarnelli, E. 1997. Structural evolution of a foreland fold-and-thrust belt: the Umbria–Marche Apennines, Italy. *Journal of Structural Geology*, **19**, 523–534.

Stress deflections around salt diapirs in the Gulf of Mexico

ROSALIND KING[1]*, GUILLAUME BACKÉ[2], MARK TINGAY[2], RICHARD HILLIS[3] & SCOTT MILDREN[4]

[1]*Centre for Tectonics, Resources and Exploration (TRaX), School of Earth and Environmental Sciences, University of Adelaide, SA 5005, Australia*

[2]*Centre for Tectonics, Resources and Exploration (TRaX), Australian School of Petroleum, University of Adelaide, SA 5005, Australia*

[3]*Deep Exploration Technologies Cooperative Research Centre, Mawson Building, University of Adelaide, SA 5005, Australia*

[4]*JRS Petroleum Research, Level 6, 28 Gawler Place, Adelaide, SA 5000, Australia*

**Corresponding author (e-mail: rosalind.king@adelaide.edu.au)*

Abstract: Delta–deepwater fold–thrust belts are linked systems of extension and compression. Margin-parallel maximum horizontal stresses (extension) on the delta top are generated by gravitational collapse of accumulating sediment, and drive downdip margin-normal maximum horizontal stresses (compression) in the deepwater fold–thrust belt (or delta toe). This maximum horizontal stress rotation has been observed in a number of delta systems. Maximum horizontal stress orientations, determined from 32 petroleum wells in the Gulf of Mexico, are broadly margin-parallel on the delta top with a mean orientation of 060 and a standard deviation of 49°. However, several orientations show up to 60° deflection from the regional margin-parallel orientation. Three-dimensional (3D) seismic data from the Gulf of Mexico delta top demonstrate the presence of salt diapirs piercing the overlying deltaic sediments. These salt diapirs are adjacent to wells (within 500 m) that demonstrate deflected stress orientations. The maximum horizontal stresses are deflected to become parallel to the interface between the salt and sediment. Two cases are presented that account for the alignment of maximum horizontal stresses parallel to this interface: (1) the contrast between geomechanical properties of the deltaic sediments and adjacent salt diapirs; and (2) gravitational collapse of deltaic sediments down the flanks of salt diapirs.

In many regions worldwide σ_H orientations are parallel or subparallel to absolute plate velocity vectors and ridge torques, for example, North America, South America and Western Europe (Richardson 1992; Zoback 1992). First-order intraplate stress patterns (wavelengths >1000 km) are therefore a result of large-scale plate boundary forces (e.g. ridge push, slab pull). The stress field generated by plate boundary forces are superimposed on major intra-plate sources of stress (gravitational forces imparted by mountain belts; lithospheric flexure) to generate the second-order stress pattern (wavelengths 100–500 km; Zoback 1992). However, recent years has seen significant advance in the understanding of short-wavelength (<100 km) third-order stress fields observed at the reservoir-, field- and basin-scale in sedimentary basins generated by local effects (e.g. topography, sediment loading, glacial rebound, elastic dislocation from large faults, overpressure generation, buckling, asperities on fault planes and lateral density contrasts; Bell 1996; Tingay *et al.* 2006; Heidbach *et al.* 2007; MacDonald *et al.* 2012). It is the relative magnitudes of the sources of stress that define the dominant stress regime in a given region (Zoback 1992; Bell 1996; Tingay *et al.* 2006). For example, a local stress source may induce large differential stresses that override the regional (far-field) stress source so that third-order stress patterns dominate in the area (e.g. Sonder 1990; Bell 1996; King *et al.* 2010*a*). Alternatively, a local stress source with low differential stresses may still affect stress orientations in regions where a layer with low shear strength (e.g. a detachment at the base of a delta system or a salt horizon) prevents the transfer of regional far-field stresses into layers where measurements are taken (Tingay *et al.* 2011; Bell 1996; King *et al.* 2010*b*).

Delta–deepwater fold–thrust belts (DDWFTBs) are linked systems of extension and compression (Morley 2003; Rowan *et al.* 2004; King *et al.* 2009; Fig. 1). Gravitational potential of accumulating sediment on the delta top generates margin-parallel maximum horizontal stress (σ_H) orientations (extension), which are marked by margin-parallel-striking normal growth faults and have listric shapes (Mandl

From: Healy, D., Butler, R. W. H., Shipton, Z. K. & Sibson, R. H. (eds) 2012. *Faulting, Fracturing and Igneous Intrusion in the Earth's Crust*. Geological Society, London, Special Publications, **367**, 141–153. http://dx.doi.org/10.1144/SP367.10

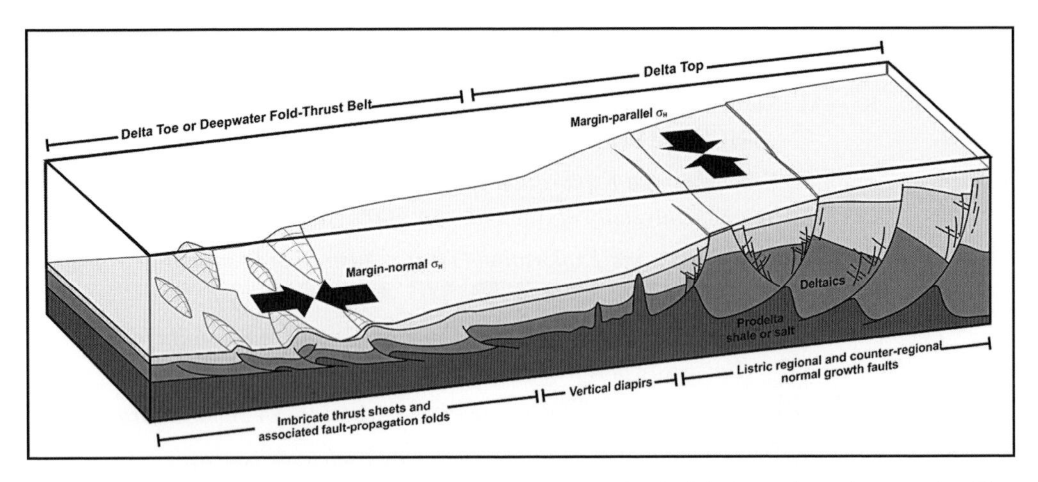

Fig. 1. Schematic diagram of a delta–deepwater fold–thrust belt illustrating the linked extension and compression. The delta top exhibits normal listric growth faults marking a margin-parallel maximum horizontal stress, and the delta toe (or deepwater fold–thrust belts) exhibits imbricate thrust sheets and associated fault-propagation folds marking a margin-normal maximum horizontal stress orientation (from King & Backé 2010).

& Crans 1981; Yassir & Zerwer 1997; King *et al.* 2009; Fig. 1). These extensional stresses drive downdip margin-normal σ_H orientations in the deepwater fold–thrust belt (or delta toe; compression), which are marked by margin-parallel-striking stacked thrust sheets and associated folds (Tingay *et al.* 2005; King *et al.* 2009; Fig. 1). However, the lobe shape of delta systems results in strike-slip stress regimes and associated structures at the outermost lateral margins of the systems (Peel *et al.* 1995). Present-day maximum horizontal stress orientations in the onshore Gulf Coast region display a clear margin-parallel trend (Tingay *et al.* 2006). However, margin-parallel σ_H orientations in the delta top region of the Gulf of Mexico, offshore Louisiana, demonstrate significant deflections from the expected margin-parallel σ_H orientations (Yassir & Zerwer 1997).

Maximum horizontal stress orientations in the Gulf of Mexico, offshore Louisiana, are third-order σ_H orientations generated by the gravitational potential of the accumulating sediment in the delta systems. However, many of these third-order σ_H orientations are deflected around salt diapirs that are considered to be caused by the contrast between the geomechanical properties of the salt and adjacent clastic deltaic sediments (Bell 1996; Yassir & Zerwer 1997). However, the Gulf of Mexico stress analysis conducted by Yassir and Zerwer (1997) was primarily a two-dimensional study using only the map pattern of σ_H orientations and does not account for the vertical changes in geomechanical properties or diapir shape. In this paper, we present 5 new σ_H orientations determined from 8 petroleum wells. We present a 3D model of the deflected σ_H orientations around a salt diapir, demonstrating changes in the orientations both laterally and vertically, and discuss the causes of these deflections.

Geological setting: the Gulf of Mexico

The Gulf of Mexico is one of the world's foremost petroleum provinces, and is located offshore the southern USA at 19–30°N and −83 to −97°W (Fig. 2a). Water depths in the Gulf of Mexico range from several metres around the coasts to more than 2000 m in the deep central parts. Much of the petroleum exploration has been focused on the shallow-water petroleum-rich delta top (Trudgill *et al.* 1999). Exploration focus shifted in the last decade to the deep water, as major discoveries in the deepwater fold–thrust belts were made (Trudgill *et al.* 1999). However, exploration and major reserve development programs continue in both the delta top and deepwater fold–thrust belt regions at present day.

The Gulf of Mexico is composed of several Upper Jurassic–Pleistocene delta systems that prograded from the north and west sourced by the Rio Grande (Galloway 1989; Fiduk *et al.* 1999; Fig. 2a). The delta systems sit on and above the Middle Jurassic Louann Salt, which is extensive across the northern Gulf of Mexico but is absent in the Mexican Ridges area (Peel *et al.* 1995; Trudgill *et al.* 1999). The Louann Salt forms the regional detachment zone beneath the deltaic sediments, with the majority of normal faults and thrust faults detaching at this level (Worrall & Snelson 1989;

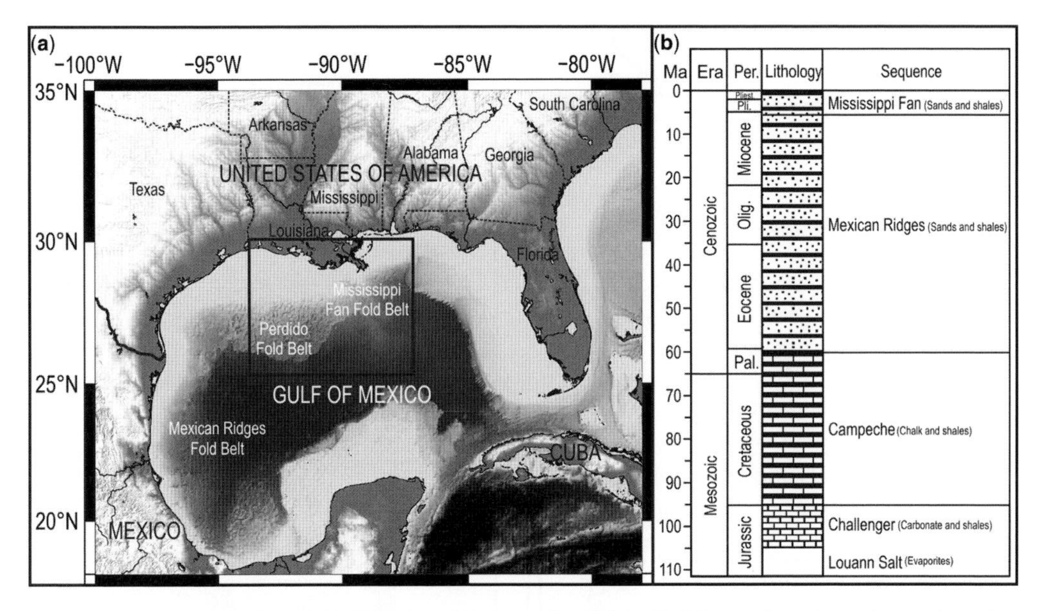

Fig. 2. (**a**) Location map of the Gulf of Mexico, offshore southern USA. Black box outlines the study area shown in Figure 4. (**b**) Mesozoic–Cenozoic tectonostratigraphy of the Gulf of Mexico (from Rowan 1997).

Wu *et al.* 1990; Rowan 1997; Fig. 1). Syn-rift sediments deposited on the rifted basement underlie the Louann Salt, but these are not well studied (Trudgill *et al.* 1999).

The post-salt stratigraphy can be divided into four broad sequences that demonstrate the evolution of the Gulf of Mexico from a shallow rifted margin in the Upper Jurassic–Early Cretaceous which continually subsided throughout the Late Cretaceous and into the Cenozoic to form the present-day shelf to basin-floor profile (Rowan 1997; Fiduk *et al.* 1999; Fig. 2). The carbonate Challenger and Campeche sequences demonstrate shallow marine environments across the Gulf of Mexico, but transition to deep marine in the present-day deepwater region during the Late Cretaceous (Weimer 1990; Feng & Buffler 1991; Rowan 1997; Fig. 2). The Palaeocene Mexican Ridges sequence is composed of deltaic sands on the present-day delta-top region and deepwater carbonates and turbidites in the present-day deepwater; the latter demonstrates bypass of deltaic sands to the basin floor (Weimer 1990; Feng & Buffler 1991; Rowan 1997; Fig. 2). The final sequence (the Palaeogene Mississippi Fan sequence) is a series of deltaic sands deposited on the present-day delta-top region and a deepwater turbidite sequence. This sequence is made up of several delta lobes (Weimer 1990; Feng & Buffler 1991; Rowan 1997; Fig. 2).

The accumulating sediment deformed by gravity sliding has resulted in a complex array of margin-parallel normal faults observed on the shelf, consistent with delta-top extension (Rowan 1997). Three fold–thrust belts generated by the updip extension are recognized in the deepwater: the Mississippi Fan Fold Belt, Perdido Fold Belt and the Mexican Ridges Fold Belt (Trudgill *et al.* 1999; Fig. 2a). It is therefore expected that σ_H orientations should rotate from margin-parallel on the delta top (marked by the margin-parallel normal faults) to margin-normal in the deepwater (marked by the imbricate thrust sheets) (e.g. King *et al.* 2009; Fig. 1).

Analysis of stress orientations in the Gulf of Mexico

Image logs and high-resolution dipmeter logs from 40 petroleum wells across the Gulf of Mexico were used to identify borehole failure and therefore determine the orientation of σ_H in 32 wells (this study; Yassir & Zerwer 1997). The orientation of σ_H can be determined from stress-induced compressive or tensile failure of the borehole wall, known as borehole breakout and drilling-induced tensile fractures (DITFs), respectively (Bell 1996; Fig. 3a).

Borehole failure

Borehole breakouts are a stress-induced elongation of the borehole cross-section. The presence of an open wellbore causes a localized perturbation of

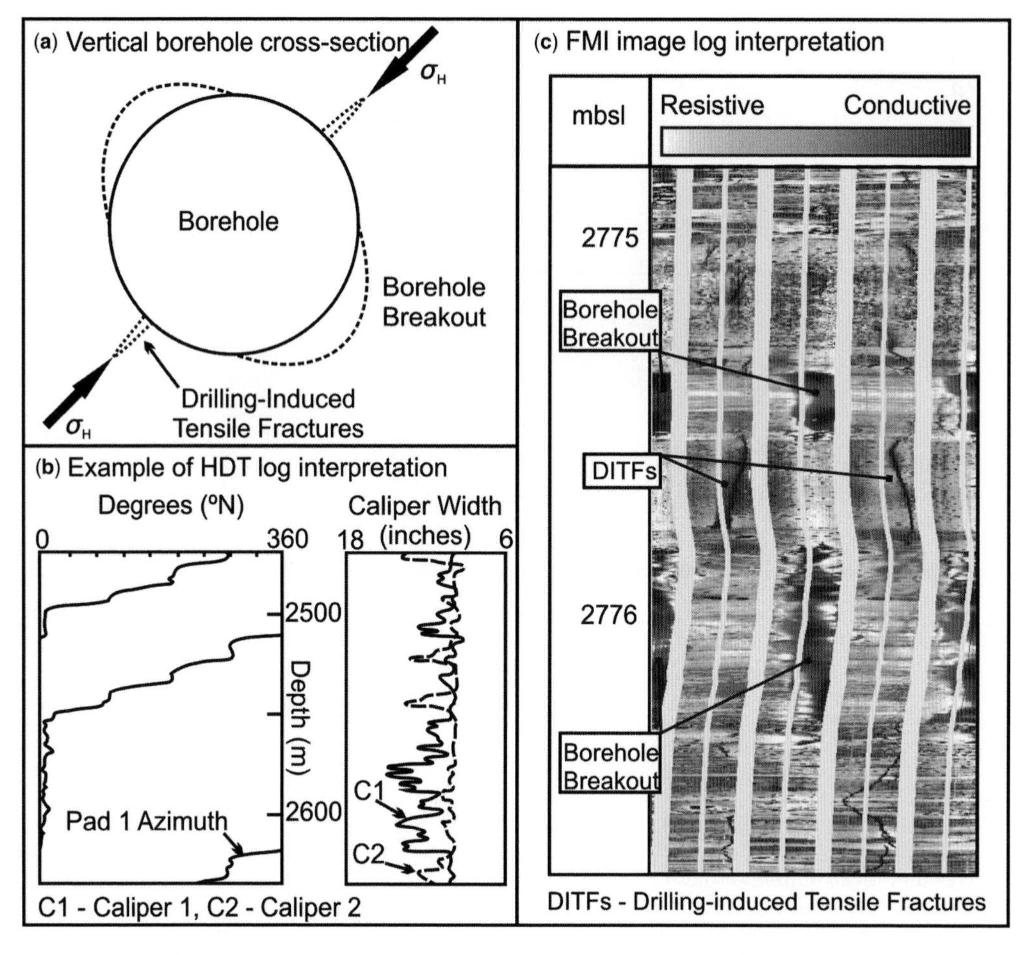

Fig. 3. (**a**) Borehole breakouts and drilling-induced tensile fractures observed on (**b**) dipmeter logs and (**c**) image logs can be used to determine the orientation of the maximum and minimum horizontal stresses (King *et al.* 2008). HDT, high-resolution dipmeter tool; FMI, formation micro-imaging tool; DITF, drilling-induced tensile fractures.

stresses in the vicinity of the borehole (Kirsch 1898). Borehole breakouts form when the maximum circumferential stress at the borehole wall exceeds the compressive rock strength, resulting in compressive failure and spalling of the borehole wall (Bell 1996). The circumferential stress is a function of the magnitude and anisotropy of σ_H and the minimum horizontal stress (σ_h) in vertical wells, with the maximum circumferential stress (and thus breakouts) developing perpendicular to the orientation of σ_H (e.g. Kirsch 1898; Bell & Gough 1979; Fig. 3a). Drilling-induced tensile fractures form due to tensile failure at the borehole wall when the minimum circumferential stress is less than the tensile strength of the borehole wall (Aadnoy & Bell 1998). Drilling-induced tensile fractures form parallel to the present-day σ_H orientation in vertical wells (Bell 1996; Brudy & Zoback 1999; Fig. 3a).

Borehole breakouts and DITFs may not directly yield the tectonic stress orientation in highly deviated boreholes due to the complex stresses that form around a borehole not oriented parallel to a principal stress (Mastin 1988; Peška & Zoback 1995). Hence, appropriate corrections to the orientations of borehole breakouts interpreted on the dipmeter or image logs were required in wells that were not vertical (after Peška & Zoback 1995).

Log interpretation

Resistivity image logs were used to interpret borehole breakout and DITF orientations in six petroleum wells. Borehole breakout orientations were analysed from four-arm high-resolution dipmeter tools (composed of two pairs of calipers) in a further two petroleum wells in this study. Eight wells were combined with breakout orientations analysed from 27 wells by Yassir & Zerwer (1997). High-resolution dipmeter tools record borehole diameters in two orthogonal directions so borehole elongations, such as borehole breakouts, can be identified (Fig. 3b). The criteria used to successfully identify borehole breakouts on dipmeter logs are as follows (after Plumb & Hickman 1985).

- The rotation of the tool stops in the zone of elongation. The tool should rotate before and after the elongation. However, in zones of several small breakouts, the rotation may terminate completely.
- The difference recorded between the two arms of the calipers is >6 mm.
- The length (along the borehole axis) of the elongation is >1.5 m.
- The largest caliper should be extended greater than the drill bit size.
- The smallest caliper should not be significantly greater than the drill bit size.

Care is required when analysing dipmeter logs, so that borehole enlargements not related to stress (e.g. washouts or key seating) are not confused with borehole breakouts (e.g. Hillis & Williams 1993). Drilling-induced tensile fractures cannot be identified on dipmeter logs because they do not generally create borehole elongation.

Formation micro-image (FMI) and formation micro-scanning tools (FMS) produce an image of resistivity contrasts at the borehole wall (Ekstrom *et al.* 1987). The images show sedimentological, lithological and structural features such as cross-bedding and natural fractures. The images also exhibit drilling-related features such as tool marks, borehole breakout and DITFs. Borehole breakouts appear on resistivity image logs (in water-based muds) as pairs of conductive (dark) poorly resolved zones separated by 180°, and typically also exhibit caliper enlargement in the breakout direction (Fig. 3c). Drilling-induced tensile fractures appear on resistivity image logs as pairs of discontinuous, vertical, conductive (dark) fractures separated by 180° (Fig. 3c; Aadnoy & Bell 1998; Barton 2000).

Quality ranking stress orientations

The σ_H orientation for each well analysed in this study has been quality ranked in accordance with the standard World Stress Map quality ranking system (Table 1). The World Stress Map ranking system is based on the total number, the standard deviation and the total length of the stress indicators observed (Sperner *et al.* 2003; Heidbach *et al.* 2010; Table 1). The ranks are from A to E, with A being the highest quality and E the lowest (Table 1). Orientations that rank A–C are considered as reliable stress orientations for investigating primary stress fields under the World Stress Map ranking system (Zoback 1992). However, more recent studies in Cenozoic basins have argued that D-quality stress orientations can also provide important information, particularly when analysing third-order stress patterns or when multiple D-quality stress orientations are observed in close proximity (Tingay *et al.* 2010). Indeed, Yassir & Zerwer (1997) themselves presented an alternative ranking system for their work in the Gulf of Mexico, which locally ranked World Stress Map D-quality indicators as B and C quality. However, all stress orientations presented in Figure 4 are ranked according to the World Stress Map criteria (Sperner *et al.* 2003; Heidbach *et al.* 2010; Table 1).

New stress orientations in the Gulf of Mexico

Eight new petroleum wells have been analysed from the Gulf of Mexico in this study. All wells are located on the shelf within the delta-top region (Fig. 4). In total, 28 borehole breakouts and no DITFs were identified in the eight wells. The mean σ_H orientations from borehole breakouts were quality ranked with B- or C-quality stress orientations observed in two wells, D-quality orientations in three wells and the remaining three wells being ranked as E-quality (Table 2). The mean east–west σ_H orientation from these wells matches the expected margin-parallel σ_H orientation on the delta top of the idealized DDWFTB model (Fig. 1; Table 3). Indeed, σ_H orientations determined from wells 1, 2, 3 and 5 are all oriented subparallel to the Louisiana coastline, suggesting a typical delta-top margin-parallel σ_H orientation (Fig. 4).

The σ_H orientations determined from five petroleum wells of the eight examined here (Table 2) have been combined with σ_H orientations determined from 27 petroleum wells presented by Yassir & Zerwer (1997; Fig. 4). Table 3 presents the regional mean σ_H orientations determined for wells ranked A–C quality (considered reliable stress indicators; Zoback 1992) and for wells ranked A–D quality. Both studies suggest that present-day σ_H orientations on the delta top are oriented margin-parallel and are consistent with the stress field generated in the shelfal sections of a DDWFTB (Figs 1 & 4). All of the stress orientations presented in this study, including those from Yassir and Zerwer (1997), exhibited significantly high standard deviations in breakout orientations within each individual well. The high degree of scatter in borehole breakout orientations within individual wells is partially due to a combination of factors, such as

Table 1. *The World Stress Map quality ranking system for four- and six-arm high-resolution dipmeter logs (Sperner* et al. *2003) and resistivity and acoustic image logs (Heidbach* et al. *2010)*

	A quality	B quality	C quality	D quality	E quality
World Stress Map quality ranking system for four- and six-arm high resolution dipmeter logs					
No. borehole breakout	≥10	≥6	≥4	<4	0
Standard deviation (°)	≤12	≤20	≤25	>25	–
Combined length (m)	>300	>100	>30	<30	–
World Stress Map quality ranking system for resistivity or acoustic image logs					
No. borehole breakout/DITFs	≥10	≥6	≥4	<4	0
Standard deviation (°)	≤12	≤20	≤25	>25	–
Combined length (m)	>100	>40	>20	<20	–

the use of poorer quality four-arm caliper data in most wells. However, it is important to note that many wells exhibiting scattered breakout orientations are located on the shelf edge and slope where many salt diapirs are located (Yassir & Zerwer 1997).

Stress deflections in the Gulf of Mexico

Yassir & Zerwer (1997) identified several σ_H orientations that were not consistent with the expected margin-parallel orientation (Fig. 4). In this study, the σ_H orientation of 036 observed in well 4 is not consistent with this margin-parallel σ_H orientation (Fig. 4). Yassir & Zerwer (1997) identified that σ_H orientations were deflected away from the regional margin-parallel orientation near the interface between a salt diapir and the adjacent deltaic sediments (Fig. 4, areas 1 and 2). They hypothesized that the contrast in geomechanical properties between the salt and adjacent deltaic sediments resulted in σ_H orientations becoming locally rotated subparallel to the interface between the salt and the sediment (Bell 1996; Yassir & Zerwer 1997). The salt–sediment boundary is a surface that can sustain only a very weak shear force; principal stress azimuths are therefore deflected subparallel/subnormal to the salt–sediment interface (Davis & Engelder 1985). Yassir & Zerwer's (1997) interpretations were made using σ_H orientations determined from well

Table 2. *Maximum horizontal stress orientations determined in this study for 8 petroleum wells from the Gulf of Mexico, including the log type used for analysis, depth of tool run and the quality rank according to the World Stress Map quality ranking system (Table 1)*

Well	Water depth (m)	Tool	Vertical well deviation	Tool Runb (m bsl)	σ_H orientation	No. of stress indicator	Total length of indicators (m)	Standard deviation (°)	Quality
1 OCS-G 01666 018 ST01 BP00	145	FMI	60° to 030	1835–2143	100	9	46	18	B
2 OCS-G 1249 B-6ST2	34	HDT	<18°	758–2192	090	7	55	08	C
3 OCS-G 00244F-1	–	FMI	<8°	2582–3531	093	6	15	28	D
4 OCS-G 026#P-40 ST1	13	HDT	<25°	1800–2169	036	3	42	12	D
5 OSC-G 0310#201	6	FMI	<9°	1975–2840	102	3	11	15	D
6 OCS-G 01959 A015 ST01BP00	48	FMI	–	1067–3147	–	0	–	–	E
7 OCS-G 0390 040 ST00 BP01	14	FMI	–	4175–4328	–	0	–	–	E
8 OCS-G 0392 #Y29	–	FMI	–	1113–3764	–	0	–	–	E

Table 3. *Mean maximum horizontal stress orientations determined for A–C and A–D quality ranked wells from this study and Yassir & Zerwer (1997)*

Quality	No. of wells	σ_H orientation	Standard deviation (°)
Data from this study			
A–C	2	095	05
A–D	5	089	24
Data from Yassir & Zerwer (1997)			
A–C	11	039	49
A–D	27	043	51
Combined Data (this study and Yassir & Zerwer 1997)			
A–C	13	061	52
A–D	32	060	51

logs at depth combined with seafloor topography (e.g. areas 1 and 2 illustrated on Fig. 4) and therefore do not appreciate the 3D nature of the system. Here, we use time slices at 1 and 2 s from the Ship Shoal 3D seismic cube to demonstrate that the hypothesized rotations in the stress field adjacent to salt diapirs are also accurate when depth is considered.

Salt diapirs in the Ship Shoal 3D seismic cube are considered to be active or passive salt diapirs

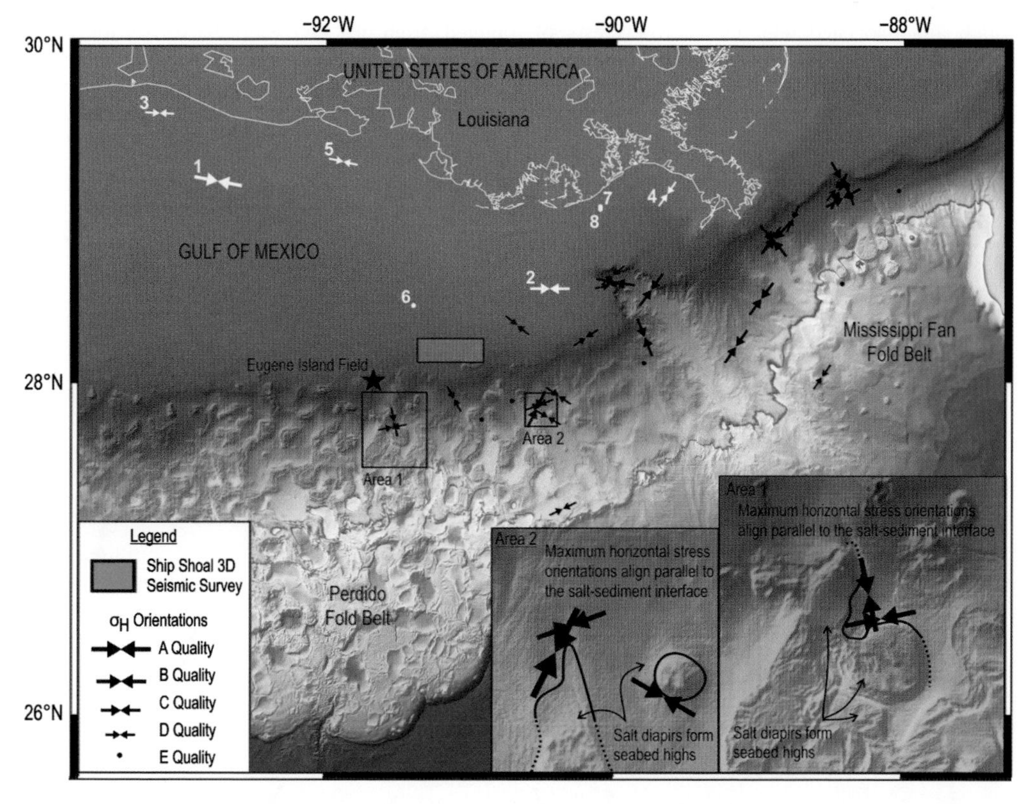

Fig. 4. Map illustrating the maximum horizontal stress orientations across the Gulf of Mexico (black arrows: Yassir & Zerwer 1997; white arrows: this study). The mean regional maximum horizontal stress orientation of 060 (standard deviation 51°) for all 27 A–D indicators is margin-parallel, consistent with the idealized model of a delta–deepwater fold–thrust belt (Fig. 1). Deflection of maximum horizontal stress orientations from margin-parallel occurs where salt diapirs pierce the deltaic sediments. The maximum horizontal stress orientations align parallel to the interface between salt and sediment, shown in areas 1 and 2.

(after Jackson *et al.* 1994). Active salt diapirs rise into overlying sediment by force, and are often associated with crestal collapse structures (Jackson *et al.* 1994). Passive diapirs rise by bouyancy into the overlying sediments and are accommodated without force; they therefore do not demonstrate any crestal collapse (Jackson *et al.* 1994). The salt diapirs have generated significant seafloor topography (Fig. 4). Time slices taken from at 1 and 2 s illustrate the shape of the salt diapirs at depth as conical structures with broad bases and narrow, domal crests (Fig. 5a, b). Figure 5 illustrates that salt diapir 1 becomes broader towards the east and west at depth and salt diapir 2 is elongated towards the north with increasing depth. Predictions of σ_H orientations at different depths around salt diapirs can be made using the time slices, following the hypothesis of Yassir & Zerwer (1997) that σ_H orientations are locally aligned subparallel to the interface between the salt diapirs and the adjacent deltaic sediments (Fig. 5a, b). At shallow depths, σ_H orientations are likely to deflect from the regional margin-parallel orientations over narrow areas which are associated with the narrow crests of salt diapirs (e.g. Fig. 5a). At greater depths, σ_H orientations are likely to deflect from the regional margin-parallel orientations over broader areas which are associated with the salt diapirs broad bases (e.g. Fig. 5b). Stress orientations in individual wells are consistent with depth and amount of lithification within deltaic sediments. This suggests that stress deflections are associated with large geomechanical contrasts in salt and clastic deltaic sediment, generating a surface that cannot sustain shear forces, and not subtle changes in the clastic deltaic sediments with depth. Furthermore, deflections of stress fields resulting in marked variations in fracture patterns around salt diapirs have also been imaged on time slices from Central Graben salt diapirs in the North Sea (e.g. Davison *et al.* 2000). Three-dimensional modelling was undertaken to better understand the geomechanical aspects controlling the observed local stress deflections near salt diapirs.

Numerical modelling of stress orientations around a salt diapir

The 3D modelling of the displacements, strain and stresses around the top of the Louann Salt, interpreted from the 3D Ship Shoal seismic cube (Fig. 6), was carried out using Poly3D, a 3D boundary element code using polygon elements and linear elasticity theory (Thomas 1993; Maerten *et al.* 2000,

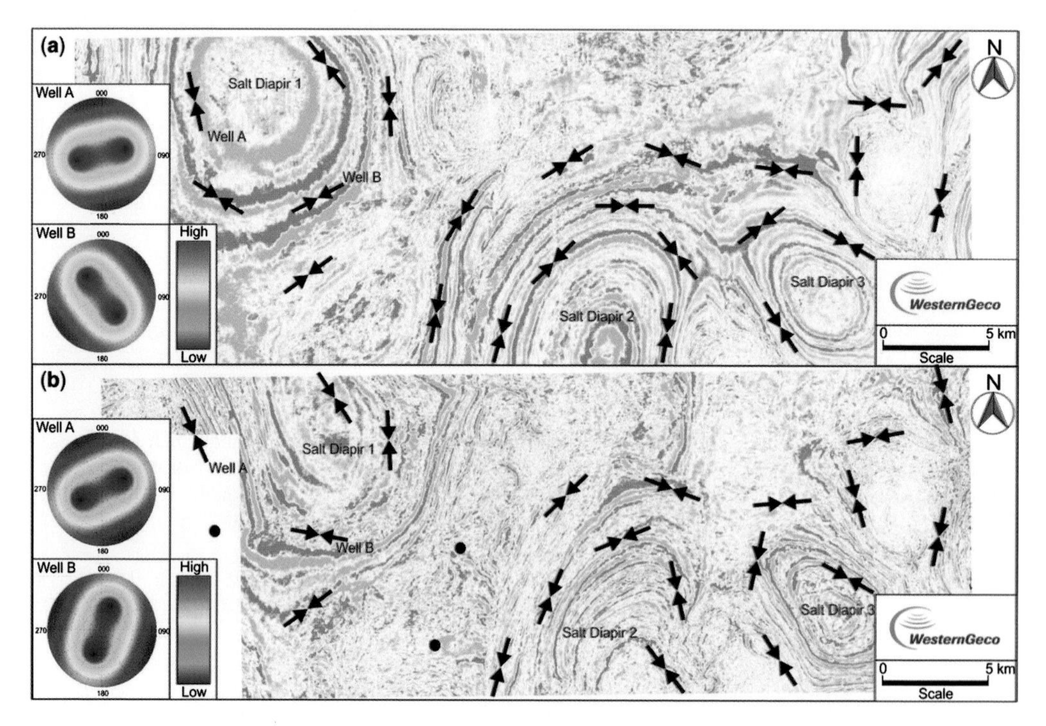

Fig. 5. Time slices taken from the Ship Shoal 3D seismic cube at (**a**) 1 s and (**b**) 2 s, demonstrating predicted maximum horizontal stress orientations (black arrows) and associated borehole stability diagrams for each orientation (vertical wells plot in the centre of the diagram, while horizontal wells plots on the circumference of the diagrams).

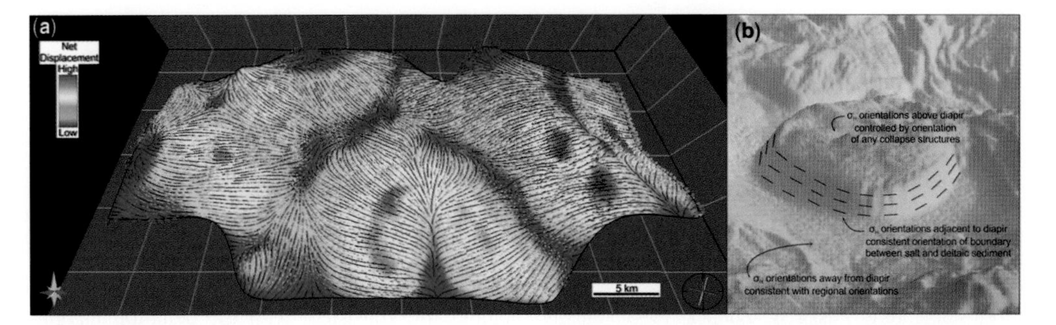

Fig. 6. (**a**) Poly3D model of the top Louann Salt horizon, interpreted from the Ship Shoal 3D seismic cube, demonstrating the displacement vectors (black lines) for the deltaic sediments at the interface between the salt diapirs and overlying sediments and the net displacement (with red being high and blue being low). Displacement vectors suggest movement will be downslope, with the greatest modelled displacement on the flanks of the salt diapirs. The predicted present-day displacement vectors are parallel to the present-day minimum horizontal stress and therefore perpendicular to the present-day maximum horizontal stress orientations (which lie in a horizontal plane) and parallel to the flanks of the salt diapirs. (**b**) A schematic 3D image of a conical salt diapir and the expected maximum horizontal stress deflections.

2002; Fiore *et al.* 2006; Maerten 2010). A boundary element method (BEM) greatly simplifies the model since only the discontinuities (such as faults or detachments) have to be taken into account and discretized. For the modelling purposes, the top Louann Salt was considered as a detachment system (i.e. a fault) and discretized with triangular elements with a 1000 m mesh size. Specific boundary conditions were applied, which govern how the elements should respond to far-field stress or strain states. The displacements were computed for a cohesionless detachment surface with boundary conditions set on the triangular boundary elements making up the faults as follows.

- The traction components parallel to the element plane in the dip direction and in the strike direction are both set to zero so the element surfaces may slip freely.
- The displacement discontinuity component normal to the element plane is set to zero to avoid opening or interpenetration of the fault surfaces.

This model examines the top Middle Jurassic Louann Salt, interpreted from the Ship Shoal 3D seismic cube (Fig. 6). The model was placed under a normal fault stress regime ($\sigma_v > \sigma_H > \sigma_h = 25$ MPa km$^{-1} > 20$ MPa km$^{-1} > 18$ MPa km^{-1}) consistent with an *in situ* stress magnitude study carried out in the Eugene Island field (Fig. 4; Finkbeiner *et al.* 2001). The resulting model suggests that the greatest net normal displacements occur on the flanks of salt diapirs, with displacement vectors for the deltaic sediments following the interface between the salt and deltaic sediments from the crest of the diapirs into the mini-basins between (black lines on Fig. 6). The model therefore implies deltaic sediments are sliding down the flanks of the salt diapirs. This gravitational collapse, under an extensional normal faulting stress regime, generates σ_H orientations that are perpendicular to the horizontal projection of the displacement vector; they will be aligned parallel to the strike of normal faults and to the flanks of salt diapirs. These observations are supported by outcrop analysis around exposed examples of diapiric flanks, which also reveals significant margin-parallel extensional faults (Alsop *et al.* 2000).

Implications for petroleum exploration and development

It has been demonstrated in previous studies that the orientations and magnitudes of present-day stresses are critical to borehole stability, water flooding, fracture stimulation and fault reactivation (e.g. Heffer & Lean 1993; Barton *et al.* 1998; Nelson *et al.* 2005; Tingay *et al.* 2009; King *et al.* 2010*c*). The evaluation of these implications can therefore become very difficult in settings such as the Gulf of Mexico, where the stress orientations are deflected away from the regional. Here we present examples of borehole stability from the Gulf of Mexico offshore Louisiana.

Boreholes can become unstable, in the form of borehole breakouts or DITFs, due to the anisotropy of the stress field. Boreholes are therefore most stable when drilled in a direction that subjects the well to the least stress anisotropy (Hillis & Williams 1993). In a normal fault stress regime on a delta top, as proposed to currently exist in the shelf region of the Gulf of Mexico, the greatest stress anisotropy occurs between the vertical stress ($\sigma_v = \sigma_1$) and the minimum horizontal stress ($\sigma_h = \sigma_3$). Horizontal wells drilled towards the regional σ_H orientation (060 and 240) would therefore be typically considered to be least stable in a normal fault stress regime as they are subject to the greatest stress anisotropy (between σ_v and σ_h; Hillis & Williams 1993). The most stable wells will be vertical wells, as they are subject to the least stress anisotropy (between σ_H and σ_h) and have the lowest applied stress (Peška & Zoback 1995). Alternatively, if σ_H is close to σ_v then wells deviated towards σ_h will be most stable (e.g. Peška & Zoback 1995; Tingay *et al.* 2009). In the reverse (or thrust) fault stress regime of a deepwater fold–thrust belt, as proposed for the slope and basin-floor region of the Gulf of Mexico, the greatest stress anisotropy occurs between σ_H (σ_1) and σ_v (σ_3). Horizontal wells drilled towards the regional σ_h orientation (150 and 330) are therefore predicted to be least stable as they are subject to the greatest stress anisotropy (between σ_H and σ_v). The most stable wells are predicted to be horizontal wells drilled towards the σ_H direction (060 and 240), as they are subject to the least stress anisotropy (between σ_h and σ_v) and/or the lowest applied stress magnitude. However, if σ_h is high and close to σ_H, then vertical wells are more stable in a reverse fault stress regime (e.g. Peška & Zoback 1995).

On the shelf region of the Gulf of Mexico (delta top), vertical wells or wells deviated towards the σ_h direction are therefore least likely to be affected by borehole breakout or DITFs. On the slope and basin-floor region (deepwater fold–thrust belt), wells drilled parallel to the regional σ_H orientation (060 and 240) or vertical wells are least likely to be affected by borehole breakout or DITFs. These therefore represent the safest drilling directions, with respect to both borehole stability and fluid losses, on the delta top and in the deepwater fold–thrust belt, respectively. However, wells drilled adjacent to salt diapirs may not follow these predictions due to the local deflection of the stress field. In these cases, vertical wells are still likely to be the most stable; if horizontal wells are required, they must be drilled with respect to the deflection of the σ_H orientations expected near the interface between the salt diapirs and adjacent deltaic sediments (Fig. 5a, b).

Conclusions

We examined present-day stress orientations in eight new petroleum wells located in the delta top of the Gulf of Mexico offshore Louisiana. In total, 28 borehole breakouts were identified from image and high-resolution dipmeter tool (HDT) logs from five wells, giving a mean σ_H orientation of wells ranked A–D of 089 with a standard deviation of 24° (Fig. 4). The stress orientations interpreted here were combined with those from a further 28 wells analysed by Yassir & Zerwer (1997) to determine a regional mean σ_H orientation from 13 wells with A–C-quality stress orientations of 061, with a standard deviation of 52° (Fig. 4). The regional mean σ_H orientation, determined from A–D-quality mean stress orientations in 32 petroleum wells, is 060 with a standard deviation of 51° (this study; Yassir & Zerwer 1997; Fig. 4). These orientations are consistent with the delta-top margin-parallel σ_H orientations expected from the idealized model of a DDWFTB (Fig. 1).

Maximum horizontal stress orientations deflect from this regional margin-parallel orientation and align parallel around salt diapirs that have pierced through the overlying deltaic sediments (Fig. 4, areas 1 and 2). These deflections are attributed to the relative geomechanical contrasts that occur between 'weaker' salt diapirs and the adjacent 'stronger' deltaic sediments (e.g. Bell 1996; Yassir & Zerwer 1997). Using the Ship Shoal 3D seismic cube, we have demonstrated that deflections of σ_H orientations vary with depth due to the conical nature of the salt diapirs. A time slice taken at 1 s demonstrates that σ_H orientations deflect over narrow areas associated with narrow crests of salt diapirs, while large areas of deltaic sediment between the crests of the salt diapirs demonstrate the regional margin-parallel σ_H orientation (Figs 4 & 5a). In contrast, a time slice taken at 2 s demonstrates that σ_H orientations deflect over broad areas associated with wide bases of salt diapirs; only small areas of deltaic sediment between the bases of the salt diapirs demonstrate the regional margin-parallel σ_H orientation (Figs 4 & 5a).

A Poly3D model was constructed of the top Louann Salt from the Ship Shoal 3D seismic cube and a normal fault stress regime was applied, consistent with that determined in the Gulf of Mexico delta top (i.e. Finkbeiner *et al.* 2001). The model predicts that σ_H orientations are aligned parallel to the interface between salt diapirs and adjacent deltaic sediments, which is consistent with σ_H orientations determined from well data (Fig. 6). The model also predicts displacements vectors, which imply that the greatest net displacements on the interface between salt and sediment occur at the flanks of the salt diapirs (Fig. 6).

The orientations and magnitudes of present-day stresses are critical to borehole stability, water flooding, fracture stimulation and fault reactivation (e.g. Heffer & Lean 1993; Barton *et al.* 1998; Nelson *et al.* 2005; Tingay *et al.* 2009). In the Gulf of Mexico the most stable well on the delta top is vertical; unless σ_H is close to σ_v, then wells deviated towards σ_h are most stable. The most stable well in the Gulf of Mexico delta toe (or deep-water fold–thrust belt) is deviated toward σ_H; unless σ_h is close to σ_H, then vertical wells are most stable. Knowledge of σ_H orientations in the Gulf of Mexico is critical for borehole stability studies due to the complex deflections of the regional margin-parallel σ_H orientations around salt diapirs. A single 'one case fits all' implied by the idealized model of a delta system (e.g. Fig. 1) does not apply for the Gulf of Mexico (Fig. 5). Careful consideration of the *in situ* stress field around salt diapirs is required.

The authors thank the reviewers for their helpful comments and the Australian Research Council for their financial support of this study. The authors would also like to thank Western Geoco for providing the Ship Shoal three-dimensional seismic cube and JRS Petroleum Research for the use of their software (JRS Suite).

References

AADNOY, B. S. & BELL, J. S. 1998. Classification of drilling induced fractures and their relationship to in situ stress directions. *The Log Analyst*, **39**, 27–42.

ALSOP, G. I., BROWN, J. P., DAVISON, I. & GIBLING, M. R. 2000. The geometry of drag zones adjacent to salt diapirs. *Journal of the Geological Society, London*, **157**, 1019–1029.

BARTON, C. A. 2000. *Discrimination of natural fractures from drilling-induced wellbore failures in wellbore image data – implications for reservoir permeability*. Society of Petroleum Engineers International Petroleum Conference and Exhibition, Mexico.

BARTON, C. A., CASTILLO, D. A., MOOS, D., PESKA, P. & ZOBACK, M. D. 1998. Characterising the full stress tensor based on observations of drilling-induced wellbore failures in vertical and inclined boreholes leading to improved wellbore stability and permeability prediction. *Australian Petroleum Production and Exploration Association Journal*, **38**, 467–487.

BELL, J. S. 1996. In situ stresses in sedimentary rocks (part 2): applications of stress measurements. *Geoscience Canada*, **23**, 135–153.

BELL, J. S. & GOUGH, D. I. 1979. Northeast-southwest compressive stress in Alberta: evidence from oil wells. *Earth and Planetary Science Letters*, **45**, 475–482.

BRUDY, M. & ZOBACK, M. D. 1999. Drilling-induced tensile wall-fractures: implications for determination of in-situ stress orientation and magnitude. *International Journal of Rock Mechanics and Mining Sciences*, **36**, 191–215.

DAVIS, D. M. & ENGELDER, T. 1985. The role of salt in fold-and-thrust belts. *Tectonophysics*, **199**, 67–88.

DAVISON, I., ALSOP, G. I. *ET AL*. 2000. Geometry and late-stage structural evolution of Central Graben salt diapirs, North Sea. *Marine and Petroleum Geology*, **17**, 499–522.

EKSTROM, M. P., DAHAN, C. A., CHEN, M. Y., LLOYD, P. M. & ROSSI, D. J. 1987. Formation imaging with microelectrical scanning arrays. *The Log Analyst*, **28**, 294–306.

FENG, J. & BUFFLER, R. T. 1991. Preliminary age determinations for new deep Gulf of Mexico Basin sequences. *Gulf Coast Association of Geological Society Proceedings*, **41**, 283–289.

FIDUK, J. C., WEIMER, P. *ET AL*. 1999. The Perdido Fold Belt, northwestern deep Gulf of Mexico, part 2: seismic stratigraphy and petroleum systems. *American Association of Petroleum Geologists Bulletin*, **83**, 578–612.

FIORE, P. E., POLLARD, D. D., CURRIN, W. R. & MINER, D. M. 2006. Mechanical and stratigraphic constraints on the evolution of faulting at Elk hills, California. *American Association of Petroleum Geologists Bulletin*, **91**, 321–341.

FINKBEINER, T., ZOBACK, M., FLEMINGS, P. & STUMP, B. 2001. Stress, pore pressure, and dynamically constrained hydrocarbon columns in the South Eugene Island 330 field, northern Gulf of Mexico. *American Association of Petroleum Geologists Bulletin*, **85**, 1007–1031.

GALLOWAY, W. E. 1989. Genetic stratigraphic sequences in basin analysis II: application to northwest Gulf of Mexico Cenozoic basin. *American Association of Petroleum Geologists Bulletin*, **73**, 143–154.

HEFFER, K. J. & LEAN, J. C. 1993. Earth Stress Orientation – A control on, and guide to, flooding directionality in a majority of reservoirs. *In*: LINVILLE, W. (ed.) *Reservoir Characterisation III*, Pennwell Becks, Tulsa, 799–822.

HEIDBACH, O., REINECKER, J., TINGAY, M., MÜLLER, B., SPERNER, B., FUCHS, K. & WENZEL, F. 2007. Plate boundary forces are not enough: second- and thrid-order stress patterns highlighted in the World Stress Map database. *Tectonics*, **26**, TC6014, doi: 10.1029/2007TC002133.

HEIDBACH, O., TINGAY, M. R. P., BARTH, A., REINECKER, J., KURFEβ, D. & MÜLLER, B. 2010. Global crustal stress pattern based on the 2008 World Stress Map database release. *Tectonophysics*, **482**, 3–15.

HILLIS, R. R. & WILLIAMS, A. F. 1993. The stress field of the North West Shelf and wellbore stability. *Australian Petroleum Production and Exploration Association Journal*, **33**, 373–385.

JACKSON, M. P. A., VENDEVILLE, B. C. & SCHULZ-ELA, D. D. 1994. Salt-related structures in the Gulf of Mexico: a field guide for geophysicists. *The Leading Edge*, **13**, 837–842.

KING, R. C. & BACKÉ, G. 2010. A balanced 2D structural model of Hammerhead Delta—Deepwater Fold-Thrust Belt, Bight Basin, Australia. *Australian Journal of Earth Sciences*, **57**, 1005–1012.

KING, R. C., HILLIS, R. R. & REYNOLDS, S. D. 2008. In situ stresses and natural fractures in the Northern Perth Basin, Australia. *Australian Journal of Earth Sciences*, **55**, 685–701.

KING, R. C., HILLIS, R. R., TINGAY, M. R. P. & MORLEY, C. K. 2009. Present-day stress and neotectonic provinces of the Baram Delta and deep-water fold-thrust belt. *Journal of the Geological Society, London*, **166**, 197–200.

KING, R. C., TINGAY, M. R. P., HILLIS, R. R., MORLEY, C. K. & CLARK, J. 2010*a*. Present-day stress orientations and tectonic provinces of the NW Borneo collisional margin. *Journal of Geophysical Research*, **115**, B10415, doi: 10.1029/2009JB006997.

King, R. C., Backé, G., Morley, C. K., Hillis, R. R. & Tingay, M. R. P. 2010*b*. Balancing deformation in NW Borneo: quantifying plate-scale v. gravitational tectonics in a Delta and Deepwater Fold-Thrust Belt Systems. *Marine and Petroleum Geology*, **27**, 238–246.

King, R. C., Hillis, R. R., Tingay, M. R. P. & Damit, A.-R. 2010*c*. Present-day stresses in Brunei, NW Borneo: superposition of deltaic and active margin tectonics. *Basin Research*, **22**, 236–247.

Kirsch, V. 1898. Die Theorie der Elastizität und die Beddürfnisse der Festigkeitslehre. *Zeitschrift des Vereines Deutscher Ingenieure*, **29**, 797–807.

Maerten, F. 2010. Adaptive cross-approximation applied to the solution of system of equations and post-processing for 3D elastostatic problems using the boundary element method. *Engineering Analysis with Boundary Elements*, **34**, 483–491.

Maerten, L., Pollard, D. D. & Karpuz, R. 2000. How to constrain 3-D fault continuity and linkage using reflection seismic data: a geomechanical approach. *AAPG Bulletin*, **84**, 1311–1324.

Maerten, L., Gillepie, P. & Pollard, D. D. 2002. Effects of local stress perturbation on secondary fault development. *Journal of Structural Geology*, **24**, 145–153.

MacDonald, J., Backé, G., King, R., Holford, S. & Hillis, R. 2012. Geomechanical modelling of fault reactivation in ther Ceduna Sub-basin, Bight Basin, Australia. *In*: Healy, D., Butler, R. W. H., Shipton, Z. K. & Sibson, R. H. (eds) *Faulting, Fracturing and Igneous Intrusion in the Earth's Crust*. Geological Society, London, Special Publications, **367**, 71–89.

Mandl, G. & Crans, W. 1981. Gravitational gliding in deltas. *In*: Price, N. J. & McClay, K. R. (eds) *Nappe and Thrust Tectonics*. Geological Society, London, Special Publications, **9**, 41–54.

Mastin, L. 1988. Effect of borehole deviation on breakout orientations. *Journal of Geophysical Research*, **93**, 9187–9195.

Morley, C. K. 2003. Mobile shale related deformation in large deltas developed on passive and active margins. *In*: van Rensbergen, P., Hillis, R. R., Maltman, A. J. & Morley, C. K. (eds) *Subsurface Sediment Mobilization*. Geological Society, London, Special Publications, **216**, 335–357.

Nelson, E. J., Hillis, R. R., Meyer, J. J., Mildren, S. D., Van Nispen, D. & Briner, A. 2005. The reservoir stress path and its implications for water flooding, Champion Southeast Field, Brunei. *Proceedings of Alaska Rocks the 40th US Symposium on Rock Mechanics*, American Rock Mechanics Association.

Peel, F. J., Travis, C. J. & Hossack, J. R. 1995. Genetic structural provinces and salt tectonics of the Cenozoic offshore U.S. Gulf of Mexico: a preliminary analysis. *In*: Jackson, M. P. A., Roberts, D. G. & Snelson, S. (eds) *Salt tectonics: A Global Perspective*. American Association of Petroleum Geologists, Tulsa, Memoir, **65**, 153–175.

Peška, P. & Zoback, M. D. 1995. Compressive and tensile failure of inclined well bores and determination of in situ stress and rock strength. *Journal of Geophysical Research*, **100**, 12 791–12 811.

Plumb, R. A. & Hickman, S. H. 1985. Stress induced borehole elongation: comparison between the four-arm dipmeter and the borehole televiewer in the Auburn geothermal well. *Journal of Geophysical Research*, **B90**, 5513–5521.

Richardson, R. M. 1992. Ridge forces, absolute plate motions, and the intra-plate stress field. *Journal of Geophysical Research*, **97**, 11 739–11 748.

Rowan, M. G. 1997. Three-dimensional geometry and evolution of a segmented detachment fold, Mississippi Fan foldbelt, Gulf of Mexico. *Journal of Structural Geology*, **19**, 463–480.

Rowan, M. G., Peel, F. J. & Vendeville, B. C. 2004. Gravity-driven fold belts on passive margins. *In*: McClay, K. R. (ed.) *Thrust Tectonics and Hydrocarbon Systems*. American Association of Petroleum Geologists, Tulsa, Memoir, **82**, 157–182.

Sonder, L. J. 1990. Effects of density contrasts on the orientation of stresses in the lithosphere: relation to principal stress directions in the Transverse ranges, California. *Tectonics*, **9**, 761–771.

Sperner, B., Müller, B., Heidbach, O., Delvaux, D., Reinecker, J. & Fuchs, K. 2003. Tectonics stress in the Earth's crust: advances in the World Stress Map project. *In*: Nieuwland, D. A. (ed.) *New Insights into Structural Interpretation and Modelling*. Geological Society, London, Special Publications, **212**, 101–116.

Thomas, A. L. 1993. *Poly3D: A three-dimensional, polygonal element, displacement discontinuity boundary element computer program with applications to fractures, faults, and cavities in the Earth's crust*. MS thesis, Stanford University.

Tingay, M., Müller, B., Reinecker, J., Heidbach, O., Wenzel, F. & Fleckenstein, P. 2005. Understanding tectonic stress in the oil patch: the world stress map project. *The Leading Edge*, **24**, 1276–1282.

Tingay, M., Müller, B., Reinecker, J. & Heidbach, O. 2006. *State and orgin of the present-day stress field in sedimentary basins: New results from the World Stress Map Project*. 41st US Symposium on Rock Mechanics, Golden Rocks 2006, Published Plenary Paper ARMA/USRMS 06–1049.

Tingay, M., Hillis, R. R., Morley, C. K., King, R. C., Swarbrick, R. E. & Damit, A. R. 2009. Present-day stress and neotectonics of Brunei: implications for petroleum exploration and production. *American Association of Petroleum Geologists Bulletin*, **93**, 75–100.

Tingay, M., Morley, C., King, R., Hillis, R., Coblentz, D. & Hall, R. 2010. Present-day stress field of southeast Asia. *In*: Heidbach, O., Tingay, M. & Wenzel, F. (eds) *Frontiers of Stress Research*. Tectonophysics, Special Issue, **482**, 92–104.

Tingay, M., Bentham, P., De Feyter, A. & Kellner, A. 2011. Present-day stress field rotations associated with evaporites in the offshore Nile Delta. *GSA Bulletin*, **123**, 1171–1180.

Trudgill, B. D., Rowan, M. G. *et al.* 1999. The Perdido Fold Belt, Northwestern Deep Gulf of Mexico, Part 1: Structural Geometry, Evolution and Regional Implications. *AAPG Bulletin*, **83**, 88–113.

WEIMER, P. 1990. Sequence stratigraphy, facies geometries, and depositional history of the Mississippi Fan, Gulf of Mexico. *AAPG Bulletin*, **74**, 425–453.

WORRALL, D. M. & SNELSON, S. 1989. Evolution of the northern Gulf of Mexico, with emphasis on Cenozoic growth faulting and the role of salt. *In*: BALLY, A. W. & PALMER, A. R. (eds) *The Geology of North America: an Overview*. Geological Society of America, Decade of North American Geology A, Geological Society of America, Boulder, CO, 97–138.

WU, S., BALLY, A. W. & CRAMEZ, C. 1990. Allochthonous salt, structure and stratigraphy of the north-eastern Gulf of Mexico, part II: structure. *Marine and Petroleum Geology*, **7**, 334–370.

YASSIR, N. A. & ZERWER, A. 1997. Stress regimes in the Gulf Coast, offshore Louisiana: data from well-bore breakout analysis. *American Association of Petroleum Geologists Bulletin*, **81**, 293–307.

ZOBACK, M. L. 1992. First- and second-order patterns of stress in the Lithosphere: the world stress map project. *Journal of Geophysical Research*, **97**, 11 703–11 728.

Evidence for non-Andersonian faulting above evaporites in the Nile Delta

MARK TINGAY[1]*, PETER BENTHAM[2], ARNOUD DE FEYTER[3,4] & AXEL KELLNER[5]

[1]*Tectonics, Resources and Exploration (TRaX), Australian School of Petroleum, University of Adelaide, SA, 5005, Australia*

[2]*BP Egypt, Road 210, Digla, Maadi, Cairo, Egypt*

[3]*International Egypt Oil Company, Road 204, Digla, Maadi, Cairo, Egypt*

[4]*Present address: OMV Yemen, Faj Attan, Beirut St., Hadda District, Sana'a, Republic of Yemen*

[5]*RWE Dea Egypt, Road 253, Digla, Maadi, Cairo, Egypt*

**Corresponding author (e-mail: mark.tingay@adelaide.edu.au)*

Abstract: This study examines present-day stress orientations from borehole breakout and drilling-induced fractures in 57 boreholes in the Nile Delta. A total of 588 breakouts and 68 drilling-induced fractures from 50 wells reveal sharply contrasting present-day maximum horizontal stress (S_{Hmax}) orientations across the Nile Delta. A typical deltaic margin-parallel S_{Hmax} exists in parts of the Nile Delta that are below or absent from evaporites (NNE–SSW in the west, east–west in the central Nile, ESE–WNW in the east). However, a largely margin-normal (NNE–SSW) S_{Hmax} is observed in sequences underlain by evaporites in the eastern Nile Delta. The margin-normal supra-salt S_{Hmax} orientations are often subperpendicular to the strike of nearby active extensional faults, rather than being parallel to the faults as predicted by Andersonian criteria. The high angle between S_{Hmax} and strike of these extensional faults represents a new type of non-Andersonian faulting that is even less-suitably oriented for shear failure than previously described anomalous faulting such as low-angle normal faults and highly oblique strike-slip faults (e.g. San Andreas). While the mechanics of these non-Andersonian faults remains uncertain, it is suggested that the margin-normal supra-salt orientation generated by basal forces imparted upon rafted blocks sliding down seawards-dipping evaporites.

The Anderson (1942, 1951) classification of faults and associated stress regimes (i.e. normal, strike-slip, thrust) has been fundamental to our understanding of faults and fault mechanics. Under the Andersonian classification of faults and stress, active fault planes are predicted to be oriented at an angle of *c.* 30° to the present-day maximum principal stress (σ_1). Furthermore, active faults are predicted to strike approximately parallel to the present-day intermediate principal stress (σ_2) in normal or thrust-faulting stress regimes, or to be vertically dipping in a strike-slip stress regime (Anderson 1942). This Andersonian relationship between active faulting and the present-day stress tensor has been observed in a wide variety of tectonic settings, including fold–thrust belts (King *et al.* 2009; Couzens-Schultz & Chan 2010), the Basin and Range province (Zoback *et al.* 1981), extrusion-related strike-slip provinces (Heidbach *et al.* 2007; Tingay *et al.* 2010), foreland basins (Reinecker *et al.* 2010) and passive margins (Bell 1996*a*; Morley *et al.* 2008). Furthermore, the commonly observed Andersonian relationship between present-day stress and structural styles forms the fundamental basis for the vast majority of palaeostress analysis techniques conducted in structural geology (Angelier 1979, 1989; Lisle 1987; Shan *et al.* 2004).

There has been significant recent interest in so called 'non-Andersonian' faults, in which the present-day stress tensor and geometry of observed active faults are inconsistent with that predicted under the Anderson model (e.g. Mount & Suppe 1992; Wernicke 1995; Zoback 2000; Collettini & Sibson 2001; Faulkner *et al.* 2006). Non-Andersonian faults examined to date can typically be classified into two categories: highly oblique strike-slip faults and low-angle normal faults. Highly oblique strike-slip faults are subvertically dipping active wrench faults that strike at a high angle (typically $\geq 60°$) to the maximum principal stress orientation in a strike-slip stress regime ($\sigma_1 = S_{Hmax}$, $\sigma_2 =$ vertical stress S_v and $\sigma_3 =$ minimum horizontal stress S_{Hmin}). Such non-Andersonian strike-slip faults have been observed in several places, most notably the San Andreas Fault and Great Sumatra Fault, which both display strikes that are oriented up to 80° to the present-day

From: Healy, D., Butler, R. W. H., Shipton, Z. K. & Sibson, R. H. (eds) 2012. *Faulting, Fracturing and Igneous Intrusion in the Earth's Crust*. Geological Society, London, Special Publications, **367**, 155–170. http://dx.doi.org/10.1144/SP367.11

maximum horizontal stress direction (Mount & Suppe 1987, 1992; Zoback *et al.* 1987; Zoback 2000). There is a therefore discrepancy of $\geq 30^\circ$ between the strike of highly oblique strike-slip faults and the strike predicted by the Anderson fault model (fault strike oriented 30° to σ_1) under the present-day stress tensor.

Low-angle normal faults are shallow-dipping ($\leq 30^\circ$) dip-slip faults that occur in a normal faulting stress regime ($S_v > S_{Hmax} > S_{Hmin}$; Wernicke 1995). Low-angle normal faults have been observed in numerous locations such as Italy (Boncio & Lavecchia 2000; Collettini & Sibson 2001), the Basin and Range province (Wernicke 1981) and Thailand (Morley 2009). Low-angle normal faults, such as highly oblique strike-slip faults, also show a $\geq 30^\circ$ discrepancy between fault orientation and the fault orientation predicted by the Andersonian model (60° dip) in the present-day stress regime.

This study provides evidence for a new and distinctly different type of non-Andersonian fault observed in the Nile Delta, offshore Egypt. Low-angle normal faults and highly oblique strike-slip faults both exhibit an anomalously high angle ($\geq 60^\circ$) between the maximum principal stress and the fault plane. In contrast, the non-Andersonian faults described herein have fault planes that are optimally oriented with respect to the maximum principal stress, but instead display anomalous fault plane orientations with respect to the intermediate and minimum principle stress. More specifically, this research describes presently active gravitationally driven normal faults in the Nile Delta that exhibit fault strikes oriented at angles of up to 90° to the intermediate principal stress (S_{Hmax}), rather than faults being parallel to S_{Hmax} as predicted by Andersonian theory. This study first describes the present-day stress field interpreted from borehole breakouts and drilling-induced tensile fractures. The present-day stress field is then examined with respect to gravity-spreading and gravity-gliding tectonics styles observed across the Nile Delta, and with faulting observed on high-resolution bathymetry and reflection seismic data, to highlight the presence of non-Andersonian normal faulting. Finally, the mechanics of this new form of non-Andersonian faulting is discussed and compared to existing suggestions for non-Andersonian faulting (low-strength fault, overpressures, local stress rotations).

Geological summary of the Nile Delta

The Nile Delta is the largest clastic accumulation in the Mediterranean Sea and has been deposited within a region of significant Cenozoic tectonic activity (Fig. 1). The Nile Delta effectively comprises two separate prograding clastic delta systems, with Jurassic–Miocene and Pliocene–Recent delta systems separated by the major Messinian unconformity (Sestini 1989; Marten *et al.* 2004). The Messinian (Late Miocene) unconformity was a consequence of the closure of the Straits of Gibraltar and the subsequent evaporation of the Mediterranean Sea, which led to both the deposition of thick evaporite sequences (primarily near the basin centre) and the formation of numerous incised valleys on the shelf and slope (Ryan & Cita 1978; Barber 1981; Aal *et al.* 2001; Loncke *et al.* 2006).

The deposition of Messinian evaporite sequences has played a major role in the structural evolution of the Pliocene–Recent Nile Delta system. The Pliocene–Recent Nile Delta displays spectacular examples of complex thin-skinned tectonics that generally sole-out into the Messinian, with zones of both gravity spreading and gravity gliding (Fig. 2; Loncke *et al.* 2006; Clark & Cartwright 2009). The sequences overlying the evaporites display typical deltaic structures such as listric growth faults and rotational block faulting, as well as structures associated with salt movement such as folds, extensional and strike-slip faults, collapsed depocentres and polygonal mini-basins (Fig. 1; Loncke *et al.* 2006). In contrast, structures in the pre-Messinian sequences are typically the result of classic gravity-spreading deltaic tectonics (e.g. proximal listric extension coupled with distal contraction near the delta toe), but also display evidence for some degree of basement-involved deformation (Sestini 1989; Aal *et al.* 2001; Shaaban *et al.* 2006).

Structures in the Nile Delta that appear to be associated with basement deformation primarily strike NW–SE and NE–SW, such as the Temsah (NW–SE oriented) and Rosetta (NE–SW) fault trends (Fig. 1; Bellaiche & Mart 1995; Aal *et al.* 2001; Shaaban *et al.* 2006). The origin, nature and influence of these basement structures remain contentious. The NW–SE basement-associated features are postulated to be linked to structures originally generated during Syrian Arc deformation or to be old crustal transforms, while the NE–SW-striking features are hypothesized to be structures related to Mesozoic rifting and the north-westwards transition from continental to oceanic crust (Bellaiche & Mart 1995; Aal *et al.* 2001; Shaaban *et al.* 2006). Shaaban *et al.* (2006) infer that the Temsah trend is a Mid–Late Miocene age uplifted horst structure. In contrast, other authors argue the basement-associated deformation is transtensional; Aal *et al.* (2001) argue for dextral motion along the Temsah trend and sinistral motion along the Rosetta trend and El-Barkooky & Helal (2002) suggest sinistral motion along the Temsah trend.

Despite the influence of basement-associated deformation on the structural evolution of the Nile Delta, there is significant evidence to suggest that

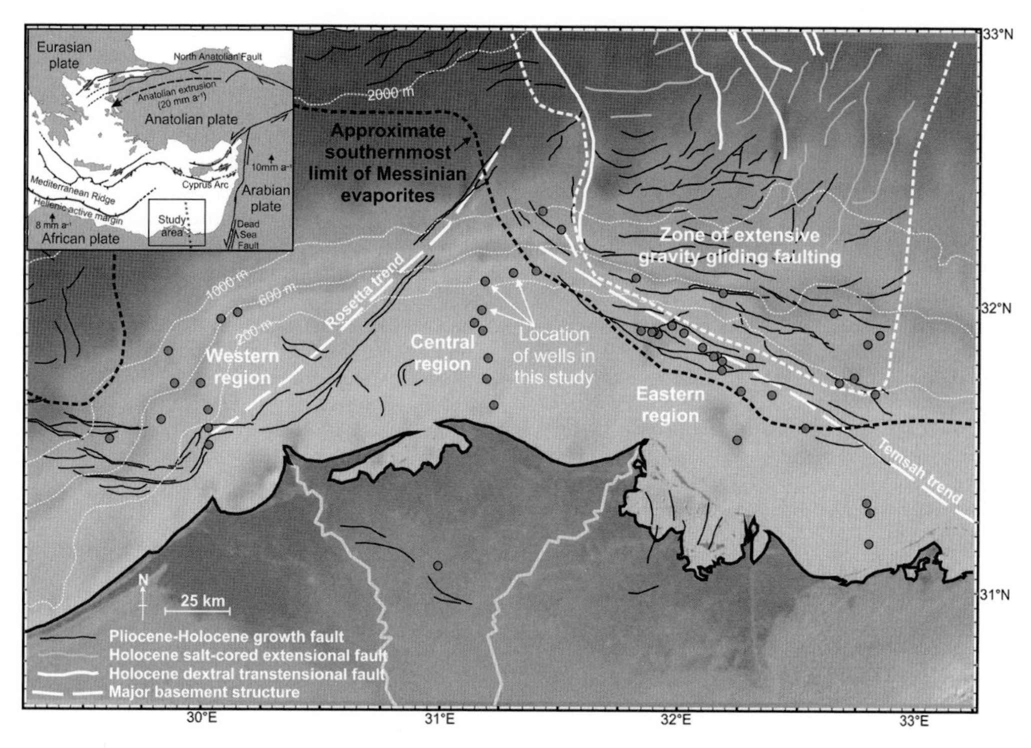

Fig. 1. Location of wells examined in this study, major Pliocene–Recent structures, basement fault trends, low-resolution bathymetry and evaporite occurrence in the Nile Delta (see inset map for location and overview of regional tectonics modified from ten Veen *et al.* 2004). The Nile Delta contains numerous Pliocene–Recent structures observed from seismic and high-resolution bathymetry data (structures from Aal *et al.* 2001; Loncke *et al.* 2006 and the author's own interpretation). However, the eastern offshore Nile Delta (highlighted by white dashed line) is characterized by an anomalous number of growth, salt-cored and transtensional faults in the sequences underlain by Messinian evaporites (evaporite sequences are located north of the grey dashed line).

there is currently little far-field or basement control on the present-day state of stress in Nile Delta sedimentary sequences. Structures in the sub-salt Nile Delta often appear to sole out into a hypothesized Oligocene–Late Cretaceous detachment horizon (most likely associated with overpressured shales), which may remove most of the influence of the basement on the stress field within the deltaic sequences (Fig. 2; Marten *et al.* 2004; Tingay *et al.* 2011). Furthermore, supra-salt structures commonly sole out on the Messinian evaporates (Loncke *et al.* 2006); an earlier study of the stress field in the eastern

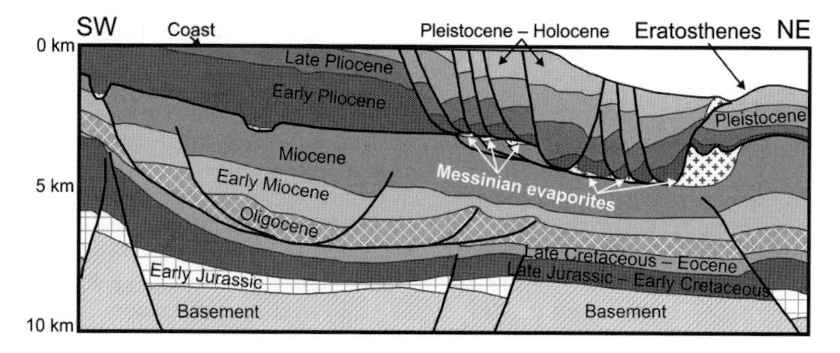

Fig. 2. Schematic cross-section of the eastern Nile Delta illustrating major structural styles, in particular the supra-salt extensional faults that sole out on the Messinian evaporites (adapted from Marten *et al.* 2004). For a comprehensive range of seismic sections across the Nile Delta, see Aal *et al.* (2001) and Loncke *et al.* (2006).

Nile Delta reveals that these evaporates decouple the supra-salt stress field from that observed in underlying sequences (Tingay *et al.* 2011). However, there is moderate seismicity in the basement underneath the Nile Delta and therefore the potential for on-going basement involvement in tectonics within the delta (Badawy & Horváth 1999; El-Sayed *et al.* 2004). Furthermore, the topography and nature of the basement (which is poorly understood) may potentially influence present-day deformation in the overlying deltaic sequences. In addition, the Nile Delta is located close to numerous plate boundary zones that may influence the stress pattern in the Nile Delta, including:

- the Cyprus Arc, particularly impingement of the Eratosthenes seamount;
- the Dead Sea Transform Fault Zone; and
- the Red Sea/Gulf of Suez Rift system (Fig. 1; ten Veen *et al.* 2004).

The Nile Delta is therefore located in an area adjacent to, and potentially overlying, active major crustal deformation that may influence both the tectonic evolution and present-day state of stress in the Nile Delta.

Determination of maximum horizontal stress orientation

Present-day S_{Hmax} orientations in the north-eastern Nile Delta were determined from borehole breakouts and drilling-induced fractures (DIFs) interpreted from resistivity image log and four-arm caliper data. Stresses become locally concentrated around any subsurface opening, such as a borehole, in response to the removal of material that was previously supporting the tectonic load (Kirsch 1898; Bell & Gough 1979). Borehole breakouts are stress-induced elongations of the wellbore and occur when the wellbore stress concentration exceeds that required to cause compressive failure of intact rock (Bell & Gough 1979). The elongation of the cross-sectional shape of the wellbore is the result of compressive shear failure on intersecting conjugate planes, which causes pieces of the borehole wall to spall off (Bell & Gough 1979). The maximum stress around a vertical borehole occurs perpendicular to the S_{Hmax} (Kirsch 1898); borehole breakouts are therefore elongated perpendicular to the S_{Hmax} direction (Bell & Gough 1979).

Drilling-induced fractures are caused by tensile failure of the borehole wall and form when the wellbore stress concentration is less than the tensile strength of the rock (Aadnoy 1990). The minimum stress around a vertical borehole occurs in the direction of the S_{Hmax} (Kirsch 1898); DIFs are therefore oriented in the S_{Hmax} direction (Aadnoy & Bell 1998).

Breakouts are interpreted here from resistivity image logs (Schlumberger formation micro imager and oil-based micro imager) and Schlumberger high-resolution dipmeter tool and oil-based dipmeter tool logs. The high-resolution dipmeter tool and oil-based dipmeter tool are four-arm caliper tools with two pairs of caliper arms at 90° to each other. Each arm has a pad on the end containing one or two resistivity 'buttons'. The resistivity data from four-arm caliper tools are processed to obtain information about the formation (primarily dip and strike of bedding) and to calculate hole volume (Schlumberger 1986). Borehole breakouts can however be interpreted from unprocessed high-resolution dipmeter tool log data. The logs used to interpret breakouts from the high-resolution dipmeter tool are the:

- borehole deviation (DEVI) and azimuth (HAZI);
- azimuth of pad one (P1AZ);
- bearing of pad one relative to the high side of the hole (RB); and
- diameter of the borehole in two orthogonal directions ['caliper one' (C1) given by arms one and three and 'caliper two' (C2) from arms two and four].

The tool tends to rotate as it is pulled up the borehole due to the lay of the cable (cable torque). However, the tool stops rotating where the cross-sectional shape of the borehole is elongated when one caliper pair becomes 'stuck' in the elongation direction (Plumb & Hickman 1985). The combined use of the six logs listed above allows the interpreter to identify zones of borehole breakout and the orientation of the elongation (Plumb & Hickman 1985). Many non-circular wellbore cross-sectional shapes are not stress-induced, such as washout and key-seating (Plumb & Hickman 1985). Borehole breakout is distinguished from other borehole elongations on four-arm or six-arm logs using a strict set of criteria outlined in Plumb & Hickman (1985) and updated and expanded in the official World Stress Map project guidelines for stress determination from caliper data (Reinecker *et al.* 2003).

Resistivity image logs evolved from the four-arm dipmeter logs and are essentially an improved version of a high-resolution dipmeter tool (Ekstrom *et al.* 1987). There are a large number of resistivity buttons on each pad of the resistivity image tool, which provide an image of the borehole wall based on resistivity contrasts (Fig. 3). Resistivity image tools also measure the hole size and all other information obtained by four-arm caliper logs, in addition to the resistivity image. The resistivity image of the wellbore wall allows for a more reliable interpretation of breakouts than can be made using dipmeter data alone (Zoback 2007; Tingay *et al.* 2008; Heidbach *et al.* 2010). Drilling-induced fractures can also be recognized on image logs (drilling-induced fractures cannot be interpreted on four-arm caliper logs). Breakouts appear on resistivity image logs as broad, parallel, often poorly resolved conductive zones separated by 180° and exhibiting caliper enlargement in the direction of the conductive zones. Drilling-induced fractures appear on image logs as narrow well-defined conductive fractures separated by 180° (Fig. 3; Zoback 2007; Tingay *et al.* 2008).

Breakouts and drilling-induced fractures can rotate in inclined boreholes and do not always directly yield the horizontal stress orientation because the assumption of the well axis being parallel to a principal stress (S_v) is no longer valid (Mastin 1988; Peska & Zoback 1995). However, it has been demonstrated that breakouts and drilling-induced fractures do not show any significant rotation in orientation and still yield the approximate S_{Hmax} orientation in boreholes with less than 20° deviation under typical stress conditions (Peska & Zoback 1995). Breakouts and drilling-induced fractures were therefore used here to estimate the S_{Hmax} direction in wellbore intervals with deviations of less than 20°.

The mean S_{Hmax} orientation from each well was given a quality ranking according to the World Stress Map project criteria with A-quality being the highest (S_{Hmax} reliable to within $\pm 15°$) and E-quality the lowest (no reliable orientation determinable; Heidbach *et al.* 2010). Table 1 lists the quality ranking criteria for breakouts and drilling-induced fractures interpreted from image and four-arm caliper logs (from Tingay *et al.* 2008; Heidbach *et al.* 2010).

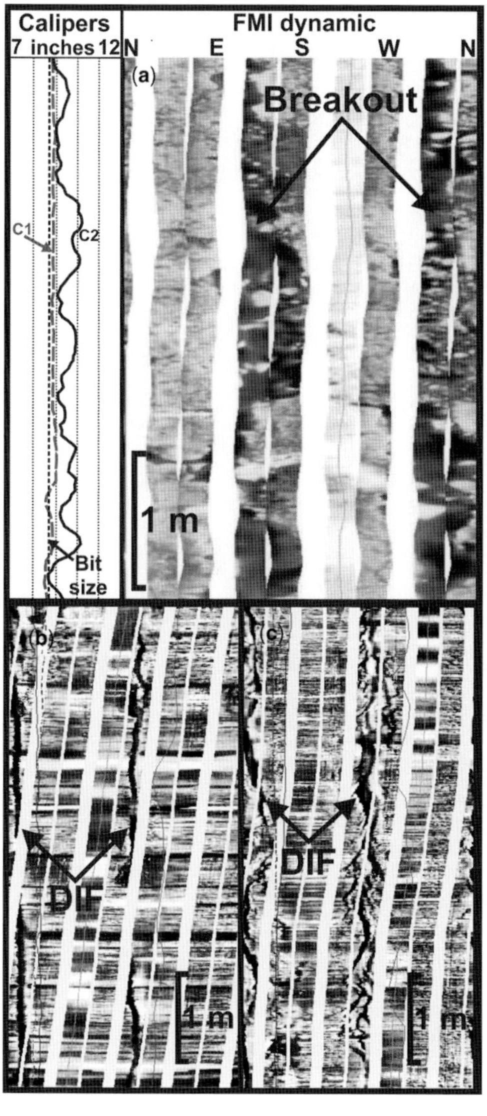

Fig. 3. Examples of borehole breakout and drilling-induced fractures observed on formation micro imager (FMI) resistivity images in the Nile Delta. FMI images present a picture of the borehole wall (as viewed from inside the wellbore) based on resistivity contrasts. FMI images are 'unrolled' and viewed clockwise from the inside of the wellbore (with north being the left and right edges of the image and south being the lateral centre of the image). (**a**) Borehole breakouts observed as broad poorly resolved conductive (dark) zones on opposite sides of the borehole and with associated caliper enlargement in one direction (caliper 2; grey line in image denotes pad 1, which is in the caliper 1 direction). Borehole breakouts are oriented approximately NNW–SSE, indicating an ENE–WSW present-day maximum horizontal stress (S_{Hmax}) orientation. (**b**) and (**c**) Drilling-induced fractures observed as narrow sharply-defined conductive (dark) features on opposite sides of the borehole. Both examples of drilling-induced fractures are oriented approximately NNE–SSW, indicating an approximately NNE–SSW present-day S_{Hmax} orientation.

Present-day stress orientation in the Nile Delta

Four-arm caliper and resistivity image logs were analysed for borehole breakout and drilling-induced fractures in 13 wells in the Nile Delta, the results of which were combined with stress orientations from 44 wells previously examined by the author in the central and eastern regions of the offshore Nile Delta (Fig. 1; Tingay *et al.* 2011). A combined total of 91 km of four-arm caliper logs and image logs were examined in the 57 wells, including 23.2 km of image log data from 30 wells. A total

Table 1. *World Stress Map (WSM) project quality ranking criteria for breakouts and drilling-induced fractures (DIFs) interpreted from four-arm caliper and image logs (Heidbach* et al. *2010). SD, standard deviation of breakout/DIF orientations*

Data Type	A-quality	B-quality	C-quality	D-quality	E-quality
Breakouts (four-arm caliper logs)	$\geq$10 breakouts with combined length $\geq$300 m and SD $\leq$12° in a single well	$\geq$6 breakouts with combined length $\geq$100 m and SD $\leq$20° in a single well	$\geq$4 breakouts with combined length $\geq$30 m and SD $\leq$25° in a single well	$<$4 breakouts or combined length $<$30 m and SD $\leq$40°	No breakouts observed or breakouts with SD $>$40°
Breakouts (image logs)	$\geq$10 breakouts with combined length $\geq$100 m and SD $\leq$12° in a single well	$\geq$6 breakouts with combined length $\geq$40 m and SD $\leq$20° in a single well	$\geq$4 breakouts with combined length $\geq$20 m and SD $\leq$25° in a single well	$<$4 breakouts or combined length $<$20 m and SD $\leq$40°	No breakouts observed or breakouts with SD $>$40°
Drilling-induced fractures (image logs)	$\geq$10 DIFs with combined length $\geq$100 m and SD $\leq$12° in a single well	$\geq$6 DIFs with combined length $\geq$40 m and SD $\leq$20° in a single well	$\geq$4 DIFs with combined length $\geq$20 m and SD $\leq$25° in a single well	$<$4 DIFs or combined length $<$20 m and SD $\leq$40°	No DIFs observed or DIFs with SD $>$40°

of 588 breakouts and 68 drilling-induced fractures with a combined length of over 3.6 km were observed in 50 wells. Breakouts and drilling-induced fractures were observed in supra-salt sequences in 19 wells, in sequences below salt in seven wells and in sequences absent from evaporites in 27 wells (Figs 3 & 4). Image and four-arm caliper logs were also examined in seven wells that either did not contain breakouts/drilling-induced fractures or were deviated by >20°; there were therefore not used here (ranked E-quality). The observed breakouts and drilling-induced fractures indicate that S_{Hmax} varies across the study region ranging from, on average, NNE–SSW in the western region, east–west in the central region, ESE–WNW in the eastern region below and absent from the Messinian evaporites and scattered but predominately NNE–SSW in sequences overlying the evaporites (Figs 4 & 5). These results show a good correlation with the S_{Hmax} orientations determined from breakouts interpreted on four-arm caliper logs in 11 wells in the onshore and shallow marine region of the Nile Delta (Fig. 4; Bosworth 2006; orientations not quality ranked).

The breakouts and drilling-induced fractures interpreted here reveal a striking variation in the regional present-day S_{Hmax} orientation in sequences above and sequences below or absent from the Messinian evaporites (Figs 4 & 5; Tingay *et al.* 2011). This variation in S_{Hmax} orientation is also observed at a local scale in all wells that penetrate through both supra- and sub-salt sequences (Fig. 6). For example, breakouts and drilling-induced fractures interpreted from image logs and four-arm caliper logs in four wells in Field A indicate that the present-day S_{Hmax} orientation in supra-salt sequences is approximately NNE–SSW (average 015°N ± 9° standard deviation; Fig. 6a). In contrast, breakouts and drilling-induced fractures interpreted from image logs below the Messinian

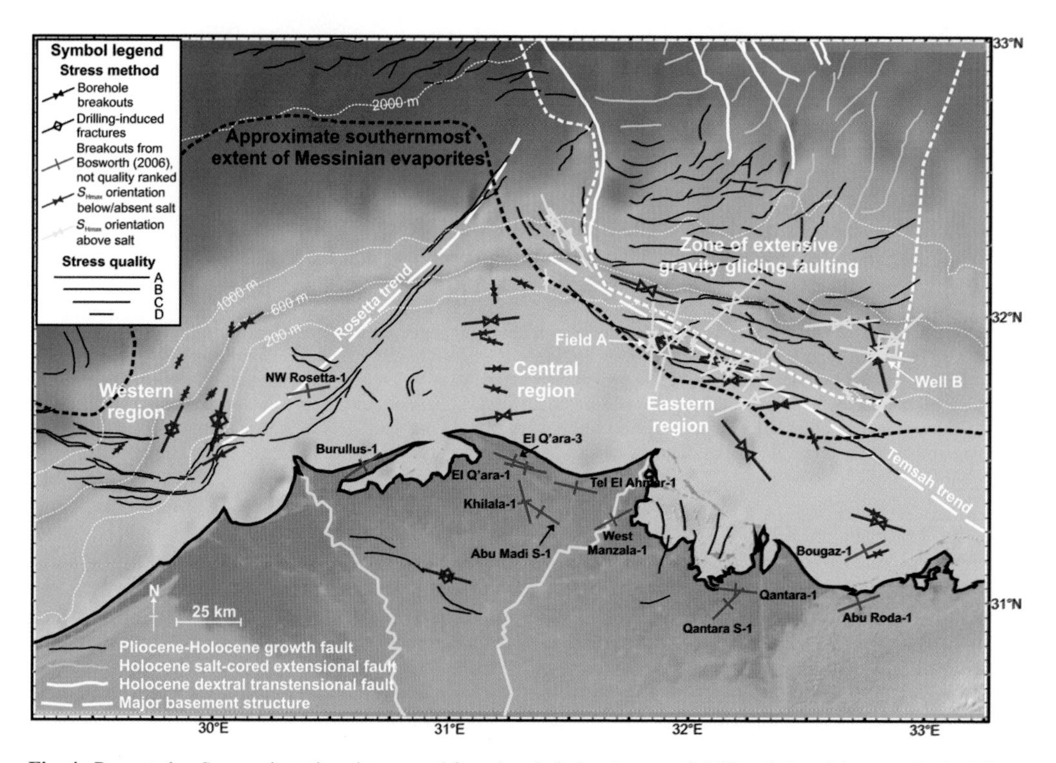

Fig. 4. Present-day S_{Hmax} orientations interpreted from borehole breakouts and drilling-induced fractures in the Nile Delta. Stress orientations below or south of the Messinian evaporites (blue symbols) indicate an approximately NNE–SSW S_{Hmax} in the western region, an east–west S_{Hmax} orientation in the central region and an approximately ESE–WNW S_{Hmax} in the eastern region of the Nile Delta. These orientations are broadly consistent with those observed in Bosworth (2006; red crosses). S_{Hmax} orientations are however scattered, but predominately NE–SW in sequences overlying Messinian evaporites (yellow symbols). The average S_{Hmax} orientation in sequences above the evaporites is therefore approximately perpendicular to that observed below the evaporites (see Fig. 5), indicating that the sub- and supra-salt sequences are mechanically detached. Furthermore, the S_{Hmax} orientations observed in supra-salt sequences are inconsistent with the predominately extensional faulting observed in this region.

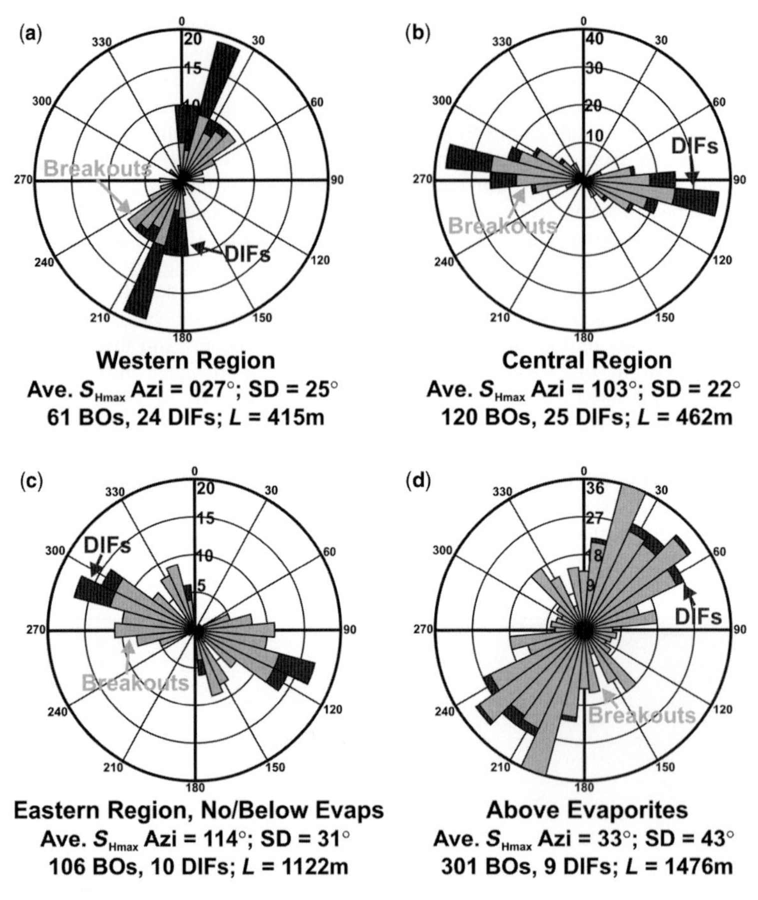

Fig. 5. Orientations of present-day S_{Hmax} indicated by borehole breakouts (light grey) and drilling-induced fractures (DIFs; dark grey). (**a**) Western offshore region outboard of the Rosetta Fault. (**b**) Central region along the Baltim trend. (**c**) Eastern offshore region in sequences below Messinian evaporites or where evaporites are absent. (**d**) Sequences overlying Messinian evaporites in the eastern offshore Nile Delta. Length of petals represents total number of stress-induced features observed in azimuthal bin, with length of light and dark grey components representing the relative number of breakouts and DIFs, respectively.

evaporites indicate an ESE–WNW S_{Hmax} orientation (average 110°N ± 5° standard deviation), almost perpendicular to the supra-salt S_{Hmax} orientation (Fig. 6a). Breakouts and drilling-induced fractures observed in Well B indicate a similar large variation in S_{Hmax} orientation above and below the evaporites, although in Well B the supra-salt S_{Hmax} orientation is approximately east–west (average 099°N ± 15° standard deviation) and the sub-salt S_{Hmax} orientation is approximately NNW–SSE (average 165°N ± 7° standard deviation; Fig. 6b). The significant variation in interpreted present-day stress orientations above and below the evaporites is considered as direct evidence that the Messinian evaporates, wherever present, act as a mechanical detachment surface in the Nile Delta (Tingay *et al.* 2011).

Present-day stress magnitudes have also been investigated by the author, but can only be summarized here due to confidentiality restrictions. Present-day stress magnitudes were determined from conventional borehole geomechanics methods, such as leak-off test and hydraulic fracture test analysis, circumferential stress modelling and frictional limit calculations (e.g. Bell 1996*b*; Zoback 2007; Tingay *et al.* 2009). The present-day state of stress in the Nile Delta typically varies with increasing pore pressure, and is constrained as being a predominately normal faulting stress regime ($S_v > S_{Hmax} > S_{Hmin}$) in normally pressured sequences, becomes borderline normal/strike-slip (S_v *c.* $S_{Hmax} > S_{Hmin}$) in moderately overpressured sequences (pore pressures 13.0–16.5 MPa km^{-1}) and trends into a strike-slip faulting stress regime

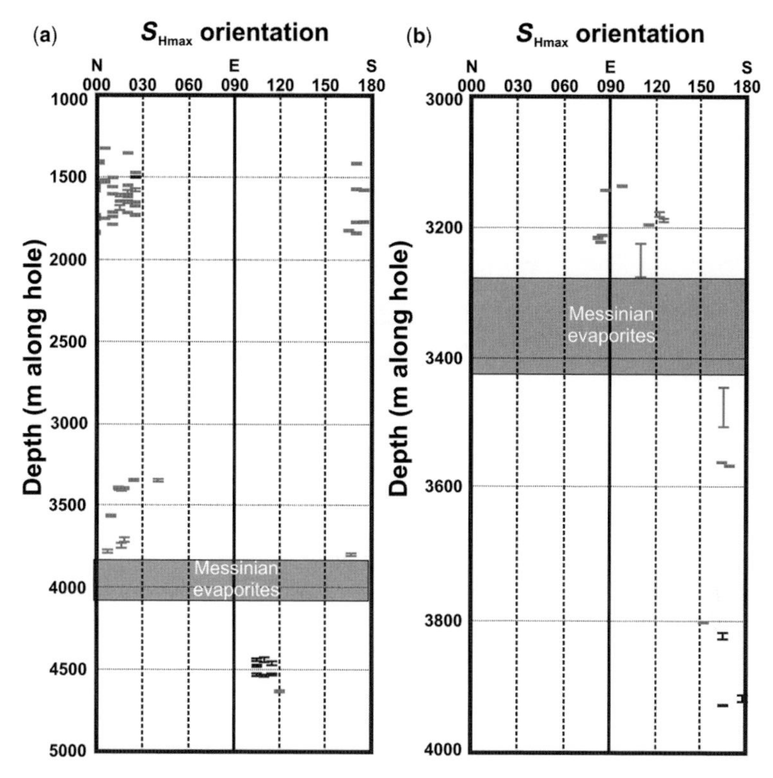

Fig. 6. Observations of S_{Hmax} orientation from breakouts (in grey) and drilling-induced fractures (DIFs; in black) versus depth (see Fig. 4 for locations). (**a**) Field A (four wells in close proximity): 68 breakouts and drilling-induced fractures indicate a NNE–SSW supra-salt S_{Hmax} orientation but a ESE–WNW sub-salt S_{Hmax} orientation. (**b**) Well B: 34 breakouts and drilling-induced fractures indicate an ESE–WNW S_{Hmax} in supra-salt sequences, but a NNE–SSE S_{Hmax} below the Messinian evaporites. The change in S_{Hmax} orientations above and below the evaporites demonstrates that the evaporites are acting as a mechanical and structural detachment horizon in the Nile Delta.

($S_{Hmax} > S_v > S_{Hmin}$) in highly overpressured sequences (pore pressure >16.5 MPa km^{-1}). No obvious variation in stress magnitudes is observed near the Messinian evaporites. Stress magnitudes in sequences overlying the evaporites suggest a predominately normal faulting stress regime, but with a borderline normal/strike-slip stress regime observed in some wells in moderately overpressured sequences adjacent to the evaporites. Furthermore, it is important to highlight that present-day horizontal stress magnitudes are anisotropic ($S_{Hmin} \neq S_{Hmax}$) and therefore not the cause of the variable stress orientations observed across the Nile Delta.

Discussion

Origin of present-day stress field in regions below or absent from evaporites

The present-day stress field on the shelf and slope areas of Cenozoic deltas, such as the Mississippi and Baram Deltas, is approximately parallel to the continental margin (Fig. 7; Yassir & Zerwer 1997; Tingay *et al.* 2005; King *et al.* 2009). This margin-parallel S_{Hmax} is thought to be generated by the shape of the clastic wedge, which promotes basinwards extension (with associated margin-parallel striking-normal faults) on the shelf and slope and basinwards contraction at the delta toe (with associated margin-parallel striking thrust faults; Fig. 7; McClay *et al.* 1998; Tingay *et al.* 2006; King *et al.* 2009).

The present-day S_{Hmax} orientations in sequences not overlying Messinian evaporites are also primarily parallel to the margin, changing from NNE–SSW in the western region, to east–west in the central region, to predominately ESE–WNW in the eastern region and to east–west in the far eastern parts of the study area directly north of the Sinai (blue and red stress orientations in Fig. 4). It is therefore suggested that a typical deltaic present-day stress field exists in Nile Delta sequences that are inboard or below the Messinian evaporites and in some supra-salt sequences.

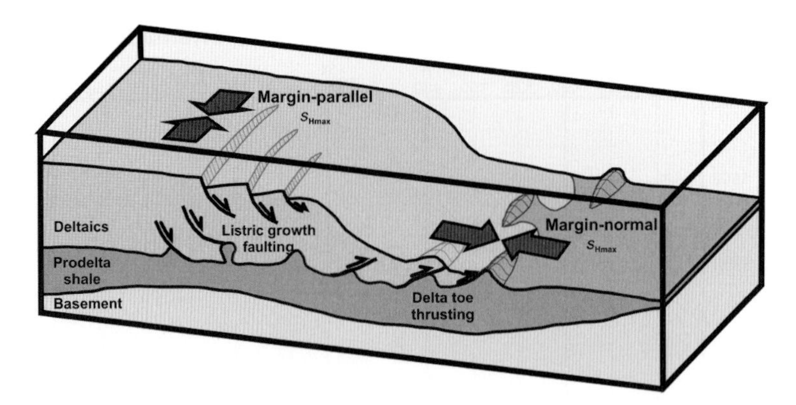

Fig. 7. Schematic relationship between present-day stress and structures in passive-margin Cenozoic deltas (after Tingay *et al.* 2006). The convex-upwards shape of the clastic wedge promotes gravity-driven extension towards the delta toe, resulting in margin-parallel maximum horizontal stress orientations on the shelf and margin-normal thrusting in the delta toe.

Structural styles observed within these areas of the Nile delta are primarily margin-parallel-striking deltaic structures such as listric growth faults, typical of those observed in passive-margin delta systems (Aal *et al.* 2001; Loncke *et al.* 2006). These structures are most likely driven by classic deltaic gravity-spreading processes, in which deformation is the result of the deltaic wedge maintaining a critical taper angle (Bilotti & Shaw 2006; Loncke *et al.* 2006; Morley 2007). The broadly margin-parallel S_{Hmax} orientations in regions of the Nile Delta that are below or absent from evaporites are therefore fully consistent with observed gravity-spreading structural styles (and the Andersonian fault model).

Present-day stress and non-Andersonian faulting above evaporites in the Nile Delta

The existence of the scattered, but predominantly NNW–SSE, S_{Hmax} orientations observed in the supra-salt sequences of the eastern Nile Delta is extremely unusual and is in stark contrast to the structural styles observed in this region, and to that expected in Cenozoic delta systems. The region of the Nile Delta underlain by evaporites contains numerous faults that are clearly imaged on high-resolution bathymetry and reflection seismic data (Mascle *et al.* 2000; Loncke *et al.* 2006). The faulting observed in much of the supra-salt Nile Delta sequences can be interpreted as typical deltaic extensional structures generated by gravity spreading, with faults commonly soling out onto the evaporites (Loncke *et al.* 2006; Tingay *et al.* 2011). However, bathymetry data also reveals a broad NNE–SSW trending corridor (approximately 100 km wide and 175 km long) of supra-salt faulting in the eastern Nile Delta with distinctly different characteristics (Fig. 8; Loncke *et al.* 2006). Faulting in this section of the supra-salt eastern Nile Delta consists of numerous margin-parallel (ESE–WNW) extensional faults near the shelf edge, trending down the slope into a zone of approximately orthogonal faults consisting of east–west-striking extensional faults coupled with north–south-striking transtensional faulting (Fig. 8). The structural features observed within this corridor of supra-salt deformation can be interpreted as being consistent with gravity-gliding tectonics, in which blocks of sediment are being translated down the slope without any significant internal deformation, along a basinwards-dipping detachment horizon (Vendeville 2005; Loncke *et al.* 2006). The structural styles observed in the supra-salt eastern Nile Delta therefore appear to be dominated by gravity gliding, in which blocks of post-Messinian sediments are rafting down a seawards-dipping evaporite detachment towards, and around, the Eratosthenes Seamount (Fig. 8).

There are no other studies known to the author of present-day stress analysis in a zone of gravity-gliding tectonics. However, the margin-normal strike of extensional faults generated by gravity gliding, as well as the geometry of such a basally detached sedimentary wedge, suggests that present-day S_{Hmax} orientations would be predicted to be margin-parallel and consistent with the Andersonian model. Breakouts and drilling-induced fractures interpreted in the supra-salt sequences of the Nile Delta however indicate an average NNE–SSW (033°N ± 43°) margin-normal present-day S_{Hmax} orientation, approximately perpendicular to the strike of extensional faults in this region and to the S_{Hmax} orientation predicted under the Anderson faulting model.

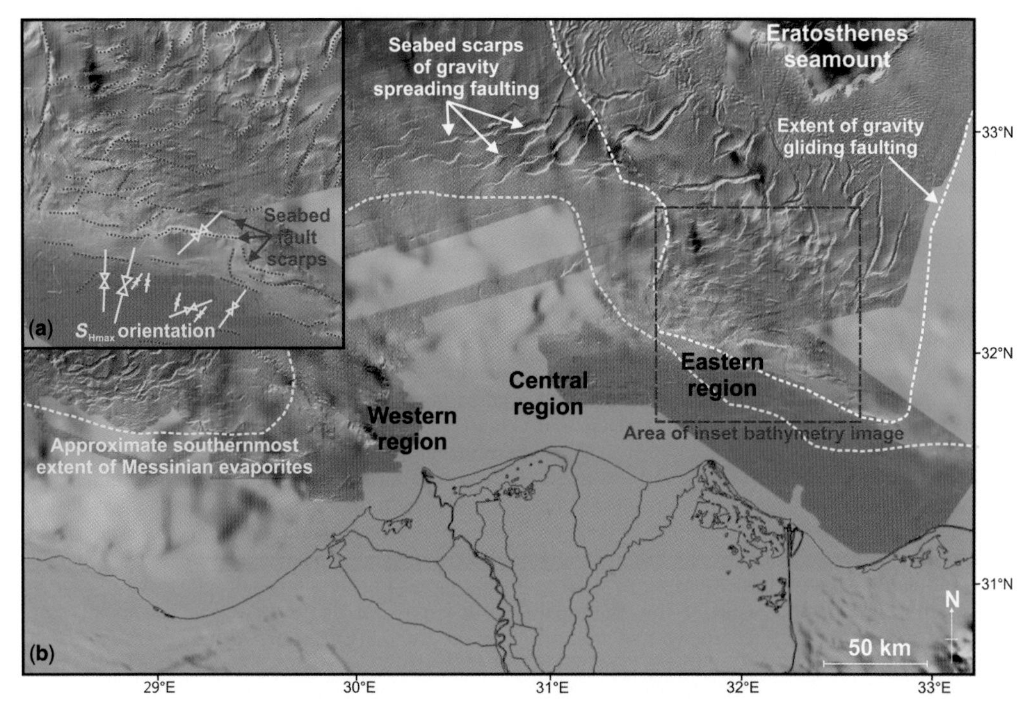

Fig. 8. (**a**) Shaded high-resolution bathymetry image of the Nile Delta and adjacent deep-sea fan region (illuminated from the north; image courtesy of BP Egypt). Numerous large seabed fault scarps are visible throughout the Nile Delta, the majority of which are margin-parallel and representative of gravity-spreading tectonics. The bathymetry also highlights a broad corridor of faulting dominated by gravity-gliding tectonics (outlined by the white dashed line), in which numerous north–south and east–west to ESE–WNW-trending seabed fault scarps are visible. (**b**) Zoomed-in view of shaded seabed bathymetry with interpreted seabed fault scarps (purple dashed lines) and present-day S_{Hmax} orientations in supra-salt sequences (yellow lines). The presence of seabed fault scarps indicates that these faults are currently or very recently active. Seabed scarps in the southern part of the inset figure are extensional faults, striking at an anomalously high angle (45–90°) to the present-day S_{Hmax} orientation. The close proximity of margin-normal present-day S_{Hmax} orientations and currently active margin-parallel striking extensional faults provides compelling evidence that these structures are a new type of non-Andersonian faulting in which the fault is anomalously oriented with respect to the intermediate (S_{Hmax}) and minimum (S_{Hmin}) principal stresses.

While the average S_{Hmax} orientation in supra-salt sequences of the eastern offshore Nile Delta is approximately NNE–SSW, it should be noted that there is a wide variation in present-day S_{Hmax} orientations within this region (Figs 4 & 5). Indeed, there are two main S_{Hmax} orientations observed in individual wells above the Messinian evaporites: a dominant north–south to NE–SW S_{Hmax} orientation (average NNE–SSE; observed in 11 wells, often of higher quality) and a secondary east–west to NW–SE S_{Hmax} orientation (average ESE–WNW; observed in 6 wells of varying quality). However, possibly the most significant observation in this study is that several wells exhibiting approximately margin-normal S_{Hmax} orientations are drilled immediately adjacent to margin-parallel-striking extensional faults. These faults cut through Pliocene–Recent units and many of these faults exhibit seabed scarps, indicating that they are currently, or very recently, active (Fig. 8; Mascle *et al.* 2000; Loncke *et al.* 2006). The close proximity of margin-normal present-day S_{Hmax} orientations and currently active margin-parallel striking extensional faults provides compelling evidence that these structures are a previously unseen form of non-Andersonian faulting in which the fault is anomalously oriented with respect to the intermediate (S_{Hmax}) and minimum (S_{Hmin}) principal stresses, rather than being misaligned with respect to the maximum principal stress as observed in highly oblique strike-slip faults and low-angle normal faults.

Mechanics of non-Andersonian faulting in the Nile Delta

It is interesting to note that the new form of non-Andersonian faulting observed in the Nile Delta is even less favourably oriented for reactivation in the present-day stress tensor than highly oblique strike-slip faults and low-angle normal faults. One method for assessing the relative likelihood of fault reactivation is the proximity of fault planes to the failure envelope on a Mohr circle shear versus normal stress diagram (Fig. 9; Morris *et al.* 1996; Jones & Hillis 2003). The shear and normal stresses acting on the non-Andersonian faults in the Nile Delta plot relatively further away from the failure envelope, and have a lower shear-to-normal stress ratio (slip tendency) than highly oblique strike-slip faults and low-angle normal faults (Fig. 9). These faults are therefore extremely poorly oriented for shear failure in the present-day stress tensor, and the mechanics of this new form of non-Andersonian faulting is uncertain.

Highly oblique strike-slip faults and low-angle normal faults have been suggested to result from a range of mechanisms. The most common hypothesis is that non-Andersonian faults are exceptionally weak, in that they have a low resistance to sliding. Weak faults may be the result of low frictional resistance to shearing (e.g. coefficients of friction <0.2), most likely due to the fault zone containing material with low shear strengths (Axen 1992; Faulkner *et al.* 2006; Collettini *et al.* 2009). An alternative hypothesis for non-Andersonian faulting having a low resistance to sliding is the presence of overpressure within a fault zone (Sibson 1992; Rice 1992; Healy 2008). High pore-fluid pressures reduce the effective normal stress that acts to resist fault slip. While overpressures are known to significantly influence the risk of fault reactivation (Sibson 1992; Tingay *et al.* 2003), it has also been suggested that overpressures are unlikely to be generated, or would be quickly dissipated, within fault zones (Townend & Zoback 2000; Fulton *et al.* 2009). Furthermore, significant overpressures are not observed in many major non-Andersonian faults, most notably the San Andreas Fault (Brace & Kohlstedt 1980; Townend & Zoback 2000; Fulton *et al.* 2009). More recently, an alternative mechanism for the development of non-Andersonian faulting has been suggested in which the maximum principal stress orientation is predicted to be locally rotated to be more favourably oriented for reactivation within and proximal to the fault zone (Faulkner *et al.* 2006; Healy 2008).

It is not clear whether the previously suggested mechanisms for non-Andersonian faulting are applicable here. While the supra-salt faulting soles out onto the presumably very weak evaporites, it seems unlikely that the fault zone in shallower sections will have exceptionally low coefficients of friction due to the inclusion of weak material. Furthermore, overpressures are not observed above 1800 m depth, and overpressures are generally mild in most of the supra-salt sequences.

It is possible that the stress field is locally perturbed by the evaporites (Tingay *et al.* 2011). Present-day S_{Hmax} orientations are often highly scattered in regions underlain by evaporites, such as in the central North Sea and Gulf of Mexico (Bell 1996*a*; Ask 1997; Yassir & Zerwer 1997). Furthermore, stress orientations can be locally deflected by contrasts in elastic rock properties with a principal stress (often S_{Hmax}) typically deflected parallel to weak materials and perpendicular to stiff materials (Bell 1996*a*; Yale 2003; Tingay *et al.* 2010). Indeed, some inclined or en echelon drilling-induced fractures were observed in one well 500 m below the evaporites, indicating that the principal stresses have been locally deflected from the

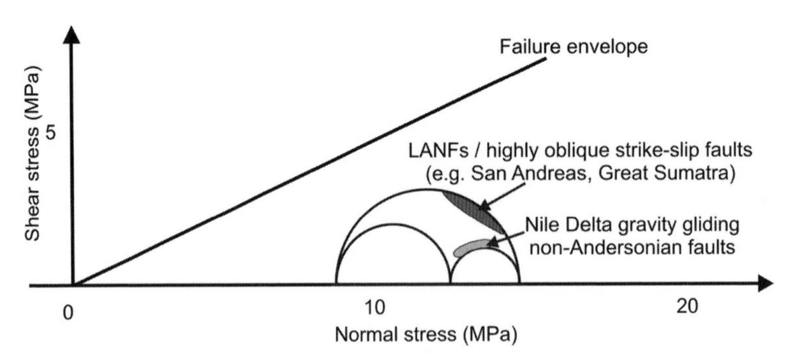

Fig. 9. Mohr circle of present-day effective stress in the supra-salt sequences of the Nile Delta at 1500 m depth. The shear and normal stresses acting on shallow supra-salt extensional faults in the Nile Delta (light-grey region) are very poorly oriented for reactivation in the observed stress regime. Indeed, the non-Andersonian faulting observed in the Nile Delta is even less-favourably oriented for failure in the present-day stress tensor than low-angle normal faults (LANFs) or highly oblique strike-slip faults (dark-grey region; plotted schematically with respect to the maximum and minimum principal stresses).

vertical and horizontal planes. It is therefore possible that the supra-salt present-day stress orientations are due to some combination of local sources of stress and small-scale stress perturbations near faults, folds and salt structures. This hypothesis is also somewhat supported by the wide variety of stress orientations and structures observed in the supra-salt sequences. The margin-normal stress orientations are however observed up to 2500 m above the salts, and not only immediately adjacent to evaporites as predicted from simple mechanical contrasts (Bell 1996*a*; Tingay *et al.* 2006). It therefore seems unlikely that the non-Andersonian faulting observed in the eastern Nile Delta is a result of localized stress rotations.

The highly margin-normal orientation of the present-day S_{Hmax} orientation, coupled with the association of the non-Andersonian faulting being located in a zone of gravity-gliding tectonics, suggests a possible alternative origin for these faults and the present-day stress field. It is interesting to note that the non-Andersonian faults and maximum principal stress orientation (S_{v}) are exactly as would be predicted for deltaic gravity-gliding tectonics – it is only the horizontal stress orientations that are anomalous. It is therefore possible that the margin-normal stress orientation is the result of forces exerted by salt withdrawal or localized gravity gliding, in which downslope movement of salt imparts margin-normal basal force on overlying rafted sedimentary sequences (Fig. 10). Such a process is analogous to the forces and motions exerted on an object sitting on a tablecloth or rug that is being pulled at one end. This hypothesis is somewhat supported by the occurrence of the large dextral transtensional faults further down the slope and bathymetric evidence for mass wasting in this region (Loncke *et al.* 2006, 2009). Indeed, it is interesting to note that the margin-normal S_{Hmax}, should it extend further offshore, is favourably oriented for the formation of these large dextral faults. It is not known to the author whether such a mechanism is plausible or whether it would generate the observed stress field and structural styles. However, this alternative mechanism suggests that these faults may originally initiate under the Andersonian model (such as is seen in other parts of the Nile Delta), with the margin-normal S_{Hmax} orientation only being generated once the faulting becomes sufficiently advanced that the now mechanically isolated sediment blocks start to raft down the slope.

It is also interesting to note that the zone of gravity gliding is proximal to the Eratosthenes seamount, with this corridor of deformation diverging around the seamount and some sequences exhibiting compressional deformation against the seamount (Aal *et al.* 2001; Loncke *et al.* 2006). It is not known to the authors whether the Eratosthenes may be influencing the state of stress in the study area, such as the transmission of compressive stresses up the slope. Such a mechanism appears unlikely however, given the significantly more complex and thicker amount of evaporates adjacent to the seamount and likely low mechanical rigidity of the sediments. Although the authors remain uncertain as to the mechanism for the non-Andersonian faulting observed in the Nile Delta, we suggest that gravity gliding and the stresses imparted on sequences

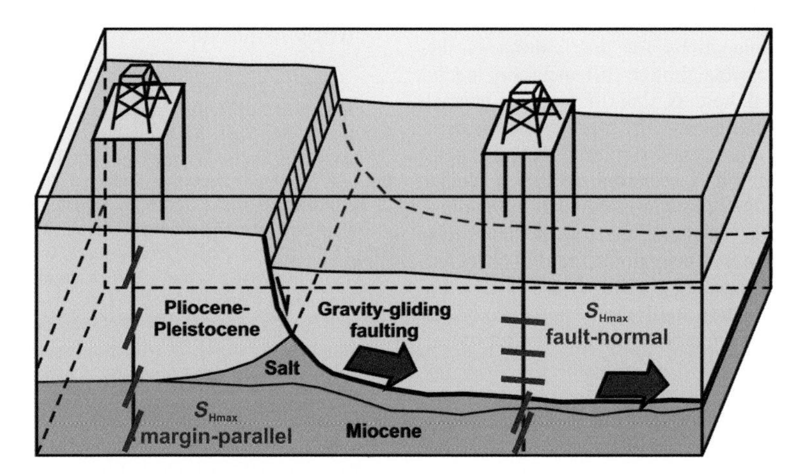

Fig. 10. A schematic model for the development of gravity gliding and observed fault-normal S_{Hmax} orientation. Once faulting initiates, the supra-salt sequences become mechanically detached and undergo gravity gliding down the seawards-dipping salt detachment. The margin-normal stress orientation is hypothesized to be the result of forces exerted by salt withdrawal, in which downslope movement of salt imparts margin-normal basal drag on overlying rafted sedimentary sequences.

rafting downslope remains the most plausible theory proposed so far.

Summary and conclusions

The present-day stress orientations in the Nile Delta highlight a range of issues concerning the structural evolution and tectonic framework of the Nile Delta. Present-day S_{Hmax} orientations in sequences not underlain by evaporites reveal a textbook example of a margin-parallel deltaic stress field and are consistent with observed gravity-spreading tectonics. The present-day S_{Hmax} orientations in the supra-salt regions are however complicated, and can be in stark contrast to the observed structural styles. The occurrence of present-day S_{Hmax} orientations approximately perpendicular to the strike of active extensional faults provides compelling evidence for an entirely new form of non-Andersonian faulting. Under this new style of non-Andersonian extensional faulting, the faults are oriented at an anomalous angle with respect to the intermediate and minimum principal stresses (rather than the 'conventional' form of non-Andersonian faulting in which faults have a dip or strike that is poorly oriented with respect to the maximum principal stress). Furthermore, the non-Andersonian faults observed in the eastern Nile Delta are even less-suitably oriented for shear failure than highly oblique strike-slip faults and low-angle normal faults. The mechanics for driving this new style of non-Andersonian faulting remains uncertain. Mechanisms suggested for other non-Andersonian faults, such as faults containing low coefficient of friction material or overpressures or the stress tensor being locally perturbed within the fault zone, appear to be unlikely explanations for the faulting in the Nile Delta. A possible, though still uncertain, mechanism suggested here is that the margin-normal S_{Hmax} orientation observed in supra-salt sequences is related to localized gravity-gliding tectonics generated by basal forces imparted on rafted blocks sliding down the basinwards-dipping evaporites. Regardless of the origin of the present-day stress orientations and non-Andersonian faulting observed in the supra-salt sequences of the eastern Nile Delta, the results of this study highlight the possible pitfalls of assuming stress from observed active fault styles. The Andersonian fault model forms the basis for most palaeostress analysis in structural geology; the present-day stress analysis discussed here therefore demonstrates that estimates of palaeostress tensors from structural styles can potentially have large errors.

The author wishes to thank BP Egypt, International Egyptian Oil Company, RWE Dea Egypt and the Egyptian Gas Company for providing the data for this study. The authors also wish to thank the two anonymous reviewers for their constructive comments. This research was funded through the Heidelberg Academy of Sciences via the World Stress Map Project and by BP Egypt, the International Egyptian Oil Company and RWE Dea Egypt. M. Tingay's contribution to this research forms TRaX record #147.

References

AADNOY, B. S. 1990. Inversion technique to determine the in-situ stress field from fracturing data. *Journal of Petroleum Science and Engineering*, **4**, 127–141.

AADNOY, B. S. & BELL, J. S. 1998. Classification of drill-induce fractures and their relationship to in-situ stress directions. *The Log Analyst*, **39**, 27–42.

AAL, A. A., EL BARKOOKY, A., GERRITS, M., MEYER, H. J., SCHWANDER, M. & ZAKI, H. 2001. Tectonic evolution of the Eastern Mediterranean Basin and its significance for the hydrocarbon prospectivity of the Nile Delta deepwater area. *GeoArabia*, **6**, 363–384.

ANDERSON, E. M. 1942. *The Dynamics of Faulting*. Oliver and Boyd, Edinburgh.

ANDERSON, E. M. 1951. *The Dynamics of Faulting and Dyke Formation*. Oliver and Boyd, London.

ANGELIER, J. 1979. Determination of the mean principal stresses for a given fault population. *Tectonophysics*, **56**, T17–T26.

ANGELIER, J. 1989. From orientation to magnitudes in paleostress determinations using fault slip data. *Journal of Structural Geology*, **11**, 37–50, doi: 10.1016/0191-8141(89)90034-5.

ASK, M. V. S. 1997. In situ stress from breakouts in the Danish Sector of the North Sea. *Marine and Petroleum Geology*, **14**, 231–243.

AXEN, G. J. 1992. Pore pressure, stress increase, and fault weakening in low-angle normal faulting. *Journal of Geophysical Research*, **97**, 8979–8991, doi: 10.1029/92JB00517.

BADAWY, A. & HORVÁTH, F. 1999. Seismicity of the Sinai subplate region: kinematic implications. *Journal of Geodynamics*, **27**, 451–468.

BARBER, P. M. 1981. Messinian subaerial erosion of the proto-Nile Delta. *Marine Geology*, **44**, 253–272.

BELL, J. S. 1996*a*. Petro Geoscience 1. In situ stresses in sedimentary rocks (part 2): applications of stress measurements. *Geoscience Canada*, **23**, 135–153.

BELL, J. S. 1996*b*. Petro Geoscience 1. In situ stresses in sedimentary rocks (part 1): measurement techniques. *Geoscience Canada*, **23**, 85–100.

BELL, J. S. & GOUGH, D. I. 1979. Northeast–southwest compressive stress in Alberta: evidence from oil wells. *Earth and Planetary Science Letters*, **45**, 475–482.

BELLAICHE, G. & MART, Y. 1995. Morphostructure, growth patterns, and tectonic control of the Rhone and Nile deep-sea fans: a comparison. *American Association of Petroleum Geologists Bulletin*, **79**, 259–284.

BILOTTI, F. & SHAW, J. H. 2006. Deep-water Niger Delta fold and thrust belt modelled as a critical-taper wedge: the influence of elevated basal fluid pressure on structural styles. *American Association of Petroleum Geologists Bulletin*, **89**, 1475–1491.

BONCIO, P. & LAVECCHIA, G. 2000. A structural model for active extension in Central Italy. *Journal of Geodynamics*, **29**, 233–244.

BOSWORTH, W. 2006. North Africa–Mediterranean Present-day Stress Field Transition and Implications for Fractured Reservoir Production in the Eastern Libyan Basins. *Proceedings of the Sedimentary Basins of Libya 3rd Symposium – Geology of East Libya.*

BRACE, W. F. & KOHLSTEDT, D. L. 1980. Limits on lithospheric stress imposed by laboratory experiments. *Journal of Geophysical Research*, **85**, 6248–6252.

CLARK, I. R. & CARTWRIGHT, J. A. 2009. Interactions between submarine channel systems and deformation in deepwater fold belts: examples from the Levant Basin, Eastern Mediterranean sea. *Marine and Petroleum Geology*, **26**, 1465–1482.

COLLETTINI, C. & SIBSON, R. H. 2001. Normal faults, normal friction? *Geology*, **29**, 927–930.

COLLETTINI, C., VITI, C., SMITH, S. A. F. & HOLDSWORTH, R. E. 2009. Development of interconnected talc networks and weakening of continental low-angle normal faults. *Geology*, **37**, 567–570, doi: 10.1130/G25645A.1.

COUZENS-SCHULTZ, B. A. & CHAN, A. W. 2010. Stress determination in active thrust belts: an alternative leak-off pressure interpretation. *Journal of Structural Geology*, **32**, 1061–1069.

EKSTROM, M. P., DAHAN, C. A., CHEN, M. Y., LLOYD, P. M. & ROSSI, D. J. 1987. Formation imaging with microelectrical scanning arrays. *The Log Analyst*, **28**, 294–306.

EL-BARKOOKY, A. N. & HELAL, M. A. 2002. Some Neogene stratigraphic aspects of the Nile Delta. *Proceedings of the 3rd Mediterranean Offshore Conference.*

EL-SAYED, A., VACCARI, F. & PANZA, G. F. 2004. The Nile Valley of Egypt: a major active graben that magnifies seismic waves. *Pure and Applied Geophysics*, **161**, 983–1002.

FAULKNER, D. R., MITCHELL, T. M., HEALY, D. & HEAP, M. J. 2006. Slip on 'weak' faults by the rotation of regional stress in the fracture damage zone. *Nature*, **444**, 922–925.

FULTON, P. M., SAFFER, D. M. & BEKINS, B. A. 2009. A critical evaluation of crustal dehydration as the cause of an overpressured and weak San Andreas fault. *Earth and Planetary Science Letters*, **284**, 447–454.

HEALY, D. 2008. Damage patterns, stress rotations and pore fluid pressure in strike-slip fault zones. *Journal of Geophysical Research*, **113**, B12407, doi: 10.1029/2008JB005655.

HEIDBACH, O., REINECKER, J., TINGAY, M., MÜLLER, B., SPERNER, B., FUCHS, K. & WENZEL, F. 2007. Plate boundary forces are not enough: Second- and third-order stress patterns highlighted in the World Stress Map database. *Tectonics*, **26**, TC6014, doi: 10.1029/2007TC002133.

HEIDBACH, O., TINGAY, M., BARTH, A., REINECKER, J., KURFEß, D. & MÜLLER, B. 2010. Global crustal stress pattern based on the world stress map database release 2008. *Tectonophysics*, **482**, 3–15.

JONES, R. M. & HILLIS, R. R. 2003. An integrated, quantitative approach to assessing fault-seal risk. *American Association of Petroleum Geologists Bulletin*, **347**, 189–215.

KING, R. C., HILLIS, R. R. & TINGAY, M. R. P. 2009. Present-day stress and neotectonic provinces of delta to deepwater fold–thrust belt systems: insights from NW Borneo. *Journal of the Geological Society, London*, **166**, 197–200.

KIRSCH, V. 1898. Die Theorie der Elastizität und die Beddürfnisse der Festigkeitslehre. *Zeitschrift des Vereines Deutscher Ingenieure*, **29**, 797–807.

LISLE, R. J. 1987. Principal stress orientations from faults: an additional constraint. *Annales Tectonicae*, **1**, 155–158.

LONCKE, L., GAULLIER, V., MASCLE, J., VENDEVILLE, B. & CAMERA, L. 2006. The Nile deep-sea fan: an example of interacting sedimentation, salt tectonics, and inherited subsalt paleotopographic features. *Marine and Petroleum Geology*, **23**, 297–315.

LONCKE, L., GAULLIER, V., DROZ, L., DUCASSOU, E., MIGEON, S. & MASCLE, J. 2009. Multi-scale slope instabilities along the Nile deep-sea fan, Egyptian margin: a general overview. *Marine and Petroleum Geology*, **26**, 633–646.

MARTEN, R., SHANN, M., MIKA, J., ROTHE, S. & QUIST, Y. 2004. Seismic challenges of developing the pre-Pliocene Akhen field offshore Nile Delta. *The Leading Edge*, **23**, 314–320.

MASCLE, J., BENKHELIL, J., BELLAICHE, G., ZITTER, T., WOODSIDE, J., LONCKE, L. & THE PRISMED II SCIENTIFIC PARTY 2000. Marine geologic evidence for a Levantine-Sinai plate, a new piece of the Mediterranean puzzle. *Geology*, **28**, 779–782.

MASTIN, L. 1988. Effect of borehole deviation on breakout orientations. *Journal of Geophysical Research*, **93**, 9187–9195.

MCCLAY, K. R., DOOLEY, T. & LEWIS, G. 1998. Analogue modelling of progradational delta systems. *Geology*, **29**, 771–774.

MORLEY, C. K. 2007. Interaction between critical wedge geometry and sediment supply in a deepwater fold belt. *Geology*, **35**, 139–142.

MORLEY, C. K. 2009. Geometry and evolution of low-angle normal faults (LANF) within a Cenozoic high-angle rift system, Thailand: Implications for sedimentology and the mechanisms of LANF development. *Tectonics*, **28**, TC5001; doi: 10.1029/2007TC002202.

MORRIS, A., FERRILL, D. A. & HENDERSON, D. B. 1996. Slip tendency analysis and fault reactivation. *Geology*, **24**, 275–278.

MORLEY, C. K., TINGAY, M., HILLIS, R. R. & KING, R. C. 2008. Relationship between structural style, overpressures and modern stress, Baram Delta province, NW Borneo. *Journal of Geophysical Research*, **113**, B09410 doi: 10.1029/2007JB005324.

MOUNT, V. S. & SUPPE, J. 1987. State of stress near the San Andreas fault: implications for wrench tectonics. *Geology*, **15**, 1143–1146.

MOUNT, V. S. & SUPPE, J. 1992. Present-day stress orientations adjacent to active strike-slip faults: California and Sumatra. *Journal of Geophysical Research*, **97**, 11 995–12 013.

PESKA, P. & ZOBACK, M. D. 1995. Compressive and tensile failure of inclined wellbores and determination of in situ stress and rock strength. *Journal of Geophysical Research*, **100**, 12 791–12 811.

PLUMB, R. A. & HICKMAN, S. H. 1985. Stress-induced borehole elongation: a comparison between the four-arm dipmeter and the borehole televiewer in the Auburn geothermal well. *Journal of Geophysical Research*, **90**, 5513–5521.

REINECKER, J., TINGAY, M. & MÜLLER, B. 2003. Borehole breakout analysis from four-arm caliper logs. *World Stress Map Project Stress Analysis Guidelines*, http://www.world-stress-map.org

REINECKER, J., TINGAY, M., MÜLLER, B. & HEIDBACH, O. 2010. Present-day stress orientation in the Molasse Basin. *Tectonophysics*, **482**, 129–138 doi: 10.1016/j.tecto.2009.07.021.

RICE, J. R. 1992. Fault stress states, pore pressure distributions, and the weakness of the San Andreas Fault. *In*: EVANS, B. & WONG, T.-F. (eds) *Fault Mechanics and Transport Properties in Rocks: A Festschrift in Honor of W.F. Brace*. Academic, San Diego, 475–503.

RYAN, W. B. F. & CITA, M. B. 1978. The nature and distribution of Messinian erosional surfaces – indicators of a several-kilometer-deep Mediterranean in the Miocene. *Marine Geology*, **27**, 193–230.

SCHLUMBERGER 1986. *Dipmeter Interpretation*. Schlumberger Limited, New York.

SESTINI, G. 1989. Nile Delta: a review of depositional environments and geological history. *In*: WHATELEY, M. K. G. & PICKERING, K. T. (eds) *Deltas: Sites and Traps for Fossil Fuels*. Geological Society, London, Special Publications, **41**, 99–127.

SHAABAN, F., LUTZ, R., LITTKE, R., BUEKER, C. & ODISHO, K. 2006. Source-rock evaluation and basin modelling in NE Egypt (NE Nile Delta and Northern Sinai). *Journal of Petroleum Geology*, **29**, 103–124.

SHAN, Y., LIN, G., LI, Z. & LI, J. 2004. A simple stress inversion of fault/slip data assuming Andersonian stress state. *Journal of Geophysical Research*, **109**, B04408, doi: 10.1029/2003JB002770.

SIBSON, R. H. 1992. Implications of fault-valve behaviour for rupture nucleation and recurrence. *Tectonophysics*, **211**, 283–293.

TINGAY, M., HILLIS, R., MORLEY, C., SWARBRICK, R. & OKPERE, E. 2003. Pore pressure/stress coupling in Brunei Darussalam – implications for shale injection. *In*: VAN RENSBERGEN, P., HILLIS, R. R., MALTMAN, A. J. & MORLEY, C. K. (eds) *Subsurface Sediment Mobilization*. Geological Society, London, Special Publications, **216**, 369–379.

TINGAY, M., HILLIS, R., MORLEY, C., SWARBRICK, R. & DRAKE, S. 2005. Present day stress orientation in Brunei: a snapshot of 'prograding tectonics' in a Tertiary delta. *Journal of the Geological Society, London*, **162**, 39–49.

TINGAY, M. R. P., MÜLLER, B., REINECKER, J. & HEIDBACH, O. 2006. State and origin of the present-day stress field in sedimentary basins: New results from the world stress map project. *Proceedings of the 41st US Rock Mechanics Symposium – ARMA's Golden Rocks 2006 – 50 Years of Rock Mechanics.*

TINGAY, M., REINECKER, J. & MÜLLER, B. 2008. Borehole breakout analysis from four-arm caliper logs. *World Stress Map Project Stress Analysis Guidelines*, http://www.world-stress-map.org

TINGAY, M., HILLIS, R., SWARBRICK, R., MORLEY, C. & DAMIT, A. 2009. Present-day stress and neotectonics of Brunei: implications for petroleum exploration and production. *American Association of Petroleum Geologists Bulletin*, **93**, 75–100.

TINGAY, M., MORLEY, C., HILLIS, R. R. & MEYER, J. J. 2010. Present-day stress orientation in Thailand's basins. *Journal of Structural Geology*, **32**, 235–248.

TINGAY, M., BENTHAM, P., DE FEYTER, A. & KELLNER, A. 2011. Present-day stress field rotations associated with evaporites in the offshore Nile Delta. *Geological Society of America Bulletin*, **123**, 1171–1180, doi:10.1130/B30185.1.

TOWNEND, J. & ZOBACK, M. D. 2000. How faulting keeps the crust strong. *Geology*, **28**, 399–402.

TEN VEEN, J. H., WOODSIDE, J. M., ZITTER, T. A. C., DUMONT, J. F., MASCLE, J. & VOLKONSKAIA, A. 2004. Neotectonic evolution of the Anaximander mountains at the junction of the Hellenic and Cyprus arcs. *Tectonophysics*, **391**, 35–65.

VENDEVILLE, B. C. 2005. Salt tectonics driven by sediment progradation: part I – mechanics and kinematics. *American Association of Petroleum Geologists Bulletin*, **89**, 1071–1079.

WERNICKE, B. 1981. Low-angle normal faults in the Basin and range province: Nappe tectonics in an extending orogen. *Nature*, **291**, 645–648.

WERNICKE, B. 1995. Low-angle normal faults and seismicity: s review. *Journal of Geophysical Research*, **100**, 20159–20174, doi:10.1029/95JB01911.

YALE, D. P. 2003. Fault and stress magnitude controls on variations in the orientation of in situ stress. *In*: AMEEN, M. S. (ed.) *Fracture and in-situ Stress Characterization of Hydrocarbon Reservoirs*. Geological Society, London, Special Publications, **209**, 55–64.

YASSIR, N. A. & ZERWER, A. 1997. Stress regimes in the Gulf Coast, offshore Louisiana: data from well-bore breakout analysis. *American Association of Petroleum Geologists Bulletin*, **81**, 293–307.

ZOBACK, M. D. 2000. Strength of the San Andreas. *Nature*, **405**, 31–32.

ZOBACK, M. D. 2007. *Reservoir Geomechanics*. Cambridge University Press, New York.

ZOBACK, M. D., ZOBACK, M. L. ET AL. 1987. New evidence on the state of stress of the San Andreas fault system. *Science*, **238**, 1105–1111.

ZOBACK, M. L., ANDERSON, R. E. & THOMPSON, G. A. 1981. Cainozoic evolution of the state of stress and style of tectonism of the basin and range province of the western United States. *In*: VINE, F. J. & SMITH, A. G. (eds) *Extensional Tectonics Associated With Convergent Plate Boundaries*. Proceedings of the Royal Society, London, 189–216.

Modelling of sediment wedge movement along low-angle detachments using ABAQUS™

A. TUITT[1,2]*, R. KING[3], T. HERGERT[4], M. TINGAY[1] & R. HILLIS[5]

[1]*Australian School of Petroleum, University of Adelaide, SA 5005, Australia*

[2]*Present address: BP Trinidad and Tobago LLC, 5a Queen's Park West, Port-of-Spain, Trinidad and Tobago*

[3]*School of Earth Sciences, University of Adelaide, SA 5005, Australia*

[4]*Karlsruhe Institute of Technology, Karlsruhe, Germany*

[5]*Deep Exploration Technologies Cooperative Research Centre, c/o University of Adelaide, SA 5005, Australia*

**Corresponding author (e-mail: adrian.tuitt@bp.com)*

Abstract: Delta–deepwater fold–thrust belts (DDWFTBs) develop over low-angle detachment faults which link extension to downslope contraction. Detachment faults have been examined in previous studies for the Amazon Fan, Niger, Nile, Angola, Baram and Bight Basin DDWFTBs. The driving mechanisms for the movement along the detachment remain uncertain, however. Previous authors have attributed the movement along detachment faults to high pore-fluid pressure, which reduces the effective normal stress acting on a fault surface thereby encouraging sliding along the fault. However, high pore-fluid pressure has not been directly confirmed in many of these faults due to a lack of well data in detachment surfaces. In this study, finite element modelling was used to test the effects of pore-fluid pressure, coefficient of friction, sediment rigidity and sediment wedge angle on sliding along the detachment. The modelling suggests that increased pore-fluid pressures and decreased coefficients of friction increase slip along a detachment. At hydrostatic pore-fluid pressures, sediment rigidity and sediment wedge angle have relatively little effect on the movement of the sediment wedge along the detachment. Modelling of these conditions using ABAQUS™ improves our understanding of the nature and mechanics of DDWFTBs and their underlying detachments.

Delta–deepwater fold–thrust belts (DDWFTBs; Fig. 1) are located on many passive continental margins and are characterized by an extensional zone coupled to downslope contraction at the delta toe via a detachment surface (Dailly 1976; Mandl & Crans 1981; Morley 2003; Rowan *et al.* 2004; Bilotti & Shaw 2005; King *et al.* 2009). These include the Amazon Fan (Cobbold *et al.* 2004), Niger (Bilotti & Shaw 2005), Nile (Tingay *et al.* 2011), Namibia (de Vera *et al.* 2010), Angola (Hudec & Jackson 2004), Baram (King *et al.* 2009) and the Bight Basin (Totterdell & Krassay 2003) DDWFTBs. Movement along detachments in DDWFTBs is suggested to be driven by a combination of gravity sliding above a seawards-dipping detachment and gravity spreading or vertical collapse of a sediment wedge under its own weight (Mandl & Crans 1981; Schultz-Ela 2001; Rowan *et al.* 2004). Far-field stresses, together with gravity tectonics, can however play a role in instigating movement of the sediment wedge along detachment surfaces as is observed in the Baram delta system in NW Borneo (Tingay *et al.* 2005; Franke *et al.* 2008; Morley *et al.* 2008; King *et al.* 2009, 2010).

Previous studies have examined the movement of a sediment wedge on a detachment surface in terms of the critical taper wedge theory (Davis *et al.* 1983; Dahlen *et al.* 1984; Dahlen 1990). Critical taper wedge theory states that a wedge of less than critical taper, when pushed along a detachment (slope = β), will deform internally to steepen its surface slope (α) (Fig. 2a) until the critical taper angle is attained. A critically tapered wedge that is not accreting new sediment is the thinnest body that can slide along its basal detachment without any internal deformation (Davis *et al.* 1983). Submarine sediment wedges have varying surface slope angles (α) and dips of detachment (β) (Fig. 2b), and at critical taper are hypothesized to be on the verge of Coulomb failure everywhere (Davis *et al.* 1983; Dahlen *et al.* 1984).

The nature of detachment surfaces varies across DDWFTBs (Rowan *et al.* 2004; Morley *et al.* 2011). Salt detachments are commonly observed

From: Healy, D., Butler, R. W. H., Shipton, Z. K. & Sibson, R. H. (eds) 2012. *Faulting, Fracturing and Igneous Intrusion in the Earth's Crust*. Geological Society, London, Special Publications, **367**, 171–183. http://dx.doi.org/10.1144/SP367.12

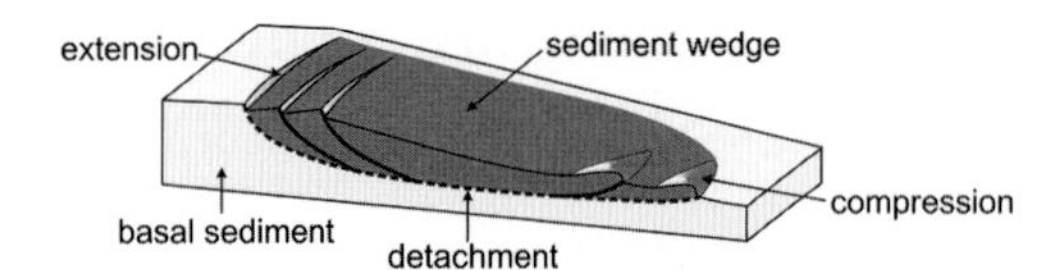

Fig. 1. Simplified diagram of extension and contraction in delta–deepwater fold–thrust belts (DDWFTBs) (Dailly 1976; Mandl & Crans 1981; Morley 2003; Rowan *et al.* 2004; Bilotti & Shaw 2005; King *et al.* 2009). Note the detachment between the sediment wedge and basal sediment. This simplified schematic is the basis for the model geometry used in this paper.

such as in the Mississippi Fan, the Perdido and Port Isabel fold–thrust belts of the northern Gulf of Mexico (Trudgill *et al.* 1999; Rowan *et al.* 2004), the Nile Delta (Rowan *et al.* 2004; Tingay *et al.* 2011) and the Kwanza Basin, Angola (Hudec & Jackson 2004). DDWFTBs can also have detachments comprising overpressured shale such as in the Niger Delta (Bilotti & Shaw 2005), the Bight Basin Delta (Totterdell & Krassay 2003), the Baram Delta (Tingay *et al.* 2009), the McKenzie Delta, Arctic Canada (Rowan *et al.* 2004) and in the Orange Basin, offshore Namibia (de Vera *et al.* 2010). Detachment surfaces are also present in other lithologies such as sandstone in the Amazon Fan DDWFTB and can exist as low-angle fault surfaces (Cobbold *et al.* 2004). The detachment surface can be basinwards- or landwards-dipping (Rowan *et al.* 2004; Morley *et al.* 2011). It remains uncertain what facilitates the gravitational movement of sediment wedges along these detachment surfaces. For salt detachments, it has been proposed that the movement of viscous salt results in the movement of the overlying sediment wedge (Rowan *et al.* 2004; Loncke *et al.* 2006; Tingay *et al.* 2012). Movement along detachments in other lithologies, such as shale and sandstone, have been ascribed to the presence of high pore-fluid pressures (Cobbold *et al.* 2004; Bilotti & Shaw 2005; de Vera *et al.* 2009).

This study investigates the movement of a sediment wedge along a detachment surface. It is clear that the nature of the detachment surface is critical to the development of DDWFTBs, as the detachment surface links processes of extension at the head to compression at the delta toe. Movement along the detachment surface is thus important for the formation of deepwater fold–thrust belts in delta systems. Our study hopes to highlight the conditions that most influence sliding along detachment surfaces.

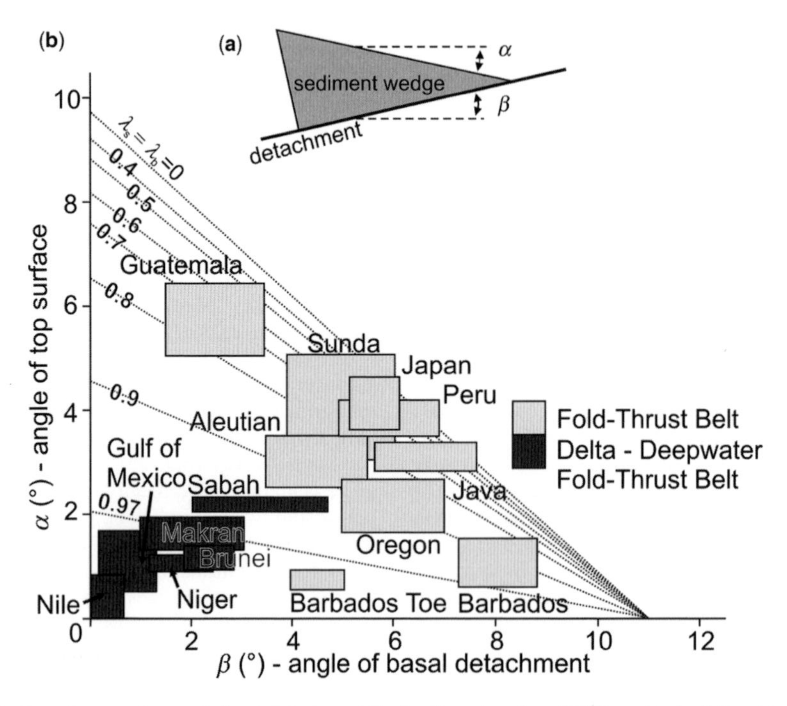

Fig. 2. (**a**) Surface slope angle α and dip of detachment β of sediment wedge. (**b**) Theoretical relationship between λ (pore pressure/lithostatic pressure) and α and β angles for accretionary prisms and DDWFTBs (modified from Davis *et al.* 1983; after Morley 2007 and Morley *et al.* 2011). Computation of λ assuming a coefficient of friction of 0.85 for the basal detachment.

In this study we use numerical modelling to construct a sediment wedge of known slope (α) and dip (β) angles and simulate the movement of this wedge along a simple detachment surface. Numerical modelling has been previously used to investigate strain and stress distributions developed within brittle faults (Ellis & Stöckhert 2004; Hetland & Hager 2006; Nüchter & Ellis 2010). Numerical models have also been used to investigate pore-fluid pressure generation in shale during the growth of a DDWFTB (Albertz *et al.* 2010; Ings & Beaumont 2010) and salt tectonics in DDWFTBs driven by sediment loading (Gemmer *et al.* 2004; Gemmer *et al.* 2005; Dirkzwager & Dooley 2008). To our knowledge however, DDWFTBs have never been numerically modelled as a sediment wedge on a brittle fault detachment as is evident in the Amazon Fan DDWFTB. This study uses numerical modelling to alter the pore-fluid pressure of the sediment, the coefficient of friction of the detachment, the rigidity of the sediments and the angle of the sediment wedge overlying a brittle detachment surface.

The aim of this study is to ascertain the conditions that encourage movement of sediment wedges along detachment surfaces. By changing these parameters we aim to gain a better understanding of the conditions that facilitate movement along detachment surfaces in DDWFTBs. This study does not take into account any external (far-field) stresses driving movement of the wedge and only uses the force of gravity to 'push' the sediment wedge along the detachment. In addition, while many DDWFTBs contain detachments comprising thick units of shale or salt, the detachment surface in this study is assumed as a brittle fault surface. The assumption of a brittle fault surface allows an additional parameter of coefficient of friction along the detachment to be examined.

Modelling method

Finite-element modelling of DDWFTBs was undertaken using the software package ABAQUS™ (Hibbitt *et al.* 2001). ABAQUS™ is a well-recognized industry standard finite-element modelling program that delivers accurate and robust high-performance solutions for simulations. The program allows models to be created and their geometries and parameters altered. Rock properties, such as density and pore-fluid pressure, can be assigned to the model that consists of individual elements. Model construction utilizes the modules of ABAQUS™: Part, Assembly, Step, Interaction, Property, Load and Mesh. The ABAQUS™ Visualization module is used to view the model results. ABAQUS™ was used to construct a simple two-part model of a DDWFTB.

Geometry and rock properties

The size and geometry of the two-dimensional (2D) model is based on the geometry of the Amazon Fan DDWFTB (Cobbold *et al.* 2004; Fig. 3a). The model consists of two parts: a sediment wedge and an underlying basal sediment block (Fig. 3b). The contact between the two parts is modelled as a frictional surface and represents the detachment in DDWFTBs. The contact surface is located at least 90 km from the sides and bottom of the model to reduce boundary effects on the contact surface. The model is 394 km long and has a maximum height of 100 km. The sediment wedge has a surface slope angle α of 1.2° and a detachment slope β of +0.8° (Fig. 3b). The β angle is positive as the detachment dips landwards in the Amazon Fan DDWFTB used as the basis for this model. The model was constructed using a quad mesh with 46,522 elements and 47,202 nodes.

The parts were assigned properties of density, pore-fluid pressure, sediment rigidity (Young's Modulus) and Poisson's ratio and were allowed to deform by a combination of linear elastic and Drucker–Prager failure. For Drucker–Prager failure criteria, the yield stresses and shear strains were based on experimental data from the Gosford sandstone (Ord *et al.* 1991). The detachment surface between the two parts was modelled as a frictional fault with a defined coefficient of friction.

Boundary conditions and stresses

Boundary conditions were applied to the model to best match the conditions present in passive-margin DDWFTBs without interference from far-field tectonic stresses. The base of the model was fixed to prevent any vertical movement of the basal sediment. In addition, the right and left sides of the basal block were prevented from moving horizontally. These conditions ensured minimal movement of the basal sediment. This is consistent with DDWFTBs in which the principal movement occurs with the sediment wedge moving above the detachment, resulting in extension at the head and compression at the toe.

Gravity was the only force applied to the model in this study. Gravity is applied as a distributed load in ABAQUS™. The procedure used in this study ensured that the model was in equilibrium with gravity and did not compact when gravity was applied ('gravitational pre-stressing'). Firstly, gravity was assigned to all elements of the model and each node of the model subjected to pore-fluid pressure. A rough interaction (infinite friction) was also applied to the detachment and each part of the model given a specific Young's Modulus. Model-derived stresses were then obtained at the

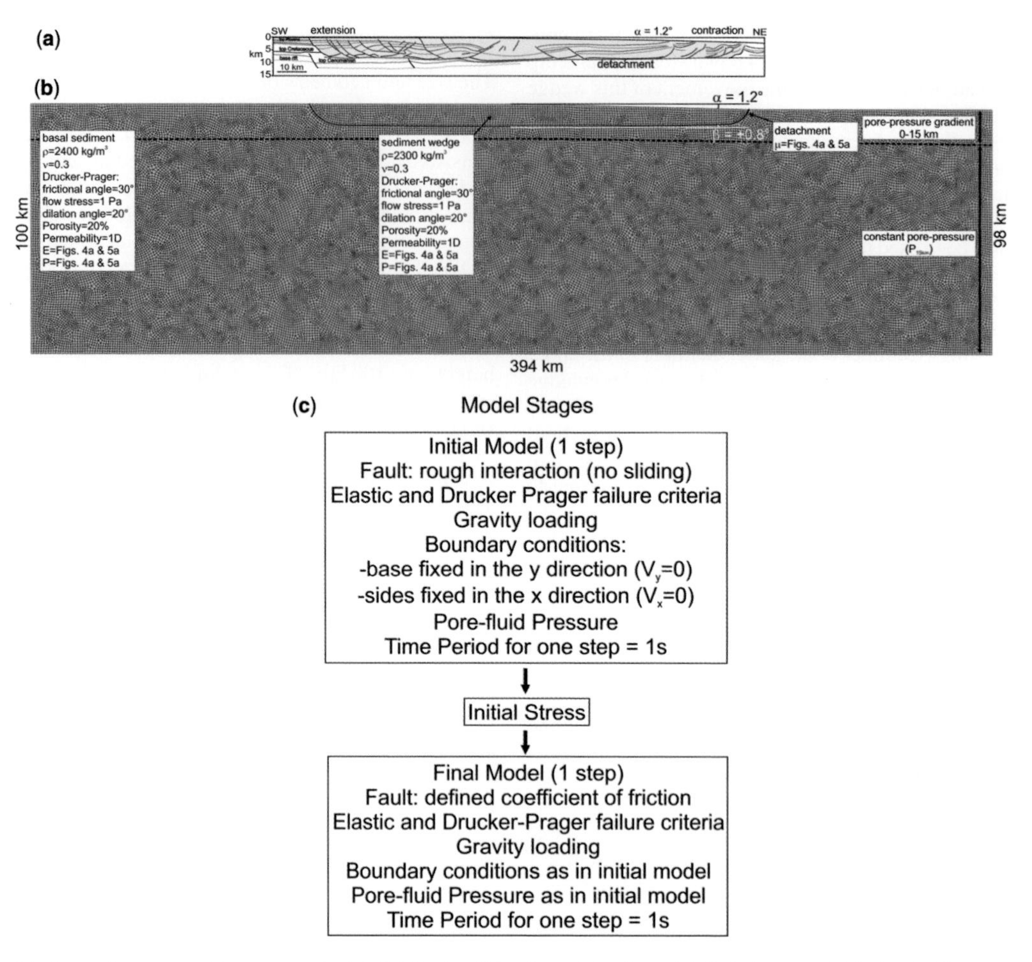

Fig. 3. (**a**) Geometry of the Amazon Fan DDWFTB (modified from Cobbold *et al.* 2004). Note the highlighted sediment wedge. (**b**) Quad mesh and geometry of ABAQUS^{TM} 2D model. The model consists of two parts: the sediment wedge and the basal sediment (μ, coefficient of friction; ρ, density; ν, Poisson's ratio; E, Young's Modulus; P, pore-fluid pressure; P_{15km}, pore-fluid pressure at 15 km depth for a given pore-fluid pressure gradient). (**c**) Stages involved in the production of models.

integration points and were applied as the initial stresses to a model containing the same pore-fluid pressures and sediment rigidities. A frictional contact was then assigned to the detachment surface of this model. Any displacement in the model when gravity is subsequently applied is therefore the result of sliding along the modelled fault rather than compaction due to gravity. The model stages are summarized in Figure 3c.

Parameters tested

The aim of the study was to quantitatively examine how different parameters influence sliding of a sediment wedge along a detachment surface. The conditions that encourage movement of sediment wedges are therefore considered to be key influences in the development of DDWFTBs. The first model built (Fig. 3b) was assigned a uniform hydrostatic pressure gradient (9.81 MPa km^{-1}), relatively low sediment rigidity and a coefficient of friction of 0.6. Hydrostatic pore-fluid pressure is the pressure exerted by a column of groundwater under the influence of gravity (Davis & Reynolds 1996) and was selected to represent the normal pore-fluid pressure of sediments containing no overpressures. Young's Moduli of 3 and 4 GPa were assigned to the sediment wedge and the basal sediment, respectively, based on the low sediment rigidities of the Baram Delta (King *et al.* 2010). A coefficient of friction of 0.6 represents the lower bound of common coefficients of friction observed in a wide variety of rock types from hundreds of sliding friction experiments (Byerlee 1978). The parameters of pore-fluid pressure, coefficient of friction, sediment rigidity and sediment wedge orientation were then systematically varied to examine the influence and sensitivity of the model to each variable.

Pore-fluid pressure

High pore-fluid pressures play a major role in the movement of faults as they offset lithostatic normal stress so that slip along faults can occur at lower stresses (Hubbert & Rubey 1959; Rice 1992). Analogue modelling has demonstrated that high pore pressures are correlated to fault weaknesses (Rice 1992). Overpressures have been directly observed in the Franciscan Complex, which hosts the San Andreas Fault (Berry 1973), and have been inferred to result in the low shear strength of this fault (Hardebeck & Hauksson 1999). Numerical modelling has predicted that the detachment fault beneath the Barbados Ridge Complex is a zone of very high pore-fluid pressure (Shi & Wang 1988). In DDWFTBs, high pore pressures have been recorded in the pro-delta shales (detachment zone) of the Baram Delta (Morley 2007; Tingay *et al.* 2007) and in sediment wedges of the Niger Delta (Bilotti & Shaw 2005) and the Amazon Fan Delta (Cobbold *et al.* 2004). However, the effects of these high pore-fluid pressures in the detachment zone beneath the sediment wedges still remain uncertain. Bilotti & Shaw (2005) predicted high pore-fluid pressures, close to lithostatic, of the basal detachment of the Niger Delta system based on the critical taper wedge theory. In this study we tested the effects of pore-fluid pressure on the movement of the sediment wedge along the detachment surface. Increased pore-fluid pressure gradients of 16.7 and 22.63 MPa km^{-1} were applied to the model to examine the effects of higher pore-fluid pressure on faulting along the detachment (Table 1, Fig. 4). The maximum pore-fluid pressure gradient (22.63 MPa km^{-1}) represents the lithostatic stress gradient. Lithostatic stress is derived from the weight of the entire column of overlying rock; here we used the commonly assumed gradient of 22.63 MPa km^{-1}, which corresponds to 1.0 psi ft^{-1} for an average sediment density of 2.3 g cm^{-3} (Dickinson 1953). Pore-fluid pressures approaching lithostatic pore pressures are thought to develop from disequilibrium compaction (Terzaghi & Peck 1948; Osborne & Swarbrick 1997), hydrocarbon generation (Osborne & Swarbrick 1997; Cobbold *et al.* 2004) and clay diagenesis (Mouchet & Mitchell 1989). For a particular pore pressure gradient, the pore-fluid pressure at each node was calculated as ρz (where ρ is pore-pressure gradient and z is depth of the node).

Sediment rigidity

Young's Modulus was used as a measure of sediment rigidity in this study. The Young's Modulus is the ratio of applied uniaxial stress to the uniaxial strain experienced by a material (Engelder & Marshak 1988; Davis & Reynolds 1996). Young's Modulus represents the stiffness of the material, or the ease at which the material undergoes strain for a given stress. It follows that soft rocks, such as sedimentary and highly fractured rocks, have a low Young's Modulus while stiff rocks, such as dense and non-fractured igneous and metamorphic rocks, have a high Young's Modulus (Gudmundsson 2004).

The rigidities (Young's Moduli) of the wedge and the basal sediment were altered from the initial rigidities of 3 and 4 GPa, respectively, to test the effect of sediment rigidity on displacement along the detachment. Values of 0.3 and 0.4 GPa

Table 1. *Model results of horizontal and vertical displacements for varying parameters of pore pressure, sediment rigidity and coefficient of friction (*P*, pore-fluid pressure gradient;* E_s*, Young's Modulus of sediment wedge;* E_b*, Young's Modulus of basal sediment;* μ_b*, coefficient of friction basal detachment)*

Parameter tested	Model	Slope angle ($\alpha°$)	P (MPa/km)	E_s (GPa)	E_b (GPa)	μ_b	Maximum horizontal displacement of sediment wedge (m)	Maximum vertical displacement at head of wedge (−m)	Maximum vertical displacement at toe of wedge (+m)
Pore-fluid pressure	1	1.2	9.81	3.0	4.0	0.6	2.0	2.9	0.0
	2	1.2	16.70	3.0	4.0	0.6	4.3	9.9	0.0
	3	1.2	22.63	3.0	4.0	0.6	624.9	614.3	400.3
Sediment rigidity	4	1.2	9.81	0.3	0.4	0.6	2.1	2.9	0.0
	5	1.2	9.81	30.0	40.0	0.6	2.0	0.3	0.0
Coefficient of friction	6	1.2	9.81	3.0	4.0	0.2	4.0	9.4	3.8
	7	1.2	9.81	3.0	4.0	0.4	2.4	4.6	0.0

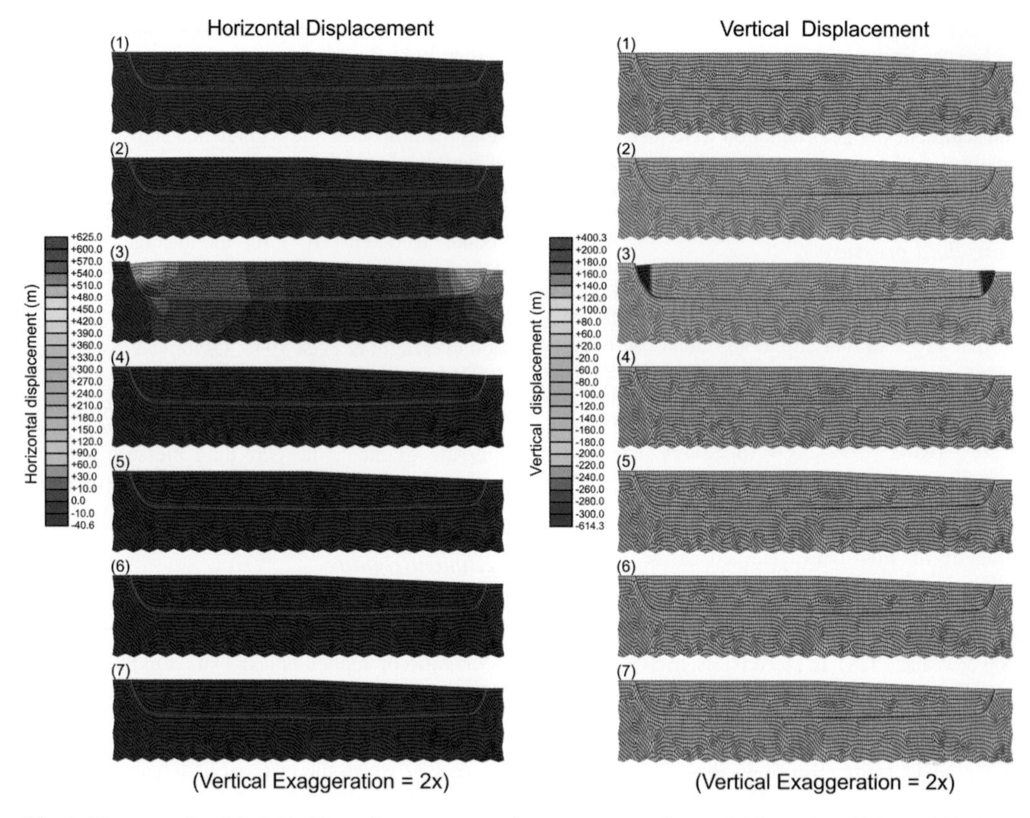

Fig. 4. Diagrams of models 1–7 with varying parameters of pore pressure, sediment rigidity and coefficients of friction and slope surface angle. Parameters for each model are listed in Table 1.

were assigned to the sediment wedge and the basal sediment, respectively, to represent softer rock. These values were increased to 30 GPa (sediment wedge rigidity) and 40 GPa (basal sediment rigidity) to represent very stiff rocks (Table 1).

Coefficient of friction

The coefficient of friction is the ratio of shear stress to normal stress required to initiate slip on an optimally oriented plane (Davis & Reynolds 1996). There is outcrop and numerical evidence to suggest that coefficient of friction plays a major role in influencing the movement of faults. Talc formed from the dehydration of serpentinite grains in a section of the San Andreas Fault in central California has been proposed as a reason for creep and low shear strength of the fault (Moore & Rymer 2007). Modelling of heat-flow measurements (d'Alessio *et al.* 2006) and stress data (Chéry *et al.* 2004) suggest an effective coefficient of friction of 0.1 and <0.1 respectively for the San Andreas Fault. This is consistent with friction experiments at room temperature that reveal a coefficient of friction in the range 0.16–0.23 for water-saturated talc gouge (Moore & Lockner 2008). Some experiments indicated that talc has a coefficient of friction <0.1 at slow strain rates at even higher temperatures of 100–400 °C (Moore & Lockner 2008).

Detachment zones in DDWFTBs are present within salt (Trudgill *et al.* 1999; Rowan *et al.* 2004), shale (Bilotti & Shaw 2005; Tingay *et al.* 2009) and sandstone (Cobbold *et al.* 2004) lithologies. However, the specific mineralogy within detachment zones of DDWFTBs remains uncertain. In this study, we model the effects of varying coefficient of friction of the modelled detachment to represent different mineralogies and lithologies. The coefficients of friction used were 0.6, 0.4 and 0.2. The minimum coefficient of friction (0.2) was selected to represent a detachment containing large thickness of gouge comprising clay minerals such as montmorillonite or vermiculite (Byerlee 1978; Suppe 1985). A coefficient of friction of 0.4 lies between a coefficient of friction (0.2) and the average coefficient of friction of 0.6 present within rock (Byerlee 1978), and thus provided a good

Table 2. *Model results of horizontal and vertical displacements for varying coefficients of friction at a pore-fluid pressure gradient of 16.7 MPa km^{-1}* (P, *pore-fluid pressure gradient;* E_s, *Young's Modulus of sediment wedge;* E_b, *Young's Modulus of basal sediment;* μ_b, *coefficient of friction basal detachment)*

Parameter tested	Model	Slope angle ($\alpha°$)	P (M Pa/km)	E_s (GPa)	E_b (GPa)	μ_b	Maximum horizontal displacement of sediment wedge (m)	Maximum vertical displacement at head of wedge (−m)	Maximum vertical displacement at toe of wedge (+m)
Coefficient of friction	8	1.2	16.70	3	4	0.6	5.2	9.9	0.0
	9	1.2	16.70	3	4	0.4	6.1	14.2	0.0
	10	1.2	16.70	3	4	0.2	13.3	27.8	25.5

average for examining coefficients of friction in our experiments.

Sediment wedge angle

We also tested the effect of changing the angle of the sediment wedge on sliding along the detachment. A new model was constructed that tilts the sediment wedge, increasing the slope angle α from 1.2° to 2.0° and changing the slope of the detachment β from +0.8° to 0° (Table 2, Fig. 5). The size and shape of the sediment wedge were maintained and the model was also allowed to deform by a combination of Drucker–Prager and elastic failure as defined in the original models. A hydrostatic pore pressure gradient was assigned to this model and the modelled detachment given a coefficient of friction of 0.6. These parameters were used to examine the effects of changing the slope of the wedge for sediments which are not overpressured.

Model results

The parameters of pore-fluid pressure, sediment rigidity, coefficient of friction and sediment wedge angle were each systematically varied in the models. The amount of movement of the sediment wedge along the detachment surface was ascertained by the degree of horizontal and vertical displacement of the wedge (Tables 1–3). Modelled negative vertical displacement at the head of the wedge represents extension while positive vertical displacement at the wedge toe represents compression. We now present the summary of the results.

Pore-fluid pressure

Models 1–3 (Fig. 4) were used to examine the effects of pore-fluid pressure on sediment wedge movement. The model results are summarized below.

(I) Higher pore-fluid pressure gradients result in greater displacements along the modelled detachment. The model with a hydrostatic pore-fluid pressure gradient, for example, predicts that no vertical movement of the wedge will occur at the toe (Model 1, Fig. 4), while lithostatic pressure gradient resulted in over 400 m of vertical displacement at the toe (Model 3, Fig. 4). The modelled maximum horizontal displacement of the sediment wedge

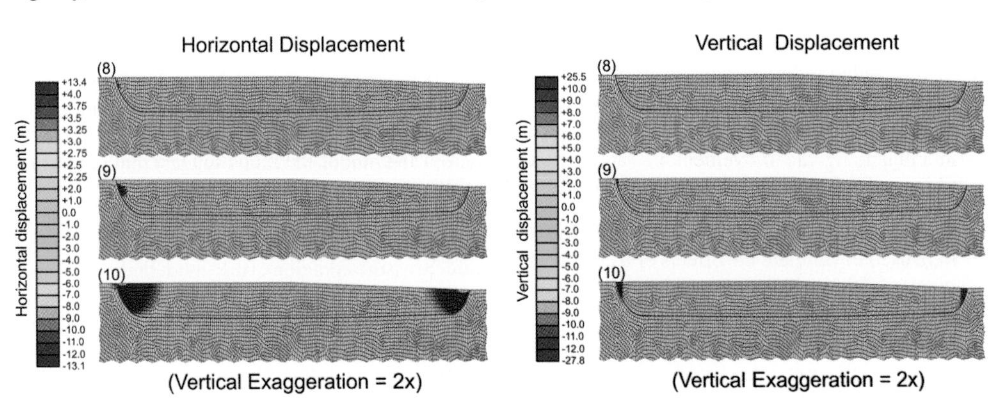

Fig. 5. Diagrams of models 8–10 with varying coefficients of friction. Parameters for each model are listed in Table 2.

Table 3. *Model results of horizontal and vertical displacements for varying sediment wedge angles at hydrostatic pore-fluid pressures (9.81 MPa km^{-1})* (P, *pore-fluid pressure gradient;* E_s*, Young's Modulus of sediment wedge;* E_b*, Young's Modulus of basal sediment;* μ_b*, coefficient of friction basal detachment)*

Parameter tested	Model	Slope angle ($\alpha°$)	Base angle ($\beta°$)	P (MPa/km)	E_s (GPa)	E_b (GPa)	μ_b	Maximum horizontal displacement of sediment wedge (m)	Maximum vertical displacement at head of wedge (−m)	Maximum vertical displacement at toe of wedge (+m)
Sediment wedge angle	1	1.2	0.8	9.81	3.0	4.0	0.6	2.0	2.9	0.0
	11	2.0	0.0	9.81	3.0	4.0	0.6	13.1	25.1	0.0

was also relatively insignificant in the hydrostatic pore-fluid pressure scenario, with just 2.0 m horizontal movement (Model 1, Fig. 4), while the lithosthatic pore-fluid pressure gradient model predicts a maximum horizontal displacement of over 625 m (Model 3, Fig. 4).

(II) The predicted amount of vertical displacement at the head (extension) exceeds the vertical displacement at the toe for all modelled pressure gradients. In the lithostatic gradient model, for example, the maximum vertical displacement of the head is 614 m, but drops to 400 m at the toe (Model 3, Fig. 4).

Sediment rigidity

Models 3–5 were used to examine the effects of sediment rigidity on predicted displacement along the detachment. The model results suggest little effect of sediment rigidity on sliding along the detachment for sediment at a hydrostatic gradient (Models 3–5, Table 1). The greatest predicted change is in the maximum vertical displacement at the head of the wedge, with an increase in Young's Modulus. With modelled Young's Moduli of 3 GPa (sediment wedge rigidity) and 4 GPa (basal sediment rigidity), the maximum vertical displacement at the head of the wedge is 2.9 m. However, this displacement decreases to 0.3 m when the modelled sediment wedge and basal sediment rigidities are increased to 30 GPa and 40 GPa, respectively (Models 3–5, Table 1).

Coefficient of friction

Model 3 and models 6–10 (Tables 1 & 2 and Figs 4 & 5) were used to simulate the effects of a change in coefficient of friction of the detachment surface. The results are summarized below:

(I) Modelled horizontal displacements along the detachment increase with decreasing coefficient of friction. For example, at a hydrostatic pore-fluid pressure gradient, a coefficient of friction of 0.2 results in a 4.0 m maximum horizontal displacement of the wedge (Model 6, Table 1) while a coefficient of friction of 0.6 only results in a displacement of 2.0 m (Model 1, Table 1). At a pore-fluid pressure gradient of 16.7 MPa km^{-1}, a coefficient of friction of 0.2 results in a maximum horizontal displacement of the wedge of 13.3 m (Model 8, Table 2) while a coefficient of friction of 0.6 predicts 5.2 m of maximum horizontal displacement (Model 10, Table 2).

(II) Maximum vertical displacements at the head of the wedge systematically increase with a decrease in coefficient of friction. This is evident at 16.7 MPa km^{-1} where a decrease in coefficient of friction from 0.6 to 0.2 results in an increase in vertical displacement at the head from 9.9 m to 27.8 m (Table 2).

(III) Maximum vertical displacement at the toe of the wedge may not change due to a decrease in coefficient of friction. At 16.7 MPa km^{-1}, for example, coefficients of friction of 0.6 and 0.4 both predict no vertical movement at the toe of the wedge (Models 9 & 10, Table 2, Fig. 5).

Sediment wedge angle

Models 1 and 11 were used to examine the effects of the dip of the sediment wedge on movement of the sediment wedge along the detachment (Table 3, Fig. 6). The modelled results suggest that increasing the slope angle and decreasing the landwards dip of the detachment yields greater vertical displacements at the head of the wedge (Table 3, Fig. 6). At a hydrostatic pressure gradient, the modelled maximum vertical displacement at the head is 25.1 m at $\alpha = 2.0°$ and $\beta = 0°$, compared to a vertical displacement of 2.0 m at the head for $\alpha = 1.2°$ and $\beta = 0.8°$ (Table 3, Fig. 6). However, the model predicts that an increase in slope angle to as much as 2.0° and a decrease in the landwards dip of the

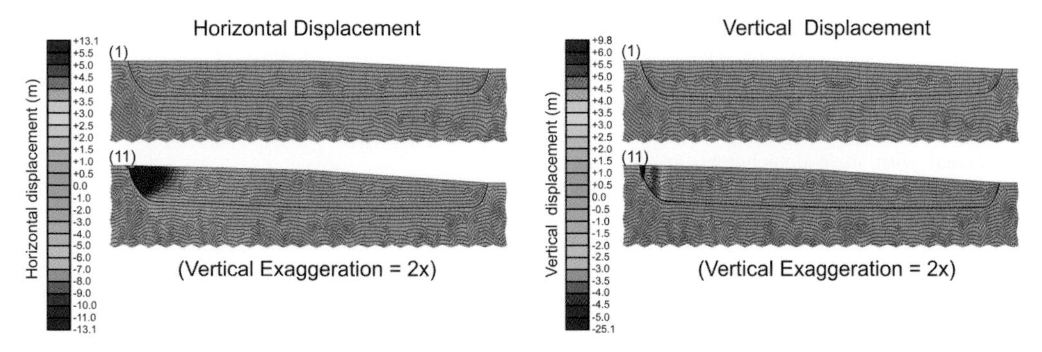

Fig. 6. Diagrams of models 1 and 11 with varying surface slopes α and dips of the detachment β. Parameters for each model are listed in Table 3.

detachment to 0° result in no horizontal displacement away from the head of the wedge and no vertical displacement at the wedge toe (Model 11, Table 3, Fig. 6). The modelled new angles of the sediment wedge has however created a 5 km wide zone of vertical uplift (up to 9.8 m uplift) downslope and adjacent to the area of maximum vertical downthrow at the head of the wedge (Model 11, Fig. 6).

Discussion

The results of the modelling suggest that movement of a sediment wedge along a detachment surface is enhanced by high pore-fluid pressures and low coefficients of friction. An increase in the surface slope of the sediment wedge and reduction in the landwards dip of the detachment has also predicted an increase in displacement at the head of the wedge at hydrostatic pressure. Changes in sediment rigidity results in relatively minimal changes in displacement of the wedge for sediment at hydrostatic pressure. These parameters are explored in real examples of DDWFTBs and other sediment wedges.

Pore-fluid pressure

The results clearly suggest that increasing pore-fluid pressure yields greater movement of the sediment wedge along the detachment surface. Movement along the basal detachment in the Amazon Fan DDWFTB, for example, has been attributed to high pore-fluid pressure due to gas generation (Cobbold *et al.* 2004). Well data in the sediment wedge of the Amazon Fan DDWFTB reveal significant overpressures of up to 18 MPa km^{-1} (Cobbold *et al.* 2004). However, there is no well data for the detachment surface. Bilotti & Shaw (2005) predicted that the basal detachment within the Akata Formation in the Niger DDWFTB is overpressured. As for the Amazon Fan DDWFTB, there is no direct evidence for overpressure in the detachment zone. However, the Akata Formation has low compressional wave speeds ($\leq$2500 m s^{-1}) at *c.* 6 km depth that are consistent with undercompaction and associated disequilibrium compaction overpressure. In the Baram DDWFTB, sublithostatic overpressures have been directly recorded by well data in the prodelta shales housing the detachment zone (Tingay *et al.* 2007, 2009). The modelling conducted here also indicates that there is greater vertical displacement at the head of the sediment wedge compared to the toe, which is consistent with observations in some DDWFTBs. Butler & Paton (2010) observed a mismatch between extension (44 km) and slip on thrusts (18–25 km) of a gravity-driven DDWFTB in the Orange Basin, offshore Namibia. They ascribe the disparity between extension and compression as the result of lateral compaction and volume loss. It should be noted, however, that some DDWFTBs show net compression. For example, compression exceeds extension in the Baram Delta and has been attributed to compression associated with far-field stresses (King *et al.* 2010).

Low coefficient of friction

Model results suggest that displacement along the detachment increases with decreasing coefficient of friction. Experiments have shown that fault surfaces containing clay minerals such as montmorillonite or vermiculite have low coefficients of friction (Byerlee 1978; Suppe 1985). Many detachments in DDWFTBs lie within clay-rich sequences, such as the Mexican Ridges and Port Isabel fold belts of the western Gulf of Mexico (Rowan *et al.* 2004), in the Segipe-Alagoas and Pará-Maranhão Basins in Brazil (Rowan *et al.* 2004), the Niger Delta (Rowan *et al.* 2004; Bilotti & Shaw 2005) and the

Baram Delta (Morley *et al.* 2008, 2011). The lower coefficients of friction associated with clay minerals may promote slip along some of these detachments. Experiments on marine sediments reveal low friction coefficients (below 0.25) for clay horizons associated with shale detachments in the Nankai and Barbados subduction thrusts (Kopf & Brown 2003). Some detachments, however, do not contain shale. The detachment of the Amazon Fan DDWFTB (upon which our model is based), for example, lies within Late Cretaceous sandstone (Cobbold *et al.* 2004); movement along this detachment may therefore be the result of other mechanisms.

Sediment rigidity (Young's Modulus)

Young's Modulus reflects the resistance of rock to elastic deformation (Davis & Reynolds 1996). In the Baram DDWFTB, log-derived Young's Moduli range from 1.12 to 6.45 GPa (King *et al.* 2010) representing sediments that are not well consolidated and thus deform relatively easily with a given stress (Gudmundsson 2004). However, the modelled results suggest no change in displacements at hydrostatic pressures with a ten-fold decrease in Young's Moduli from 3 and 4 GPa to 0.3 and 0.4 GPa of the sediment wedge and the basal sediment, respectively. The modelled results may indicate that, although softer sediments may deform more easily due to gravity, sliding along the detachment is mainly governed by pore-fluid pressure and coefficient of friction as these properties directly affect brittle faulting.

Sediment wedge angle

The model results predict higher displacements at the head of the wedge with an increased slope angle and a decrease in the landwards dip of the detachment. The tilting of the wedge may have brought the fault at the head of the wedge at a more favourable orientation for Andersonian faulting (based on Anderson 1951), thereby increasing the amount of slip at the head of the wedge. Our model results also indicate a zone of uplift adjacent to the area of maximum vertical displacement at the head of the wedge upon tilting. It is also evident that tilting of the wedge still resulted in no movement along the low-angle detachment and no displacement at the toe of the wedge at hydrostatic pressure. It is therefore our belief that while the tilt of the wedge resulted in the favourable orientation at the top end of the fault, the other areas of the detachment fault were still not favourably oriented to result in movement at hydrostatic pressures. This may have created a buttress against the maximum vertical displacement of the well-oriented section of the fault at the head of the wedge to result in adjacent compression and subsequent uplift. This may not be observed in DDWFTBs as there is movement along the detachment surface to prevent any buttressing effect.

Implications for critical taper wedge theory

Critical taper wedge theory states that a wedge of less-than-critical taper when pushed along a detachment will deform internally to steepen its surface slope α until the critical taper angle is attained (Davis *et al.* 1983). Based on this theory, higher pore-fluid pressures are predicted in detachments underlying wedges with lower slope angles (α) (Fig. 2b; Davis *et al.* 1983). Pore-fluid pressure observations in the Taiwan Accretionary Prism and the Amazon Fan DDWFTB are consistent with the critical taper wedge theory. The Taiwan Accretionary Prism has a steeper slope angle α of *c.* 3° and a dip of the detachment β of *c.* 6° (Davis *et al.* 1983) compared to the Amazon Fan DDWFTB with a slope angle α of 1.2° and a detachment dip β of 0.8°. In the Taiwan Accretionary Prism, the Chinshui Shale detachment in the Chenglungpu Fault exhibits only hydrostatic pore-fluid pressure (Yue 2007; cited in Suppe 2007) while well data in the sediment wedge of the Amazon Fan DDWFTB indicate that the sediments at 4 km depth (*c.* 5 km above the detachment) are overpressured (Cobbold *et al.* 2004). Our modelled results are also consistent with the critical taper wedge theory in that significant movement of a sediment wedge with a relatively small slope angle (1.2°) along the detachment is achieved only at high pore-fluid pressures.

In addition to pore-fluid pressure, coefficient of friction of the detachment also plays a role in determining the shape of the sediment wedge. Experiments conducted by Huiqui *et al.* (1992), for example, show that low coefficients of friction of the detachment result in the formation of sediment wedges with lower slope angles (α). Our modelled results suggests that low coefficients of friction of the detachment are needed for significant movement of an overlying sediment wedge with a relatively small slope angle (1.2°), in agreement with the critical taper wedge theory.

In our study a DDWFTB was modelled as a simple wedge on a basal detachment. In reality, however, deformation takes place within the sediment wedge with extensional faults at the head and reverse faults at the toe as observed in DDWFTBs worldwide (Dailly 1976; Mandl & Crans 1981; Morley 2003; Rowan *et al.* 2004; Bilotti & Shaw 2005; King *et al.* 2009). According to critical taper wedge theory, the sediment wedge deforms internally to attain critical taper (Davis *et al.* 1983; Dahlen *et al.* 1984). The surface slope α produced drives gravitational spreading of the sediment wedge along the detachment surface (Rowan *et al.* 2004). In our models, the lack of internal deformation within the sediment wedge is therefore a limitation to application to critical taper wedge theory. Our modelling however shows that, despite internal deformation, the wedge is actively attempting to produce lower slope angles at higher pore-fluid pressure and lower coefficients of friction with greater movements along the detachment surface.

Conclusion

Gravity forces have acted on continental margins to result in the formation of DDWFTBs. These systems are characterized by extension at the head and compression at the toe of a sediment wedge. In this study we have constructed finite element models using ABAQUS™ to investigate the conditions influencing slip on basal detachments in DDWFTB systems. The model results suggest that the development of DDWFTBs is best when sediments (and by extension the detachment surface) have high pore-fluid pressures. This is consistent with results derived from numerical and analogue models by previous authors (Hubbert & Rubey 1959; Rice 1992) and direct measurements within sediment wedges of DDWFTBs (Cobbold *et al.* 2004; Tingay *et al.* 2009). The model results also suggest that a decrease in coefficient of friction also increases the displacement of the sediment wedge along the detachment surface. Previous work on outcrops has shown that low coefficients of friction have resulted in easier fault movement (Moore & Rymer 2007). An increase in sediment rigidity and tilting of sediment wedge under hydrostatic pressure conditions have relatively little effect on sliding along the detachment, emphasizing the major roles of higher pore-fluid pressures and lower coefficients of friction to sediment wedge movement. Our model results indicate that the sediment wedge attempts to decrease slope surface angle with increasing pore-fluid pressures and decreasing coefficients of friction consistent with critical taper wedge theory. The lack of internal deformation in the sediment wedge, however, limits some application of the models to the critical taper wedge theory. This shortcoming may be overcome by introducing more complex structures, such as normal and reverse faults, to the sediment wedge. In addition, future models of sediment wedges from different DDWFTBs and accretionary prisms can be compared to further understand critical taper wedge theory.

The modelled increase in sediment wedge movement under conditions of higher pore-fluid pressures and lower coefficients of friction is critical in our understanding of the origin and nature of DDWFTBs. DDWFTBs are the sites of major hydrocarbon exploration (Morley *et al.* 2011). The largest producing deepwater asset in the world is the Thunder Horse (1999) reserve of the Mississippi Fan Fold Belt, with an estimated 1 billion barrels of oil equivalent (bboe) (Cossey 2004). An understanding of the structure and origin of DDWFTBs is therefore important for the migration, accumulation and exploration of hydrocarbons. Modelling using ABAQUS™ has yielded possible conditions for such systems, and faults in general, to propagate in nature.

We are grateful to the Australian Research Council (ARC) for funding this project. This work represents TRaX contribution #212. We also thank Simulia for providing ABAQUS™.

References

ALBERTZ, M., BEAUMONT, C. & INGS, S. J. 2010. Geodynamic modeling of sedimentation-induced overpressure, gravitational spreading, and deformation of passive margin mobile shale basins. *In*: WOOD, L. (ed.) *Shale Tectonics*. American Association of Petroleum Geologists, Tulsa, Memoir, **93**, 29–62.

ANDERSON, E. M. 1951. *The Dynamics of Faulting and Dyke Formation with Applications to Britain*. Oliver & Boyd, Edinburgh.

BERRY, F. 1973. High fluid potential in California Coast Ranges and their tectonic significance. *American Association of Petroleum Geology Bulletin*, **57**, 1219–1249.

BILOTTI, F. & SHAW, J. H. 2005. Deepwater Niger Delta fold and thrust belt modelled as a critical-taper wedge: the influence of elevated basal fluid pressure on structural styles. *American Association of Petroleum Geology Bulletin*, **89**, 1475–1491.

BUTLER, R. W. H. & PATON, D. A. 2010. Evaluating lateral compaction in deepwater fold and thrust belts: how much are we missing from 'nature's sandbox'? *Geological Society of America Today*, **20**, 4–10.

BYERLEE, J. D. 1978. Friction of rocks. *Pure and Applied Geophysics*, **116**, 615–626.

CHÉRY, J., ZOBACK, M. D. & HICKMAN, S. 2004. A mechanical model of the San Andreas fault and SAFOD Pilot Hole stress measurements. *Geophysical Research Letters*, **31**, L15S13, doi: 10.1029/2004GL019521.

COBBOLD, P. R., MOURGUES, R. & BOYD, K. 2004. Mechanism of thin-skinned detachment in the Amazon Fan: assessing the importance of fluid overpressure and hydrocarbon generation. *Marine Petroleum Geology*, **21**, 1013–1025.

COSSEY, S. P. J. 2004. Deepwater discoveries toasted over 30 years: AAPG Explorer, http://www.aapg.org/explorer/2004/09sep/gom_history.cfm.

DAHLEN, F. A. 1990. Critical taper model of fold-and-thrust belts and accretionary wedges. *Annual Review Earth Planet Science*, **18**, 55–99.

DAHLEN, F. A., SUPPE, J. & DAVIS, D. 1984. Mechanics of fold-and-thrust belts and accretionary wedges: cohesive coulomb theory. *Journal of Geophysical Research*, **89**, 10 087–10 101.

DAILLY, G. C. 1976. A possible mechanism relating progradation, growth faulting, clay diapirism and overthrusting in the regressive sequence of sediments. *Bulletin of Canadian Petroleum Geology*, **24**, 92–116.

D'ALESSIO, M. A., WILLIAMS, C. F. & BÜRGMANN, R. 2006. Frictional strength heterogeneity and surface heat flow: implications for the strength of the creeping San Andreas fault. *Journal of Geophysical Research*, **111**, B05410, doi: 10.1029/2005JB003780.

DAVIS, D., SUPPE, J. & DAHLEN, F. A. 1983. Mechanics of fold-and-thrust belts and accretionary wedges. *Journal of Geophysical Research*, **88**, 1153–1172.

DAVIS, G. H. & REYNOLDS, S. J. 1996. *Structural Geology of Rocks and Regions*. 2nd edn. John Wiley and Sons, Inc., New York.

DE VERA, J., GRANADO, P. & MCCLAY, K. 2010. Structural evolution of the Orange basin gravity-driven system, offshore Namibia. *Marine Petroleum Geology*, **27**, 233–237.

DICKINSON, G. 1953. Geological aspects of abnormal reservoir pressures in Gulf Coast Louisiana. *American Association of Petroleum Geologists Bulletin*, **37**, 410–432.

DIRKZWAGER, J. & DOOLEY, T. P. 2008. In-situ stress modeling of a salt-based gravity driven thrust belt in a passive margins setting using physical and numerical modeling. *Proceedings of the 42nd US Rock Mechanics Symposium*. San Francisco, 29th June–2nd July 2008. American Rock Mechanics Association paper number 08–270.

ELLIS, S. & STÖCKHERT, B. 2004. Elevated stresses and creep rates beneath the brittle–ductile transition caused by seismic faulting in the upper crust. *Journal of Geophysical Research*, **109**, B05407, doi: 10.1029/2003JB002744.

ENGELDER, T. & MARSHAK, S. 1988. Analysis of data from rock-deformation experiments. *In*: MARSHAK, S. & MITRA, G. (eds) *Basic Methods of Structural Geology*. Prentice-Hall, Englewood Cliffs, NJ, 193–212.

FRANKE, D., BARCKHAUSEN, U., HEYDE, I., TINGAY, M. & RAMLI, N. 2008. Seismic images of a collision zone offshore NW Sabah/Borneo. *Marine Petroleum Geology*, **25**, 606–624.

GEMMER, L., INGS, S. J., MEDVEDEV, S. & BEAUMONT, C. 2004. Salt tectonics driven by differential sediment loading: stability analysis and finite-element experiments. *Basin Research*, **16**, 199–218.

GEMMER, L., BEAUMONT, C. & INGS, S. J. 2005. Dynamic modelling of passive margin salt tectonics: effects of water loading, sediment properties and sedimentation patterns. *Basin Research*, **17**, 383–402.

GUDMUNDSSON, A. 2004. Effects of Young's modulus on fault displacement. *Comptes Rendus Geoscience*, **336**, 85–92.

HARDEBECK, J. L. & HAUKSSON, E. 1999. Role of fluids in faulting inferred from stress field signatures. *Science*, **285**, 236–239.

HETLAND, E. A. & HAGER, B. H. 2006. Interseismic strain accumulation: Spin-up, cycle invariance, and irregular rupture sequences. *Geochemistry Geophysics Geosystems*, **7**, Q05004, doi: 10.1029/2005GC001087.

HIBBITT, D., KARLSSON, B. & SORENSEN, P. 2001. *Abaqus/Standard User's Manual, vol. 1 and 2, version 6.2*. Hibbitt, Karlsson and Sorensen Inc., Pawtucket, Rhode Island.

HUBBERT, M. K. & RUBEY, W. W. 1959. Role of fluid pressure in mechanics of overthrust faulting. *Geological Society of America Bulletin*, **70**, 115–166.

HUDEC, M. R. & JACKSON, M. P. A. 2004. Regional restoration across the Kwanza Basin, Angola: salt tectonics triggered by repeated uplift of a metastable passive margin. *American Association of Petroleum Geology Bulletin*, **88**, 971–990.

HUIQUI, L., MCCLAY, K. R. & POWELL, C. M. 1992. Physical models of thrust wedges. *In*: MCCLAY, K. R. (ed.) *Thrust Tectonics*. Chapman and Hall, New York, 71–81.

INGS, S. J. & BEAUMONT, C. 2010. Continental margin shale tectonics: preliminary results from coupled fluid-mechanical models of basin scale delta instability. *Journal of Geological Society, London*, **167**, 571–582.

KING, R. C., HILLIS, R. R., TINGAY, M. R. P. & MORLEY, C. K. 2009. Present-day stress and neotectonic provinces of the Baram Delta and deep-water fold–thrust belt. *Journal of Geological Society, London*, **166**, 1–4.

KING, R. C., BACKÉ, G., MORLEY, C. K., HILLIS, R. & TINGAY, M. 2010. Balancing deformation in NW Borneo: Quantifying plate-scale vs gravitational tectonics in a delta and deepwater fold–thrust belt system. *Marine Petroleum Geology*, **27**, 238–246.

KOPF, A. & BROWN, K. M. 2003. Friction experiments on saturated sediments and their implications for the stress state of the Nankai and Barbados subduction thrusts. *Marine Geology*, **202**, 193–210.

LONCKE, L., GAULLIER, V., MASCLE, J., VENDEVILLE, B. & CAMERA, L. 2006. The Nile deep-sea fan: An example of interacting sedimentation, salt tectonics, and inherited subsalt paleotopographic features. *Marine and Petroleum Geology*, **23**, 297–315.

MANDL, G. & CRANS, W. 1981. Gravitational gliding in deltas. *In*: MCCLAY, K. R. & PRICE, N. J. (eds) *Thrust and Nappe Tectonics*. Geological Society, London, Special Publications, **9**, 41–54.

MOORE, D. E. & RYMER, M. J. 2007. Talc-bearing serpentinite and the creeping section of the San Andreas fault. *Nature*, **448**, 795–797.

MOORE, D. E. & LOCKNER, D. A. 2008. Talc friction in the temperature range 25°–400°C: Relevance for fault-zone weakening. *Tectonophysics*, **449**, 120–132.

MORLEY, C. K. 2003. Mobile shale related deformation in large deltas developed on passive and active margins. *In*: VAN RENSBERGEN, P., HILLIS, R. R., MALTMAN, A. J. & MORLEY, C. K. (eds) *Subsurface Sediment Mobilization*. Geological Society, London, Special Publications, **216**, 335–357.

MORLEY, C. K. 2007. Interaction between critical wedge geometry and sediment supply in a deepwater fold belt, NW Borneo. *Geology*, **35**, 139–142.

MORLEY, C. K., TINGAY, M., HILLIS, R. & KING, R. 2008. Relationship between structural style, overpressures, and modern stress, Baram Delta Province, north west Borneo. *Journal of Geophysical Research*, **113**, doi: 10.1029/2007JB005324.

Morley, C. K., King, R., Hillis, R., Tingay, M. & Backe, G. 2011. Deepwater fold and thrust belt classification, tectonics, structure and hydrocarbon prospectivity: A review. *Earth-Science Reviews*, **104**, 41–91.

Mouchet, J. P. & Mitchell, A. 1989. *Abnormal pressures while drilling*. Manuels techniques, elf acquitaine.

Nüchter, J.-A. & Ellis, S. 2010. Complex states of stress during the normal faulting seismic cycle: Role of midcrustal postseismic creep. *Journal of Geophysical Research*, **115**, B12411, doi: 10.1029/2010JB007557.

Ord, A., Vardoulakis, I. & Kajewski, R. 1991. Shear band formation in Gosford Sandstone. *International Journal of Rock Mechanics and Mining Sciences and Geomechanics Abstracts*, **28**, 397–409.

Osborne, M. J. & Swarbrick, R. E. 1997. Mechanisms for generating overpressure in sedimentary basins: A reevaluation. *American Association of Petroleum Geology Bulletin*, **81**, 1023–1041.

Rice, J. R. 1992. Fault stress states, pore pressure distributions, and the weakness of the San Andreas fault. *In*: Evans, B. & Wong, T.-F. (eds) *Fault Mechanics and Transport Properties in Rocks*. Academic Press, San Diego, 475–503.

Rowan, M. G., Peel, F. J. & Vendeville, B. C. 2004. Gravity-driven fold belts on passive margins. *In*: McClay, K. R.(ed.) *Thrust Tectonics and Hydrocarbon Systems*. American Association of Petroleum Geologists, Tulsa, Memoir, **82**, 157–182.

Schultz-Ela, D. D. 2001. Excursus on gravity gliding and gravity spreading. *Journal of Structural Geology*, **23**, 725–731.

Shi, Y. & Wang, C.-Y. 1988. Generation of high pore pressure in accretionary prisms: inferences from the Barbados Subduction Complex. *Journal of Geophysical Research*, **93**, 8893–8910.

Suppe, J. 1985. *Principles of Structural Geology*. Prentice-Hall, Englewood Cliffs, NJ.

Suppe, J. 2007. Absolute fault and crustal strength from wedge tapers. *Geology*, **35**, 1127–1130.

Terzaghi, K. & Peck, R. B. 1948. *Soil Mechanics in Engineering Practice*. John Wiley, New York.

Tingay, M., Hillis, R., Morley, C., Swarbrick, R. & Drake, S. 2005. Present-day stress orientation in Brunei: a snapshot of 'prograding tectonics' in a Tertiary delta. *Journal of the Geological Society London*, **162**, 39–49.

Tingay, M., Hillis, R., Swarbrick, R., Morley, C. & Damit, A. 2007. Vertically transferred overpressures in Brunei: evidence for a new mechanism for the formation of high magnitude overpressures. *Geology*, **35**, 1023–1026.

Tingay, M., Hillis, R., Swarbrick, R., Morley, C. K. & Damit, A. 2009. Origin of overpressure and pore pressure prediction in the Baram Delta Province, Brunei. *American Association of Petroleum Geology Bulletin*, **93**, 51–74.

Tingay, M., Bentham, P., de Feyter, A. & Kellner, A. 2011. Present-day stress-field rotations associated with evaporites in the offshore Nile Delta. *Geological Society of America Bulletin*, **123**, 1171–1180.

Tingay, M., Bentham, P., de Feyter, A. & Kellner, A. 2012. Evidence for non-Andersonian faulting above evaporites in the Nile Delta. *In*: Healy, D., Butler, R. W. H., Shipton, Z. K. & Sibson, R. H. (eds) *Faulting, Fracturing and Igneous Intrusion in the Earth's Crust*. Geological Society, London, Special Publications, **367**, 155–170.

Totterdell, J. M. & Krassay, A. A. 2003. The role of shale deformation and growth faulting in the Late Cretaceous evolution of the Bight Basin, offshore southern Australia. *In*: Van Rensbergen, P., Hillis, R. R., Maltman, A. J. & Morley, C. K. (eds) *Subsurface Sediment Mobilization*. Geological Society, London, Special Publications, **216**, 429–442.

Trudgill, B. D., Rowan, M. G. *et al.* 1999. The perdido fold belt, Northwestern deep Gulf of Mexico, Part 1: Structural Geometry, evolution and regional implications. *American Association of Petroleum Geology Bulletin*, **83**, 88–113.

Yue, L. F. 2007. *Active structural growth in central Taiwan in relationship to large earthquakes and pre-fluid pressures*. PhD thesis, Princeton University.

On the nucleation of non-Andersonian faults along phyllosilicate-rich mylonite belts

ANDREA BISTACCHI[1]*, MATTEO MASSIRONI[2], LUCA MENEGON[2,3], FRANCESCA BOLOGNESI[1] & VALERIANO DONGHI[1]

[1]*Dipartimento di Scienze Geologiche e Geotecnologie, Università degli Studi di Milano Bicocca, Piazza della Scienza 4, 20126 Milano, Italy*

[2]*Dipartimento di Geoscienze, Università degli Studi di Padova, Via Gradenigo 6, 35131 Padova, Italy*

[3]*Institutt for Geologi, Universitetet i Tromsø, Dramsveien 201, 9037 Tromsø, Norway*

**Corresponding author (e-mail: andrea.bistacchi@unimib.it)*

Abstract: The weakness of fault zones is generally explained by invoking an elevated fluid pressure or the presence of extremely weak minerals in a continuous fault gouge horizon. This allows for faults to slip under an unfavourable normal to shear stress ratio, in contrast to E. M. Anderson's theory of faulting. However, these mechanisms do not explain why faults should nucleate in such an orientation as to make them misoriented and non-Andersonian. Here we present a weakening mechanism, involving the mechanical anisotropy of phyllosilicate-bearing mylonite belts, which is likely to influence the nucleation of faults in addition to their subsequent activity. Considering three natural examples from the Alps (the Simplon, Brenner and Sprechenstein-Mules fault zones) and a review of laboratory tests on anisotropic rocks, we apply anisotropic slip tendency analysis and show that misoriented weak faults can nucleate along a sub-planar phyllosilicate-rich mylonitic foliation, constituting a large-scale mechanical anisotropy belt and preventing the development of Andersonian optimally oriented faults.

A growing number of contributions deal with the problem of 'weak' faults, which are faults that show a weakness with respect to the surrounding (stronger) crust or faults (characterized by particularly weak materials) showing a lower resistance with respect to the standard brittle crust, as described by the empirical Byerlee law (Byerlee 1978). These two forms of weakness are referred to as 'relative' and 'absolute' weakness, respectively (e.g. Rice 1992). Focusing on relatively weak faults, the most common tool to assess their weakness is to demonstrate that they are unfavourably oriented with respect to the regional stress field, and thus experience an unfavourable tangential to normal stress ratio. These faults are also indicated as 'misoriented' (e.g. Sibson 1985). On the other hand, faults optimally oriented with respect to the active stress field are referred to as 'Andersonian' faults, after E. M. Anderson's theory of faulting (Anderson 1951). Typical misoriented faults are low-angle normal faults (LANFs; Wernicke 1981), high-angle reverse faults (Cox 1995; Butler *et al.* 2008) and strike-slip faults forming a high angle to the most compressional stress axis σ_1 (e.g. the San Andreas Fault; Townend & Zoback 2004).

Different weakening mechanisms have been proposed which generally solve the problem of unlocking and promoting slip along a pre-existing misoriented fault. They fall into two main classes: (1) mechanisms involving layers of particularly weak fault gouge, containing minerals with extremely low friction coefficients (e.g. Moore & Rymer 2007; Collettini *et al.* 2009; Smith & Faulkner 2010); and (2) mechanisms involving elevated fluid pressure (e.g. Byerlee 1990; Axen 1992; Rice 1992; Faulkner & Rutter 2001; Collettini *et al.* 2006). A third class of mechanisms predict a rotation of principal stress axes within the fault zone which may favour slip (e.g. Rice 1992). However, this rotation is due to a contrast in mechanical properties (the absolute weakness of the fault zone) so it might be better described as a secondary effect, enhancing relative weakness in presence of an absolute weakness.

In any case, most of these mechanisms fail in predicting *why* a misoriented fault should nucleate instead of an Andersonian fault. For instance, elevated fluid pressure conditions would decrease to the same degree the effective normal stress acting on an optimally oriented or on a misoriented fault. Hence, the nucleation and development of an optimally oriented fault will still be favoured. Following this reasoning, the occurrence of non-Andersonian faults should be limited to cases where the stress field first produces an Andersonian fault and then

From: Healy, D., Butler, R. W. H., Shipton, Z. K. & Sibson, R. H. (eds) 2012. *Faulting, Fracturing and Igneous Intrusion in the Earth's Crust.* Geological Society, London, Special Publications, **367**, 185–199. http://dx.doi.org/10.1144/SP367.13

the stress field or the fault is re-oriented in such a way as to make the fault misoriented. However, misoriented faults are quite common.

An additional weakening mechanism may be related to the mechanical anisotropy of foliated rocks, which has been characterized in triaxial tests since the 1960s (e.g. Paterson & Wong 2005, p. 38) but never quantitatively considered in the study of large-scale tectonic faults. In this contribution we quantitatively investigate this fourth weakening mechanism which is likely to influence the nucleation stage of brittle faults, making the occurrence of misoriented faults more probable than that of Andersonian faults. This mechanism is likely to operate under some common geological conditions which are encountered in collisional belts, rifts and also continental transform margins, where regional-scale mylonite belts are present. Using a modified slip tendency analysis (see Morris *et al.* 1996 for the original slip tendency analysis), we will show that a sub-planar phyllosilicate-rich metamorphic foliation (resulting in a large-scale mechanical anisotropy) can be preferentially activated in the brittle field although it is unfavourably oriented, and that this prevents the development of Andersonian optimally oriented faults. Compared to the common qualitative understanding that fabric anisotropy may influence different aspects of brittle faulting (e.g. Hobbs *et al.* 1976; Butler *et al.* 2008), in this contribution we will quantitatively show how, and under which conditions, the nucleation of brittle faults along misoriented anisotropy surfaces is more likely to occur than that of Andersonian faults.

Large-scale weak brittle faults in metamorphic terrains

In this section we characterize three case studies from the metamorphic core of the Alps. Two of these (the Simplon and Brenner lines) can be classified as typical LANFs while the third (the Sprechenstein-Mules fault) is a transpressional misoriented fault. All these faults developed in the last stages of the Europe–Adria continental collision during tectonic phases where the over-thickened orogen experienced a mixture of extension and strike-slip deformations during continuing plate convergence (Dal Piaz *et al.* 2003). The units involved in this kind of deformation are mainly those of the metamorphic core of the Alps (Argand 1916): Austroalpine Units of Adriatic origin (upper plate) and European Penninic Units (lower plate).

The Simplon Line

The Simplon Fault Zone (Bearth 1956) is a late-collisional low-angle normal fault of the Western Alps, cropping out at the boundary between Italy and Switzerland (Fig. 1). It is characterized by a 1-km-thick mylonitic horizon (called the Simplon Fault Zone or SFZ; Mancktelow 1985) exposed in the footwall, which evolved due to differential exhumation from conditions at the boundary between the amphibolite and greenschist facies towards lower temperatures (Mancktelow 1985, 1992; Mancel & Merle 1987; Steck 1987, 2008; Campani *et al.* 2010*a*). Structures related to subsequent stages of this deformation path, showing different mineral assemblages and characteristic microstructures, define a regional-scale shear zone that tends to localize in a progressively narrower horizon during cooling and exhumation, thus preserving higher-temperature mylonites at the periphery and showing rocks reworked at increasingly lower temperatures approaching the core (Fig. 1b). This makes the SFZ a classic example of a crustal-scale LANF (Wernicke 1981). A transition to brittle deformation mechanisms is observed at the lower boundary of the greenschist facies (*c.* 250–300 °C), resulting in the nucleation and propagation of cataclasite layers along the mylonitic foliation. These cataclasites mark the boundary between the footwall and the hanging wall and are indicated as the Simplon Line (SL, Fig. 1b; Bearth 1956; Mancktelow 1985, 1992). Regarding the hanging wall, which was exhumed before the main normal-fault activity of the SFZ, two interpretations have been published: Mancktelow (1985, 1992) holds that no ductile fabric directly related to the SFZ can be found here, while Steck (1987, 2008) interpreted the foliations in the hanging wall as related to an earlier dextral-extensional phase of a generalized Simplon shear zone. In any case (of relevance for our study), it must be noted that metamorphic fabrics in the hanging wall are generally at a high angle to the SL.

Mylonites of the SFZ (exposed in the footwall) result from the polyphase deformation of pelitic–psammitic and granitoid protoliths with subordinate carbonate rocks, belonging to the Lower Penninic Nappes of the Ossola–Tessin tectonic window. This corresponds to the lowermost structural level in the Alps, and has been exhumed in the last stages of the continental collision between *c.* 26 and 3 Ma. The final phases of this exhumation in the western part of the window took place along the SFZ and consist of a differential uplift of *c.* 15 km (Grasemann & Mancktelow 1993; Campani *et al.* 2010*b*).

The mylonites of the SFZ show a SW-dipping low-angle composite foliation, resulting from a very well-developed and penetrative SCC′ fabric. On average, dip/dip direction for S planes is 20/240, 30/235 for C planes and 45/230 for C′ planes (Fig. 1c). The foliation is characterized by alternating quartz-feldspar and phyllosilicate-rich layers, plus accessory phases, occurring in different

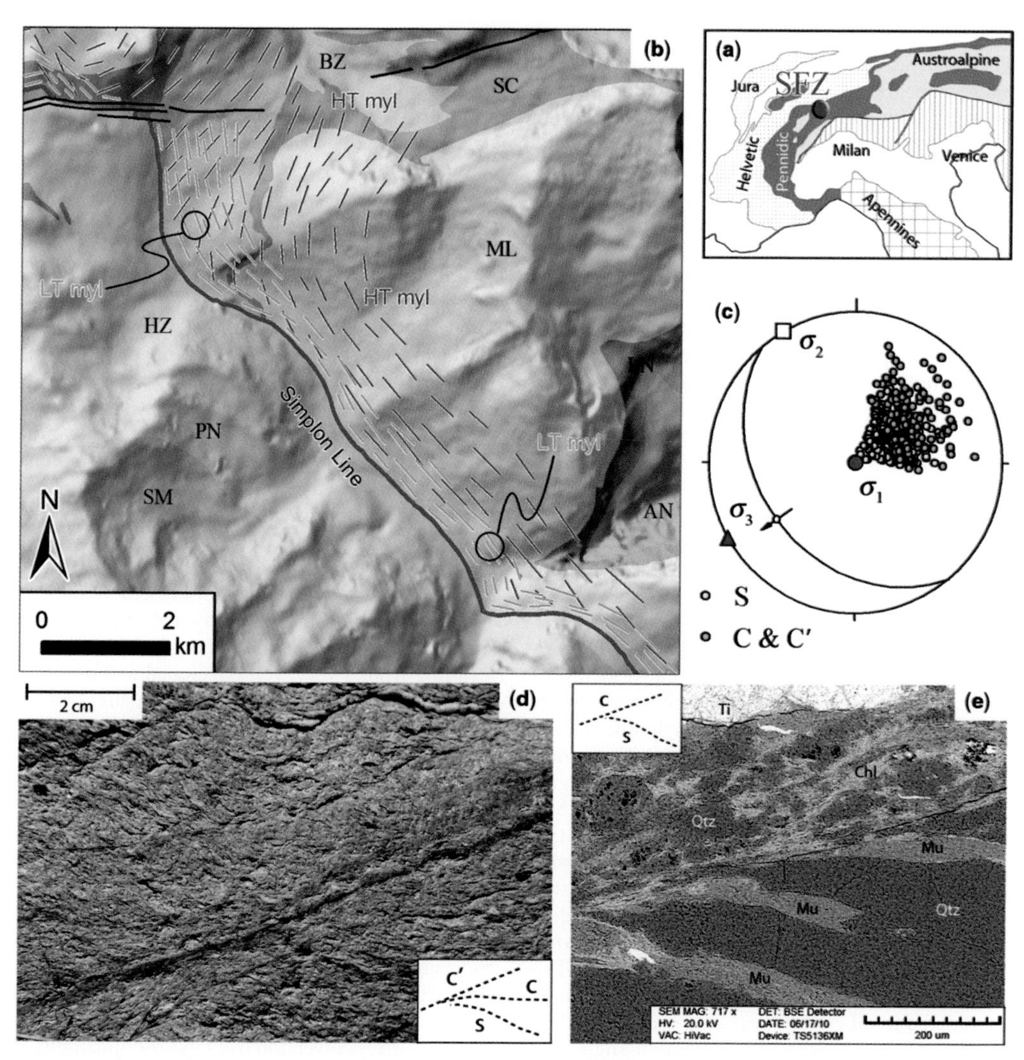

Fig. 1. The Simplon Fault Zone (SFZ). (**a**) Location of the SFZ in the Alps. (**b**) Tectonic sketch of the Simplon Pass area. Main tectonic units modified after Steck *et al.* (1999). The trace of the Simplon Line (in red) differs from Bearth (1956) and other works only in the area to the NW, where the SL is crosscut by younger east–west normal faults (in black), not evidenced by previous authors. Note parallelism between the low-temperature SFZ mylonitic foliation (LT myl) and the brittle SL trace. The higher-temperature mylonites (HT myl) are gently folded in the Simplon Pass area. Tectonic units as follows. Middle Penninic (Grand San Bernhard System): Siviez-Mischabel (SM), Pontis Nappe (PN), Houillère Zone (HZ), Berisal Zone (BZ). Outer Penninic: Sion-Courmayeur Zone (SC). Lower Penninic (Ossola–Tessin window): Monte Leone Nappe (ML), Lebendun Nappe (LN), Antigorio Nappe (AN). (**c**) Equal-area lower-hemisphere plot showing the average SL attitude and kinematics (great circle with slip vector), poles to the composite footwall mylonitic foliation (yellow, S; green, C and C′), and mean palaeostress tensor principal axes (σ_1, σ_2, σ_3). (**d**) C′ shear band reactivated as an ultracataclasite seam, evidenced by black colour in the SFZ, *c.* 20 m below the SL. (**e**) Thin ultracataclasite seam (*c.* 150 μm) developed along a C shear band in the SFZ mylonitic foliation *c.* 15 m below the SL; Chl, chlorite; Mu, muscovite; Qtz, quartz; Ti, titanite.

proportions in orthogneiss, paragneiss and mica-schists. Phyllosilicates are white mica and biotite in the outer higher-temperature portion of the SFZ and white mica and chlorite in the inner lower-temperature zone (marked as HT and LT respectively in Fig. 1b). The amount of phyllosilicates in different representative mylonites has been estimated with image-analysis on optical microscopy images (appendix) and varies between 20–25 and 40–45%, being lower for orthogneiss and progressively higher for paragneiss and micaschists. This analysis also allowed a visual estimate of the interconnection of phyllosilicate layers (appendix), which is very high even in orthogneiss due to the strong shape-preferred orientation (SPO) of mica flakes and their organization in discrete and continuous layers (lepidoblastic fabric).

The first evidence of brittle deformation in the SFZ rocks during progressive cooling is represented by thin (*c.* 200 μm) ultracataclasite seams which nucleate along S, C and C′ planes (Fig. 1d, e). This kind of reactivation can be found in a *c.* 100 m thick zone, approaching the SL from the footwall. Contrastingly, these structures cannot be generally found in the hanging wall where the SFZ mylonites are not observed. The spacing of cataclasite seams varies from almost 1 m to a few centimetres as the fault core is approached. The fault core itself (corresponding to the mapped SL) is represented by a *c.* 5 m thick cataclasite and ultracataclasite horizon, where less-deformed clasts are composed of foliated mylonites with evidence of the same kind of reactivation seen in the footwall. It must be emphasized that both the ultracataclasite seams (developed in the SFZ mylonites) and the cataclasite/ultracataclasite core zone (the SL) conform to the original low-angle attitude of the mylonitic SFZ.

Palaeostress inversion performed at the regional scale from mesoscopic faults in both the hanging wall and footwall, at some distance from the SFZ (Grosjean *et al.* 2004, and data collected during this study; Fig. 1c), point to a NE–SW extension during the activity of the SL with σ_1 almost vertical, σ_3 almost horizontal directed NE–SW (*c.* 00/241) and a stress tensor shape ratio $\phi = (\sigma_2 - \sigma_3)/(\sigma_1 - \sigma_3) \approx 0.6$ (as defined by Angelier 1990). This means that the low-angle misoriented SL can be considered as a LANF.

It is also notable that almost no high-angle optimally oriented Andersonian fault can be found both at the meso- and micro-scale in volumes where the low-angle mylonitic foliation is reactivated, although they are present out of this zone. This can be interpreted in terms of the presence of low-angle misoriented faults inhibiting the development of Andersonian faults, and we will come back to this in the section on Slip Tendency Analysis.

The Brenner Line

The Brenner Fault Zone (BFZ, Behrmann 1988) is a greenschist facies mylonitic belt with a variable thickness of 400 m to 1 km, which can be traced continuously for *c.* 25 km (Fig. 2). According to estimates by Behrmann (1988) and Selverstone (1988), this low-angle normal fault has accommodated between 5 and 8 km of differential uplift, evidenced by the juxtaposition of Penninic Nappes of the Tauern window in the footwall and Upper Austroalpine Nappes in the hanging wall (Fig. 2b). A considerable jump of cooling ages across the fault is consistent with its extensional activity in the Miocene. Late Cretaceous micas in the hanging wall contrast with Oligocene–Mid-Miocene ages within the Tauern window (Thöni & Hoinkes 1987; von Blanckenburg *et al.* 1989; Christensen *et al.* 1994; Genser *et al.* 1996; Elias 1998). Similarly, zircon and apatite fission tracks yield Late Cretaceous–Oligocene ages in the hanging wall and Mid–Late Miocene ages in the footwall (Grundmann & Morteani 1985; Fügenschuh *et al.* 1997; Elias 1998).

Mylonites of the BFZ are both more varied in composition and generally more rich in phyllosilicates than those of the SFZ (for this reason we did not perform the image analysis as for the Simplon). They develop mainly from mica schists, phyllonites and calcschists of the uppermost unit in the Tauern Window (Fig. 2b), the Glockner Nappe (also referred to as the Upper Schieferhülle) and marginally from other cover units (Lower Schieferhülle).

The BFZ is characterized by a W-dipping low-angle low-T greenschist facies composite foliation, resulting from a very well-developed and penetrative SCC′ fabric. On average, dip/dip direction for S planes is 10/270 while it is 25/270 for C planes and 40/270 for C′ planes (Fig. 2c). Either C or C′ shear bands are locally preferentially developed in different lithologies, but in general both are present.

Approaching the Brenner Line (BL: the boundary to the hanging wall, using the same terminology as for the SFZ/SL), this footwall mylonitic foliation is pervasively reactivated by thin cataclastic seams (Fig. 2d, e) which become more frequent and thicker, eventually resulting in foliated gouges and (ultra)cataclasites at the fault core (Behrmann 1988; Axen *et al.* 1995; Bistacchi *et al.* 2003).

Palaeostress inversion from mesoscopic faults in both the hanging wall and footwall (Bistacchi *et al.* 2010, and some newer data collected for this study; Fig. 2c) point to an east–west extension during the activity of the BL with σ_1 almost vertical, σ_3 almost horizontal directed E–W (*c.* 05/093) and a stress tensor shape ratio $\phi = (\sigma_2 - \sigma_3)/(\sigma_1 - \sigma_3) \approx 0.6$–$0.7$ (as defined by Angelier 1990). This high value is in accordance with frequent $\sigma_1 - \sigma_2$ permutations leading to repeated switches between normal faulting (vertical σ_1) and strike-slip (north–south horizontal σ_1) regimes in the footwall. In any case, this means that the low-angle BL can be considered both misoriented and a LANF.

As for the Simplon, no high-angle Andersonian faults can be found both at the meso- and micro-scale in volumes where the low-angle mylonitic foliation is reactivated. Once again, this can be interpreted in terms of the presence of low-angle misoriented faults inhibiting the development of Andersonian faults.

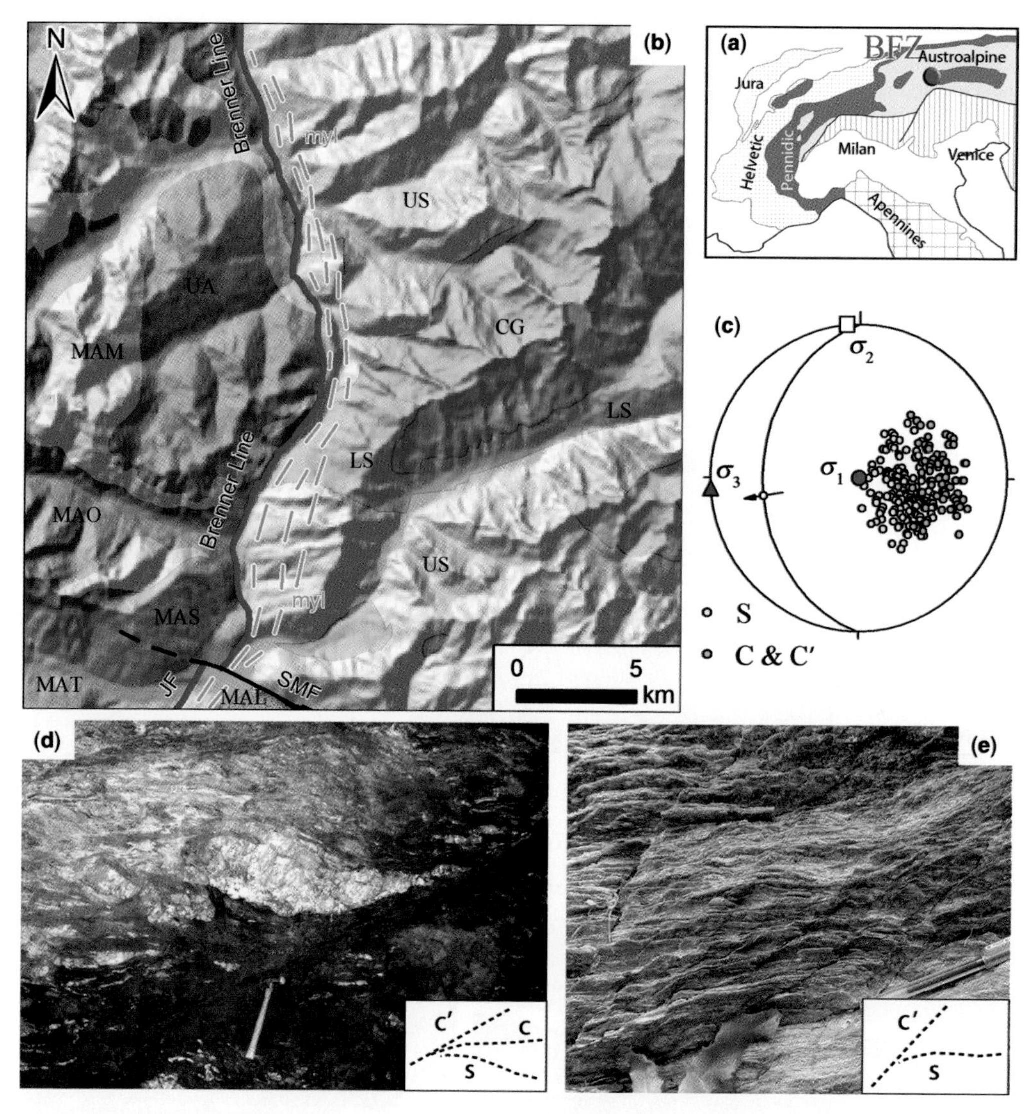

Fig. 2. The Brenner Fault Zone (BFZ). (**a**) Location of the BFZ in the Alps. (**b**) Tectonic sketch of the Brenner Pass area. Base geology modified after Baggio *et al.* (1971), Baggio *et al.* (1982), Behrmann (1988), Bistacchi *et al.* (2003), Bistacchi *et al.* (2008). Note parallelism between the BFZ mylonitic foliation (myl) and the brittle BL trace (in red). Tectonic units as follows. Upper Austroalpine (UA). Middle Austroalpine: Mesozoic cover (MAM), Ötztal Nappe (MAO), Schneeberg Complex (MAS), Texel Unit (MAT), St. Leonard Unit (MAL). Penninic (Tauern Window): Upper Schieferhülle (US), Lower Schieferhülle (LS), Central Gneiss (CG). (**c**) Equal-area lower-hemisphere plot showing the average BL attitude and kinematics (great circle with slip vector), poles to the composite footwall mylonitic foliation (yellow, S; green, C and C′) and mean palaeostress tensor principal axes (σ_1, σ_2, σ_3). (**d**) S, C, and C′ planes overprinted by ultracataclasite layer, evidenced by black colour, in the BFZ, *c*. 40 m below the BL. Hammer as scale. (**e**) Pervasively reactivated SC′ fabric of the BFZ mylonitic foliation, *c*. 25 m below the BL. Pen as scale.

The Sprechenstein-Mules Fault Zone

Unlike the Simplon and Brenner detachments, the WNW-trending dextral-reverse Sprechenstein-Mules fault (SMF) did not accommodate a relevant differential exhumation but still shows clear evidence of brittle reactivation of a pre-existent highly foliated mylonitic fabric. This fault belongs to the 700-km-long Periadriatic fault system, which is the tectonic divide between the Alpine orogenic wedge and its southern buttress, the Southern Alps (e.g. Castellarin *et al.* 2006). The major element of the Periadriatic fault system in the Eastern sector of the Alps is the subvertical east–west Pusteria Fault, which is connected to the Brenner detachment via the 20-km-long SMF (Fig. 3; Bistacchi *et al.* 2010). The SMF is an array of dextral fault strands (with a minor reverse component), interconnected by contractional jogs cutting the Austroalpine units sandwiched between the Tauern Window and the Pusteria Fault.

In more detail, the SMF cuts (from north to south) through three units (Fig. 3b): (1) the Austroalpine basement; (2) the so-called Mules Tonalitic Lamella (an Oligocene intrusive sheet); and (3) the South Alpine Bressanone Granite (Bistacchi *et al.* 2010). While the Bressanone Granite never experienced temperatures exceeding 200–250 °C during deformation, and hence does not show any metamorphic fabric, the Tonalitic Lamella and Austroalpine basement underwent progressive cooling from ≥450 °C to *c.* 250 °C during the

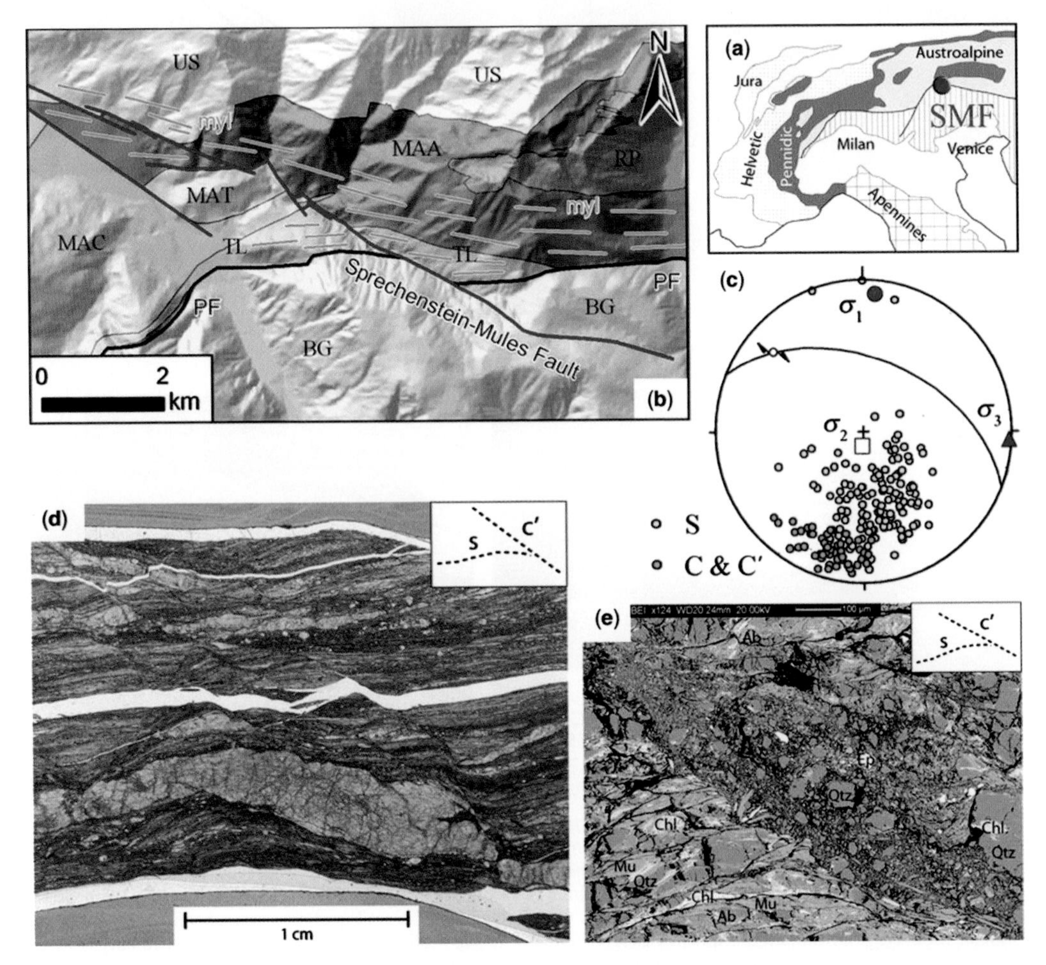

Fig. 3. The Sprechenstein-Mules Fault Zone (SMF). (**a**) Location of the SMF in the Alps. (**b**) Tectonic sketch of the Mules Valley, after Bistacchi *et al.* (2010). Note parallelism between the mylonitic foliation (myl) in the units to the north of the fault and the brittle SMF trace. Tectonic units as follows. Penninic (Tauern Window): Upper Schieferhülle (US). Middle Austroalpine: Mules Trias (MAT), Cervina Unit (MAC), Mules-Anterselva Unit (MAA). Southern Alps: Bressanone Granite (BG). Tertiary intrusives: Tonalitic Lamella (TL), Rensen Pluton (RP). (**c**) Equal-area lower-hemisphere plot showing the average SMF attitude and kinematics (great circle with slip vector), poles to the composite footwall mylonitic foliation (yellow, S; green, C and C′) and mean palaeostress tensor principal axes (σ_1, σ_2, σ_3). (**d**) SCC′ composite mylonitic foliation in the Austroalpine Basement, pervasively reactivated by thin ultracataclasite seams, *c.* 40 m to the north of the SMF core zone. (**e**) C′ shear band reactivated as a 100-μm-thick ultracataclasite seam in the mylonitic Tonalitic Lamella, *c.* 15 m to the south of the SMF core zone; Ab, albite; Chl, chlorite; Ep, epidote; Mu, muscovite; Qtz, quartz.

Oligocene–Miocene activity of the Periadriatic Lineament, thus developing mylonitic to phyllonitic fabrics in the higher temperatures range (Bistacchi *et al.* 2010). The brittle SMF fault array nucleated just within the inherited N–NNE steeply dipping composite foliation of the Austroalpine and Tonalitic Lamella phyllonites and mylonites. These are composed of 40–60% phyllosilicates (white mica and subordinate chlorite) and show a penetrative SCC′ composite foliation with average dip/dip direction for S, C and C′ planes of 56/353, 52/005 and 51/027, respectively (Fig. 3c). Due to the high phyllosilicate content, we did not perform the image analysis in this case as for the Simplon.

The penetrative brittle reactivation of the mylonitic fabric is particularly developed in Austroalpine phyllonites, where thin ultracataclasite seams can be found along S, C and C′ ductile planes at about 500 m from the fault core. This reactivation becomes more and more intense approaching the fault cores (in places the fault is duplicated across jogs), eventually resulting in continuous ultracataclasite and fault gouge layers *c.* 2 m thick.

Results of stress inversion in the footwall and hanging wall of the SMF (Fig. 3c) are similar to those already presented for the Brenner Fault (the distance between the two faults is less than 20 km and their activity overlaps), but σ_1 and σ_2 show a permutation. In this case σ_1 is not so unfavourably oriented, but σ_3 is at a very low angle to the fault and the inferred stress tensor has a high shape ratio (about 0.7). This results in a very unfavourable normal to tangential stress ratio on the SMF. This fault should therefore also be considered misoriented and non-Andersonian.

As well as the Simplon and Brenner, in this case no optimally oriented Andersonian faults can be found at the meso- or micro-scale in volumes where the phyllonitic foliation is reactivated. Once again, this can be interpreted in terms of the presence of reactivated mylonitic foliation inhibiting the development of Andersonian faults.

A review of mechanical anisotropy data

Following Paterson & Wong (2005, p. 38), mechanical anisotropy of foliated rocks is expressed as a variation of fracture stress with orientation of the principal stress axes relative to the fabric. In these rocks shear fractures tend to nucleate along the pre-existing foliation, therefore (re)activating it; the Mohr–Coulomb failure envelope is no more unique (e.g. Hobbs *et al.* 1976, p. 328). Indeed, two different envelopes can be defined: (1) one for failure along the foliation (acting as a very penetrative and generalized ‘plane of weakness’) and (2) one for general failure at a high angle to the foliation (not activated in this case). The latter resembles the failure envelope of an isotropic rock with standard internal friction coefficients, while the first one is more depressed in a normal versus shear stress Mohr plot (Fig. 4). When failure takes place along the foliation, the differential stress is always lower than when the foliation is not activated (Fig. 4); this must therefore be considered an absolute weakening mechanism.

In order to assess the level of mechanical anisotropy that may be expected due to a foliation marked by phyllosilicates, we have reviewed data published in several papers reporting on triaxial tests (Donath 1961, 1972; Walsh & Brace 1964; Attewell & Sandford 1974; McCabe & Koerner 1975; Bell & Coulthard 1997; Duveau *et al.* 1998) and direct shear tests (Collettini *et al.* 2009). Triaxial tests were performed changing systematically the orientation of foliation with respect to the imposed σ_1 axis from 0° to 90°. This results in a marked drop in failure stress for angles of *c.* 25–45°, due to nucleation of shear fractures along the foliation (Paterson & Wong 2005, fig. 16). Direct shear tests have been carried out on ‘rock wafers’ (hexahedral samples with one side parallel to foliation) and result in realistic fracturing and shearing along the foliation (Collettini *et al.* 2009). In Table 1 we summarize the results of tests performed on different rocks, focusing on the different phyllosilicate content and on the minimum and maximum internal friction coefficient, corresponding to shear fracturing along the foliation and fracturing across the foliation respectively. In Figure 5 we plot minimum/maximum internal friction coefficients versus phyllosilicate content, and we show reference values of friction coefficients for isotropic rocks ($\mu = 0.6$–0.75; e.g. Jaeger *et al.* 2007) and phyllosilicate single crystals or pure aggregates ($\mu = 0.3$–0.4; Kronenberg *et al.* 1990; Moore & Lockner 2004). While maximum friction coefficients of foliated rocks (fracture across foliation) show values similar to those for isotropic rocks, minimum friction coefficients show that reactivation of a foliation is a relevant weakening mechanism for phyllosilicate contents as low as 25–30%. In other words, for rocks like a gneiss or a micaschist, the internal friction coefficient for fracturing along the foliation can be in the range 0.3–0.4. Rocks with particularly weak phases such as talc may show even lower values (Collettini *et al.* 2009). The pronounced ‘knee’ shown by minimum friction coefficient values at phyllosilicate contents of about 25–30% (Fig. 5) is probably due to the appearance of an interconnected network of phyllosilicates which controls the overall mechanical behaviour (Niemeijer & Spiers 2005; Numelin *et al.* 2007).

Anisotropic slip tendency analysis

Theory and implementation

Slip tendency analysis has been developed by Morris *et al.* (1996) in order to quantitatively predict the tendency of faults to undergo slip (be reactivated) in a given stress field, assuming a general frictional behaviour with a constant, homogeneous and isotropic friction coefficient (e.g. given by Byerlee Law). This tendency depends uniquely on the tangential to normal stress ratio, hence slip tendency is defined as $T_s = \tau/\sigma_n$ where τ and σ_n are the tangential and normal stress, respectively. Since the maximum tangential stress on a surface with friction coefficient μ is $\tau_{critical} = \mu\sigma_n$, maximum slip tendency is $\max(T_s) = \tau_{critical}/\sigma_n = \mu$. Building on this, Lisle & Srivastava (2004) introduced the normalized slip tendency: $NT_s = T_s/\max(T_s) = T_s/\mu$. This varies between 0 (least slip probability) and 1 (maximum proneness to slip).

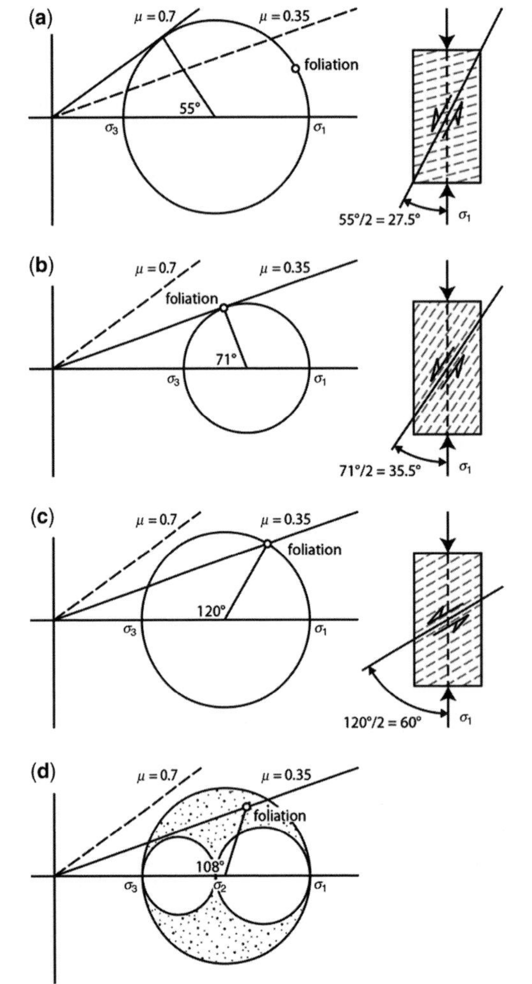

Fig. 4. Non-uniqueness of Mohr–Coulomb failure envelopes for anisotropic foliated rocks. Two different envelopes can be defined: one for failure along the foliation and one for general failure at a high angle to the foliation. The latter resembles the failure envelope of an isotropic rock with standard internal friction coefficients ($\mu = 0.7$ in the examples), while the former shows a lower internal friction coefficient ($\mu = 0.35$ in the examples). Failure takes place either when any plane reaches the 'general failure' envelope or when a foliation plane reaches the envelope for failure along foliation. Three examples of triaxial tests ($\sigma_2 = \sigma_3$) and the general case of a natural fault in foliated rocks ($\sigma_2 > \sigma_3$) are shown in (a–d), respectively. (**a**) Failure across foliation produces Andersonian fractures. (**b**) Failure along foliation with an optimal orientation produces fractures with an orientation similar to Andersonian, but also an extreme weakening (very low differential stresses at failure). (**c**) For failure along a misoriented foliation, differential stresses are lower with respect to the Andersonian case (a), thus the reactivation of a weak foliation is an absolute weakening mechanism. (**d**) Failure along a misoriented foliation with a general orientation relative to the principal stress axes, as it should be expected in natural cases, with differential stresses lower than the Andersonian case (a).

The input stress tensor for slip tendency analysis can be expressed either in terms of its absolute values or as a reduced stress tensor (e.g. Angelier 1990) where four parameters are considered: the orientation of the three principal stresses and the stress tensor shape ratio $\phi = (\sigma_2 - \sigma_3)/(\sigma_1 - \sigma_3)$. These parameters, and particularly the shape ratio, are very critical in defining the slip tendency. On the other hand, the absolute magnitude of stresses is not relevant since slip tendency is defined as a ratio between tangential and normal stresses.

In order to account for the anisotropic behaviour outlined in the previous section we generalize the slip tendency concept to fault nucleation, considering a frictional Mohr–Coulomb failure criterion in its cohesionless form for simplicity. If we consider an anisotropic foliated material, the internal friction angle will vary with the orientation relative to the foliation. Hence, with respect to an external reference frame, μ_A (μ anisotropic) will be a function of the attitude of the surface being investigated: $\mu_A = \mu(\text{dip, dip direction})$.

Building on this concept, we can introduce the anisotropic normalized slip tendency, defined as $ANT_s = T_s/\mu(\text{dip, dip direction}) = \tau/\sigma_n\mu_A$. As we will see in the following section, this markedly changes the prediction on the orientation of the fault that is most likely to nucleate in a given stress field. The analysis has been performed with a Matlab® toolbox (available to readers upon request to the corresponding author).

Table 1. *Mechanical anisotropy data from triaxial tests. See discussion in the main text and Figure 5*

Reference	Lithology	Phyllos. %	Min μ	Max μ	Test	*N* in Figure 5
Attewell & Sandford (1974)	Slate	25.00	0.50	0.80	Triaxial	a
Bell & Coulthard (1997)	Laminated clay	65.00	0.34	0.47	Triaxial	b
Donath (1961)	Slate	35.00	0.40	0.53	Triaxial	c
Donath (1972)	Phyllite	30.00	0.36	0.68	Triaxial	d
Donath (1972)	Phyllite	30.00	0.35	0.65	Triaxial	d
Duveau *et al.* (1998)	Chl-Ms schist	50.00	0.29	0.70	Triaxial	e
Duveau *et al.* (1998)	Chl-Ms schist	50.00	0.30	0.80	Triaxial	e
McCabe & Koerner (1975)	Mica schists	25.00	0.34	0.53	Triaxial	f
Walsh & Brace (1964)	Slate	35.00	0.29	0.70	Triaxial	g
Collettini *et al.* (2009)	Phyllonite	57.00	0.31		Direct shear on wafers	h
Collettini *et al.* (2009)	Phyllonite	61.00	0.25		Direct shear on wafers	h

Results

In Figure 6 the results of anisotropic slip tendency analysis on the Simplon, Brenner and Sprechenstein–Mules faults are shown. For each fault we compare normalized slip tendency (NT_s; first two columns) and anisotropic normalized slip tendency (ANT_s; last two columns) plotted on an equal-angle lower-hemisphere stereoplot and in the adimensional Mohr space (consistent with the reduced stress tensor). In the middle column the anisotropic internal friction coefficient μ_A is plotted in an equal-angle lower-hemisphere stereoplot.

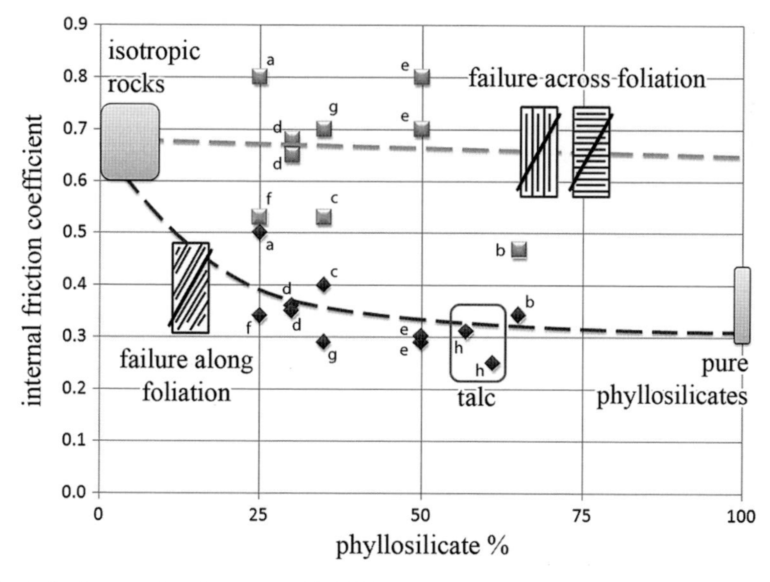

Fig. 5. Mechanical anisotropy data. Data reported in Table 1 are summarized in a phyllosilicate content versus internal friction coefficient plot. Minimum internal friction coefficient values (diamonds) are for failure along the foliation. Maximum internal friction coefficient values (squares) are for failure across the foliation, which in this case is not reactivated. Typical values for isotropic rocks and pure phyllosilicates (tested as single crystals or aggregates) are shown as end members. Direct shear test data by Collettini *et al.* (2009), performed on talc-bearing rocks, labelled as 'talc'. Numbers shown beside data points refer to Table 1.

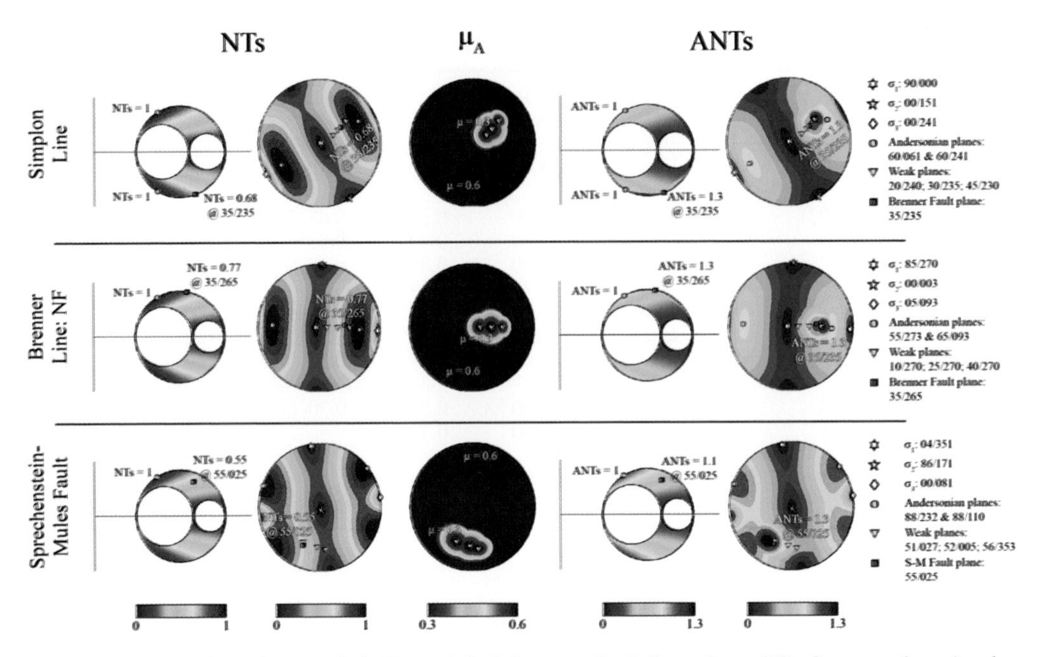

Fig. 6. Anisotropic slip tendency analysis. For each fault the normalized slip tendency (NT_s; first two columns) and anisotropic normalized slip tendency (ANT_s; last two columns) are compared in equal-angle lower-hemisphere stereoplots and in adimensional Mohr plots. In the middle column the anisotropic internal friction coefficient μ_A is plotted in an equal-angle lower-hemisphere stereoplot. Orientation data for the fault zone, mylonitic foliation and palaeostress axes are listed in the legend to the right.

NT_s is calculated based on palaeostress conditions defined for the different case studies (principal axes shown with white symbols in all stereoplots). All three faults (red squares) show low NT_s values between 0.55 (Sprechenstein-Mules) and 0.77 (Brenner); their activation is therefore unlikely in the isotropic case. Moreover, theoretical Andersonian planes (blue dots) have markedly different attitudes.

The anisotropic internal friction coefficient μ_A is calculated in all three cases considering a coefficient varying between 0.65 for failure across the foliation and a coefficient of 0.35 for failure along the composite SCC′ mylonitic foliation. Poles to planes representing the mean values of S, C and C′ in the three mylonite belts are represented with yellow triangles.

In the anisotropic analysis columns (Fig. 6, right), we see that ANT_s is very much elevated with respect to NT_s (left columns) along the 'weak orientations' associated with the SCC′ mylonitic foliations. Poles to planes representing the mean attitude of the Simplon, Brenner and Sprechenstein-Mules faults (red squares) fall in high ANT_s fields and show ANT_s values higher than those attributed to the theoretical Andersonian planes (blue dots). In this kind of analysis, this means that faults oriented subparallel to the mylonitic foliation would have a higher tendency or probability to nucleate (and then accumulate deformation) than the Andersonian planes. This happens because the anisotropic rocks are weaker parallel to the foliation, and because this weakness compensates for the unfavourable normal to shear stress ratio experienced by the misoriented faults.

Discussion and conclusion

Triaxial and direct shear test on foliated phyllosilicate-bearing rocks indicate that these rocks show a relevant mechanical anisotropy for phyllosilicate contents that could be as low as 25–30%. Anisotropic slip tendency analysis demonstrates that nucleation and continuing activity of misoriented brittle faults, which exploit a pre-existing suitable mechanical anisotropy, could be more likely to occur than the nucleation of Andersonian faults. The presence of regional-scale phyllosilicate-rich mylonitic belts could therefore represent an absolute weakening mechanism for the brittle crust and, at the same time, explain the nucleation of large faults with non-Andersonian orientations. In the remaining we will discuss whether this mechanism may apply (and to what extent) to large-scale faults in the upper crust.

A first critical issue is whether foliated phyllosilicate-rich rocks would show the same kind of anisotropy, evidenced by room-temperature triaxial and direct shear test and at temperatures of about 250–300 °C at the base of the brittle crust. Different kinds of experiments investigate the mechanical behaviour of phyllosilicates at temperatures up to 700–800 °C (e.g. Kronenberg *et al.* 1990; Moore & Lockner 2004; Mariani *et al.* 2006). Even if mechanical anisotropy is not always explicitly considered in such experiments, the related datasets show that the mechanical behaviour of phyllosilicates is not subject to important variations between room temperature and *c.* 300–400 °C. Crossing this boundary, metamorphic reactions start to significantly influence the mechanical behaviour. This means (at least at a first-order approximation) that the level of mechanical anisotropy detected at room temperature can also be postulated at lower greenschist facies conditions (250–300 °C) where large-scale faults nucleate. Moreover, other deformation processes are likely to occur in phyllosilicates and further reduce their strength under low-grade metamorphic conditions such as stress corrosion (e.g. Scholz 2002), solution transfer (e.g. Niemeijer & Spiers 2005) and dislocation glide on (001) planes (Wintsch *et al.* 1995). Thus the room temperature anisotropy may even be amplified at lower greenschist facies conditions.

Also relevant is the critical amount of phyllosilicates above which the mechanical anisotropy strongly controls fault nucleation. Triaxial test data (Fig. 5) point to a critical transition at about 25–30% of phyllosilicates. Below this value, most rocks show an isotropic behaviour and in many cases conform to Byerlee law. Above this critical phyllosilicate content a marked anisotropy appears, and at higher values we observe a 'plateau' where the ratio between the along-foliation and across-foliation friction coefficients does not show important variations. Shea & Kronenberg (1993) and Rawling *et al.* (2002) also evidenced a similar transition in triaxial tests, and proposed that it may be related to the interconnection of phyllosilicate layers which is realized at about 25–30% of phyllosilicate content. A similar result was obtained by Logan & Rauenzahn (1987) for friction experiments on synthetic clay-quartz gouge with varying composition. This means that, in the presence of distinct layers of phyllosilicates with a relevant SPO (lepidoblastic fabric), a 25–30% phyllosilicate content is sufficient to control the overall mechanical behaviour of a polymineralic aggregate, as shown in independent micromechanical models by Rawling *et al.* (2002) and Niemeijer & Spiers (2005). This condition is very likely to be met in phyllosilicate-bearing mylonites developed from protoliths of a wide range of compositions.

Returning to the field examples described in the first part of this contribution, we can describe how two clues point to an anisotropy-related weakening. The first is the widespread occurrence of metamorphic foliations 'reactivated' under brittle conditions which demonstrates nucleation, growth, coalescence and continuing activity of (ultra)cataclastic horizons exploiting the phyllosilicate-rich layers. The second (not less important in our opinion) is the almost complete absence of Andersonian conjugate fault sets in volumes that are quite obviously intensely deformed in the brittle field. This indicates that the development of Andersonian faults is inhibited by the activation of phyllosilicate-rich layers, which in mechanical terms means that the differential stress level is buffered by the phyllosilicates to levels not allowing the 'isotropic' strength envelope to be reached (Fig. 4).

A similar observation was made by Butler *et al.* (2008) in the hangingwall of the Liachar Thrust (Nanga Parbat Massif), where brittle faults preferentially nucleate along a high-grade metamorphic foliation. In this case the resulting faulting pattern is geometrically and topologically simple, with most of the faults developed parallel to each other and to the foliation, and almost no cross-cutting conjugate faults. According to Butler *et al.* (2008), this contrasts to more complex faulting patterns including Andersonian and even more complicate conjugate sets developed in other parts of the same massif (where the foliation was not preferentially activated).

In conclusion, it seems to be more than just a coincidence that LANFs and other weak faults are common in continental deformation belts and plate margins. The association between weak faults and phyllosilicates organized along regional-scale metamorphic mylonites is evidenced by many examples in continental deformation belts, both in collisional environments (e.g. Cox 1995; Collettini *et al.* 2009; Bistacchi *et al.* 2010), extensional environments and rifts (e.g. Wernicke 1981; Numelin *et al.* 2007; Smith & Faulkner 2010) and large-scale strike-slip or transform faults (e.g. Stewart *et al.* 2000; Holdsworth *et al.* 2011). The frequent presence of regional-scale phyllosilicate-rich mylonite belts, due to the polyphase ductile deformation history along plate margins and intraplate deformation belts (e.g. Storti *et al.* 2003), is likely to influence the deformation style and overall strength of the whole brittle crust in active regions.

The importance of mechanical anisotropy in fault weakening is a concept that has been developed with equal contributions by AB, MM and LM. Fieldwork and meso-scale structural analysis was carried out by AB, FB and VD on the SFZ, by AB and MM on the BFZ and by AB, MM and LM on the SMF. Microstructural analysis by LM, FB, AB and MM. The review on mechanical anisotropy data was carried out by FB and VD. The anisotropic slip tendency analysis concept and Matlab® toolbox were developed by AB, who also wrote most of this text; MM, LM and FB contributed to different sections of the manuscript. We thank C. Bond and E. Mariani for constructive reviews, D. Healy for the editorial handling and S. Smith for a last-minute friendly review.

Appendix: Estimate of phyllosilicate content in mylonites

The amount of phyllosilicates in mylonites from the Simplon Fault Zone was estimated by analysis of optical microscopy images. The same analysis was not performed for the Brenner and Sprechenstein-Mules Faults, where the mylonitic micaschists and phyllonites are generally much more rich in phyllosilicates and at the same time show a high variability in phyllosilicate content between mica-rich and quartz-feldspar-rich layers.

Colour digital photos of representative portions of a thin section, cut perpendicular to foliation and parallel to stretching lineation, have been captured with a 12 megapixel digital camera attached to a standard polarizing microscope. In order to highlight phyllosilicates with respect to quartz and feldspars, we have collected crossed-polar images with foliation trending at about 45° (Fig. 7) so that micas show higher interference colours with respect to other minerals. Magnification was in general kept low, in order to capture large areas of the thin sections.

Images have been processed with a low-pass digital filter in order to eliminate noise and small patches of bright pixels related to grain boundaries.

The filtered images were classified with a supervised maximum likelihood algorithm (e.g. Richards 1999) which, after a selection of training areas in the images, allows classified images to be generated with just three classes: phyllosilicates (red in Fig. 7), quartz and feldspar (green in Fig. 7) and not classified pixels (black in Fig. 7).

The relative amount of phyllosilicates in every image is finally obtained as area proportion in pixels, excluding not classified pixels. These are always between 5 and 10% in this kind of analysis, so we estimate that our analysis is accurate to about 5%.

This kind of analysis also allowed us to visually estimate the interconnection of phyllosilicate layers, which is quite impressive even in orthogneiss with phyllosilicate content as low as 24% (Fig. 7a, c) due to the strong shape-preferred orientation (SPO) of mica flakes and their organization in distinct and continuous layers (lepidoblastic fabric).

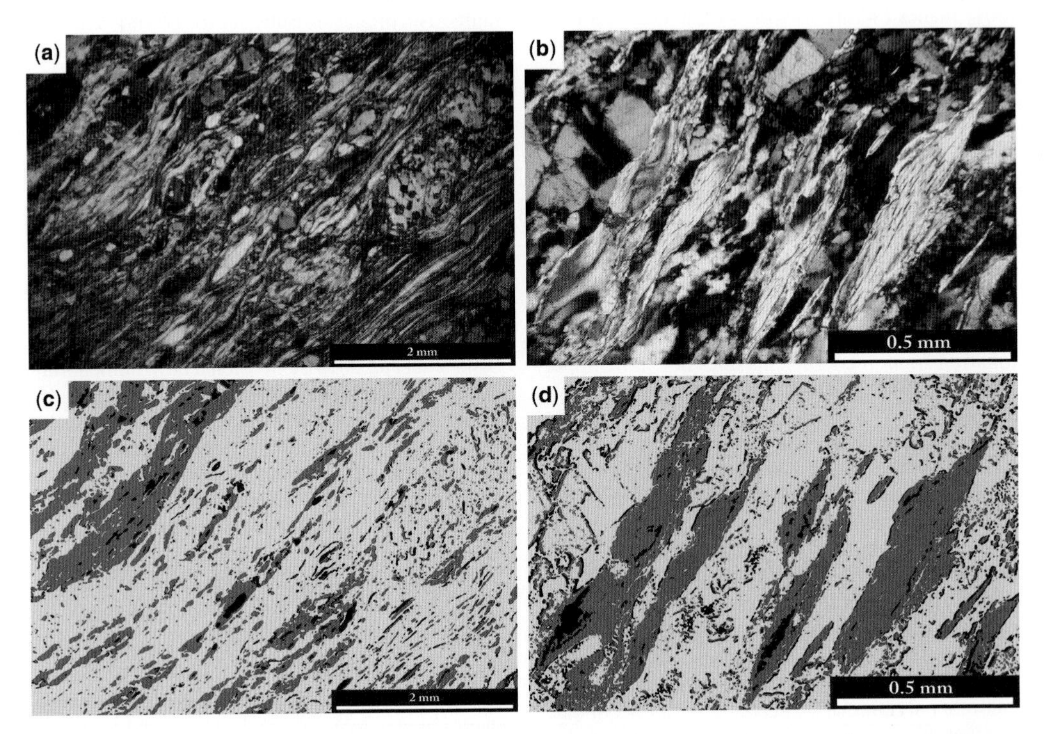

Fig. 7. Image analysis with optical microscopy digital photos. (**a**, **b**) Colour photos of representative areas in thin sections (crossed polars). (**c**, **d**) Results of maximum likelihood image classification for (a) and (b), respectively. Red, phyllosilicates; green, quartz and feldspar; black, not classified pixels. Results of classification for the examples shown here: (a, c) phyllosilicates 23.8%, quartz and feldspar 70.8%, not classified 5.4%; (b, d) phyllosilicates 30.3%, quartz and feldspar 59.9%, not classified 9.7%.

References

Anderson, E. M. 1951. *Dynamics of Faulting and Dyke Formations with Application to Britain*. Oliver & Boy, Edinburgh.

Angelier, J. 1990. Inversion of field data in fault tectonics to obtain the regional stress. III. A new rapid direct inversion method by analytical means. *Geophysical Journal International*, **103**, 363–376.

Argand, E. 1916. Sur l'arc des Alpes occidentales. *Eclogae Geologicae Helvetiae*, **14**, 145–191.

Attewell, P. B. & Sandford, M. R. 1974. Intrinsic shear strength of a brittle, anisotropic rock I: experimental and mechanical interpretation. *International Journal of Rock Mechanics and Mining Science & Geomechanics Abstracts*, **11**, 423–430.

Axen, G. 1992. Pore pressure, stress increase, and fault weakening in low-angle normal faulting. *Journal of Geophysical Research*, **97**, 8979–8991.

Axen, G., Bartley, J. & Selverstone, J. 1995. Structural expression of a rolling hinge in the footwall of the Brenner Line normal fault, eastern Alps. *Tectonics*, **14**, 1380–1392.

Baggio, P., Friz, C. et al. 1971. *Carta Geologica d'Italia 1:100.000, F. 4, Merano*. Servizio Geologico Italiano.

Baggio, P., De Vecchi, G. P. & Mezzacasa, G. 1982. Carta geologica della media ed alta Valle di Vizze e regione vicine (Alto Adige). *Bollettino della Società Geologica Italiana*, **101**, 89–116.

Bearth, P. 1956. Zur Geologie der Wurzelzone östlich des Ossolatales. *Eclogae Geologicae Helvetiae*, **49**, 267–278.

Behrmann, J. H. 1988. Crustal scale extension in a convergent orogen: the Sterzing–Steinach mylonite zone in the Eastern Alps. *Geodynamica Acta*, **2**, 63–73.

Bell, F. G. & Coulthard, J. M. 1997. A survey of some geotechnical properties of the Tees Laminated Clay of central Middlesbrough, North East England. *Engineering Geology*, **48**, 117–133.

Bistacchi, A., Dal Piaz, G. V., Dal Piaz, G., Massironi, M., Monopoli, B. & Schiavo, A. 2003. Carta Geologica e note illustrative del transetto Val di Vizze-Fortezza (Alpi Orientali). *Memorie di Scienze Geologiche, Padova*, **55**, 169–189.

Bistacchi, A., Massironi, M., Menegon, L., Zanchetta, S. & Zanchi, A. 2008. Late Alpine fault systems in the central-eastern Alps between Merano and Vipiteno: preliminary results. *Rendiconti della Società Geologica Italiana*, **6**, Nuova Serie, 31–36.

Bistacchi, A., Massironi, M. & Menegon, L. 2010. Three-dimensional characterization of a crustal-scale fault zone: the Pusteria and Sprechenstein fault system (Eastern Alps). *Journal of Structural Geology*, **32**, 2022–2041.

Butler, R. W. H., Bond, C. E., Shipton, Z. K., Jones, R. R. & Casey, M. 2008. Fabric anisotropy controls faulting in the continental crust. *Journal of the Geological Society, London*, **165**, 449–452.

Byerlee, J. D. 1978. Friction of rocks. *Pure and Applied Geophysics*, **116**, 615–626.

Byerlee, J. D. 1990. Friction, overpressure and fault-normal compression. *Geophysical Research Letters*, **17**, 2109–2112.

Campani, M., Mancktelow, N., Seward, D., Rolland, Y., Müller, W. & Guerra, I. 2010*a*. Geochronological evidence for continuous exhumation through the ductile–brittle transition along a crustal-scale low-angle normal fault: simplon Fault Zone, central Alps. *Tectonics*, **29**, TC3002.

Campani, M., Herman, F. & Mancktelow, N. 2010*b*. Two- and three-dimensional thermal modeling of a low-angle detachment: exhumation history of the Simplon Fault Zone, central Alps. *Journal of Geophysical Research*, **115**, B10420.

Castellarin, A., Vai, G. B. & Cantelli, L. 2006. The Alpine evolution of the Southern Alps around the Giudicarie faults: a Late Cretaceous to Early Eocene transfer zone. *Tectonophysics*, **414**, 203–223.

Christensen, J. N., Selverstone, J., Rosenfeld, J. L. & De Paolo, D. J. 1994. Correlation by Rb–Sr geochronology of garnet growth histories from different structural levels within the Tauern Window, Eastern Alps. *Contributions to Mineralogy and Petrology*, **118**, 1–12.

Collettini, C., De Paola, N. & Goulty, N. R. 2006. Switches in the minimum compressive stress direction induced by overpressure beneath a low-permeability fault zone. *Terra Nova*, **18**, 224–231.

Collettini, C., Niemeijer, A., Viti, C. & Marone, C. 2009. Fault zone fabric and fault weakness. *Nature*, **462**, 907–911.

Cox, S. F. 1995. Faulting processes at high fluid pressures: an example of fault–valve behaviour from the Wattle Gully Fault, Victoria, Australia. *Journal of Geophysical Research*, **100**, 12 841–12 859.

Dal Piaz, G. V., Bistacchi, A. & Massironi, M. 2003. Geological outline of the Alps. *Episodes*, **26**, 175–180.

Donath, F. A. 1961. Experimental study of shear failure in anisotropic rocks. *Geological Society of America Bulletin*, **72**, 985–990.

Donath, F. A. 1972. Effects of cohesion and granularity on deformational behavior of anisotropic rocks. *Geological Society of America, Memoirs*, **135**, 95–128.

Duveau, G., Shao, J. F. & Henry, J. P. 1998. Assessment of some failure criteria for strongly anisotropic materials. *Mechanics of Cohesive-Frictional Materials*, **3**, 1–26.

Elias, J. 1998. The thermal history of the Ötztal-Stubai complex (Tyrol; Austria/Italy) in the light of the lateral extrusion model. *Tübinger Geowissenschaftliche Arbeiten, Reihe A*, **42**, 1–172.

Faulkner, D. R. & Rutter, E. H. 2001. Can the maintenance of overpressured fluids in large strike-slip fault zones explain their apparent weakness? *Geology*, **29**, 503–506.

Fügenschuh, B., Seward, D. & Mancktelow, N. 1997. Exhumation in a convergent orogen: the western Tauern window. *Terra Nova*, **9**, 213–217.

Genser, J., van Wees, J. D., Cloething, S. & Neubauer, F. 1996. Eastern Alpine tectonometamorphic evolution: constraints from two-dimensional P–T–t modeling. *Tectonics*, **15**, 584–604.

Grasemann, B. & Mancktelow, N. S. 1993. Two-dimensional thermal modelling of normal faulting: the Simplon Fault Zone, Central Alps, Switzerland. *Tectonophysics*, **225**, 155–165.

Grosjean, G., Sue, C. & Burkhard, M. 2004. Late Neogene extension in the vicinity of the Simplon

Fault Zone (central Alps, Switzerland). *Eclogae Geologicae Helvetiae*, **97**, 33–46.

GRUNDMANN, G. & MORTEANI, G. 1985. The young uplift and thermal history of the central Eastern Alps (Austria/Italy), evidence from apatite fission track ages. *Geologisches Jahrbuch*, **128**, 197–216.

HOBBS,B. E., MEANS, W. D. & WILLIAMS, P. F. 1976. *An Outline of Structural Geology*. Wiley, New York.

HOLDSWORTH, R. E., VAN DIGGELEN, E. W. E., SPIERS, C. J., DE BRESSER, J. H. P., WALKER, R. J. & BOWEN, L. 2011. Fault rocks from the SAFOD core samples: implications for weakening at shallow depths along the San Andreas Fault, California. *Journal of Structural Geology*, **33**, 132–144.

JAEGER, J. C., COOK, N. G. W. & ZIMMERMANN, R. W. 2007. *Fundamentals of Rock Mechanics*. 4th edn. Blackwell, Oxford.

KRONENBERG, A. K., KIRBY, S. H. & PINKSTON, J. 1990. Basal slip and mechanical anisotropy of biotite. *Journal of Geophysical Research*, **95**, 19 257–19 278.

LISLE, R. J. & SRIVASTAVA, D. C. 2004. Test of the frictional reactivation theory for faults and validity of fault-slip analysis. *Geology*, **32**, 569–572.

LOGAN, J. M. & RAUENZAHN, K. A. 1987. Frictional dependence of gouge mixtures of quartz and montmorillonite on velocity, composition and fabric. *Tectonophysics*, **144**, 87–108.

MANCEL, P. & MERLE, O. 1987. Kinematics of the northern part of the Simplon line (Central Alps). *Tectonophysics*, **135**, 265–275.

MANCKTELOW, N. S. 1985. The Simplon Line: a major displacement zone in the western Lepontine Alps. *Eclogae Geologicae Helvetiae*, **78**, 73–96.

MANCKTELOW, N. S. 1992. Neogene lateral extension during convergence in the central Alps: evidence from interrelated faulting and backfolding around Simplonpass (Switzerland). *Tectonophysics*, **215**, 295–317.

MARIANI, E., BRODIE, K. H. & RUTTER, E. H. 2006. Experimental deformation of muscovite shear zones at high temperatures under hydrothermal conditions and the strength of phyllosilicate-bearing faults in nature. *Journal of Structural Geology*, **28**, 1569–1587.

MCCABE, W. M. & KOERNER, R. M. 1975. High pressure shear strength investigation of an anisotropic mica schist rock. *International Journal of Rock Mechanics and Mining Science & Geomechanics Abstracts*, **12**, 219–228.

MOORE, D. E. & LOCKNER, D. A. 2004. Crystallographic controls on the frictional behavior of dry and water-saturated sheet structure minerals. *Journal of Geophysical Research*, **109**, B03401.

MOORE, D. E. & RYMER, M. J. 2007. Talc-bearing serpentinite and the creeping section of the San Andreas fault. *Nature*, **448**, 795–797.

MORRIS, A., FERRIL, A. & HENDERSON, D. B. 1996. Slip tendency analysis and fault reactivation. *Geology*, **24**, 275–278.

NIEMEIJER, A. R. & SPIERS, C. J. 2005. Influence of phyllosilicates on fault strength in the brittle–ductile transition: insights from rock analogue experiments. *In*: BRUHN, D. & BURLINI, L. (eds) *High-Strain Zones: Structure and Physical Properties*. Geological Society, London, Special Publications, **245**, 303–327.

NUMELIN, T., MARONE, C. & KIRBY, E. 2007. Frictional properties of natural fault gouge from a low-angle normal fault, Panamint Valley, California. *Tectonics*, **26**, TC2004.

PATERSON, M. S. & WONG, T.-F. 2005. *Experimental Rock Deformation – The Brittle Field*. 2nd edn. Springer, Berlin.

RAWLING, G. C., BAUD, P. & WONG, T.-F. 2002. Dilatancy, brittle strength and anisotropy of foliated rocks: experimental deformation and micromechanical modelling. *Journal of Geophysical Research*, **107**, 2234.

RICE, J. R. 1992. Fault stress states, pore pressure redistributions, and the weakness of the San Andreas fault. *In*: EVANS, B. & WONG, T.-F. (eds) *Fault Mechanics and Transport Properties of Rocks*. Academic Press, London, 476–503.

RICHARDS, J. A. 1999. *Remote Sensing Digital Image Analysis*. Springer, Berlin.

SCHOLZ, C. H. 2002. *The Mechanics of Earthquakes and Faulting*. 2nd edn. Cambridge University Press, Cambridge.

SELVERSTONE, J. 1988. Evidence for east–west crustal extension in the Eastern Alps: implications for the unroofing history of the Tauern window. *Tectonics*, **7**, 87–105.

SHEA, W. T. & KRONENBERG, A. K. 1993. Strength and anisotropy of foliated rocks with varied mica contents. *Journal of Structural Geology*, **15**, 1097–1121.

SIBSON, R. H. 1985. A note on fault reactivation. *Journal of Structural Geology*, **7**, 751–752.

SMITH, S. A. F. & FAULKNER, D. R. 2010. Laboratory measurements of the frictional strength of a natural low-angle normal fault. *Journal of Geophysical Research*, **115**, B02407.

STECK, A. 1987. Le massif du Simplon – Réflexions sur la cinématique des nappes de gneiss. *Schweizerische Mineralogische und Petrographische Mitteilungen*, **67**, 27–45.

STECK, A. 2008. Tectonics of the Simplon massif and Lepontine gneiss dome: deformation structures due to collision between the underthrusting European plate and the Adriatic indenter. *Swiss Journal of Geosciences*, **101**, 515–546.

STECK, A., BIGIOGGERO, B., DAL PIAZ, G. V., ESCHER, A., MARTINOTTI, G. & MASSON, H. 1999. *Carte géologique des Alpes de Suisse occidentale, 1: 100 000, Carte géologique spéciale N° 123*. Service Hydrologique et Géologique National.

STEWART, M., HOLDSWORTH, R. E. & STRACHAN, R. A. 2000. Deformation processes and weakening mechanisms within the frictional–viscous transition zone on major crustal-scale faults: insights from the Great Glen Fault Zone, Scotland. *Journal of Structural Geology*, **22**, 543–560.

STORTI, F., HOLDSWORTH, R. & SALVINI, F. 2003. Intraplate strike-slip deformation belts. *In*: STORTI, F., HOLDSWORTH, R. & SALVINI, F. (eds) *Intraplate Strike-Slip Deformation Belts*. Geological Society, London, Special Publications, **210**, 1–14.

THÖNI, M. & HOINKES, G. 1987. The southern Ötztal basement: geochronological and petrological consequences of Eoalpine metamorphic overprinting. *In*: FLÜGEL, H. W. & FAUPL, P. (eds) *Geodynamics of the Eastern Alps*. Deuticke Verlag, Deuticke, Wien, 200–213.

TOWNEND, J. & ZOBACK, M. D. 2004. Regional tectonic stress near the San Andreas fault in central and southern California. *Geophysical Research Letters*, **31**, L15S11.

VON BLANCKENBURG, F., VILLA, I., BAUR, H., MORTEANI, G. & STEIGER, R. 1989. Time calibration of a PT-path from the Western Tauern window, Eastern Alps: the problem of closure temperatures. *Contributions to Mineralogy and Petrology*, **101**, 1–11.

WALSH, J. B. & BRACE, W. F. 1964. A fracture criterion for brittle anisotropic rock. *Journal of Geophysical Research*, **69**, 3449–3456.

WERNICKE, B. 1981. Low-angle normal faults in the Basin and Range province: 3D nappe tectonics in an extending orogen. *Nature*, **291**, 645–648.

WINTSCH, R. P., CHRISTOFFERSEN, R. & KRONENBERG, A. K. 1995. Fluid–rock reaction weakening of fault zones. *Journal of Geophysical Research*, **100**, 13021–13032.

Anisotropic poroelasticity and the response of faulted rock to changes in pore-fluid pressure

DAVID HEALY

School of Geosciences, King's College, University of Aberdeen, Aberdeen AB24 3UE UK (e-mail: d.healy@abdn.ac.uk)

Abstract: The Law of Effective Stress has found wide application in structural geology, rock mechanics and petroleum geology. The commonly used form of this law relies on an assumption of isotropic porosity. The porosity in and around fluid-saturated fault zones is likely to be dominated by tectonically induced cracks of various shapes and sizes. Previously published field and laboratory data show that these cracks occur in distinct patterns of preferred orientation, and that these patterns vary around the fault zone. This paper uses the more general form of the Law of Effective Stress which incorporates anisotropic poroelasticity to model the geomechanical response of fault zones surrounded by patterns of oriented cracks. Predictions of fault stability in response to fluid pressure changes are shown to depend on both the nature (or symmetry) of the crack pattern and the orientation of the crack patterns with respect to the in situ stress. More complete data on the porosity of natural fault zones will enable more accurate predictions of fault stability in the subsurface.

Accurately predicting the mechanical response of faulted and saturated rocks is a key requirement for society in general (e.g. earthquake, landslide and volcano hazard mitigation) and for those agencies involved in the management of resources in the subsurface (e.g. aquifers, CO_2 repositories and hydrocarbon reservoirs). In faulted and saturated rocks, the pore fluid occupies cracks of various sizes and shapes. It has been established from field and laboratory studies that cracks frequently occur in patterns of preferred orientations arranged more or less systematically around larger fault zones, with crack density increasing exponentially towards principal fault slip surfaces (e.g. Flinn 1977; Moore & Lockner 1995; Vermilye & Scholz 1998; Wilson *et al.* 2003). Critically, the individual cracks are not equant voids, but have aspect ratios which depart significantly from unity and are often characterized as penny-shaped discs or oblate spheroids (Hallbauer *et al.* 1973; Tapponnier & Brace 1976; Reches & Lockner 1994; Healy *et al.* 2006*a*, *b*). The presence of locally intense brittle fabrics composed of inequant cracks means that the porosity in these rocks is anisotropic and not isotropic. The cracks are where the pore fluid resides and provide the key mechanical link between the pore fluid, and therefore any changes in pore-fluid pressure, and the bulk rock.

Assessments of strength in saturated rock have traditionally been made using the Law of Effective Stress, originally developed for soils (Terzaghi 1943). The law in this original form assumes isotropic porosity and the effective stress σ' in an isotropic saturated material is written

$$\sigma' = \sigma - P_f \tag{1}$$

where σ is the applied stress and P_f is the pore-fluid pressure. This equation forms the basis of several classical analyses in structural geology (e.g. Hubbert & Rubey 1959; Secor 1965; Phillips 1972). The inherent assumption of isotropic poroelasticity in this equation implies that the pore fluid resides in equant (e.g. spherical) pores; numerous field and laboratory studies have however shown that the porosity of faulted saturated rock is dominated by inequant cracks. Better predictions of strength and stability in faulted rocks can be obtained by incorporating anisotropic poroelasticity into the Law of Effective Stress.

This paper develops the idea originally put forward by Chen & Nur (1992) that the predicted changes in effective stress with anisotropic poroelasticity can be significantly different from predictions made using the more common isotropic form. The implications of the anisotropic form of the effective stress law for the core zones of so-called 'weak' faults have already been explored for strike-slip faults (Healy 2008) and low-angle normal faults (Healy 2009). This paper describes the geomechanical consequences of anisotropic poroelasticity for Andersonian (Anderson 1951) normal, strike-slip and thrust faults. The novelty in this paper lies in the application of the previously published equations (Chen & Nur 1992) to the

From: Healy, D., Butler, R. W. H., Shipton, Z. K. & Sibson, R. H. (eds) 2012. *Faulting, Fracturing and Igneous Intrusion in the Earth's Crust*. Geological Society, London, Special Publications, **367**, 201–214. http://dx.doi.org/10.1144/SP367.14

setting of Andersonian fault zones and, in particular, the geomechanical response of their fluid-saturated damage zones in the approach to failure.

Throughout this paper the following conventions are used. The maximum principal stress σ_1 is taken as compressive, with $\sigma_1 \geq \sigma_2 \geq \sigma_3$. The term 'cracks' is used to include all porosity in a deformed rock, including the original pores and the tectonically induced micro- and macro-scale brittle damage. The models presented in this paper are all based on the poroelastic response of Berea sandstone. The intact elastic properties of this rock are listed in Table 1 (Scott *et al.* 1993). The outline of the paper is as follows. The patterns of brittle damage recorded in faulted rocks are briefly reviewed from the published literature and then simplified into three patterns used in the models presented in this paper. The related concepts of brittle fabrics and crack tensors are also discussed. The theoretical basis for anisotropic poroelasticity is described, complete with a description of the modelling of poroelastic properties in cracked rocks using Effective Medium Theory. The geomechanical consequences of the modified Law of Effective Stress are first explored in a framework of pore pressure/stress coupling. The influences of crack pattern, stress field orientation and crack density are described. The implications for fault stability are then quantified in terms of fracture susceptibility, that is, the change in pore-fluid pressure required to cause fault slip. The Discussion section revisits the key issues raised by the predictive models and is followed by a brief Summary.

Brittle fabrics in faulted rocks

Field, laboratory and theoretical studies have shown that natural faults are composite shear fractures formed through the interaction and coalescence of many smaller, dominantly tensile, microcracks (e.g. Peng & Johnson 1972; Hallbauer *et al.* 1973; Tapponnier & Brace 1976; Wong 1982; Lockner *et al.* 1991; Reches & Lockner 1994; Vermilye & Scholz 1998; Wibberley *et al.* 2000; Healy *et al.* 2006*a*, *b*). Microstructural analyses of field and laboratory samples has shown that the pattern of microcracks, that is, their average preferred orientation over a sampled spatial domain, varies systematically with position around the larger fault zone (Moore & Lockner 1995; Vermilye & Scholz 1998; Wibberley *et al.* 2000). Seismic shear waves have also been used to demonstrate that patterns of fluid-filled cracks are common in the faulted upper crust, with most cracks aligned in the direction of the *in situ* maximum horizontal stress (Zatsepin & Crampin 1997).

Tensile microcracks are predicted to form with their poles parallel to the direction of σ_3 (the least compressive stress). In the relatively intact rock around the fault tip line, a homogeneous state of stress should promote the formation of parallel crack patterns (uniaxial clusters) and this is matched by observations from those field and laboratory samples where the fault tip lines are well constrained. Alongside the fault surface, repeated slip events in response to stress cycling combined with stress field heterogeneity due to an irregular fault surface will cause local stress rotations and the dominance of axial crack patterns (uniaxial

Table 1. *Elastic constants of intact Berea sandstone used in this study (Scott* et al. *1993). The full stiffness matrix (c_{ij}) is shown in Voigt format and units of GPa. For isotropic Berea sandstone, this gives values of 37.8 GPa for Young's Modulus and 0.11 for Poisson's ratio*

	c_{i1}	c_{i2}	c_{i3}	c_{i4}	c_{i5}	c_{i6}
c_{1j}	38.83	4.80	4.80	0.00	0.00	0.00
c_{2j}	4.80	38.83	4.80	0.00	0.00	0.00
c_{3j}	4.80	4.80	38.83	0.00	0.00	0.00
c_{4j}	0.00	0.00	0.00	17.02	0.00	0.00
c_{5j}	0.00	0.00	0.00	0.00	17.02	0.00
c_{6j}	0.00	0.00	0.00	0.00	0.00	17.02

Table 2. *Elastic stiffness matrices of cracked Berea sandstone, calculated using the equations of Effective Medium Theory (Guéguen & Sarout 2009) with a random set of cracks at 3 different crack densities of 0.1, 0.4 and 0.7*

	c_{i1}	c_{i2}	c_{i3}	c_{i4}	c_{i5}	c_{i6}
$\rho = 0.1$						
c_{1j}	32.62	3.35	3.35	0.00	0.00	0.00
c_{2j}	3.35	32.62	3.35	0.00	0.00	0.00
c_{3j}	3.35	3.35	32.62	0.00	0.00	0.00
c_{4j}	0.00	0.00	0.00	14.63	0.00	0.00
c_{5j}	0.00	0.00	0.00	0.00	14.63	0.00
c_{6j}	0.00	0.00	0.00	0.00	0.00	14.63
$\rho = 0.4$						
c_{1j}	22.12	1.51	1.51	0.00	0.00	0.00
c_{2j}	1.51	22.12	1.51	0.00	0.00	0.00
c_{3j}	1.51	1.51	22.12	0.00	0.00	0.00
c_{4j}	0.00	0.00	0.00	10.30	0.00	0.00
c_{5j}	0.00	0.00	0.00	0.00	10.30	0.00
c_{6j}	0.00	0.00	0.00	0.00	0.00	10.30
$\rho = 0.7$						
c_{1j}	16.76	0.86	0.86	0.00	0.00	0.00
c_{2j}	0.86	16.76	0.86	0.00	0.00	0.00
c_{3j}	0.86	0.86	16.76	0.00	0.00	0.00
c_{4j}	0.00	0.00	0.00	7.95	0.00	0.00
c_{5j}	0.00	0.00	0.00	0.00	7.95	0.00
c_{6j}	0.00	0.00	0.00	0.00	0.00	7.95

Table 3. *Elastic stiffness matrices of cracked Berea sandstone, calculated using the equations of Effective Medium Theory (Guéguen & Sarout 2009) with a perfectly uniaxial or parallel set of cracks at 3 different crack densities of 0.1, 0.4 and 0.7*

	c_{i1}	c_{i2}	c_{i3}	c_{i4}	c_{i5}	c_{i6}
$\rho = 0.1$						
c_{1j}	38.62	4.59	3.08	0.00	0.00	0.00
c_{2j}	4.59	38.62	3.08	0.00	0.00	0.00
c_{3j}	3.08	3.08	24.93	0.00	0.00	0.00
c_{4j}	0.00	0.00	0.00	13.67	0.00	0.00
c_{5j}	0.00	0.00	0.00	0.00	13.67	0.00
c_{6j}	0.00	0.00	0.00	0.00	0.00	17.01
$\rho = 0.4$						
c_{1j}	38.42	4.39	1.49	0.00	0.00	0.00
c_{2j}	4.39	38.42	1.49	0.00	0.00	0.00
c_{3j}	1.49	1.49	12.02	0.00	0.00	0.00
c_{4j}	0.00	0.00	0.00	8.61	0.00	0.00
c_{5j}	0.00	0.00	0.00	0.00	8.61	0.00
c_{6j}	0.00	0.00	0.00	0.00	0.00	17.02
$\rho = 0.7$						
c_{1j}	38.36	4.33	0.98	0.00	0.00	0.00
c_{2j}	4.33	38.36	0.98	0.00	0.00	0.00
c_{3j}	0.98	0.98	7.92	0.00	0.00	0.00
c_{4j}	0.00	0.00	0.00	6.28	0.00	0.00
c_{5j}	0.00	0.00	0.00	0.00	6.28	0.00
c_{6j}	0.00	0.00	0.00	0.00	0.00	17.02

Table 4. *Elastic stiffness matrices of cracked Berea sandstone, calculated using the equations of Effective Medium Theory (Guéguen & Sarout 2009) with a perfectly axial (or girdle) pattern of cracks at 3 different crack densities of 0.1, 0.4 and 0.7*

	c_{i1}	c_{i2}	c_{i3}	c_{i4}	c_{i5}	c_{i6}
$\rho = 0.1$						
c_{1j}	30.29	2.94	3.66	0.00	0.00	0.00
c_{2j}	2.94	30.29	3.66	0.00	0.00	0.00
c_{3j}	3.66	3.66	38.58	0.00	0.00	0.00
c_{4j}	0.00	0.00	0.00		0.00	0.00
c_{5j}	0.00	0.00	0.00	0.00		0.00
c_{6j}	0.00	0.00	0.00	0.00	0.00	
$\rho = 0.4$						
c_{1j}	18.29	1.08	2.13	0.00	0.00	0.00
c_{2j}	1.08	18.29	2.13	0.00	0.00	0.00
c_{3j}	2.13	2.13	38.24	0.00	0.00	0.00
c_{4j}	0.00	0.00	0.00	11.43	0.00	0.00
c_{5j}	0.00	0.00	0.00	0.00	11.43	0.00
c_{6j}	0.00	0.00	0.00	0.00	0.00	8.61
$\rho = 0.7$						
c_{1j}	13.11	0.56	1.50	0.00	0.00	0.00
c_{2j}	0.56	13.11	1.50	0.00	0.00	0.00
c_{3j}	1.50	1.50	38.10	0.00	0.00	0.00
c_{4j}	0.00	0.00	0.00	9.17	0.00	0.00
c_{5j}	0.00	0.00	0.00	0.00	9.17	0.00
c_{6j}	0.00	0.00	0.00	0.00	0.00	9.17

girdles) in the damage zones of Andersonian faults. Cataclasis and accumulated fault slip will tend to obliterate preferred orientations in the fault core and, in the absence of phyllosilicate minerals, the tectonically induced porosity is likely to be randomly oriented (isotropic).

Spatially varying idealized crack patterns around Andersonian faults are shown in Figures 1–3 for normal, strike-slip and thrust faults, respectively. The crack patterns are shown in schematic block diagrams with any symmetry attributes indicated, and then as stereonets of poles to microcracks. In natural examples, a degree of scatter (noise) is to be expected but the models presented in this paper use these end-member orientation distributions to explore the effects of anisotropic brittle damage around saturated faults.

The presence of preferred orientations in the cracks observed around fault zones leads to the idea of quantifying these patterns. Several workers in engineering and medicine have suggested the use of a fabric tensor, a symmetric second-rank tensor which characterizes the arrangement of components in a multiphase or porous system (Oda 1982; Kanatani 1984; Cowin 1985). In a porous rock, the fabric tensor can be considered as the second-order measure of porosity, with the scalar value of porosity (*sensu stricto*) as the relative volume fraction of pores. The benefit of the tensorial approach is that the orientations of the pores, or inequant cracks in the case of a faulted rock, can be quantified. A similar approach has been taken by Woodcock (1977), who used the eigenvalues of an orientation tensor to capture and then display the key attributes of any distribution of directional data on a modified Flinn plot (Flinn 1962; Ramsay 1967). A modified Flinn plot for brittle fabrics is shown in Figure 4, using the idealized crack patterns from Figures 1–3. Two useful parameters derived from the eigenvalues of the orientation distribution are K and C, defined:

$$K = \frac{\log_e(e_1/e_2)}{\log_e(e_2/e_3)} \qquad (2)$$

$$C = \log_e(e_1/e_3). \qquad (3)$$

where e_1, e_2 and e_3 are the normalized eigenvalues of the orientation tensor. K measures the kind of pattern (i.e. cluster or girdle), whereas C measures the strength of the pattern. K values of 0 and + infinity define the x-axis (uniaxial girdle) and y-axis (uniaxial cluster) respectively. A C value of 0 designates a random pattern (Fig. 4), and these fabrics plot at the origin. Brittle fabrics intermediate between uniaxial clusters and uniaxial girdles fall

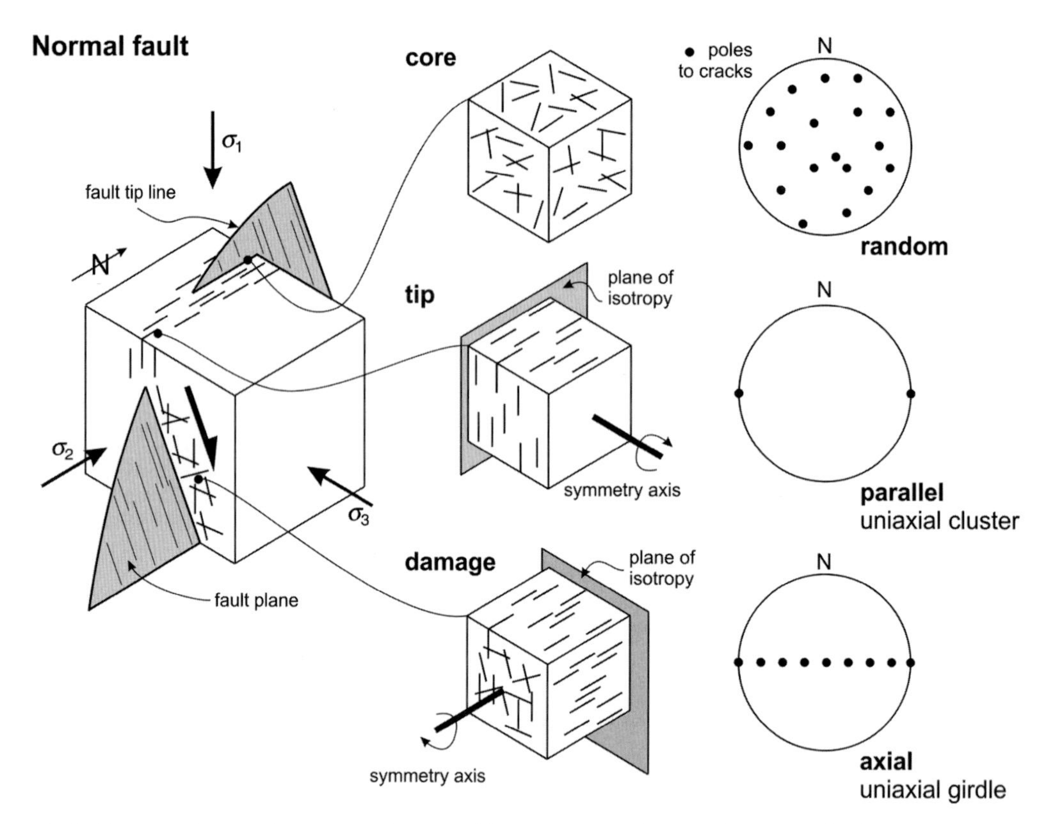

Fig. 1. Schematic diagram of an Andersonian normal fault and the associated crack patterns. The fault core is characterized by a random pattern, the damage zone has an axial (or radial) pattern with the poles to cracks in a vertical plane and the tip zone has a vertical parallel pattern. The axial and parallel patterns correspond to uniaxial girdles and uniaxial clusters, respectively.

in the middle of the plot. Woodcock (1977) pointed out that this plot is strictly only valid for fabrics with orthorhombic symmetry and above. For monoclinic and triclinic fabrics, the eigenvalues of the orientation tensor do not coincide with the maximum and minimum moments of the distribution. Work is in progress to provide a quantitative formal assessment of brittle fabrics from fault zones, including their symmetry.

Theoretical basis of anisotropic poroelasticity

The concept of effective stress is attributed to Terzaghi (1943) who experimentally validated an empirical law for saturated soils [Equation (1)]. Skempton (1960) explored further and assessed the law using experimental data for soils, rocks and concrete. He found the fit was good for soils, but poor for rocks and concrete. Handin *et al.* (1963) conducted the first systematic rock deformation experiments to measure effective stress and failure in the laboratory for a range of rock types, loads and confining pressures. The underlying theoretical basis for the Law of Effective Stress lies in poroelasticity (Biot 1955). For isotropic materials, a simple formulation has been provided by Rice & Cleary (1976), while the solution for anisotropic materials has been formulated by several authors (Nur & Byerlee 1971; Carroll 1979; Thompson & Willis 1991). The generalized Law of Effective Stress incorporating anisotropic poroelasticity is

$$\sigma'_{ij} = \sigma_{ij} - \beta_{ij} P_{\mathrm{f}} \tag{4}$$

where σ' is the effective stress tensor and β is a second-rank tensor variously termed the effective stress, or Biot, coefficient. In this paper, β will be referred to as the Biot tensor and is defined as

$$\beta_{ij} = \delta_{ij} - c_{ijkl} s'_{klmm} \tag{5}$$

where δ is the Kronecker delta, c is the elastic stiffness of the cracked rock and s' is the elastic

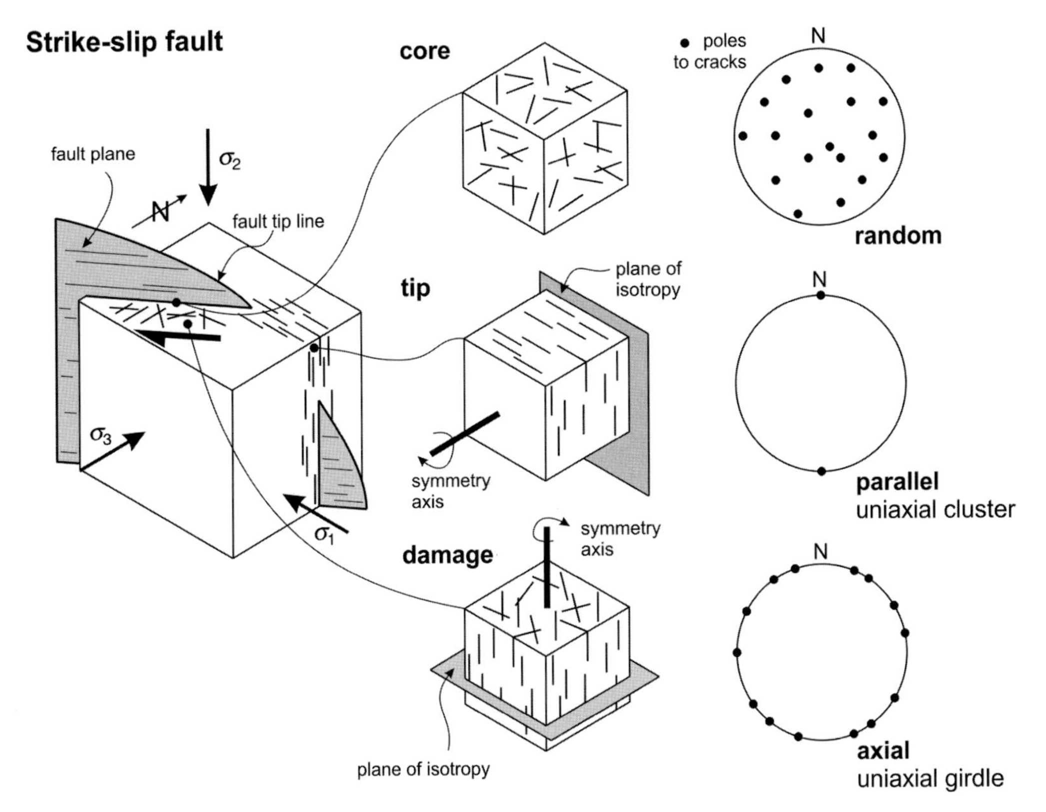

Fig. 2. Schematic diagram of an Andersonian strike-slip fault and the associated crack patterns. The fault core is characterized by a random pattern, the damage zone has an axial (or radial) pattern with the poles to cracks in a horizontal plane and the tip zone has a vertical parallel pattern. The axial and parallel patterns correspond to uniaxial girdles and uniaxial clusters, respectively.

compliance of the intact rock (Carroll 1979). These simple explicit equations mean that if we can measure or calculate the elastic properties of the intact and cracked rock, we can predict the geomechanical response of fractured rock according to the general form of the Law of Effective Stress.

Rock deformation experiments show that changes in the proportion of cracks alter the bulk elastic properties of a rock (Brace *et al.* 1966). While increases in confining pressure lead to small increases in elastic stiffness, increasing deviatoric stress towards the yield strength generates marked decreases in elastic stiffness (Scott *et al.* 1993; Sayers & Kachanov 1995). Conceptually, these effects can be understood in terms of the opening and closing of cracks along the loading path. Effective Medium Theory (EMT) provides a rigorous and quantitative basis for linking changes in crack density to changes in elastic properties (Sayers & Kachanov 1995; Guéguen & Sarout 2009). The crack pattern is defined as a second-rank tensor in EMT, analogous to the fabric and orientation tensors described above which incorporates the effects of any orientation distribution in the crack population. The quantitative comparison of data from rocks deformed under controlled laboratory conditions (e.g. Sayers & Kachanov 1995; Schubnel *et al.* 2006) with theoretical predictions using EMT confirms that the theory accurately captures the effects of crack damage in a tractable form. The models presented in this paper are based on calculations of effective stress using Equation (4), where the Biot tensor has been calculated using Equation (5) and the elastic properties for different patterns of cracks in Berea sandstone have been derived from the previously published equations of EMT (see Healy 2008, 2009 for further details).

Pore-pressure/stress coupling in fractured, anisotropic rocks

Conventional models for fault stability have assumed that pore-fluid pressure and stress are independent (e.g. Hubbert & Rubey 1959; Handin

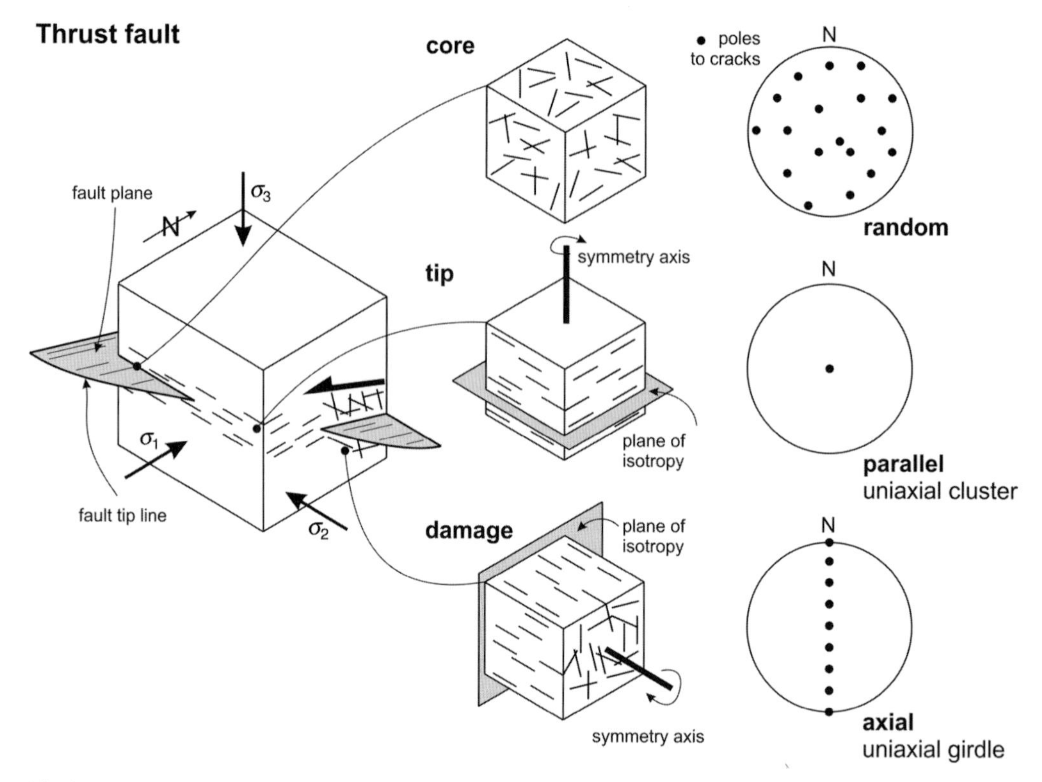

Fig. 3. Schematic diagram of an Andersonian thrust fault and the associated crack patterns. The fault core is characterized by a random pattern, the damage zone has an axial (or radial) pattern with the poles to cracks in a vertical plane and the tip zone has a horizontal parallel pattern. The axial and parallel patterns correspond to uniaxial girdles and uniaxial clusters, respectively.

et al. 1963). However, data from hydrocarbon fields and basins documents a systematic relationship between pore-fluid pressure and the minimum horizontal stress σ_h (typically σ_3 in these dominantly extensional settings). At the scale of individual oil fields, pore pressure depletion from oil extraction is associated with a decrease in the minimum horizontal stress magnitude (Salz 1977; Teufel *et al.* 1991; Tingay *et al.* 2009). In addition, at the scale of sedimentary basins, higher values of σ_h are observed in overpressured sequences (Breckels & van Eekelen 1982; Addis 1997; Hillis 2001). This relationship between changes in pore-fluid pressure and minimum horizontal stress is called pore-pressure/stress coupling and is a renewed focus of interest in extending the life of mature fields; it is also relevant for predicting the effects of CO_2 sequestration (Vidal-Gilbert *et al.* 2009). Previous analyses of pore-pressure/stress coupling have adopted a poroelastic approach with a vertical uniaxial strain boundary condition to reflect sedimentary compaction and the assumption of elastic isotropy in both the rock framework and the pore space (Teufel *et al.* 1991; Engelder & Fischer 1994; Hillis 2001; Goulty 2003).

This paper develops the concept of pore-pressure/stress coupling in a framework of anisotropic poroelasticity without the uniaxial strain boundary condition, as this is believed to be more applicable to fractured rocks in and around saturated fault zones. The assumption of isotropic poroelasticity used in previous work has the corollary that any macroscopic fractures in a saturated rock mass (i.e. the probable loci for geomechanical risk) exist in isolation from any patterns of micro- to meso-scale cracks. This is contradicted by data from field and laboratory studies which unequivocally show that (a) the intensity of brittle crack damage, at all scales, increases exponentially towards fault zones and (b) this damage has distinct patterns. Even low densities of inequant cracks and pores reduce the elastic stiffness of rocks and preferred orientations of inequant cracks produce elastic anisotropy (Brace *et al.* 1966; Scott *et al.* 1993; Crampin & Chastin 2003).

The idealized crack patterns described in relation to Andersonian faults (Figs 1–3) generate different

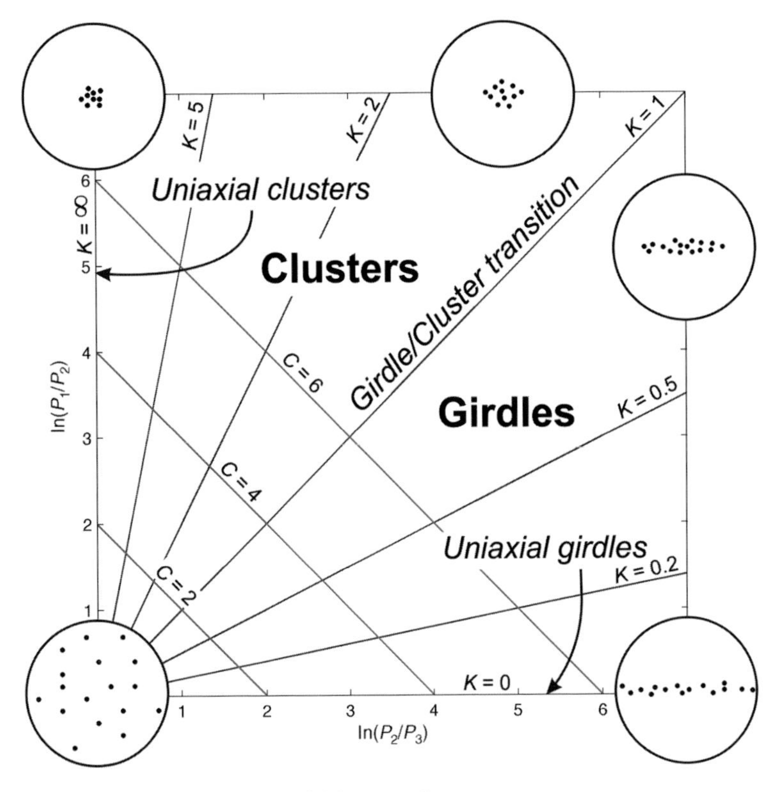

Fig. 4. Schematic modified Flinn plot of brittle fabrics or crack patterns.

geomechanical behaviour in response to changes in pore-fluid pressure. In the case of random cracks (Fig. 5a), the bulk porosity is isotropic and the predicted changes in effective stress mimic the predictions from the simplified isotropic form of the Law of Effective Stress (Equation (1)). An increase in pore-fluid pressure drives the stress state to the left on a Mohr diagram (i.e. towards failure; Fig. 5b), whereas a decrease in fluid pressure drives the stress state to the right (i.e. away from failure; Fig. 5c). Note that the effective differential stress σ'_d $(=\sigma'_1 - \sigma'_3)$ remains constant (the diameter of the Mohr circle for the stress state is fixed) and therefore the maximum shear stress does not vary.

A pattern of parallel cracks (a uniaxial cluster) predicts different behaviour (Fig. 5d), with changes in the effective differential stress due to the anisotropic effects of pore pressure in the inequant pores (Fig. 5e, f). For a parallel crack pattern oriented with the poles to the cracks parallel to σ_3, the effect of changes in pore-fluid pressure are 'felt' more strongly in the direction of σ'_3 and less strongly in the directions of σ'_1 and σ'_2. Note how the values of σ'_1 and σ'_2 barely change in the Mohr plots (Fig. 5e, f), while the value of σ'_3 moves significantly: this is pore-pressure/stress coupling. The change in effective differential stress means that the rock mass approaches failure (pore pressure increase) or stability (pore pressure decrease) more rapidly than the isotropic case. For a pattern of axial cracks (a uniaxial girdle) oriented with the poles to the cracks confined to the $\sigma_1 - \sigma_3$ principal plane, the effects are more subtle. The effective differential stress $\sigma'_1 - \sigma'_3$ remains constant, but the value of σ'_2 is different in comparison to the isotropic (random pattern) case (compare Fig. 5h, i with Fig. 5b, c). This is explained by the fact that effective stresses in the $\sigma'_1 - \sigma'_3$ principal plane 'feel' an isotropic pattern of cracks, whereas effective stresses in the σ'_2 direction barely register fluid pressure changes along crack tips (Fig. 5g). Note that in these cases the effective stress ratio R' is changing with respect to the stress ratio R, where these terms are defined by Lisle *et al.* (2006) as:

$$R' = \frac{(\sigma'_2 - \sigma'_3)}{(\sigma'_1 - \sigma'_2)} \tag{6}$$

$$R = \frac{(\sigma_2 - \sigma_3)}{(\sigma_1 - \sigma_2)}. \tag{7}$$

Stresses and fluid pressures can change over time; while many faults may form under more or

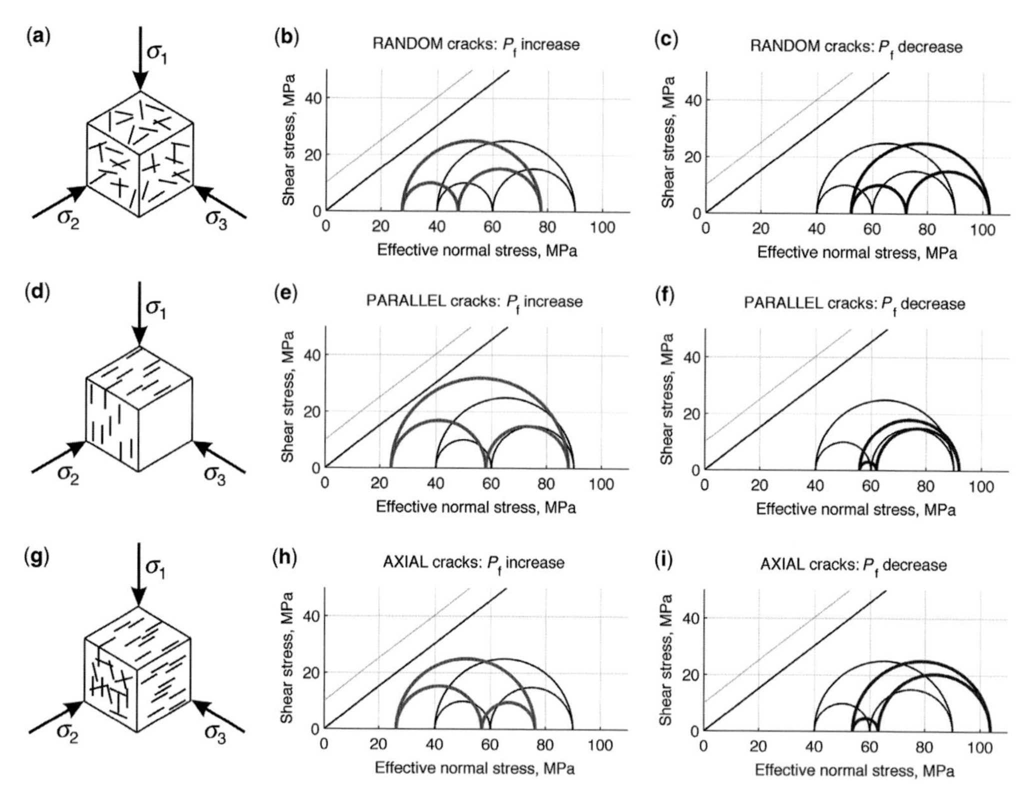

Fig. 5. Mohr diagrams showing the interaction between crack pattern, fluid pressure change and effective stress. Two failure envelopes are shown, one for brittle failure (*sensu stricto*) with a finite cohesion, and one for frictional sliding with no cohesion. (**a–c**) For a random pattern of cracks, the porosity is isotropic and the changes in effective stress due to an increase or decrease in fluid pressure obey the familiar isotropic version of the law of effective stress. (**d–f**) For a parallel pattern of cracks oriented with their normals parallel to σ_3, the change in effective stress due to an increase or decrease in fluid pressure is markedly different from the random pattern (isotropic poroelasticity) case. Note how the differential stress σ'_d ($\sigma'_1 - \sigma'_3$) changes for both an increase and decrease in fluid pressure, and therefore the maximum shear stress is different in comparison to the isotropic case. (**g–i**) For the axial pattern of cracks oriented with their poles in the $\sigma_1 - \sigma_3$ principal plane, the change in effective stress appears similar to that for the random pattern case. In detail though, there is a clear impact on the value of σ'_2. In all cases, the elastic constants of the intact rock are those of Berea sandstone (Table 1) and the modelled crack density is 0.7 (see Tables 2, 3 and 4).

less Andersonian conditions, the ambient stress field may become reoriented. In such cases, the crack patterns surrounding the fault and containing the pore fluid may become 'misoriented' with respect to the local stress. *In situ* stress measurements reveal that many extensional rift basins dominated by normal faults are now in a state of Andersonian strike-slip or thrust faulting, for example, North Sea, Norwegian Sea and offshore Australia (Hillis & Nelson 2005; King *et al.* 2008).

An example of a 'misoriented' crack pattern in a stress field different from that in which it formed is shown in Figure 6c, with σ_1 now perpendicular to the cracks. In the case of intact anisotropic rocks, such as shales, the porosity can be defined by inequant grain boundary cracks. This bedding-parallel anisotropy will also be normal to σ_1 in a normal fault regime (Fig. 6d). Parallel crack patterns, with their poles oriented parallel to σ_1 (Fig. 7a), lead to counter-intuitive effective stress predictions as first noted by Chen & Nur (1992).

Figure 7b, c show the Mohr stress states for an increase and decrease in pore-fluid pressure, respectively. An increase in pore-fluid pressure can now cause a reduction in effective differential stress and therefore a reduction in geomechanical risk; the final stress state in Figure 7b is further from failure than the initial state. Conversely, a decrease in pore-fluid pressure can push the stress state closer to failure (Fig. 7c) with an increase in effective

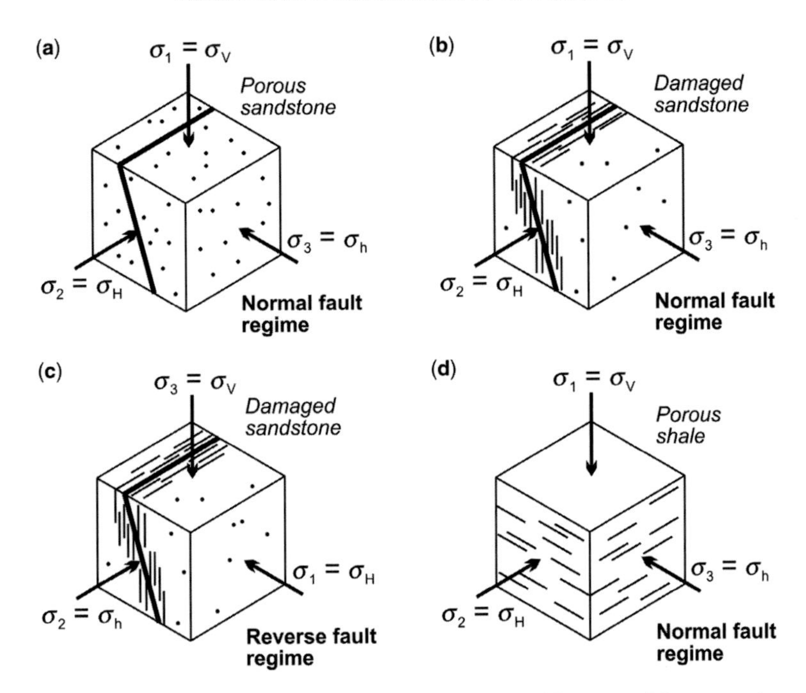

Fig. 6. Schematic illustrations of saturated porous and cracked rocks under different possible stress regimes. (**a**) Porous sandstone in an Andersonian normal fault regime. In the absence of any brittle damage, a common assumption is that the rock mass is poroelastically isotropic. (**b**) Brittle damage around normal faults can be approximated as arrays of cracks around the fault plane. (**c**) Pre-existing normal fault now in an Andersonian reverse (thrust) fault regime. In this case, the maximum compressive stress (σ_1) is oriented perpendicular to the pre-existing brittle damage in the normal fault zone. (**d**) Anisotropic shale in an Andersonian normal fault regime. Porosity in shales is often dominated by grain boundary cracks and pores aligned parallel to the laminations, and therefore perpendicular to the maximum compressive stress (σ_1).

differential stress. The predictions for a 'misoriented' axial crack pattern, oriented with crack poles distributed in the $\sigma_2 - \sigma_3$ principal plane (Fig. 7d), also lead to differences in comparison to the behaviour shown for the standard case (compare Fig. 7e, f with Fig. 5h, i).

One key point to emerge from these simple models is that, in the presence of anisotropic porosity (e.g. example inequant cracks around a fault zone), the geomechanical response of faulted saturated rock will depend on (a) the pattern of cracks (i.e. the brittle fabric) and (b) the orientation of the local stress system with respect to this pattern. A rich variety of possible behaviours exist, including counter-intuitive relationships between fault stability and changes in fluid pressure. Scope exists to map and quantify brittle fabrics in much more detail to provide more accurate and reliable estimates of fault stability in saturated rocks.

The models shown so far have all used a single and constant value of crack density (0.7). The value of crack density in relation to progressive brittle rock deformation has been calibrated from laboratory experiments (Guéguen *et al.* 1997), with a value of 0.15 at the percolation threshold (i.e. connected pores, onset of bulk permeability) and 1.0 at whole rock failure. In natural fault zones, crack density varies with distance from the principal slip surface and it is important to constrain the effect of this parameter in the model predictions. Figure 8 shows the same crack patterns, applied stresses and pore-fluid pressures used in Figure 5, but each Mohr plot now shows predicted effective stress states for three separate crack density values: 0.1, 0.4 and 0.7. In all cases, a non-linear evolution is observed: with increasing crack density the change in stress state slows, but for all positive increments of crack density the behaviour is consistent and mirrors that shown in Figure 5 for a single crack density value.

Implications for fault mechanics

Several quantitative measures of fault and fracture stability have been proposed: slip tendency (Morris *et al.* 1996); dilatation tendency (Ferrill *et al.* 1999); and fracture susceptibility (Mildren

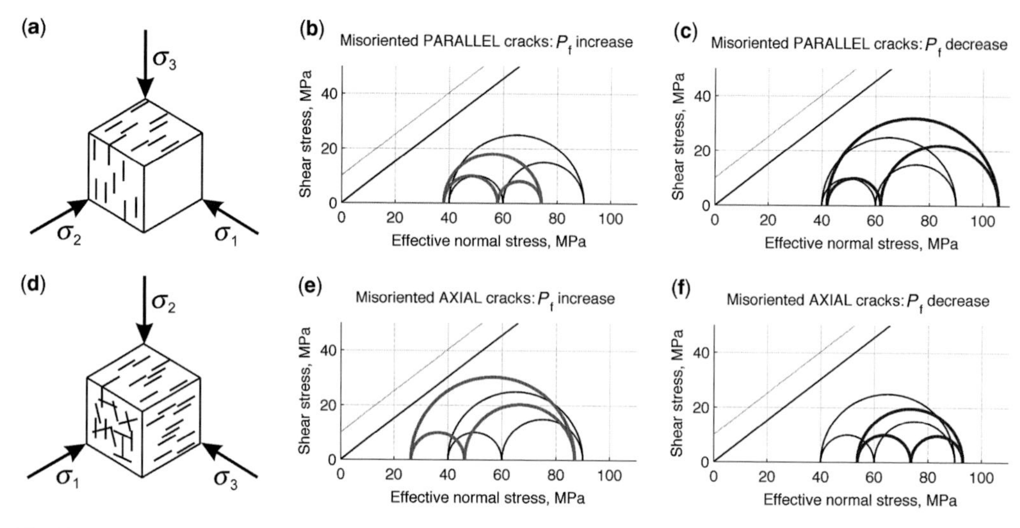

Fig. 7. Mohr diagrams for misoriented crack patterns. (**a–c**) For the parallel crack pattern, with the poles to the cracks oriented parallel to σ_1 instead of σ_3, the difference in predicted effective stresses for fluid pressure changes are marked in comparison to the 'correctly oriented' case (see Fig. 5d–f). Increases in P_f move the effective stress state *away* from failure and decreases in P_f move the effective stress state *towards* failure. (**d–f**) For the axial pattern with the poles to cracks distributed in the $\sigma_2 - \sigma_3$ principal plane, the difference in effective stress with respect to the 'correctly oriented' case is again striking (see Fig. 5g–i). Increases in P_f move the effective stress state much closer towards failure and decreases in P_f move the effective stress state further from failure.

et al. 2002). The published expressions for slip and dilatation tendency do not include a term for pore-fluid pressure, and should therefore only be applied to dry rocks. Fracture susceptibility measures the change in pore-fluid pressure necessary to cause fault reactivation (for cohesive or non-cohesive rocks) and incorporates the isotropic form of the Law of Effective Stress. The predictions are plotted as stereonets to poles of fracture planes coloured by pore-fluid pressure. Low values of fracture susceptibility imply that only a small increase in pore-fluid pressure is necessary to reactivate a fault of that orientation.

The published expression for fracture susceptibility can easily be modified to include anisotropic poroelasticity (Equation (4)). Figure 9 shows the effects of anisotropic poroelasticity due to different crack patterns on the predicted fracture susceptibility for Andersonian faults. For each fault orientation, the parallel and axial crack patterns are oriented in the correct orientation for the applied stresses (see Figs 1–3 for orientations). For random crack patterns (Fig. 9, top row) the predicted fracture susceptibility mirrors the inverse of slip tendency: low values of fracture susceptibility (small increase in pore-fluid pressure) correspond to high slip tendency. For example, in an Andersonian normal fault stress regime (Fig. 9, left column) the most likely slip planes are arranged at *c.* 30° to σ_1 and their poles lie close to σ_3 (Fig. 9, blue regions). The most unlikely slip planes, needing the highest increases in pore-fluid pressure, are those with fault normals arranged subvertically (Fig. 9, red colours).

As shown in the Mohr diagrams in Figure 5, the effect of an axial crack pattern on fracture susceptibility is barely noticeable (Fig. 9, bottom row) in comparison to the random crack pattern case. This is explained by the orientation of the random component in the axial pattern, with the poles to the cracks confined to the $\sigma_1 - \sigma_3$ principal plane for all fault types. Fault stability critically depends on σ_1' and σ_3', and these stresses 'see' an isotropic distribution of cracks in both the random and axial pattern cases. For parallel crack patterns the predicted fracture susceptibility is markedly different from the comparable random (or isotropic) case (Fig. 9, middle row). For a given orientation of potential slip, the magnitude of pore-fluid pressure is different; note that the orientation of preferred slip planes also changes. For the normal, strike-slip and thrust-fault regimes, the effect of a parallel crack pattern is to promote slip on non-Andersonian fault planes. This is most clearly seen for the strike-slip case, where the minima in the fracture susceptibility define four distinct clusters (Fig. 9, central stereonet). Compare this to the corresponding case for a random crack pattern (Fig. 9, top row, centre) where the fault orientations most likely to fail (small increase in pore-fluid pressure, blue

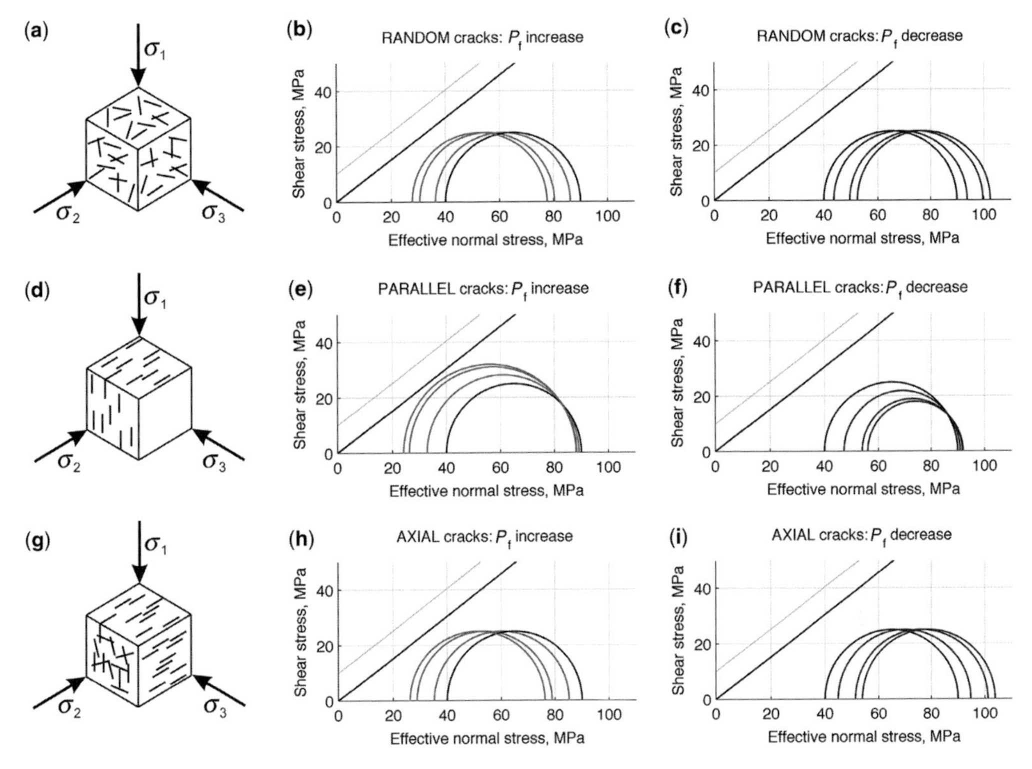

Fig. 8. Mohr diagrams showing predicted effective stresses for different crack patterns and varying crack density for (**a–c**) a random pattern of cracks; (**d–f**) a parallel pattern of cracks oriented with their normals parallel to σ_3; and (**g–i**) the axial pattern of cracks oriented with their poles in the $\sigma_1 - \sigma_3$ principal plane. In all cases, the elastic constants of the intact rock are those of Berea sandstone (Table 1) and the modelled crack densities are 0.1, 0.4 and 0.7 (see Tables 2, 3 and 4).

colours) define two pole clusters straddling the horizontal stereonet primitive. In the random crack pattern case, predictions of fracture susceptibility fit the standard Andersonian conjugate model. For parallel crack patterns, predictions of fracture susceptibility suggest a quadrimodal fracture pattern.

Discussion

The model predictions presented in this paper are based on three simplified crack patterns: random (with orientation tensor values of $C = 0$, $K = 0$); perfectly uniaxial/parallel ($C = +$infinity, $K = +$ infinity); and perfectly axial/girdle ($C = +$infinity, $K = 0$). How well these idealized patterns represent natural distributions of fault-related cracks is hard to quantify for two reasons. Firstly, there is a relative paucity of brittle fabric data, especially in contrast to their ductile (or plastic) fabric equivalents (i.e. lattice or crystallographic-preferred orientations). This may be partly explained by the labour-intensive nature of crack orientation measurements with an optical microscope fitted with a Universal stage. Work is in progress to develop a rapid, consistent and automated method to address this key issue. Secondly, several datasets are severely incomplete with thin section measurements of microcrack orientation that do not span the full three dimensions or are limited to cracks measured in a single mineral phase in a polymineralic rock. Given these limitations, the best-published datasets lend support to the idea of random, parallel and axial crack patterns as useful end-member distributions of natural cracks around fault zones (Anders & Wiltschko 1994; Vermilye & Scholz 1998; Wilson *et al.* 2003).

Theoretical considerations and field data suggest that the interaction and coalescence of finite tensile cracks will not produce strictly Andersonian fault orientations even under conditions of homogeneous stress. Quadrimodal and polymodal fault orientation distributions are expected and observed in the field and the laboratory (Aydin & Reches 1982; Reches & Dieterich 1983; Healy *et al.* 2006*a*, *b*) and a predicted discordance between the orientations of the tensile cracks and the

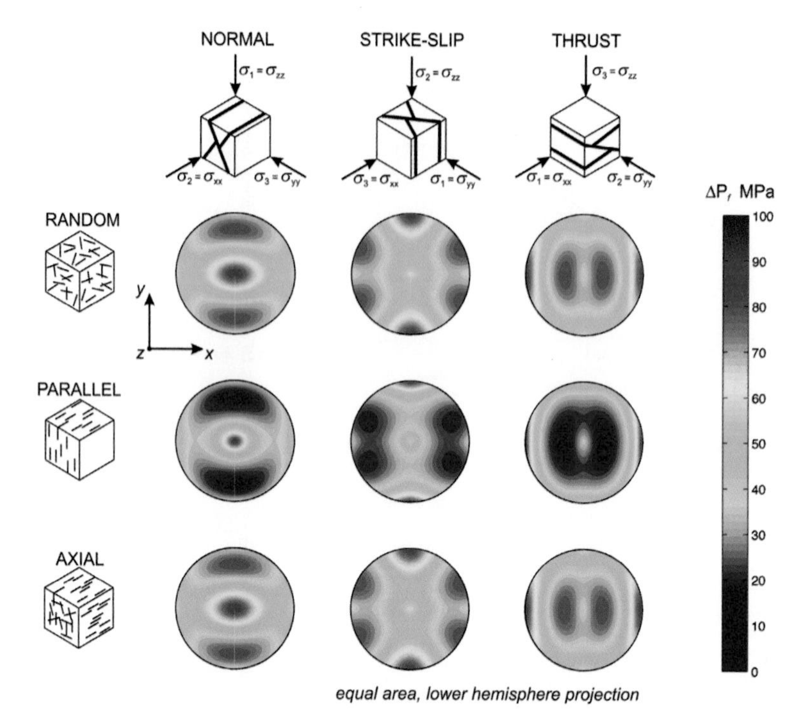

Fig. 9. Stereonets of fracture susceptibility for different Andersonian fault regimes showing the effect of changing the crack pattern. All stereonets are plotted as equal-area lower hemispheres. Fracture susceptibility is expressed as the change in pore-fluid pressure required to cause frictional slip on a cohesionless fault with a friction coefficient of 0.76. The applied stress is $\sigma_1 = 90$, $\sigma_2 = 60$ and $\sigma_3 = 40$ MPa. The intact rock elastic constants are those of Berea sandstone (Table 1) and the crack density used to model the poroelastic response is 0.7 (see Tables 2, 3 and 4).

composite shear fractures is also observed in the data (Blenkinsop 2008). To better understand the patterns of cracks developed around faults, quantitative crack patterns from natural faults and laboratory deformation experiments will in future be plotted on the modified Flinn plot for brittle fabrics (Fig. 4) to assess the presence or absence of symmetry in their distributions, together with a detailed assessment of their orientation with respect to the larger fault structure.

Faulting in the Earth's crust involves a wide range of processes and deformation mechanisms; models restricted to poroelasticity, whether isotropic or anisotropic, are therefore at best incomplete. However, any deformation driven by, or related to, natural fluid pressure variations will involve a poroelastic response along some of the load path. Isolating the purely poroelastic response, and identifying the important role anisotropies can play in this response, remains a valid and useful first step. Work in progress aims to couple this approach with reaction kinetics to provide further insight into the stability of natural fault zones (e.g. Hacker 1997). Further research could address the limits of poroelastic behaviour during brittle failure through careful laboratory experiments, and provide key insights into the physical nature of the Biot tensor at the scale of grains (and their cements), pores and cracks.

Summary

Porosity around fault zones is dominated by tectonically induced cracks. The shapes of individual cracks and their preferred orientations render the bulk rock anisotropic. Better predictions of strength or stability can be made for fluid-saturated fault zones by incorporating anisotropic poroelasticity into the Law of Effective Stress. Model predictions of the geomechanical response of fractured sandstone show that the difference between the commonly assumed isotropic form of the Law of Effective Stress and the modified anisotropic form are significant for likely combinations of crack pattern, orientation and density. Increases in pore-fluid pressure can lead to increased stability (or reduced risk) and decreases in fluid pressure can lead to fault reactivation (or increased risk). There is a large volume of previous work addressing the permeability of fractured rock. However, the measurement and characterization of porosity in faulted rocks, and the potentially significant mechanical consequences, seem to have been overlooked.

The author thanks the following people for discussion: D. Dewhurst (CSIRO Perth); B. Gurevich (Curtin University, Perth); R. Hillis (DET CRC, Adelaide); G. Lloyd (University of Leeds); P. Ruelland (Total); P. Schutjens (Shell); Z. Shipton (Strathclyde University); C. Wibberley (Total). The comments of three reviewers (I. Main and 2 anonymous) also helped to clarify the arguments made in this paper. This work has been part-funded by a NERC New Investigator award NE/I001743/1, which is gratefully acknowledged.

References

ADDIS, M. 1997. The stress–depletion response of reservoirs. *SPE Annual Technical Conference and Exhibition, 5–8 October 1997*. San Antonio, Texas.

ANDERS, M. & WILTSCHKO, D. 1994. Microfracturing, paleostress and the growth of faults. *Journal of Structural Geology*, **16**, 795–815.

ANDERSON, E. 1951. *The Dynamics of Faulting and Dyke Formation with Applications to Britain*. Oliver & Boyd, Edinburgh and London.

AYDIN, A. & RECHES, Z. 1982. Number and orientation of fault sets in the field and in experiments. *Geology*, **10**, 107.

BIOT, M. 1955. Theory of elasticity and consolidation for a porous anisotropic solid. *Journal of Applied Physics*, **26**, 182–185.

BLENKINSOP, T. 2008. Relationships between faults, extension fractures and veins, and stress. *Journal of Structural Geology*, **30**, 622–632.

BRACE, W., PAULDING, B. JR. & SCHOLZ, C. 1966. Dilatancy in the fracture of crystalline rocks. *Journal of Geophysical Research*, **71**, 3939–3953.

BRECKELS, I. & VAN EEKELEN, H. 1982. Relationship between horizontal stress and depth in sedimentary basins. *Journal of Petroleum Technology*, **34**, 2191–2199.

CARROLL, M. 1979. An effective stress law for anisotropic elastic deformation. *Journal of Geophysical Research*, **84**, 7510–7512.

CHEN, Q. & NUR, A. 1992. Pore fluid pressure effects in anisotropic rocks: mechanisms of induced seismicity and weak faults. *Pure and Applied Geophysics*, **139**, 463–479.

COWIN, S. 1985. The relationship between the elasticity tensor and the fabric tensor. *Mechanics of Materials*, **4**, 137–147.

CRAMPIN, S. & CHASTIN, S. 2003. A review of shear wave splitting in the crack critical crust. *Geophysical Journal International*, **155**, 221–240.

ENGELDER, T. & FISCHER, M. 1994. Influence of poroelastic behavior on the magnitude of minimum horizontal stress, Sh in overpressured parts of sedimentary basins. *Geology*, **22**, 949–952.

FERRILL, D., WINTERLE, J., WITTMEYER, G., SIMS, D., COLTON, S., ARMSTRONG, A. & MORRIS, A. 1999. Stressed rock strains groundwater at Yucca Mountain, Nevada. *GSA Today*, **9**, 1–8.

FLINN, D. 1962. On folding during three-dimensional progressive deformation. *Quarterly Journal of the Geological Society*, **118**, 385.

FLINN, D. 1977. Transcurrent faults and associated cataclasis in Shetland. *Journal of the Geological Society*, **133**, 231–247.

GOULTY, N. 2003. Reservoir stress path during depletion of Norwegian chalk oilfields. *Petroleum Geoscience*, **9**, 233–241.

GUÉGUEN, Y. & SAROUT, J. 2009. Crack-induced anisotropy in crustal rocks: predicted dry and fluid-saturated Thomsen's parameters. *Physics of the Earth and Planetary Interiors*, **172**, 116–124.

GUÉGUEN, Y., CHELIDZE, T. & LE RAVALEC, M. 1997. Microstructures, percolation thresholds, and rock physical properties. *Tectonophysics*, **279**, 23–35.

HACKER, B. 1997. Diagenesis and fault valve seismicity of crustal faults. *Journal of Geophysical Research*, **102**, 24459–24468.

HALLBAUER, D., WAGNER, H. & COOK, N. 1973. Some observations concerning the microscopic and mechanical behaviour of quartzite specimens in stiff, triaxial compression tests. *International Journal of Rock Mechanics and Mining Sciences & Geomechanics Abstracts*, **10**, 713–726.

HANDIN, J., HAGER, R. JR, FRIEDMAN, M. & FEATHER, J. 1963. Experimental deformation of sedimentary rocks under confining pressure: pore pressure tests. *AAPG Bulletin*, **47**, 717–755.

HEALY, D. 2008. Damage patterns, stress rotations and pore fluid pressures in strike-slip fault zones. *Journal of Geophysical Research*, **113**, B12407.

HEALY, D. 2009. Anisotropy, pore fluid pressure and low angle normal faults. *Journal of Structural Geology*, **31**, 561–574.

HEALY, D., JONES, R. & HOLDSWORTH, R. 2006*a*. Three-dimensional brittle shear fracturing by tensile crack interaction. *Nature*, **439**, 64–67.

HEALY, D., JONES, R. & HOLDSWORTH, R. 2006*b*. New insights into the development of brittle shear fractures from a 3-D numerical model of microcrack interaction. *Earth and Planetary Science Letters*, **249**, 14–28.

HILLIS, R. 2001. Coupled changes in pore pressure and stress in oil fields and sedimentary basins. *Petroleum Geoscience*, **7**, 419–425.

HILLIS, R. & NELSON, E. 2005. In situ stresses in the North Sea and their applications: petroleum geomechanics from exploration to development. *In*: *Petroleum Geology Conference series 6*. Geological Society of London, **551**.

HUBBERT, M. K. & RUBEY, W. W. 1959. Role of fluid pressure in mechanics of overthrust faulting. *Geological Society of America Bulletin*, **70**, 167.

KANATANI, K. 1984. Distribution of directional data and fabric tensors. *International Journal of Engineering Science*, **22**, 149–164.

KING, R., HILLIS, R. & REYNOLDS, S. 2008. In situ stresses and natural fractures in the Northern Perth Basin, Australia. *Australian Journal of Earth Sciences*, **55**, 685–701.

LISLE, R., ORIFE, T. O., ARLEGUI, L., LIESA, C. & SRIVASTAVA, D. C. 2006. Favoured states of palaeostress in the

Earth's crust: evidence from fault slip data. *Journal of Structural Geology*, **28**, 1051–1066.

Lockner, D., Byerlee, J., Kuksenko, V., Ponomarev, A. & Sidorin, A. 1991. Quasi-static fault growth and shear fracture energy in granite. *Nature*, **350**, 39–42.

Mildren, S., Hillis, R. & Kaldi, I. 2002. Calibrating predictions of fault seal reactivation in the Timor Sea. *APPEA Journal*, **42**, 187–202.

Moore, D. & Lockner, D. 1995. The role of microcracking in shear-fracture propagation in granite. *Journal of Structural Geology*, **17**, 95–111.

Morris, A., Ferrill, D. & Henderson, D. 1996. Slip-tendency analysis and fault reactivation. *Geology*, **24**, 275–278.

Nur, A. & Byerlee, J. 1971. An exact effective stress law for elastic deformation of rock with fluids. *Journal of Geophysical Research*, **76**, 6414–6419.

Oda, M. 1982. Fabric tensor for discontinuous geological materials. *Soils Found*, **22**, 96–108.

Peng, S. & Johnson, A. 1972. Crack growth and faulting in cylindrical specimens of Chelmsford granite. *International Journal of Rock Mechanics and Mining Sciences & Geomechanics Abstracts*, **9**, 37–42.

Phillips, W. 1972. Hydraulic fracturing and mineralization. *Geological Society of London Journal*, **128**, 337–359.

Ramsay, J. 1967. *Folding and Fracturing of Rocks*. McGraw-Hill, New York.

Reches, Z. & Dieterich, J. 1983. Faulting of rocks in three-dimensional strain fields I. Failure of rocks in polyaxial, servo-control experiments. *Tectonophysics*, **95**, 111–132.

Reches, Z. & Lockner, D. 1994. Nucleation and growth of faults in brittle rocks. *Journal of Geophysical Research*, **99**, 18159–18173.

Rice, J. & Cleary, M. 1976. Some basic stress diffusion solutions for fluid-saturated elastic porous media with compressible constituents. *Reviews of Geophysics*, **14**, 227–241.

Salz, L. 1977. Relationship between fracture propagation pressure and pore pressure. *SPE Annual Fall Technical Conference and Exhibition, 9–12 October 1977*. Denver, Colorado.

Sayers, C. & Kachanov, M. 1995. Microcrack-induced elastic wave anisotropy of brittle rocks. *Journal of Geophysical Research*, **100**, 4149–4156.

Schubnel, A., Benson, P., Thompson, B., Hazzard, J. & Young, R. 2006. Quantifying damage, saturation and anisotropy in cracked rocks by inverting elastic wave velocities. *Pure and Applied Geophysics*, **163**, 947–973.

Scott, T. Jr, Ma, Q. & Roegiers, J. 1993. Acoustic velocity changes during shear enhanced compaction of sandstone. *International Journal of Rock Mechanics and Mining Sciences & Geomechanics Abstracts*, **30**, 763–769.

Secor, D. 1965. Role of fluid pressure in jointing. *American Journal of Science*, **263**, 633–646.

Skempton, A. 1960. Effective stress in soils, concrete and rocks. *In*: International Society of Soil Mechanics and Foundation Engineering (ed.) *Pore Pressure and Suction in Soils*. Butterworths, London.

Tapponnier, P. & Brace, W. 1976. Development of stress-induced microcracks in Westerly granite. *International Journal of Rock Mechanics and Mining Sciences & Geomechanics Abstracts*, **13**, 103–112.

Terzaghi, K. 1943. *Theoretical Soil Mechanics*. John Wiley & Sons, New York.

Teufel, L., Rhett, D. & Farrell, H. 1991. *Effect of Reservoir Depletion and Pore Pressure Drawdown on in situ Stress and Deformation in the Ekofisk Field, North Sea*. Sandia National Labs, Albuquerque, NM (USA).

Thompson, M. & Willis, J. 1991. A reformation of the equations of anisotropic poroelasticity. *Journal of Applied Mechanics*, **58**, 612–616.

Tingay, M., Hillis, R., Swarbrick, R., Morley, C. & Damit, A. 2009. Origin of overpressure and pore-pressure prediction in the Baram province, Brunei. *AAPG Bulletin*, **93**, 51–74.

Vermilye, J. & Scholz, C. 1998. The process zone: a microstructural view of fault growth. *Journal of Geophysical Research*, **103**, 12223–12237.

Vidal-Gilbert, S., Nauroy, J. & Brosse, E. 2009. 3D geomechanical modelling for CO_2 geologic storage in the Dogger carbonates of the Paris Basin. *International Journal of Greenhouse Gas Control*, **3**, 288–299.

Wibberley, C., Petit, J. & Rives, T. 2000. Micromechanics of shear rupture and the control of normal stress. *Journal of Structural Geology*, **22**, 411–427.

Wilson, J., Chester, J. & Chester, F. 2003. Microfracture analysis of fault growth and wear processes, Punchbowl Fault, San Andreas system, California. *Journal of Structural Geology*, **25**, 1855–1873.

Wong, T. 1982. Micromechanics of faulting in Westerly granite. *International Journal of Rock Mechanics and Mining Sciences & Geomechanics Abstracts*, **19**, 49–64.

Woodcock, N. 1977. Specification of fabric shapes using an eigenvalue method. *Geological Society of America Bulletin*, **88**, 1231–1236.

Zatsepin, S. V. & Crampin, S. 1997. Modelling the compliance of crustal rock – I. Response of shear-wave splitting to differential stress. *Geophysical Journal International*, **129**, 477–494.

The dilatancy–diffusion hypothesis and earthquake predictability

IAN G. MAIN[1]*, ANDREW F. BELL[1], PHILIP G. MEREDITH[2], SEBASTIAN GEIGER[3] & SARAH TOUATI[1]

[1]*University of Edinburgh, School of GeoSciences, West Mains Road, Edinburgh*

[2]*Department of Earth Sciences, University College London, Gower Street, London*

[3]*Heriot-Watt University, Institute of Petroleum Engineering, Riccarton, Edinburgh*

**Corresponding author (e-mail: ian.main@ed.ac.uk)*

Abstract: The dilatancy–diffusion hypothesis was one of the first attempts to predict the form of potential geophysical signals that may precede earthquakes, and hence provide a possible physical basis for earthquake prediction. The basic hypothesis has stood up well in the laboratory, where catastrophic failure of intact rocks has been observed to be associated with geophysical signals associated both with dilatancy and pore pressure changes. In contrast, the precursors invoked to determine the predicted earthquake time and event magnitude have not stood up to independent scrutiny. There are several reasons for the lack of simple scaling between the laboratory and the field scales, but key differences are those of scale in time and space and in material boundary conditions, coupled with the sheer complexity and non-linearity of the processes involved. 'Upscaling' is recognized as a difficult task in multi-scale complex systems generally and in oil and gas reservoir engineering specifically. It may however provide a clue as to why simple local laws for dilatancy and diffusion do not scale simply to bulk properties at a greater scale, even when the fracture system that controls the mechanical and hydraulic properties of the reservoir rock is itself scale-invariant.

Ernest Masson Anderson developed his theory for the structure of faults and fractures primarily from matching observations in nature made by pioneers such as Hutton and Lyell to hypotheses developed by Navier, Coulomb and Mohr in the 19th century, citing some early controlled experiments on analogue materials such as layered clay or mastic (Anderson 1905). Experiments on actual rocks in compression were available in the mining engineering literature by the time of the publication of his book (Anderson 1942), but largely corroborated the inferences already made. Anderson extrapolated these results more or less linearly to the crustal scale (Fig. 1). Some features scale remarkably well, notably the typical orientation of the angle of deformation in shear (controlled by the internal frictional properties) and in tension (opening against the least resistance or minimum principal stress). More recently, a much broader range of structural properties of populations of faults and fractures have been shown to scale remarkably well from laboratory failure to crustal scales; this has been observed directly in field outcrop (Bonnet *et al.* 2001) or inferred from the scaling of earthquake stress drops and frequency-magnitude scaling (Main 1996). Given this *structural* scaling, it might at first glance seem natural to assume that other aspects of the *physics* of catastrophic failure will scale linearly from the lab to natural earthquakes. Is this appealing notion really how nature works, however?

Early papers on the role of dilatancy in the earthquake cycle were based on geological observation (Mead 1925; Frank 1965). The dilatancy–diffusion hypothesis itself was developed from the observation of changes in geophysical properties associated with dilatant strain in laboratory tests (Nur 1972). It was one of the first to be put forward as a physical basis for purported earthquake precursors, also assuming (implicitly) linear scaling of the physics involved from lab to field (Scholz *et al.* 1973). This paper contains what now seem like wildly over-optimistic statements on the existence of earthquake precursors ('occur before many, and perhaps all, earthquakes') and the prospects for earthquake predictability ('the mechanism of premonitory changes appears to lead to prediction which is deterministic rather than probabilistic') given subsequent experience. In science however, we often learn more from hypothesis failure than confirmation. In a classical example, Einstein's special theory of relativity followed the failure of the hypothesis of the 'ether' as a fixed reference frame for the propagation of light in the Michaelson–Morley experiment. In this paper we re-examine the failure of the dilatancy–diffusion hypothesis with the benefit of hindsight, and suggest new

From: Healy, D., Butler, R. W. H., Shipton, Z. K. & Sibson, R. H. (eds) 2012. *Faulting, Fracturing and Igneous Intrusion in the Earth's Crust*. Geological Society, London, Special Publications, **367**, 215–230. http://dx.doi.org/10.1144/SP367.15

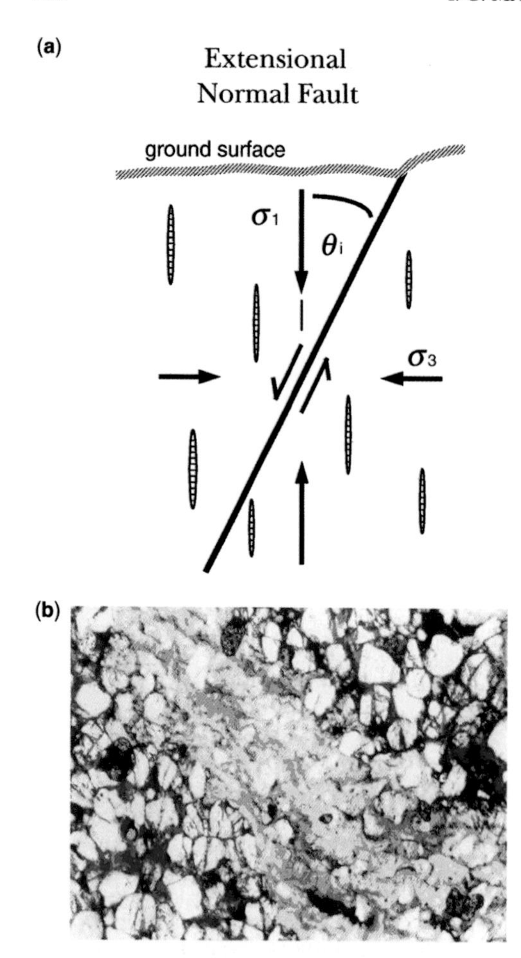

Fig. 1. Upper: shear fault and extensional fracture orientations predicted by E.M. Anderson's model for failure in the brittle crust, in the case of a vertical maximum principal stress (after Sibson 2001, image provided by Richard Sibson). Lower: orientations of a shear band and local tensile microcracks on the grain scale (around 300 microns) in a porous sandstone, also in the case of a vertical maximum principal stress (after Mair *et al.* 2000).

areas to explore in constraining the physics of earthquakes and the prospects for predictability.

The dilatancy–diffusion hypothesis has not yet been validated at a crustal scale, primarily due to the general absence of the predicted dilatancy-related precursors (including seismic velocity, seismicity, electrical conductivity or radon release, e.g. Jordan *et al.* 2011). Scholz (1997) has argued instead that the search for dilatancy-related precursors has become biased, with the mainstream community too ready to dismiss or not look for evidence of precursors. Data recorded in real time since then, even at well-monitored borehole sites near the 2004 Parkfield earthquake rupture (Bakun *et al.* 2005) and at other significant events in California (e.g. Loma Prieta, Northridge, Landers, Superstition Hills), have however all failed to show any direct evidence for detectable precursory behaviour.

Among this predominantly negative evidence, Niu *et al.* (2008) observed two large excursions in the travel-time data that are coincident with two earthquakes (magnitudes 3 and 1) that are among those predicted to produce the largest coseismic stress changes at the San Andreas Fault Observatory at Depth (SAFOD) drilling site (Ellsworth *et al.* 2005). The two excursions started *c.* 10 and 2 hours before the events, respectively. Niu *et al.* (2008) suggested that they may be related to pre-rupture stress-induced changes in crack properties, as observed in early laboratory studies. More recently, satellite interferometry has confirmed more directly the absence of any significant precursory strain recorded at the Earth's surface in the case of the 2009 L'Aquila earthquake in Abruzzo, Italy (Amoruso & Crescentini 2010).

While the search for precursors continues despite this, we take a different tack here and ask instead how such rigorous 'negative' observations may nevertheless be used in a positive way as a significant physical constraint on the actual physics of the process involved. In particular, a careful 'upscaling' exercise remains to be done for the dilatancy–diffusion hypothesis, that is, to take account of differences in loading and sample boundary conditions, spatial and temporal scale and the material, structural, mechanical and hydraulic complexities involved. For example, Nur (1975) pointed out that various forms of dilatancy (microcrack, existing fractures, granular) could be expected in the Earth's seismogenic crust, and they would be expected to have different stress sensitivities. The hydrofracture dilatancy reported in the vicinity of some fault zones (Sibson 1981) requires pore pressure in excess of the minimum principal stress ($p > \sigma_3$), for example. This can only be achieved under low levels of differential stress, that is, $\sigma_1 - \sigma_3 < 4T$ where T is the tensile failure stress. This is in direct contrast with the high levels of differential stress required for microcrack dilatancy in the laboratory, and possibly also the high pore pressure (Sibson 2009) required for microfracture in nature.

Laboratory tests typically utilize intact uniform samples of rock in order to produce as uniform a stress field as possible in a controlled test. This introduces a kind of 'sample bias', or epistemic error not accounted for in linear scaling arguments, because it is not representative even of the small-scale heterogeneity in the Earth. In contrast, it is clear that the majority of moderate-to-large crustal earthquakes involve repeated reactivation of existing faults (e.g. Holdsworth *et al.* 1997) which may have very different properties to those of an intact rock sample.

Another potential source of epistemic error in the application of traditional laboratory-scale experiments to the Earth is the laboratory testing protocol itself, which typically involves increasing the axial stress σ_1 on a right-cylindrical specimen at constant strain rate under axisymmetric compression provided by a hydraulic fluid at constant pressure ($P = \sigma_2 = \sigma_3$). Any dilatancy can be accommodated readily by a reduction in fluid volume at constant pressure, whereas in nature dilatancy must require an increase in the confining stresses from the surrounding solid rock. During the test the mean stress and the breaking shear stress are both increasing in time, corresponding to load-strengthening behaviour. In nature σ_2 and σ_3 are neither fixed nor equal in the general case. During crustal extension for example, the loading of a normal fault to failure involves progressive reduction of σ_3 while the vertical stress σ_1 stays fixed. In this case the mean stress and fault strength are decreasing while shear stress on the fault and the differential stress are both increasing (load-weakening behaviour). This may help to account for the observation that foreshock activity is more commonly associated with normal faults than with reverse (Abercrombie & Mori 1996). Recognizing the importance of E. M. Anderson's inferences, some laboratories have examined the effect of 'true' triaxial stresses ($\sigma_1 > \sigma_2 > \sigma_3$) on rock strength (Haimson & Chang 2000) and its effect on geophysical properties such as shear-wave birefringence due to aligned microcracks (Crawford *et al.* 1995).

In addition to these mechanical and spatial scaling arguments, recent laboratory 'creep' tests carried out under constant load conditions have demonstrated a systematic decrease in bulk sample dilatancy as strain rates are lowered towards more realistic values for crustal-scale deformation (Heap *et al.* 2009). The results are consistent with the absence of strong dilatancy-related precursors associated with large earthquakes. A further suite of even slower deformation experiments is planned to test the extrapolation, to fill in an important gap in our understanding of the temporal scaling of brittle-field rheology.

The dilatancy–diffusion hypothesis

The hypothesis was based on solid and repeatable evidence of primarily mechanical and geophysical precursors to failure in the laboratory, associated with measured changes in sample volume after the yield point in crystalline rocks. This typically occurs at around or above half of the ultimate strength of the rock sample: dilatancy is a 'high-stress' phenomenon. Such bulk dilatancy, due to microcracking of the type shown also in sedimentary rocks (Fig. 1), was associated in the laboratory with changes in seismic velocity, electrical resistivity and acoustic emission event rate and the scaling of event size, expressed by the exponent b (the 'b-value') in the Gutenberg–Richter relation for the frequency F of events of magnitude m, of the form $F(m) = a - bm$. We might then expect such sample dilatancy to affect pore-fluid volume and/or pressure, depending on the permeability of the medium, the local strain rate and the experimental boundary conditions.

At sufficiently high volumetric strain rates dilatancy in a relatively impermeable crystalline rock in the Earth's subsurface would be initially expected to produce a local decrease in pore pressure and a concomitant increase in the effective normal stress, resulting in material hardening and delaying failure (Paterson & Wong 2005). Implicit in this scenario is that the rate of dilatancy (volumetric strain rate) must remain higher than that which will allow pore water to diffuse into the new cracks to restore the pore pressure and induce concomitant material softening. In practice, the low strain rates at the onset of dilatancy means there will be a finite lag time between the onset of dilatancy and local pore pressure reduction. Assuming a supply of fluid from outside the dilatant zone and a deceleration in the rate of dilatancy associated with the hardening effect, the drop in pore pressure would be followed by a slow recovery by fluid flow from the surrounding undilated region. This recovery would ultimately trigger dynamic failure.

Scholz *et al.* (1973) presented no direct measurements of fluid pressure variations from the laboratory; instead, they inferred such a decrease then recovery, solving a simple diffusion law for transient pressure recovery in a spatially uniform medium with a constant diffusivity to estimate the duration of the recovery time. The hypothesis predicted systematic *qualitative* changes in geophysical signals associated with stages of elastic loading, dilatant yield, pore pressure recovery, dynamic failure and post-seismic relaxation as illustrated in Figure 2.

The basic coupled process has been replicated to some extent (though not exactly) in the laboratory, including contemporary measurement of actual pore pressure change and its impact on seismic (acoustic emission) precursors under constant strain rate loading. Figure 3 (from Sammonds *et al.* 1992) shows an example of two tests, one nominally 'dry' and one completely saturated in a constant volume of water, held at pressure in the sample under 'undrained' conditions with sample boundaries sealed to fluid flow in either direction. The dry sample shows an acceleration in event

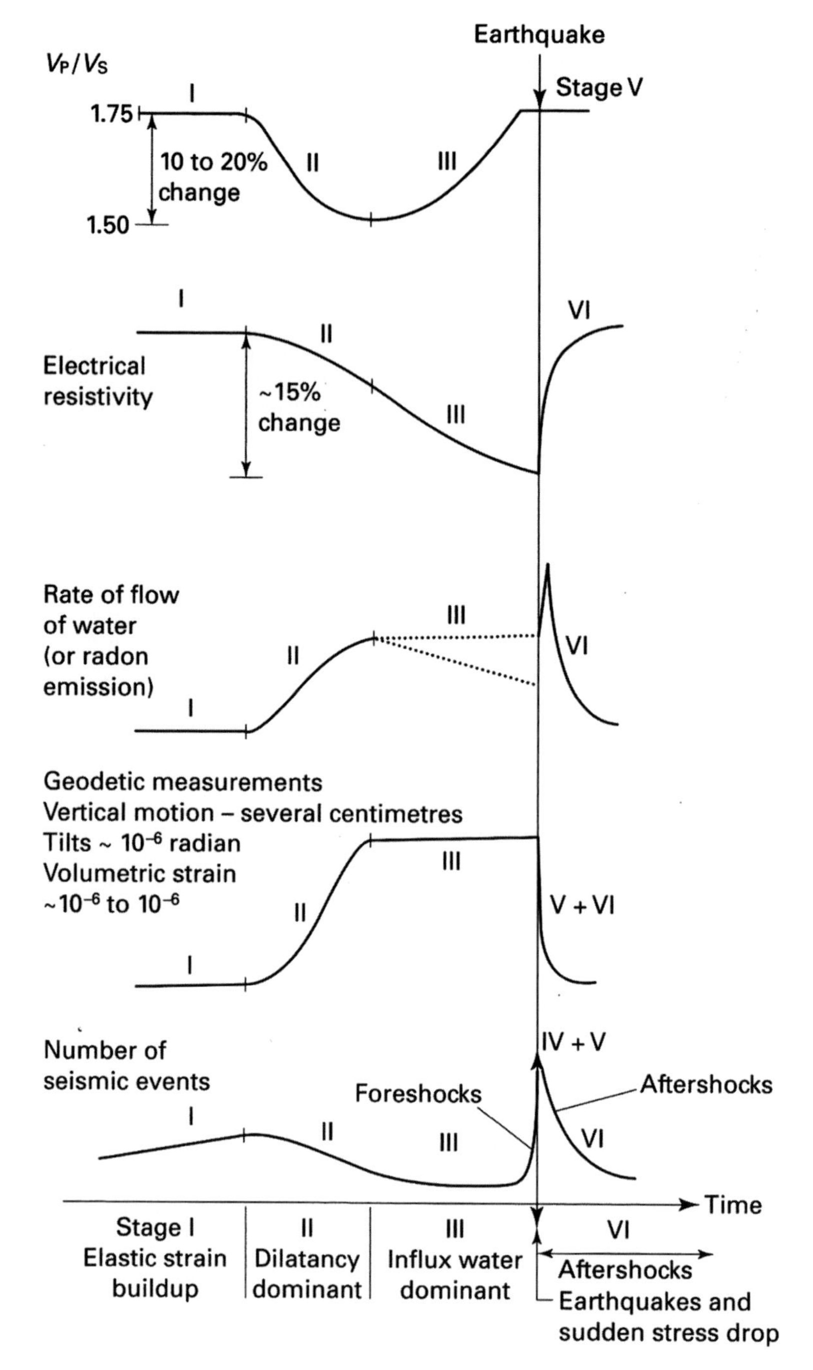

Fig. 2. Predictions of anomalies in geophysical signals associated with elastic loading, dilatancy, diffusion, earthquake and post-seismic periods (after Scholz *et al.* 1973, 2002).

rate (related to a) and a decrease in the seismic b-value associated with an increase in stress. Similar behaviour is seen in a 'drained' test held at constant boundary pore pressure, allowing fluid flow at the sample boundary (Fig. 3c). In the undrained test of Figure 3b the pore pressure, measured at the sample boundary, first increases due to crack and pore closure associated with an

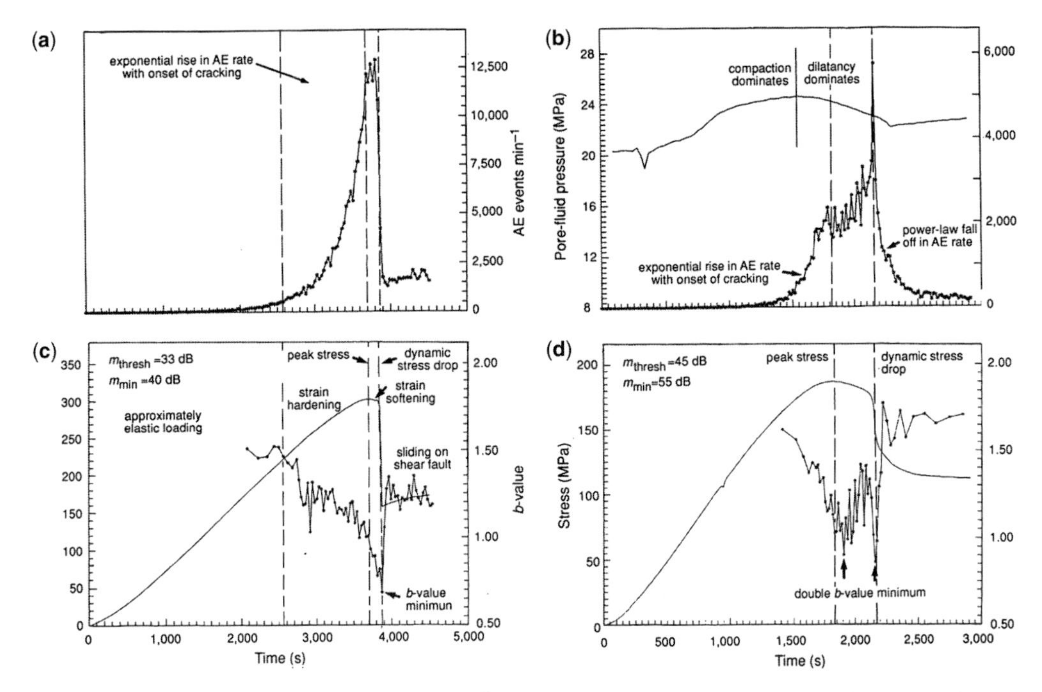

Fig. 3. Acoustic emission event rate (AE events min^{-1}) and pore-fluid pressure (in MPa) (upper diagrams) and variations in the differential stress (in MPa) and the Gutenberg–Richter *b*-value (lower diagrams) for (**a**) nominally dry and (**b**) water-saturated samples of Darley–Dale sandstone (after Sammonds *et al.* 1992).

increase in mean stress, and then decreases up to the failure time due to shear-enhanced dilatancy. The inferred dilatancy hardening with zero-permeability boundary conditions does indeed significantly extend the post-peak stress deformation phase and delay the failure time. No pore pressure recovery is seen because of the sealed boundary, but the event rate flattens off and the *b*-value recovers in an extended strain-softening phase before dropping to a minimum at the final stage near dynamic failure. The controlled laboratory experimental conditions used to obtain the results of Figure 3 are ideal end-members to the more intermediate conditions that may apply in the general case in the Earth. Such conditions would be difficult to replicate, and to the authors' knowledge, have not yet been applied in the laboratory. The 'diffusion' or final pore pressure recovery phase has yet to be demonstrated in such an open system in the laboratory.

The dilatancy–diffusion hypothesis is based on the assumption of a finite-sized 'preparation zone', within which microcrack damage is occurring, of size related to the eventual size of the mainshock. A preparation zone is well defined in the laboratory by the sample boundaries but remains elusive in field data: behaviour identified after the earthquake as anomalous is often available only at one site (or at most a handful of selected sites) even for spatially very extensive mainshocks. In some cases this has been argued to be a consequence of low instrument density.

To overcome this problem, Scholz *et al.* (1973) estimated the size of the upcoming event (based on the extent of the aftershock zone) from its correlation with and the duration of the reported precursor, based on the literature then available. Interestingly, this correlation could be explained by the duration of the inferred diffusive pore pressure recovery phase in a uniform medium, albeit with an inferred diffusivity higher than a typical laboratory test for a low-porosity crystalline rock under similar pressure conditions. The hypothesis remains unproven because the predicted precursors failed to materialize convincingly in a consistent and reliable way in field evidence (Wyss & Booth 1997; Bakun *et al.* 2005).

At this point it is useful to note that the notion of dilatancy–diffusion does have an important bearing on dynamic failure processes. Rudnicki & Chen (1988) developed a coupled model to explain how rapid frictional slip may be stabilized on an otherwise weakening fault by dilatancy hardening. Under constant flow rate boundary conditions the same coupled model predicts a dynamic 'suction pump' effect, where fluids are actively channelled into the dilating fault zone. The results of a

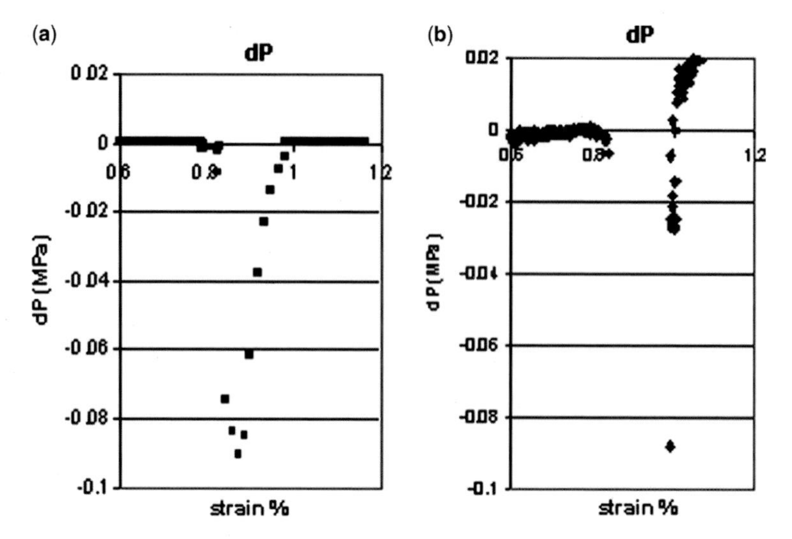

Fig. 4. The dilatant 'suction pump' in action: (**a**) a dynamic model and (**b**) observation for pore pressure drop during dynamic failure of a sample of Clashach sandstone under constant input fluid flow rate, using unreactive oil as a permeant (after Grueschow *et al.* 2003).

numerical model for this dynamic effect compare favourably with those of a laboratory experiment under similar conditions (Grueschow *et al.* 2003). The suction generated by dynamic dilatancy in the fault zone is manifest by a drop in the inlet pressure required to push fluid in at a constant rate at the sample boundary (Fig. 4). Such seismic pumping, repeated over many cycles, is consistent with the observation of mineral deposits formed by episodic channelling of hydrothermal fluids along faults and fractures in meso-thermal conditions (Sibson *et al.* 1975), but this interpretation is not unique (e.g. Sibson 1981, 2001). In any case, such *dynamic* effects appear to scale better between the laboratory and the brittle Earth than the *quasi-static* loading phase where precursors might be expected. However, this dynamic coupling is consistent with short-lived dilatancy concentrated very near the fault zone, rather than the longer-term quasi-static regional dilatancy invoked by Scholz *et al.* (1973) or related theories based on seismic anisotropy and extensive fracture dilatancy outside the nominal mainshock 'preparation zone' (Crampin *et al.* 1984).

The flawed search for earthquake precursors

There has been much discussion of this issue in the literature, and only a brief summary can be given here. An excellent and accessible summary of the repeated conflict between an otherwise reasonable hypotheses and data, along with an interesting and very relevant discussion of the social, human and even political dimensions that are very much part of the story, is given by Hough (2009). Following the most comprehensive study to date by an expert panel convened by the International Association for Seismology and Physics of the Earth's Interior (IASPEI), Wyss & Booth (1997) concluded that there were no candidate precursors that satisfied all of the criteria set by the panel for a physically and statistically reasonable precursory signal (e.g. any anomaly must be seen at more than one site to be acceptable). This means the 'precursor' durations used to determine the magnitude correlation by Scholz *et al.* (1973), and hence the inferred quantitative value of the fluid pressure diffusion constant, were based on questionable published data.

The correlation between a reported fluctuation in a geophysical parameter and a subsequent earthquake itself may have other more mundane causes, for example, retrospective selection bias in a noisy signal (Mulargia 2001) as illustrated in the example shown as a tutorial in Figure 5. Figure 5c reproduces a figure from Scholz *et al.* (1973) cited as evidence for changes in event rate a and in the scaling exponent b prior to a magnitude 3 earthquake. At first glance this selected data seem to show a convincing minimum and recovery in event rate prior to the magnitude 3 mainshock time identified on the diagram, consistent with the predictions of the dilatancy–diffusion model of Figure 2. For reference, Figure 5a shows a parent random Poisson process for one year, with Figure 5b sampling the parent in a similar way to the real data of Figure 5c. Figure 5a illustrates the large relative fluctuations expected from simple counting errors of the

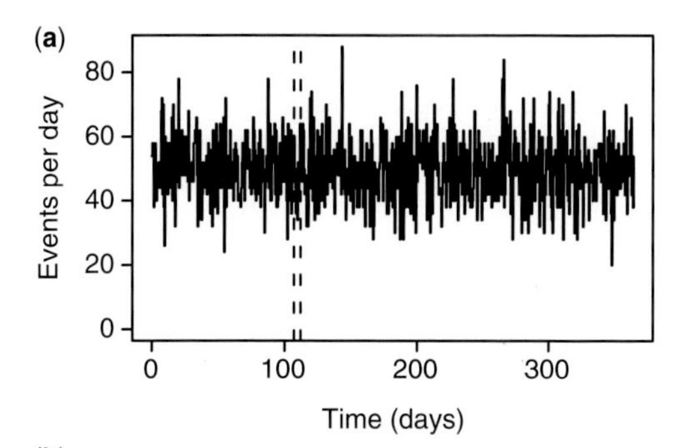

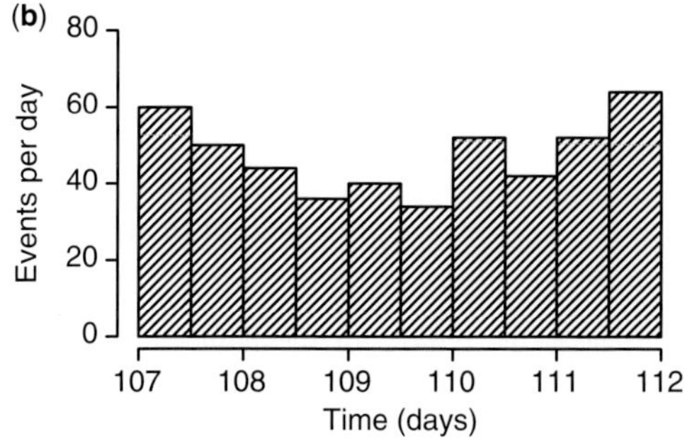

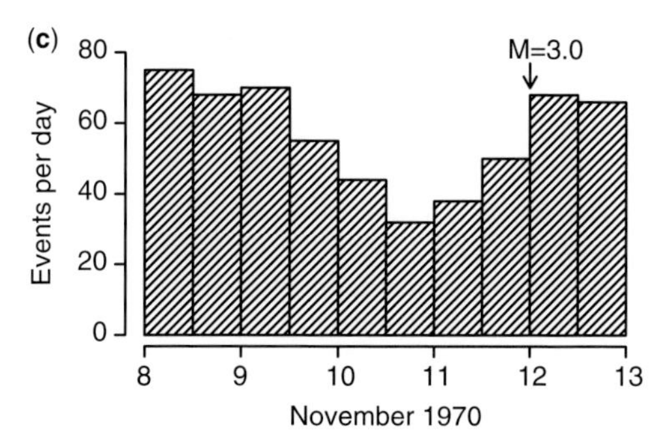

Fig. 5. (**a**) Fluctuations in a random (Poisson) process with an average daily event rate of 50, sampled at 12 hour intervals. (**b**) Blow-up of one of the minima between the two vertical dashed lines in (a). (**c**) Variation in seismic event rate per day over a similar timescale as (b), also sampled in 12 hour intervals, with a similar average event rate as in (a), after Scholz *et al.* 1973.

number of events in a random process with this average, and the tendency to cluster rather than produce the flat graph expected for an infinitely sampled process. Figure 5b is a blow-up of one of the minima in Figure 5a, illustrating how minima such as Figure 5c could occur simply by finite sampling of a random process.

In another example, Main *et al.* (2008) showed that the non-linear statistics of seismicity (exemplified by the Gutenberg–Richter law and exacerbated by earthquake triggering not considered in Fig. 5) can lead to very large samples (several thousand) being required even to get a stable value of average total event rate and its standard deviation. The simplest interpretation consistent with the data of Figure 5a is therefore a finite (small) sample of a random process, with one of several candidate magnitude 3 earthquakes selected retrospectively within a time window that effectively introduces two additional free parameters (start time and end time) to the search.

Most candidate precursors fail as potential predictors because of poor hypothesis testing protocols, notably examining and selecting data in retrospect and not accounting for the resulting sample bias in assessing the significance of any correlation identified. Such data selection is perfectly valid in developing a hypothesis, but not in testing or validation. As a consequence, clinicians developed the prospective 'double-blind' test as the standard and only acceptable method of testing in medical sciences (Modell & Houde 1958). Ultimately any hypothesis must be put at risk in a situation where the outcome is not known *a priori*. In our case this means actual forecasting in real time is needed to evaluate fully the significance of any precursor and its quantitative impact on earthquake predictability. This aspect has now been fully embraced by the global seismological community, with a range of regional testing centres now set up by the Collaboratory for the Study of Earthquake Predictability (http://www.cseptesting.org/).

Using negative evidence as a constraint

The dilatancy–diffusion hypothesis was based on direct evidence of dilatant strain in laboratory samples. In the laboratory, dilatant strain can be measured by strain gauges placed directly on the sample, or more recently through changes in the volume of the pore fluid or the fluid-confining medium. Modern satellite interferometric data utilizing synthetic aperture radar imagery (InSAR) now provide extremely sensitive measurements of strain at the Earth's surface. For example, during the two years before the M_W 6.3 earthquake struck the city of L'Aquila, Italy on 6 April 2009, no anomalous precursory strain larger than a few tens of nano-strain units is visible; this limits the volume of the possible earthquake preparation zone to less than 100 km^3 (Amoruso & Crescentini 2010) or a linear dimension of 4.6 km. This is much smaller than the 10–20 km or so rupture length for a magnitude 6.3 mainshock, calling into question the generality of the notion of a preparation zone similar to the sample dimensions of a laboratory test. Even seconds before the 2009 L'Aquila earthquake, 'strain is stable at the 10^{-12} level and pre-rupture nucleation slip in the hypocentral region is constrained to have a moment less than 2×10^{12} Nm, that is 0.00005% of the main shock seismic moment' (Amoruso & Crescentini 2010). Assuming a scale-invariant strain change with a typical stress drop of 30 bar for a continental earthquake or 10^{-4} units of strain for the earthquake itself, the nucleation zone is restricted to a scale length of at most 100 m (likely at typical earthquake nucleation depths of 10 km or so). At this localized scale, likely related to re-fracture of a healed locked asperity, we might expect to see the same physics as we observe in a laboratory test but this is going to be hard to detect. Clearly the nucleation zone is much, much smaller than the eventual rupture, and the two need not be directly related.

While precursory dilatant strain has not yet been observed directly and systematically for continental earthquakes, there is evidence that post-seismic strain relaxation is clearly visible: for example, following the 26 December 2003 Bam earthquake in Iran (Fielding *et al.* 2009). Using satellite-based InSAR observations, and after accounting for poroelastic effects, they identify a localized zone of dilatant strain recovery near (within 200 m or so) the centre of the mapped fault trace where the co-seismic slip was greatest. Such dilatancy is therefore much more likely to be due to co-seismic dilatancy of the type modelled by Rudnicki & Chen (1988) and observed in the laboratory by Grueschow *et al.* (2003), rather than any residual memory of precursory dilatant strain. This confirms the inference that actual precursory dilatant strain is quantitatively much less and/or much more highly localized than the bulk behaviour of a laboratory test.

While more difficult to measure and subject to much debate, the inferred shear stresses involved in crustal loading prior to earthquake rupture could also provide a constraint on the type and amount of dilatancy that might be expected in the seismogenic crust (e.g. Brune & Thatcher 2003). One view holds that the ambient effective differential stresses are low (of the order 100 bars or less), implying almost total stress relaxation during rupture. Another holds that shear stresses are high (of the order 1 kbar or more), comparable to those where dilatancy is seen in crystalline rocks in the laboratory. The absence of a clear dilatancy signal from microcracking around seismogenic faults is consistent with relatively low-stress (shear stress <100 bar) rupturing on existing, relatively weak, structures.

In summary, direct observation of dilatant strain implies that the dilatancy–diffusion process does apply well (and on a large scale) to the co-seismic and post-seismic phases, and may apply to earthquake nucleation on a very small scale-up to a few hundred metres. This geodetic constraint is supported by recent seismic evidence (Bouchon *et al.* 2011) of an accelerating signal concentrated on a very localized zone, identified by cross-correlation techniques prior to the 1999 M_w 7.6 Izmit (Turkey) earthquake. The signal consisted of a succession of small foreshocks in the form of repetitive seismic bursts, accelerating with time in the 2 min preceding the event, and increased low-frequency seismic noise in the 44 min preceding the event. Any one of these foreshocks is located within 20 m or less of the majority of the other events, comparable to the size of the largest events (25 m). These results confirm a very short-duration highly localized but nevertheless detectable nucleation phase for this event. Modern techniques of data assimilation applied to continuously recorded broadband seismic data will be required to confirm the generality or otherwise of this intriguing observation.

Upscaling of a complex system in space and time

Ultimately the reason for the failure of the dilatancy–diffusion hypothesis to scale simply to crustal processes is due to the complexity and non-linearity of the processes and large differences in space and time between the laboratory and the field case. We address these issues separately below.

Complexity and predictability

In a laboratory test such as illustrated in Figure 3 the sample is initially chosen for its uniformity. It is loaded first by increasing the isotropic stress (axial stress and confining pressure) to a given level, and then by increasing the axial stress alone at a constant strain rate to change the differential stress from zero. In this sense, the sample is loaded from a very sub-critical state (zero differential stress) to a more critical (high-stress) state near the dynamic failure time. In the Earth, the spatial structure is however highly heterogeneous. The tectonic stress perpetually maintains the system in a state much nearer its critical value than the starting conditions of such laboratory tests, making the system much more sensitive to small stress perturbations.

Among other drivers, such complexity has led to a completely alternative view on earthquake mechanics proposed by Bak & Tang (1989) who postulated that earthquakes occurred in a state of self-organized criticality. This hypothesis neatly explained much of the phenomenology of earthquakes including the Gutenberg–Richter law, the scale-invariant distribution of faults, the relatively low and constant stress drop, the ease with which small natural and man-made stress perturbations can induce earthquakes and the long-term stationarity inferred for seismic hazard calculation (Main 1995, 1996). Unfortunately, this came at the expense of degraded predictability; the size of an event in a near-critical system is determined by small details of the avalanche-like response, so that event size would be primarily determined during (and not before) the event (Main 1995). This is consistent with the small size of the nucleation patch inferred by Amoruso & Crescentini (2010) and Bouchon *et al.* (2011) described above.

These inferences and observations are all consistent with the relatively low correlation between magnitudes estimated from the early part of the seismogram and the eventual magnitude of the earthquake used in earthquake 'early-warning' systems, including the recent M_w 9.0 tsunamigenic earthquake in north-eastern Japan (Cyranoski 2011). The notion of self-organized criticality, or near-criticality, implies that any hope for deterministic prediction of earthquakes is remote (Main 1997). Nevertheless, the finite (albeit small) stress drop of earthquakes implies a slightly subcritical system where a small but finite degree of forecasting power might be expected, albeit of a probabilistic nature. (See Nature website debate at http://www.nature.com/nature/debates/earthquake/equake_frameset.html and Main & Naylor 2008 for a discussion of the role of dissipation on maintaining a near-but-subcritical state.)

Such self-organized near-but-strictly-*sub*-criticality is also consistent with recent data from earthquake repeat times in palaeoseismic data from the San Andreas fault (Scharer *et al.* 2010). These data show to first order the temporally random recurrence of a purely critical system, but a second-order quasi-periodic component to the stress renewal process expected from a system with finite stress drop. Any probability gain due to this effect is therefore extremely subtle. In retrospective mode quasi-periodic renewal models have been suggested to provide a factor 2–5 probability gain over a temporally random process (Imoto 2004), but this is likely to be an upper bound to a true prospective forecasting scenario. This marginal probability gain has led to its effectiveness as an operational tool being questioned both in California (Chui 2009) and in Japan (Geller 2011).

Much higher probability gains (over background) are possible with space–time clustering associated with earthquake triggering in a system operating near its critical point. For example, the long-term background seismic risk (of more than 100 fatalities) in the L'Aquila area is of the order 10^{-6} per day (van Stiphout *et al.* 2010). From the clustering properties of swarm activity preceding the 2009 M_w 6.3 L'Aquila earthquake, they estimated this probability was increased by a factor 30 or so prior to the mainshock. To put these numbers into perspective, the typical estimated probability of dying in an earthquake for an individual person in the next 24 hour was temporarily elevated to 10^{-9}, whereas the average probability of dying in a car accident in Italy in any 24 hour period is 2.7×10^{-9}. Taking this a step further, van Stiphout *et al.* (2010) developed a quantitative cost–benefit analysis to demonstrate that the negative consequences of an evacuation (integrated over all such swarms, including the many false alarms, and funds diverted from other potential life-saving activities) outweighed the positive benefits. Dealing with such high probability gain but still low-probability forecasts remains a generic subject of research at the interface between the natural and social sciences (http://www.protezionecivile.it/cms/attach/ex_sum_finale_eng1.pdf).

Scaling in space

Like resistivity, hydraulic diffusivity (or the related permeability) spans a huge number of scale ranges. For example, Figure 6a shows calculations of the flow velocity in a real fracture network mapped in detail at the surface where the fractures are 1 mm wide (Geiger & Emmanuel 2010). These range from 10^{-16} to 10^{-4} m s^{-1}. A reservoir engineer must then estimate a representative single permeability from a block of this size (shown in Fig. 6b) typically comprising a horizontal resolution of *c.* 100 m and a vertical one of *c.* 10 m.

Because it is computationally not feasible to perform reservoir simulations at the resolution of individual fractures and/or sedimentary layers, major research efforts have been dedicated to develop best practices for computing effective coarse-scale permeabilities that preserve the average fine-scale flow behaviour through individual sedimentary beds (cf. Christie 1996, 2001; Renard & de Marsily 1997). Common analytical methods in reservoir engineering include arithmetic permeability averaging (for flow parallel to layering), harmonic permeability averaging (for flow perpendicular to layering) and geometric permeability averaging (for flow in randomly correlated permeability fields). Flow-based permeability averaging can be used to compute the effective permeability of more complex geological structures. Here, a steady-state pressure field is computed for a subsection of the reservoir model using the fine-scale permeability field and known boundary conditions to obtain the total volumetric flux through the model. Using the total flux through the model and the known boundary conditions, the average permeability can be computed straightforwardly from Darcy's law although the final value will be sensitive to the applied boundary condition (e.g. no-flow versus leaky boundaries parallel to the main flow direction).

It has become increasingly common to estimate the error introduced by upscaling *a priori* by

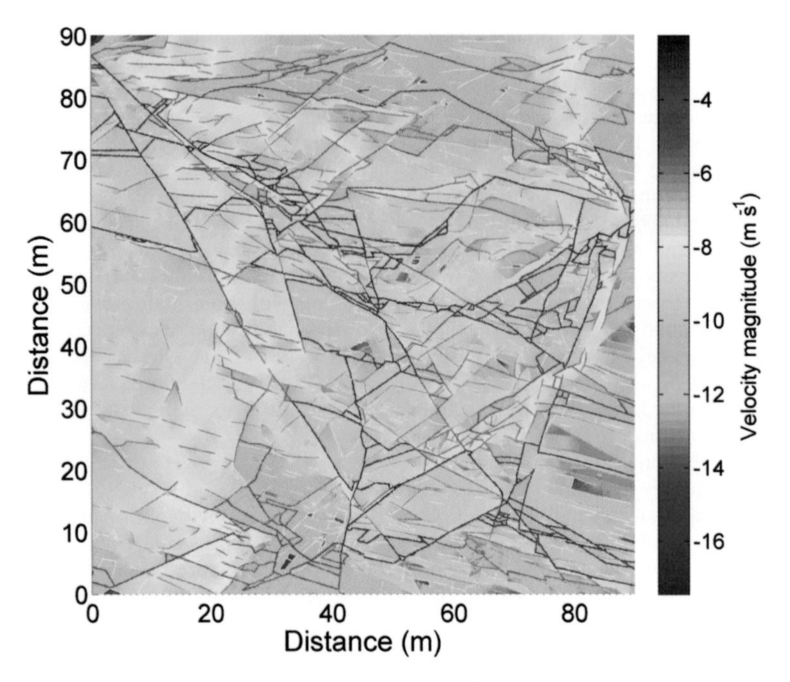

Fig. 6. Illustration of spatial complexity in material properties on different scales. (**a**) Spatial variations in computed flow rates, expressed as the magnitude of the Darcy velocity (note the $\log_{10}$ scale) on a heavily fractured mapped outcrop (after Geiger & Emmanuel 2010).

computing a measure for heterogeneity (usually containing permeability, porosity and flow rate) in all sedimentary layers comprising the geological model and comparing it on a layer-by-layer basis (King *et al.* 2006). This allows the reservoir engineer to generate optimized simulation grids that non-uniformly group different geological layers of similar heterogeneity while preserving others that have a major impact on flow. Coarsening the detailed geological model uniformly, for example by grouping every ten vertical layers of the geological model into one single layer for the flow simulation model, provides little control over the upscaling error. Other methods to validate the quality of upscaling include streamline comparisons between the original fine-scale geological model and coarse-scale flow simulation model (Samier *et al.* 2002).

The common view among reservoir engineers is therefore that sedimentary heterogeneities in clastic rocks can be upscaled reliably and classical benchmark studies such as the 10th SPE Comparative Solution Project (Christie & Blunt 2001) appear to confirm this view. Here it has been demonstrated that a fine-scale geological model of a fluvial North Sea reservoir containing over 10^6 cells can be simulated equally well with a wide range of upscaled flow models containing between 10^3 and 7×10^4 cells. It needs to be pointed out that the upscaled flow models used vastly different upscaling methods and different simulators, which allowed an upscaled flow model to be fine-tuned for a certain simulator; if on the other hand different upscaling methods and simulators are applied to a flow model with a fixed number of grid cells, then results vary significantly.

Upscaling however remains a fundamental problem if fractures are present in the reservoir and pose the philosophical question: can something as inherently discrete as the fracture network of Figure 6a, which is embedded in a permeable rock matrix, be described by a 'representative elemental volume' that can be modelled by a continuum theory with averaged parameters such as fluid pressure diffusion, such that all relevant time- and length scales (spanning several orders of magnitude) are retained? Formally the answer is no – the correlation length of the fractures is much larger than the block size and may even approach crustal scales according to evidence from borehole logs (Dolan *et al.* 1998; Berkowitz 2002).

Still, for practical purposes, fracture networks are upscaled in reservoir simulations using the so-called Oda's method, which attempts to compute an effective permeability tensor of the fracture network based on the aperture and connectivity

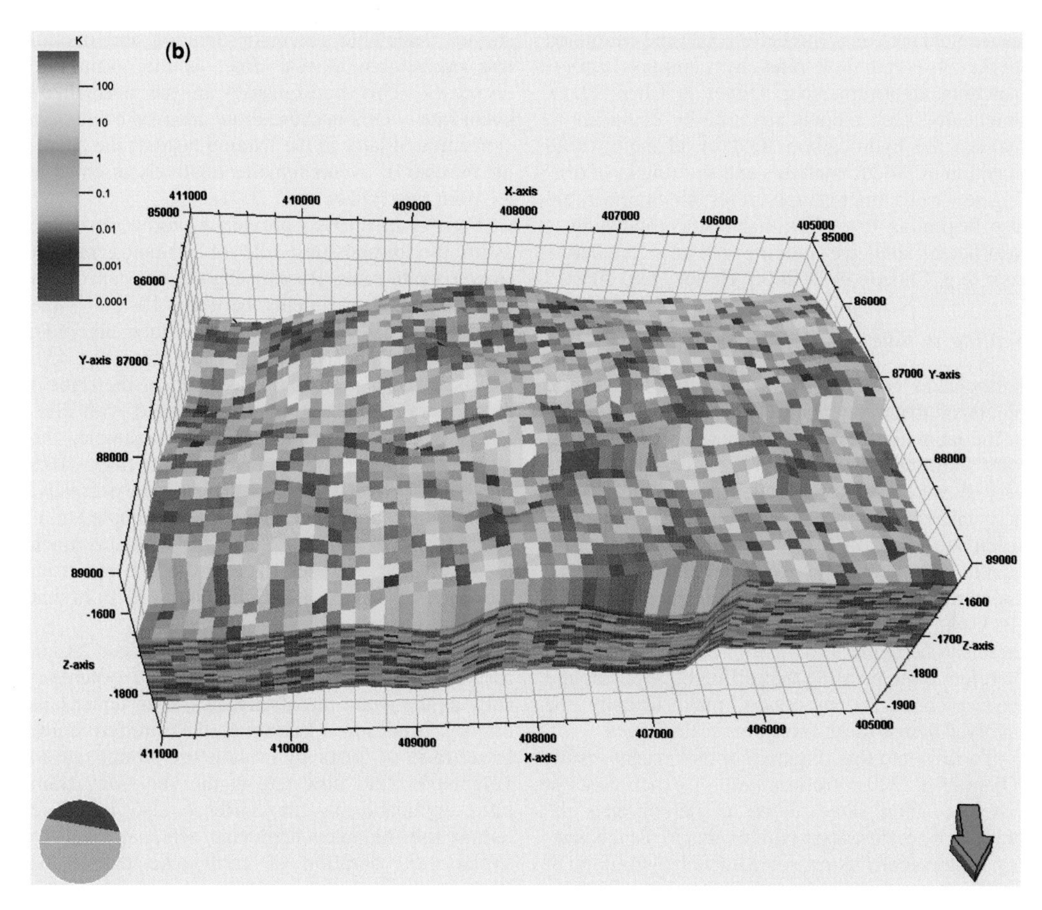

Fig. 6. (*Continued*) (**b**) A typical reservoir model where the spatial extent of a single voxel is also of the order 100 m and the colour-coding denotes the variation in the reservoir property (courtesy of Viswa Chandra, Heriot-Watt University).

of the individual fractures (Dershowitz *et al.* 2000). The effective fracture permeability is then employed in a dual-porosity simulation, which assumes that the fractures comprise the flowing domain of the porous media while the rock matrix is stagnant and provides the fluid storage (cf. Warren & Root 1963). Fluid exchange between fracture and matrix is modelled using a transfer function, which attempts to account for the physics of the fracture-matrix fluid transfer and the geometry of the fracture network. The dual-porosity approach can be extended to simulate geomechanical effects where changes in fracture permeability and matrix porosity are computed individually (Bagheri & Settari 2008).

Defining the appropriate scale (i.e. grid cell size in the flow simulation model) to compute the effective fracture permeability tensor remains a major challenge, however; it must preserve the connectivity and permeability of the original fracture network, which evolves if the fractured porous media is deforming. Dershowitz *et al.* (2000) therefore concluded that 'if there is no grid cell scale that can reproduce the connectivity of the (discrete fracture network), (dual porosity) continuum simulation results will need to be treated with caution'.

The upscaling problem is exacerbated by the fact that the detailed information from the total geological exposure of a fracture system observed at the Earth's surface and modelled in Figure 6a is not available, and an initial estimate of bulk permeability must be made from very limited and quasi-1D data available from core samples and logs in wellbores. As a result, the upscaled permeability estimate for a reservoir block in Figure 6b is often obtained empirically by combining the qualitative geological and structural interpretation of the reservoir with quantitative geo-statistical simulation of the flow field (i.e. permeability and porosity fields), conditioned to the sparse field data (e.g. Strebelle 2002) and calibrated to the observed flow rates by complex history-matching algorithms (e.g. Oliver & Chen 2011). Practically, such models are used by engineers to manage the hydrocarbon field by changing fluid injection or production rates and shutting in or drilling new wells. Interestingly, reservoir engineers are also beginning to realize that history matching is insufficient and are moving towards predictive tests (e.g. Christie *et al.* 2006; Heffer *et al.* 2010).

Scaling in time

Earthquakes occur at strain rates that are several orders of magnitude lower than those achievable in the laboratory. A typical laboratory test at constant strain rate loading is of the order $10^{-5}\,s^{-1}$; a very slow 'creep' test (loaded at constant stress) may take a few weeks or months, slowing the strain rate down to $10^{-8}\,s^{-1}$ or so. In contrast, earthquake strain rates in continental zones occur under regional strain rates of the order $10^{-15}\,s^{-1}$ (Jackson & McKenzie 1988) although locally these can be higher ($10^{-12}\,s^{-1}$; Sibson 1982). It is therefore quite possible that different physical, and physico-chemical, processes may actually be involved across these enormous scale ranges.

To illustrate this, Figure 7 shows recent results (Ojala *et al.* 2004) from a suite of experiments at different strain rates aimed at determining the process of acceleration to failure due to the mechanism of stress corrosion associated with dissolution of silica in a sandstone sample (Ojala *et al.* 2003). In the quasi-static phase the acoustic emission event rate shows a systematic acceleration to failure of the near-asymptotic form $a = a_0(t_m + c - t/t_m)^{-p'}$, where t_m is the mainshock (dynamic failure) time, p' is a positive exponent and c is a characteristic time that keeps the event rate finite at the main event time. This form is consistent with the predictions from accelerated stress corrosion cracking (Main 1999, 2000), where $p' = 1 - 2/(n - 2)$ for event rate (assuming event rate is proportional to crack growth rate) and the stress corrosion index n is defined by the empirical observation that the velocity of subcritical crack growth scales as the nth power of the stress intensity (Charles' law: Meredith & Atkinson 1983).

Figure 7 shows that the absolute time for a warning that an acceleration has started (in real time) decreases systematically as the strain rate decreases, from a few minutes at a strain rate of $10^{-5}\,s^{-1}$ to a few tens of seconds at $10^{-8}\,s^{-1}$. This may be due to variations in the rate-limiting step for the process (e.g. the diminishing effect of the slow transport rate of reactive fluid to the fresh crack tip, Atkinson 1987). The net effect is that the system becomes more non-linear, with a shorter detectable precursor duration and overall less predictable in real time, as the strain rate decreases. This non-linearity in the normalized event rate occurs because more acoustic events are concentrated later in the loading history; the absolute number of events remains relatively insensitive to strain rate (Ojala *et al.* 2004).

Fujii *et al.* (1998) provide a more direct clue as to the diminishing role of dilatant strain at slower strain rates. By carrying out constant strain rate loading experiments between $10^{-8}\,s^{-1}$ and $10^{-3}\,s^{-1}$ on Kimachi sandstone (their fig. 13), Inada granite and Noboribetsu tuff (their fig. 15), they showed a systematic decrease in the critical tensile strain (most strongly associated with dilatancy) with lower strain rate. For example, the tensile strain decreases in Inada granite from 0.075 to 0.060% between 10^{-8} and $10^{-4}\,s^{-1}$ respectively, albeit with a large uncertainty due to sample variability of 0.5%. This is consistent with the much lower dilatant volumes that can be inferred from the result of Amoruso & Crescentini (2010) in that single example.

In summary, the signals from dilatant strain associated with microcracking diminish systematically as the strain rate diminishes, even under laboratory conditions. Likewise, the inferred much lower ratio of dilatancy (volumetric strain) rate to volumetric fluid flow rate at the very low strain rates applicable in the Earth is also likely to reduce the dilatancy hardening effect invoked to explain the duration of earthquake precursors. Both processes may contribute significantly to the lack of scaling of the dilatancy–diffusion process from the laboratory to the field case, with associated degradation of predictability.

Conclusion

The dilatancy–diffusion hypothesis as originally proposed failed for several reasons, largely associated with the validity of the assumptions on which it was based. Primarily, the anecdotal data used as evidence of geophysical precursors similar to those observed in the laboratory has not stood up to subsequent more rigorous testing: systematic precursors remain elusive, and the community has now moved on to true prospective testing of earthquake forecasting, specifically in order to avoid the retrospective selection bias inherent in previous literature. The hypothesis assumed a linear scaling (after renormalization of the parameters) of the physics from tests of small uniform lab samples with well-defined boundaries, loaded from zero to critical shear stress intensities; the Earth is however much more complex, has no such clear boundaries and is maintained by plate tectonics in a near-critical

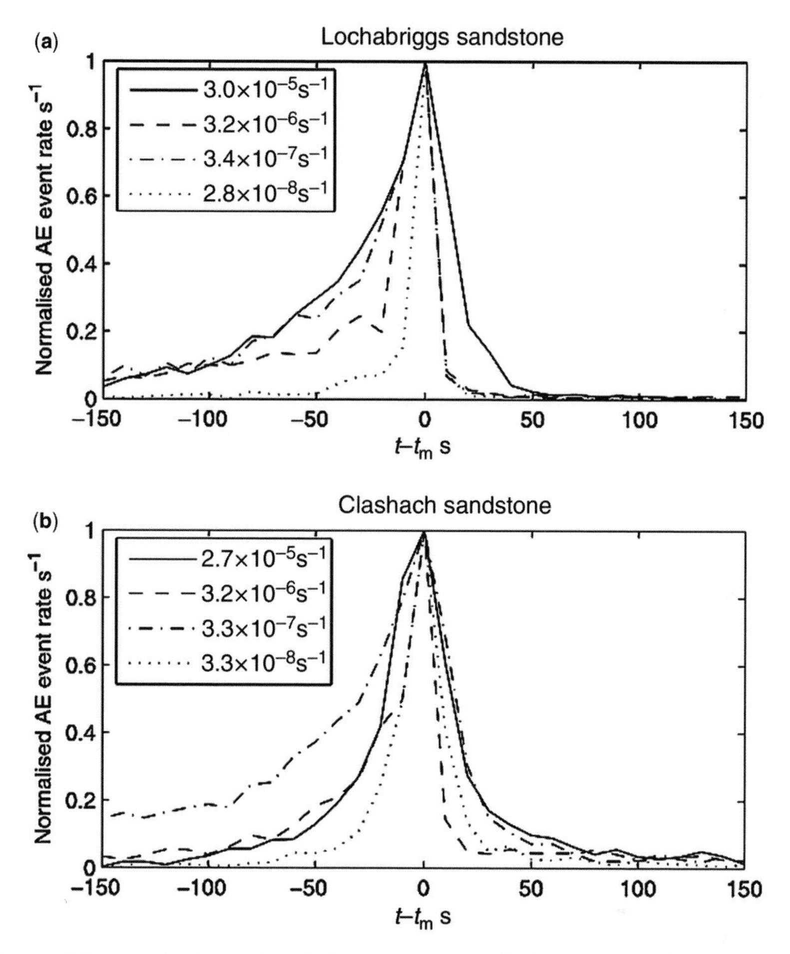

Fig. 7. Evolution of the normalized acoustic emission event rate a for the four tests at strain rates between 10^{-5} and 10^{-8} s^{-1}. (**a**) Locharbriggs and (**b**) Clashach sandstones, all run at a temperature of 80 °C.

effective stress state with relatively small stress fluctuations between events and a strong sensitivity to even smaller stress perturbations. Nevertheless, quasi-periodic stress 'renewal' models can provide a statistically significant (though small in absolute terms) probability gain over a purely random process.

The concept of a large-scale 'preparation zone', indicating the likely magnitude of a future event, remains as ethereal as the ether that went undetected in the Michaelson–Morley experiment. There appears to be little correlation even of aspects of the early part of rupture with the eventual magnitude of an event, consistent with the complexity and the critical or domino-like cascade of the rupture process. In contrast, recent geodetic and seismic data reveal in some cases the existence of a small but finite nucleation zone of around a few hundred metres at earthquake nucleation depths. Perhaps the basic features of precursory dilatancy and diffusion do occur locally on this scale, but are essentially impractical to detect reliably with current methods.

A significant upscaling exercise in space and time is needed to account for the multi-scale physics involved in extrapolating from laboratory tests to crustal scales, involving large-scale computational simulation to handle the several orders of magnitude differences in spatial and temporal scales. New laboratory tests at very slow strain rates are needed to bridge the gap to natural strain rates, and to explore the effect of the ratio of volumetric strain rate to volumetric fluid flow rate on the coupled behaviour in the precursory phase.

Finally, the hypothesis is not a complete failure; coupled dilatancy–diffusion processes remain a prime candidate for coseismic and post-seismic processes localized on or near the fault rupture plane and validated by geodetic and geological observation, and perhaps for localized nucleation processes inferred or constrained by geodetic or seismic data. These are interesting topics for study, irrespective of their implications for earthquake forecasting.

During the writing of this paper, I. Main was funded by a Royal Society of Edinburgh Research Fellowship, A. Bell by NERC grant NE/H02297X/1 and S. Touati by an EPSRC studentship. We are grateful to two anonymous reviewers for their constructive and insightful comments on a previous manuscript.

References

ABERCROMBIE, R. E. & MORI, J. 1996. Occurrence patterns of foreshocks to large earthquakes in the western United States. *Nature*, **381**, 303–307.

AMORUSO, A. & CRESCENTINI, L. 2010. Limits on earthquake nucleation and other pre-seismic phenomena from continuous strain in the near field of the 2009 L'Aquila earthquake. *Geophysical Research Letters*, **37**, L10307.

ANDERSON, E. M. 1905. The dynamics of faulting. *Transactions of the Edinburgh Geological Society*, **8**, 387–402.

ANDERSON, E. M. 1942. *The Dynamics of Faulting*. Oliver & Boyd, Edinburgh.

ATKINSON, B. K. 1987. Introduction to fracture mechanics and its applications. *In*: ATKINSON, B. K. (ed.) *Fracture Mechanics of Rock*. Academic Press, London, 1–23.

BAGHERI, M. & SETTARI, A. 2008. Modeling of geomechanics in naturally fractured reservoirs. *SPE Reservoir Evaluation and Engineering*, **11**, 108–118.

BAK, P. & TANG, C. 1989. Earthquakes as a self-organized critical phenomenon. *Journal of Geophysical Research*, **94**, 15 635–15 637.

BAKUN, W. H., AAGAARD, B. *ET AL.* 2005. Implications for prediction and hazard assessment from the 2004 Parkfield earthquake. *Nature*, **437**, 969–974.

BERKOWITZ, B. 2002. Characterizing flow and transport in fractured geological media: a review. *Advances in Water Resources*, **25**, 861–884.

BONNET, E., BOUR, O., ODLING, N. E., DAVY, P., MAIN, I., COWIE, P. & BERKOWITZ, B. 2001. Scaling of fracture systems in geological media. *Reviews of Geophysics*, **39**, 347–383.

BOUCHON, M., KARABULUT, H., AKTAR, M., ÖZALAYBEY, S., SCHMITTBUHL, J. & BOUIN, M.-P. 2011. Extended nucleation of the 1999 Mw 7.6 Izmit Earthquake. *Science*, **331**, 877–880.

BRUNE, J. N. & THATCHER, W. 2003. Strength and energetics of active fault zones. *In*: LEE, W.H., KANAMORI, H., JENNINGS, P. C. & KISSLINGER, C. (eds) *International Handbook of Earthquake and Engineering Seismology*. Academic Press, Amsterdam, **81A**, 569–588.

CHRISTIE, M. A. 1996. Upscaling for reservoir simulation. *Journal of Petroleum Technology*, November, 1004–1010.

CHRISTIE, M. A. 2001. Flow in porous media – scale up of multiphase flow. *Current Opinion in Colloid & Interface Science*, **6**, 236–241.

CHRISTIE, M. A. & BLUNT, M. J. 2001. Tenth SPE comparative solution project: a comparison of upscaling techniques. *SPE Reservoir Evaluation and Engineering*, **4**, 308–317.

CHRISTIE, M. A., DEMYANOV, V. & ERBAS, D. 2006. Uncertainty quantification for porous media flows. *Journal of Computational Physics*, **217**, 143–158.

CHUI, G. 2009. Shaking up earthquake theory. *Nature*, **461**, 870–872.

CRAMPIN, S., EVANS, R. & ATKINSON, B. K. 1984. Earthquake prediction: a new physical basis. *Geophysical Journal of the Royal Astronomical Society*, **76**, 147–156.

CRAWFORD, B. R., SMART, B. G. D., MAIN, I. G. & LIAKOPOULOU-MORRIS, F. 1995. Strength characteristics and shear acoustic anisotropy of rock core subjected to true triaxial compression. *International Journal of Rock Mechanics and Mining Sciences*, **32**, 189–200.

CYRANOSKI, D. 2011. Japan faces up to failure of its earthquake preparations. *Nature*, **471**, 556–557.

DERSHOWITZ, B., LAPOINTE, P., EIBEN, T. & WEI, L. 2000. Integration of discrete fracture network methods with conventional simulator approaches. *SPE Reservoir Evaluation and Engineering*, **3**, 165–170.

DOLAN, S. S., BEAN, C. J. & RIOLLET, B. 1998. The broadband fractal nature of heterogeneity in the upper crust from petrophysical logs. *Geophysical Journal International*, **132**, 489–507.

ELLSWORTH, W., HICKMAN, S. *ET AL.* 2005. Observing the San Andreas Fault at depth. *Eos, Transactions, American Geophysical Union*, **86**, 52.

FIELDING, E. J., LUNDGREN, P. R., BÜRGMANN, R. & FUNNING, G. J. 2009. Shallow fault-zone dilatancy recovery after the 2003 Bam earthquake in Iran. *Nature*, **458**, 64–68.

FRANK, D. C. 1965. On dilatancy in relation to seismic sources. *Reviews of Geophysics and Space Physics*, **3**, 485–503.

FUJII, Y., KIYAMA, T., ISHIJIMA, Y. & KODAMA, J. 1998. Examination of a rock failure criterion based on circumferential tensile strain. *Pure and Applied Geophysics*, **152**, 551–577.

GEIGER, S. & EMMANUEL, S. 2010. Non-fourier thermal transport in fractured geological media. *Water Resources Research*, **46**, W07504.

GELLER, R. J. 2011. Shake-up time for Japanese seismology. *Nature*, **472**, 407–409, doi: 10.1038/nature10105.

GRUESCHOW, B., KWON, O., MAIN, I. G. & RUDNICKI, J. W. 2003. Observation and modelling of the suction pump effect during rapid dilatant slip. *Geophysical Research Letters*, **30**, 1226.

HAIMSON, B. & CHANG, C. 2000. A new true triaxial cell for testing mechanical properties of rock, and its use to determine rock strength and deformability of Westerly granite. *International Journal of Rock Mechanics and Mining Sciences*, **37**, 285–296.

HEAP, M. J., BAUD, P., MEREDITH, P. G., BELL, A. F. & MAIN, I. G. 2009. Time-dependent brittle creep in Darley Dale sandstone. *Journal of Geophysical Research*, **114**, B07203–1.

HEFFER, K. J., GREEHOUGH, J. & MAIN, I. G. 2010. The accuracy of production forecasts: conventional models v. a statistical model for short-term predictions allied to reservoir monitoring and characterization. *In*: *Proceedings, Petex 2010*, London, 23–25 November (extended abstract).

HOLDSWORTH, R. E., BUTLER, C. A. & ROBERTS, A. M. 1997. The recognition of reactivation during continental deformation. *Journal of the Geological Society*, **154**, 73–78.

HOUGH, S. 2009. *Predicting the Unpredictable: The Tumultuous Science of Earthquake Prediction*. Princeton University Press, Princeton.

IMOTO, M. 2004. Probability gains expected for renewal process models. *Earth Planets Space*, **56**, 563–571.

JACKSON, J. & MCKENZIE, D. 1988. The relationship between plate motions and seismic moment tensors, and the rates of active deformation in the Mediterranean and middle-east. *Geophysical Journal*, **93**, 45–73.

JORDAN, T., CHEN, Y.-T. *ET AL*. 2011. Operational earthquake forecasting: State of knowledge and guidelines for utilization. *Annals of Geophysics*, **54**, 316–391.

KING, M. J., BURN, K. S., WANG, P., MURALIDHARAN, V., ALVARADO, F., MA, X. & DATTA-GUPTA, A. 2006. Optimal coarsening of 3D reservoir models for flow simulation. *SPE Reservoir Evaluation and Engineering*, **9**, 317–334.

MAIN, I. G. 1995. Earthquakes as critical phenomena: implications for probabilistic seismic hazard analysis. *Bulletin of the Seismological Society of America*, **85**, 1299–1308.

MAIN, I. G. 1996. Statistical physics, seismogenesis, and seismic hazard. *Reviews of Geophysics*, **34**, 433–462.

MAIN, I. G. 1997. Earthquakes – long odds on prediction. *Nature*, **385**, 19–20.

MAIN, I. G. 1999. Applicability of time-to-failure analysis to accelerated strain before earthquakes and volcanic eruptions. *Geophysical Journal International*, **139**, F1–F6.

MAIN, I. G. 2000. A damage mechanics model for power-law creep and earthquake aftershock and foreshock sequences. *Geophysical Journal International*, **142**, 151–161.

MAIN, I. G. & NAYLOR, M. 2008. Maximum entropy production and earthquake dynamics. *Geophysical Research Letters*, **35**, L19311.

MAIN, I. G., LI, L., MCCLOSKEY, J. & NAYLOR, M. 2008. Effect of the Sumatran mega-earthquake on the global magnitude cut off and event rate. *Nature Geoscience*, **1**, 142.

MAIR, K., MAIN, I. G. & ELPHICK, S. C. 2000. Sequential development of deformation bands in the laboratory. *Journal of Structural Geology*, **22**, 25–42.

MEAD, W. J. 1925. The geologic role of dilatancy. *Journal of Geology*, **33**, 685–698.

MEREDITH, P. G. & ATKINSON, B. K. 1983. Stress corrosion and acoustic emission during tensile crack propagation in Whin sill dolerite and other basic rocks. *Geophysical Journal of the Royal Astronomical Society*, **75**, 1–21.

MODELL, W. & HOUDE, R. W. 1958. Factors influencing clinical evaluation of drugs – with special reference to the double-blind technique. *Journal of the American Medical Association*, **167**, 2190–2199.

MULARGIA, F. 2001. Retrospective selection bias (or the benefit of hindsight). *Geophysical Journal International*, **146**, 489–496.

NIU, F., SILVER, P. G., DALEY, T. M., CHENG, X. & MAJER, E. L. 2008. Preseismic velocity changes observed from active source monitoring at the Parkfield SAFOD drill site. *Nature*, **454**, 204–208.

NUR, A. 1972. Dilatancy, pore fluids, and premonitory variations of ts/tp travel times. *Bulletin of the Seismological Society of America*, **62**, 1217–1222.

NUR, A. 1975. A note on the constitutive law for dilatancy. *Pure and Applied Geophysics*, **113**, 197–206.

OJALA, I., NGWENYA, B. T., MAIN, I. G. & ELPHICK, S. C. 2003. Correlation of microseismic and chemical properties of brittle deformation in Locharbriggs sandstone. *Journal of Geophysical Research*, **108**, 2268.

OJALA, I. O., MAIN, I. G. & NGWENYA, B. T. 2004. Strain rate and temperature dependence of Omori law scaling constants of AE data: implications for earthquake foreshock–aftershock sequences. *Geophysical Research Letters*, **31**, L24617.

OLIVER, D. S. & CHEN, Y. 2011. Recent progress on reservoir history matching: a review. *Computational Geosciences*, **15**, 185–221.

PATERSON, M. S. & WONG, T.-F. 2005. *Experimental Rock Deformation – The Brittle Field*. 2nd edn. Springer, Berlin, Heidelberg.

RENARD, P. & DE MARSILY, G. 1997. Calculating equivalent permeability: a review. *Advances in Water Resources*, **20**, 253–278.

RUDNICKI, J. W. & CHEN, C.-H. 1988. Stabilization of rapid frictional slip on a weakening fault by dilatant hardening. *Journal of Geophysical Research*, **93**, 4745–4757.

SAMIER, P., QUETTIER, L. & THIELE, M. 2002. Applications of streamline simulations to reservoir studies. *SPE Reservoir Evaluation and Engineering*, **5**, 324–332.

SAMMONDS, P. R., MEREDITH, P. G. & MAIN, I. G. 1992. Role of pore fluids in the generation of seismic precursors to shear fracture. *Nature*, **359**, 228–230.

SCHARER, K. M., BIASI, G. P., WELDON, R. & FUMAL, T. E. 2010. Quasi-periodic recurrence of large earthquakes on the southern San Andreas Fault. *Geology*, **38**, 555–558.

SCHOLZ, C. H. 1997. Whatever happened to earthquake prediction? *Geotimes*, **42**, 16–19.

SCHOLZ, C. H. 2002. *The Mechanics of Earthquakes and Faulting*. Cambridge University Press, Cambridge.

SCHOLZ, C. H., SYKES, L. R. & AGGARWAL, Y. P. 1973. Earthquake prediction – a physical basis. *Science*, **181**, 803–810.

SIBSON, R. H. 1981. Controls on low-stress hydro-fracture dilatancy in thrust, wrench and normal-fault terrains. *Nature*, **289**, 665–667.

SIBSON, R. H. 1982. Fault zone models, heat flow, and the depth distribution of earthquakes in the continental crust of the Unites states. *Bulletin of the Seismological Society of America*, **72**, 151–163.

SIBSON, R. H. 2001. Seismogenic framework for hydrothermal transport and ore deposition. *Reviews in Economic Geology*, **14**, 25–50.

Sibson, R. H. 2009. Rupturing in overpressured crust during compressional inversion – the case from NE Honshu, Japan. *Tectonophysics*, **473**, 404–416.

Sibson, R. H., Moore, J. Mc. M. & Rankin, A. H. 1975. Seismic pumping – a hydrothermal fluid transport mechanism. *Journal of the Geological Society, London*, **131**, 653–659.

Strebelle, S. 2002. Conditional simulation of complex geological structures using multi-point statistics. *Mathematical Geology*, **34**, 1–21.

Van Stiphout, T., Wiemer, S. & Marzocchi, W. 2010. Are short-term evacuations warranted? Case of the 2009 L'Aquila earthquake. *Geophysical Research Letters*, **37**, L06306.

Warren, J. E. & Root, P. J. 1963. The behaviour of naturally fractured reservoirs. *Society of Petroleum Engineers Journal*, **3**, 245–255.

Wyss, M. & Booth, D. C. 1997. The IASPEI procedure for the evaluation of earthquake precursors. *Geophysical Journal International*, **131**, 423–424.

XLII. *The Dynamics of Faulting.* By ERNEST M. ANDERSON, M.A., B.Sc., H.M. Geological Survey.

(Read 15th March 1905.)

IT has been known for long that faults arrange themselves naturally into different classes, which have originated under different conditions of pressure in the rock mass. The object of the present paper is to show a little more clearly the connection between any system of faults and the system of forces which gave rise to it.

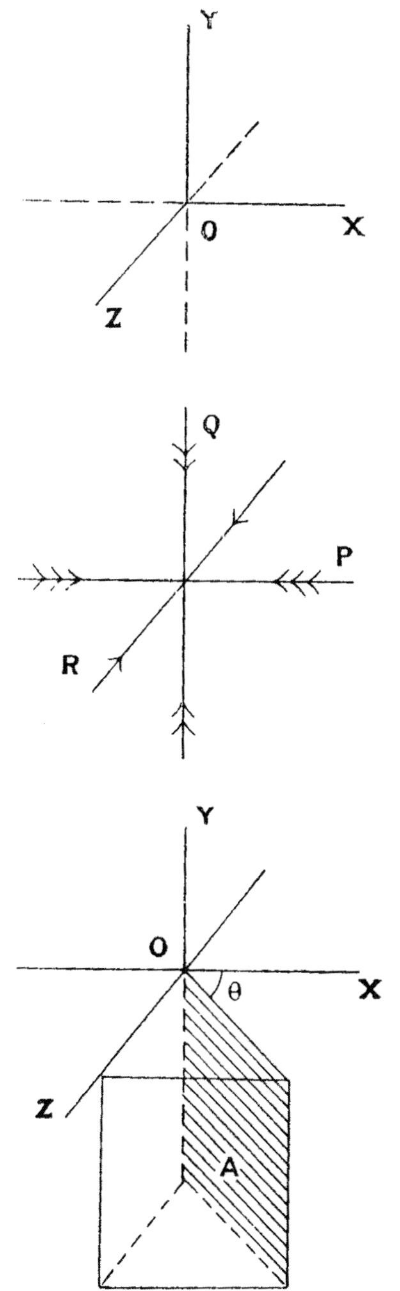

It can be shown mathematically that any system of forces, acting within a rock which for the time being is in equilibrium, resolves itself at any particular point into three pressures or tensions (or both combined), acting across three planes which are at right angles to one another.

Across these particular planes there is no tangential stress, but there will be tangential stress at that point across any other plane which may be drawn through it. There will evidently be positions of this hypothetical plane for which the tangential stress will be a maximum. It is evident that these maximum positions of the plane will have much to do with determining the directions of faults in the rock. We will therefore take the general case and investigate what the positions are. Suppose O to be any point in a rock, and let the three directions along which the pressures or tensions act (the directions perpendicular to the three planes mentioned above)

From: HEALY, D., BUTLER, R. W. H., SHIPTON, Z. K. & SIBSON, R. H. (eds) 2012. *Faulting, Fracturing and Igneous Intrusion in the Earth's Crust*. Geological Society, London, Special Publications, **367**, 231–246.

First published: *Transactions of the Edinburgh Geological Society*, **8** (1905), 387–402.

be OX, OY, OZ. Let the pressures, or tensions, acting along these three directions be P, Q, R, which we will suppose positive when they denote pressures, and negative when they denote tensions. Suppose further that P is the greatest pressure, or the least tension in the case where there are only tensions, and that R is the greatest tension, or the least pressure in the case where there are only pressures, so that P, Q, R are algebraically in descending order of magnitude. Then it can easily be shown that the planes of greatest tangential stress are parallel to the line OY, and inclined at an angle to the directions OX and OZ. That is to say, they are parallel to the direction of intermediate pressure, and inclined at certain angles to the directions of greatest and least pressure in the rock.

To determine what these angles are, suppose A to be a plane parallel to OY, and making an angle θ with the direction of OX. Then by a simple proof it can be shown that the tangential stress across A is $\frac{P-R}{2}\sin 2\theta$.

This proof is a well-known theorem, and can be found in any book which deals with the subject of stresses in solid bodies. The method, which I shall merely indicate, is to consider the forces acting on a right triangular prism, having its edges parallel to OY, one of its faces parallel to A, and two others parallel to the planes XOY and YOZ. This prism we suppose to exist in the rock, somewhat as the statue exists beforehand in the block of marble, and by considering the forces acting on it we are led to the above result, namely—

$$\text{Tangential stress} = \frac{P-R}{2}\sin 2\theta.$$

It is evident that this force will vanish when $P=R$. That is, there can be no tangential stress when the pressures in the two directions are equal. It will be large when P and R are of opposite sign; *i.e.* when one represents a pressure and the other a tension.

For any given system, however, the tangential stress is greatest when $\sin 2\theta = 1$, $2\theta = 90$, $\theta = 45°$. It is evident that it will be equally great when $\theta = -45°$, and thus we shall have two series of planes across which tangential stress is a maximum. The one set will be parallel to the plane which passes through OY and bisects the angle XOZ; the other set will be parallel to the plane passing through OY and bisecting the angle XOZ. Across any plane belonging to either of these series the tangential stress will be $\frac{P-R}{2}$.

Now, as we shall afterwards see, the planes of faulting in any ck do not follow exactly the directions of maximum tangential stress, but deviate from these positions in a more or less determinate manner. In endeavouring to explain this I have been led to suppose that the forces which hinder rupture from taking place in any rock are not the same in every direction. If we suppose that the resistance which any solid (otherwise isotropic) offers to being broken by shearing along any plane consists of two parts, one part being a constant quantity and the other part proportional to the pressure across that plane, we shall arrive at results which agree very well with the observed geological facts.

The second force will have an effect somewhat similar to friction, and I shall use the symbol μ in this connection, while by no means assuming that we are dealing with the same phenomenon. The effect of this force will be to make faulting more difficult along planes across which there is great pressure.

Supposing, as before, that P and R are the greatest and least pressures at any point. Then, as we found, the tangential stress across a plane parallel to the direction of Q, and inclined at an angle θ to the direction of P, is $\frac{P-R}{2}\sin 2\theta$, while the *pressure* across such a plane may easily be shown to be $P\sin^2\theta + R\cos^2\theta$.

Now, supposing a plane crack had actually formed in this direction, and that movement were just about to begin along it, the resistance to this movement due to friction would be $\mu(P\sin^2\theta + R\cos^2\theta)$, or $\mu\left(\frac{P+R}{2} - \frac{P-R}{2}\cos 2\theta\right)$.

If we assume the existence of the second force above referred to, then instead of considering the maxima of $\frac{P-R}{2}\sin 2\theta$, we must subtract from this quantity one of like form to that given above. We are supposing now that μ is the ratio which the variable part of the resistance to breakage bears to the pressure across the plane considered. We then get the following quantity, $\frac{P-R}{2}(\sin 2\theta + \mu\cos 2\theta) - \mu\frac{P+R}{2}$; and this will be a maximum in the directions in which faulting will be the most likely to occur.

For a maximum, it follows from the principles of the Differential Calculus that $\cos 2\theta - \mu\sin 2\theta = 0$, $\tan 2\theta = \frac{1}{\mu}$

This gives us $\theta = 45°$ for $\mu = 0$

$\theta = 30°$ for $\mu = \frac{1}{\sqrt{3}}$ or ·577.

$\theta = 22\frac{1}{2}°$ for $\mu = 1$

It is difficult to form any estimate of what value must be assigned to μ for any particular rock, but we see what will be the general result. The planes of faulting, instead of bisecting the angles between the directions of greatest and least pressure, will deviate from these positions so as to form smaller angles with the direction of greatest pressure. This result agrees very well with the recorded facts.

I shall next consider a little more fully what takes place under (1) an increase and (2) a relief of lateral pressure in any rock. It is important to notice that it does not follow that because there is an increase of pressure in one horizontal direction, there will necessarily be so in all. On the other hand, it is quite possible that there may be an increase of pressure in one horizontal direction along with a relief of pressure in a horizontal direction at right angles to the first. This will form a third case to be treated separately.

(1) Suppose there is an increase of pressure in all horizontal

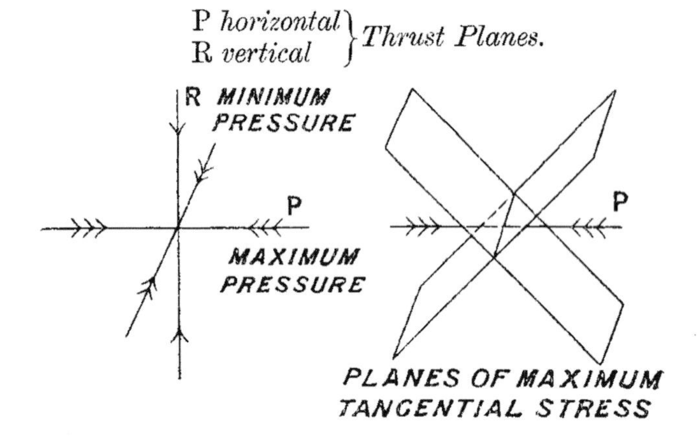

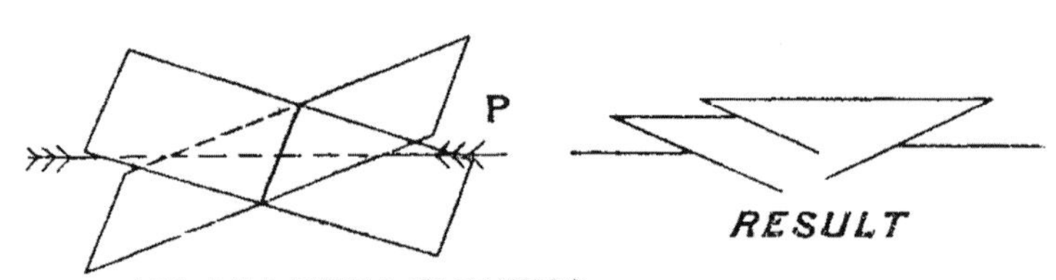

directions. Then it may possibly happen that the pressure in all horizontal directions is equal. This will form a special case to be treated later on. For the meanwhile we shall assume, what is far more likely to happen in fact, that the different horizontal pressures are not exactly equal, but that there is one horizontal direction along which pressure is greatest. This maximum pressure we shall, as before, denote by P. Then R, the minimum pressure, will be vertical, and the intermediate principal pressure Q will be in a horizontal direction perpendicular to P.

Then there will be two sets of planes across which tangential stress will be a maximum. Both sets will have their "strike" parallel to Q and perpendicular to P. Both sets will "dip" at an angle of 45°, but they will dip in opposite directions.

Suppose now that the stresses are so great as to lead to actual rupture. Then the planes of faulting should strike in the same direction, but they should, as we have seen, be less inclined to the direction of greatest pressure, which in the present case is horizontal. Thus we should have a double series of fault-planes inclined to the horizontal at angles of less than 45° and striking perpendicularly to the direction of greatest pressure. Motion would take place along any of these planes in such a way as to relieve the pressure, that is, in the form of overthrust.

(2) Suppose next that there is a relief of pressure in all

P *vertical* }
R *horizontal* } *Normal Faults.*

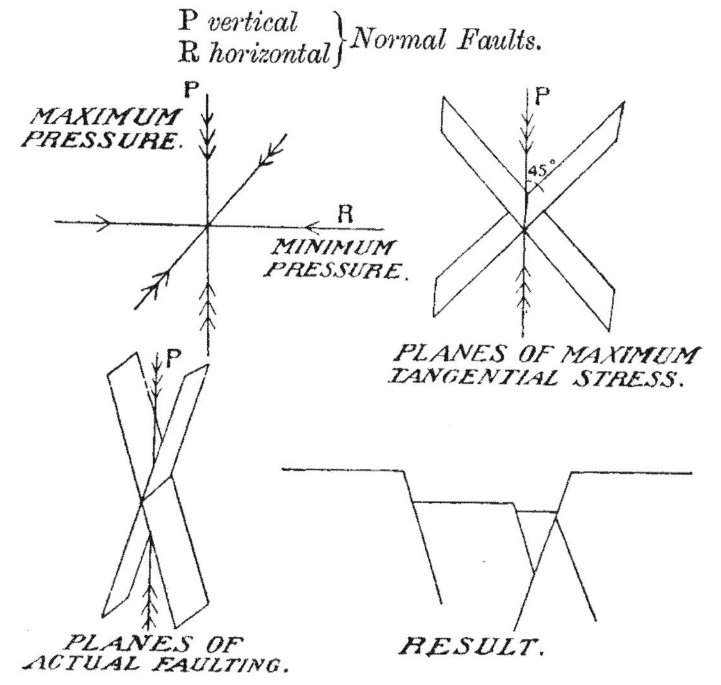

horizontal directions, so that P, the greatest pressure, will be the vertical pressure due to gravity. Then it can only happen very rarely that the pressures, or tensions, in all horizontal directions, will be equal. In the general case there will be one horizontal direction for which the pressure is a minimum. Taking this as the direction of R, then Q, the intermediate principal pressure, will be in a horizontal direction perpendicular to R.

In this case the planes of maximum tangential stress will strike parallel to Q and perpendicular to R; while they will dip in opposite directions at angles of 45° as before.

The planes of actual faulting will deviate from these positions so as to form smaller angles with P, the vertical pressure. The result will be a double series of fault-planes dipping in opposite directions at angles of *more* than 45°, and striking perpendicularly to that direction in which the relief of pressure is the greatest. Motion will take place along these planes in the normal manner. I shall try to show later on how such faulting will tend to equalise the pressures.

(3) We have next to consider the case in which there is an

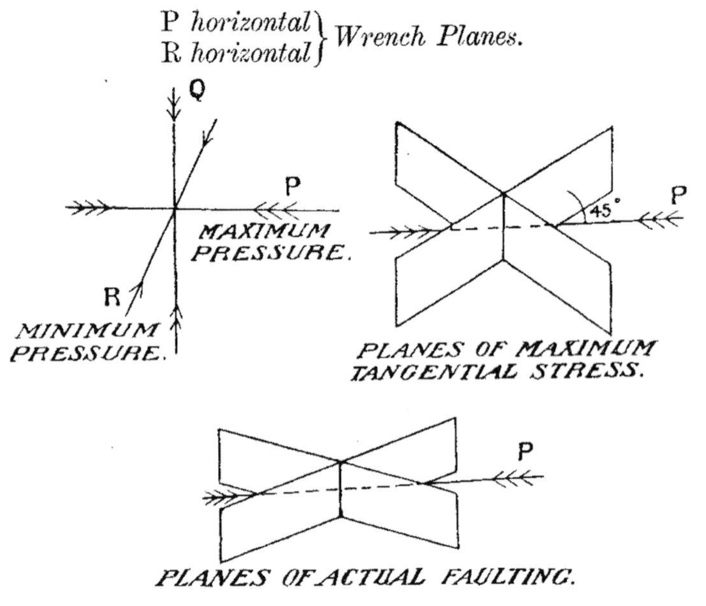

increase of pressure in one horizontal direction, together with a decrease of pressure in a horizontal direction at right angles to the first. In this case the maximum pressure P is horizontal; the intermediate pressure Q is vertical; while the third principal direction, which may correspond to a tension, or to the smallest pressure, is horizontal and at right angles to the direction of P.

Then the planes of maximum tangential stress are vertical, and inclined at angles of 45° to the directions of P and R. The planes of actual faulting will deviate from these positions so as to form smaller angles with the direction of P, the maximum pressure.

They might, in fact, form an arrangement not unlike that of the cleavage-planes of a hornblende crystal, supposing the crystal to be placed with its prism axis vertical. Motion would take place along any one of these planes in a horizontal manner. A plane of dislocation along which motion has actually taken place in a horizontal direction is sometimes called a "wrench-plane." Thus we see that under the system of forces last considered a network of such wrench-planes may possibly be developed. Each of these would hade vertically, and the two systems would cross each other at acute angles.

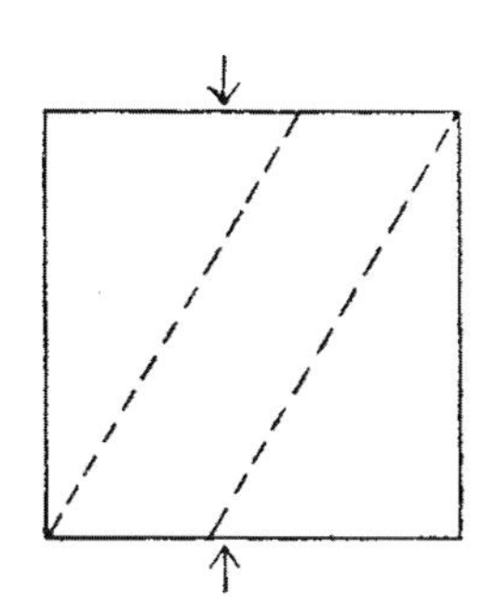

In this last case it is not particularly evident in what manner a single dislocation tends to relieve the stress. If, however, we take a number of such wrench-planes, crossing one another, it becomes optically evident how the stress will be relieved.

The accompanying diagram shows how a square area of land is deformed by a double system of wrench-planes; the particular case chosen being that in which the greatest pressure is from N. to S. and the least pressure from E. to W. The joint result is to decrease the N. to S. dimensions of the area, and to increase its dimensions from E. to W. Thus we see how the stress will tend to be relieved. The diagram also indicates in what direction motion will take place along a fault of either series.

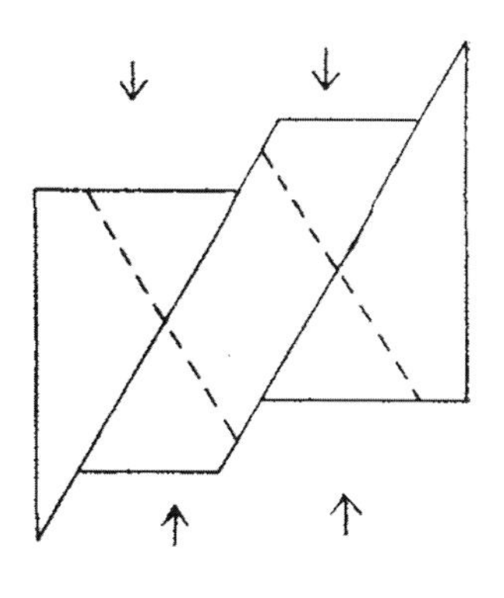

Regarded as a vertical section, the same diagram will do to illustrate the case of normal faults. As a matter of fact, however, it is almost impossible to tell, from surface indications, whether so complete a network ever does exist in this case. Two systems of faults like those in the diagram, originating under the same system of pressures, might very well be called "complementary" systems.

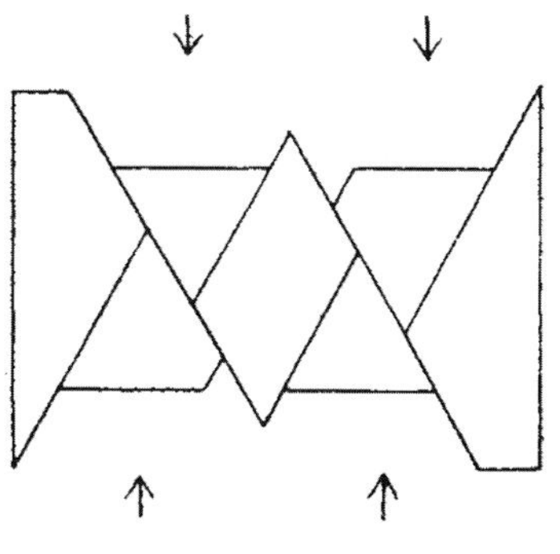

There remain to be considered only the cases in which the stresses in two principal directions are equal; while the stress in the third principal direction is not equal to the other two. In any such case shearing stress will reach its maximum across planes having an indefinite number of directions, but parallel to the tangential planes of a cone having its axis in the direction of unequal pressure. Thus faulting might take place in an indefinite number of directions. This is a case, however, which is very unlikely to arise in reality, as the odds are very great against the pressure in two principal directions being exactly equal.

I shall next consider in how far these theoretical conclusions are borne out by the recorded facts.

With regard to the first of the three cases, it is a well-known fact that reversed faults and thrust-planes do in general dip at a low angle; this is especially the case in connection with the great series of thrust-planes occurring in the N.W. Highlands. These are a series of dislocations dipping originally at a low angle to the east; motion has taken place along them by the country to the east being pushed over the country to the west. The pressure which produced them was obviously from east to west. There must at the same time have been a slightly more than normal pressure from north to south, while the least pressure would be in a nearly vertical direction.

In this case our theory would have led us to expect a complementary series of faults, dipping at a low angle to the west, and along which the country to the west had been pushed over the country to the east. In Scotland no such complementary series exists.

It is, however, very instructive to compare the Scottish system of thrust-planes with that which has been made out to exist in Scandinavia. In this peninsula rocks of Silurian age are overlain through more than 10° of latitude by rocks of what is called the Seve group; possibly Cambrian, and undoubtedly older than the rocks beneath them. The labours of many geologists, and notably Törnebohm (" Grundragen af det Centrala Skandinaviens Bergbyggnad"), have shown that this is due to thrusting on an enormous scale. In the case of the larger thrusts, the country to the west has been pushed over the country to the east. Törnebohm, however, states that in the case of some of the smaller thrusts this rule is reversed, and thrusting has taken place of east over west, as in Scotland.

In any case it seems probable that the thrust-planes in Scandinavia owe their origin to the same great series of pressures as produced those in Scotland, and that thus a pressure acting from east to west over a wide stretch of country may produce in one part thrusting of west over east, and in another of east over west.

When we come to consider angles, however, the results of observation are less in accordance with our theory. Thus in Scotland the prevailing dip of the thrust-planes to the east is too low to have been produced by any purely horizontal pressure. In Scandinavia the planes of dislocation have themselves been affected by subsequent movements, and they may dip west, east, or in any direction. If the original dip was to the west, however, it would seem to have been an extremely low one. The accompanying diagram is a rough attempt to give mathematical form to an idea of Dr Peach's with regard to the low angle of dip. What is intended to be expressed is as follows. Supposing a mountain chain was being produced by the same series of pressures that caused the thrusts, and suppose that it ran (as it naturally would) in a north and south direction. Suppose further that the west of Sutherland lay on the western margin of this mountain-chain, and that the district in Scandinavia where the thrusts have been produced lay on its eastern margin, or on the eastern margin of some parallel chain. Then the effect of the declivity of the ground

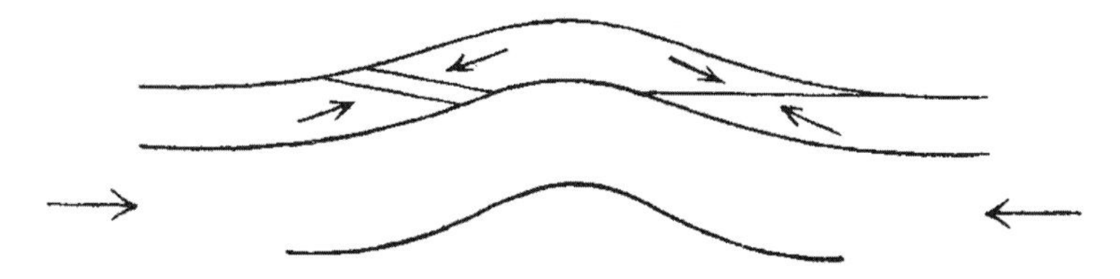

on either side would be to tilt the directions of greatest pressure, on either side, as indicated in the diagram. The planes of most likely thrusting would thus also be tilted, so as to become more nearly horizontal.

With regard to the second of our two cases, it is well known that normal faults have in general a dip much steeper than 45°. In the central lowlands of Scotland there is a great system of normal faults striking approximately east and west. Of these faults a certain number hade towards the north, and have a downthrow on that side; others have their hade and downthrow towards the south. We have thus here a beautiful example of a double series of faults produced by the same system of forces, and in this case we may regard it as proven—

(1) That at the same time when these faults were formed there was a relief of pressure in all horizontal directions.

(2) That this relief was greatest in the direction from north to south.

I have spoken everywhere above only of a relief of pressure, but it is possible that in some cases a relief of pressure may go so far as to amount to an actual tension. If this tension became so great as to produce actual rupture by pulling the rocks asunder, it seems likely that cracks would be formed perpendicular to the direction of tension. They would run parallel to the set of faults likely to be produced by a smaller tension, but their hade would most likely be vertical, and they would not necessarily produce any displacement of the strata on either side. In this connection it is extremely interesting to notice that sets of intrusive dykes are in some cases known to accompany systems of normal faults, and to run in the same direction.

Thus along with the east and westerly faults of the central valley of Scotland, occur, as is well known, a set of east and westerly dykes. Sets of parallel faults also accompany the north-westerly dykes of Tertiary age in some parts of the British Islands. Thus in the Cowal district Mr Clough has observed a series of north-westerly faults along with these dykes, and in Anglesea Mr Greenley has found a similar series accompanying dykes of the same age, which he knows to be normal faults. Although in a case like this the faults and the dykes may not be strictly contemporaneous, it seems almost certain that they owe their origin to the same great series of forces; and it may be possible that the dykes occupy fissures formed while the tension was at its strongest, by the actual pulling asunder of the strata.

With regard to the third class of faults, there is as yet much less recorded material with which to compare our conclusions. Planes of dislocation do occur, along which there has been much horizontal movement, but it is difficult to show in any particular case that this has not been accompanied by an equal or greater amount of vertical displacement.

In the Fassa Monzoni district of the Tyrol a double series of faults has been described by Mrs Gordon under the name Judicarian.[1] These form two sets running N.N.E. and N.N.W., and Mrs Gordon believes them to be strictly contemporaneous. On this assumption the only possible method of explaining their production is to suppose them to be wrench-planes which have formed under the influence of a great pressure in the north and south direction, accompanied by a relief of pressure in the east and west direction. If this explanation be the correct one, these faults form a very striking example of a double system originating under the same set of forces.

To come nearer home, a good example of faults accompanied by lateral wrench occurs in certain parts of the Highlands of Scotland. It is now well known that there is a great series of north-easterly or north-north-easterly faults traversing the Highlands, and giving rise to well-marked physical features. South of the Great Glen are four faults which have been named the Loch

[1] "The Geological Structure of Monzoni and Fassa." *Trans. Ed. Geol. Soc.*,

Tay fault, the Killin fault, the Glen Fine or Tyndrum fault, and the Loch Awe fault. North of the Glen occurs the Inbhir-Chorainn fault, which extends perhaps from the southern part of Skye to near Ben Wyvis in Ross-shire.

These are all faults with lateral displacement, and they run in the direction already mentioned. It is noticeable that the ground on the east side of these faults has in every case been shifted north-eastwards with regard to the ground on the west; the amount of displacement often amounts to two or three miles. From this fact we are justified in the following conclusion, that at the time when these faults were formed there was an increase of horizontal pressure in the north-south direction, accompanied by a relief of pressure in the direction from east to west.

It is just possible that the Loch Maree fault, which is a wrench running in the N.W. direction, may be complementary to the series above described, as the movement along it indicates a pressure which was greatest in a direction only a little W. of N. (and E. of S.). Otherwise we must suppose that we have a series like that of the thrust-planes which occur in another part of the Highland area, where for some reason only one set of a possible double set of faults has been developed.

According to the theory, a fault of the kind we are discussing should have a nearly vertical hade. I have seen this verified in the case of some small faults with horizontal slickensides, which occur in the valley of the Allt Coire Rainich in easter Ross-shire. Mr Clough has observed the Inbhir-Chorainn fault to hade in different directions in different parts of its course. It is possible that the *general* hade of this fault may be nearly vertical, and that what Mr Clough has noticed may be due to a sort of slickensiding or corrugating of the fault plane on a large scale.

In the above discussion I have everywhere assumed that one of the principal directions of pressure is vertical. If it were not, we might have systems of faults intermediate in character between the classes described above. Thus we might have a single fault, or a system of faults, each member of which was partly a wrench-plane and partly a normal fault. That such intermediate cases do occur seems certain, from the number of cases in which faults are accompanied by slickensides with a direction intermediate between the horizontal and vertical.

Or again, although, as we have seen, the majority of thrust-planes and normal faults do follow the rules above laid down, individual cases do occur in which a thrust is met with, with a steeper dip than 45°, or a normal fault which has a less dip than the above figure. These may very likely be explained in the same way, by a departure from the perpendicular and horizontal directions of the principal axes of pressure.

I have in the preceding part of this paper used the terms strike and dip, making them apply to fault-planes in the same sense as they do to planes of bedding. I have by this means avoided the use of the word "hade," which sometimes gives rise to ambiguity.

It is very difficult to estimate what amount of tangential force will be necessary in order to produce actual rupture and so lead to faulting. In Professor Ewing's book on "The Strength of Materials," figures are given for the amount of force necessary to produce crushing in prismatic blocks of various materials, the force being applied to the ends of the prisms. For prisms 1″ in section, the following are the results for a few common rocks

Granite	6-10 tons.
Basalt	8-10 tons.
Slate	5-10 tons.
Sandstone . . .	2-5 tons.

In § 9 of the same book occurs this sentence, which is of some significance in connection with the present subject:—

"When a bar is pulled asunder, or a block is crushed by pressure applied to two opposite faces, it frequently happens that yielding takes place wholly or in part by shearing on surfaces inclined to the direction of pull or thrust."

Now, in the case of a block crushed by pressure applied to two opposite faces, we are dealing with a single pressure, which we may denote by P; Q and R being nearly zero. Assuming, then, that yielding does take place by shearing to begin with (and it is difficult to imagine what else could happen), the amount of tangential stress necessary to produce this shearing cannot be greater than $\frac{P}{2}$

In granite this may be as much as 5 tons per square inch; in hard sandstone it would amount to $2\frac{1}{2}$ tons per square inch; in soft sandstone to only 1 ton; in shale, and in soft rocks of Tertiary formation, it would probably be even less.

If we suppose this tangential stress to be the result of a single pressure, the amount of such a pressure necessary to produce faulting is indicated by the figures already quoted from Professor Ewing.

For hard sandstone the pressure, if a single pressure, must amount to 5 tons per square inch. Now, supposing the S.G. of the rock we are dealing with to be 2·65 (the S.G. of quartz), a pressure of this amount will be caused by the weight of the superincumbent rock at a depth of $1\frac{5}{6}$ miles (1·844).

We are thus led to enquire what it is that prevents faulting taking place incessantly at this and all greater depths. We see from our formula that supposing P to be the vertical pressure, a lateral pressure of amount R diminishes the tendency to shear from P/2 to (P−R)/2. Thus the answer to the above question is that there must be lateral pressure in all directions. It is easily seen, too, that at the critical depth above mentioned, the lateral pressure cannot exceed twice the vertical, or else (P−R)/2 would become a negative quantity greater numerically than $2\frac{1}{2}$ tons per square inch, which we are taking for the critical amount of stress.

As we go further down in the substance of the earth's crust, the lateral pressures must increase along with the vertical, as the difference between the vertical pressure and the pressure in any horizontal direction, even for a rock as hard as basalt, can never exceed 10 tons per square inch. At a depth of, say, 25 miles, the vertical pressure will be something very much greater than 10 tons per square inch, and so at this depth the differences between the pressures must be small quantities when compared to the pressures themselves. Thus there must be a condition of things, at great depths, similar in one respect to fluid pressure.

The question next arises, what is to produce this lateral pressure, which we see is necessary to preserve equilibrium, altogether apart from the production of anticlines or thrust-planes. The question may be answered by considering the case of an arch consisting of a single layer of bricks. If we take the brick at the summit of the arch, it is easy to show by drawing a triangle of forces that the horizontal forces acting on the brick, and due to the pressure between it and its next neighbours, are great in comparison with the vertical force acting on it due to gravity. This will be the case even when the arch is loaded by the weight of further material resting on it. For a brick in this position, then, the horizontal pressure is not equal, but much greater than the vertical.

The same would be the case if all the extraneous forces, instead of being directed in parallel lines, were directed towards the centre of the arch ; and it is easy to see that the statement also applies to the case of a hollow globe, acted on by a system of forces tending towards its centre. Under a force as great as that of gravity at the earth's surface, however, a hollow globe of the size of the earth could not exist, at least if composed of any known rock. The horizontal pressure would necessarily be so great as to cause shearing, being uncompensated by any vertical pressure of corresponding magnitude.

Thus it is impossible to look on the earth as being a series of independent, self-supporting, concentric shells. At the same time, if, by a mathematical fiction, we suppose the earth divided into a series of concentric shells, it is not the case that each shell will have to bear the whole weight of those above it.

Each of the latter acts *to a small extent* as an arch, and so, as it were, bears part of its own weight. This part will only be a small fraction of the whole In fact, if A denote the lateral pressure which would exist in any such shell, supposing it to be entirely unsupported, and to act as an arch, and if H denote the actual mean horizontal pressure, then, roughly, H/A will denote the fraction of its own weight borne by such a shell.

The same will hold in the case of a liquid globe, in equilibrium under its own attraction ; only in this case the problem will be far more definite, as the pressure must be the same in all directions at any point.

I have brought in this discussion to show how we might account for a lateral pressure even much greater than that which actually exists. We have seen that the actual lateral pressure existing in each layer of the earth's crust is fixed within certain limits, so long as equilibrium is to be maintained. I shall try to show how it is that adjustment takes place whenever the lateral pressure transgresses these limits.

Suppose that during any period the outer surface of the earth is not contracting so quickly as the earth's interior. The result will be that the outer crust ceases to have so much support from the underlying layers (whether liquid or solid) as it would normally have. It has thus to support a greater than usual fraction of its own weight, and the result is greater lateral pressure.

If the lateral pressure only exceeds the vertical by so small a quantity as to lie within the above mentioned limits, it may result in the gradual formation of anticlines or isoclinal folds. If, however, the lateral pressure exceeds the vertical by a greater quantity than the rigidity of the rocks concerned will allow, a series of reversed faults or thrust-planes will result. In either case the result is the same, the circumference of the surface layer is diminished ; it thus settles down on the layers below it, and ceases to bear a more than normal fraction of its weight, and so lateral pressure is diminished to within the before mentioned limits.

Suppose, on the other hand, that during some geological period the surface layer is contracting more rapidly than the more central portions. The result will be that the latter are compressed, and that there is an increase of pressure in the central portions which tends to set up a tension in the surface layer. In this way lateral pressure in the surface layer will be partially, or it may be wholly, relieved.

As soon as the lateral pressure falls short of the vertical at any point, by more than a certain amount, a series of normal faults will result. These increase the earth's circumference so that the arch-like condition of the surface layer is restored, and lateral pressure is increased to within the required limits.

When considering the formation of faults, we have generally to deal with systems of forces which existed in past geological time. In a few cases the contrary may be the case. Thus in Japan and other countries which are liable to earthquakes, there is no doubt that faults are being produced or are growing in magnitude even at the present day. To a less extent this may be the case in our own country; at least we know that movement is going on in some districts along faults which have already been formed. This brings us to the interesting question whether it may be possible, by studying the direction of fault-lines along which motion is taking place, to arrive at some conclusion with regard to the system of forces at present at work in any given area.

From a consideration of the facts published by Mr Davison, we see that in Scotland there is a tendency for earthquakes to occur along faults which run in a north-easterly direction. This may perhaps be an accident due to the fact that the largest faults in the country do happen to run in that direction. At the same time the fact is worth noticing; it might be connected with a slight increase of horizontal pressure in a direction from south-east to north-west; it might, however, be due to other forces, and we are very far from being in a position to form any definite conclusion on the subject. In England, earthquakes seem to take place along faults running in every direction of the compass; perhaps there may be a very slight tendency in some parts to select faults which run north-eastwards, but it would be very unsafe to base any hypothesis on this supposition. In a case like this it is impossible to say whether we are dealing with a simple system of pressures or not. On the other hand the extent of country over which roughly parallel folding may often be found is a proof that at certain past geological epochs the same sets of forces must have extended over wide areas. The length which certain mountain chains extend in what are practically straight lines may be taken as additional evidence to the same effect.

It has been observed that fault breccias occur more frequently in connection with normal faults and wrench-planes than they do in connection with thrusts. This may be easily explained as follows. The normal pressure due to gravity in a rock mass is a fairly constant quantity at a given depth. It may therefore be taken as a standard with which to compare our other pressures.

Now, in the case of normal faults both horizontal principal pressures are less than the vertical; in the case of wrench-planes one is greater and one is less; while in the case of thrusts the pressure in all horizontal directions is greater than the vertical. If we take the formula $P \sin^2\theta + R \cos^2\theta$, which for a certain value of θ denotes the pressure *across* a fault plane, we see that in the first case the pressure must be less than the vertical. In the second it may be greater or it may be less, while in the third case, that of thrusts, it is bound to be greater. The explanation seems therefore to be that fault-breccias form more readily along faults across which the pressure is not very intense.

Some interesting conclusions are suggested by what has already been brought forward as to the ultimate strength of rocks. Thus it is impossible to have a sheer cliff face of sandstone more than $1\frac{5}{6}$ miles in height, this figure expressing what we have before denoted as the "critical depth" for hard sandstone. If the cliff exceeded this height, fracture would immediately be set up at the base of the cliff.

Before I conclude, I may as well briefly summarise the results to which I have been led in this paper. They are as follows:—

Faults may be grouped roughly into the three classes, known as reversed faults, normal faults, and wrench-planes, but varieties intermediate in character between these three types also occur.

(*a*) Reversed faults and thrust-planes originate when the greatest pressure in the rock mass is horizontal, and the least pressure vertical. They "strike" in a direction perpendicular to that of greatest pressure, and dip in either direction at angles of less than 45°.

(*b*) Normal faults originate when the greatest pressure is vertical, and the least pressure in some horizontal direction. They "strike" in a direction perpendicular to that of least pressure, and dip in either direction at angles of more than 45°.

(*c*) The third type of faults, to which the name of wrench-planes has been applied, originate when the greatest pressure is in one horizontal direction, and the least pressure in another horizontal direction, necessarily at right angles to the first. They "strike" in two possible directions, forming acute angles which are bisected by the direction of greatest pressure; their hade is theoretically vertical.

Index

Page numbers in *italics* denote figures. Page numbers in **bold** denote tables.